ORAL-BASED DIAGNOSTICS

ANNALS OF THE NEW YORK ACADEMY OF SCIENCES

Volume 1098

ORAL-BASED DIAGNOSTICS

Edited by Daniel Malamud and R. Sam Niedbala

Published by Blackwell Publishing on behalf of the New York Academy of Sciences
Boston, Massachusetts
2007

Library of Congress Cataloging-in-Publication Data

Oral-based diagnostics / edited by Daniel Malamud and R. Sam Niedbala.
p. ; cm. – (Annals of the New York Academy of Sciences, ISSN 0077-8923 ; v. 1098)
"This volume is the result of a conference entitled Oral-based Diagnostics held October 10–13, 2006 at Emerald Point Resort, Lake Lanier Islands, Georgia."
Includes bibliographical references.
ISBN-13: 978-1-57331-661-3 (alk. paper)
ISBN-10: 1-57331-661-X (alk. paper)
1. Saliva–Examination–Congresses. I. Malamud, Daniel. II. Niedbala, R. Sam. III. New York Academy of Sciences. IV. Series.
[DNLM: 1. Diagnosis, Oral–Congresses. 2. Saliva–Congresses. W1 AN626YL v.1097 2006 / WU 141 O628 2006]

RB52.5.O73 2006
617.6075–dc22

2006102099

The *Annals of the New York Academy of Sciences* (ISSN: 0077-8923 [print]; ISSN: 1749-6632 [online]) is published 28 times a year on behalf of the New York Academy of Sciences by Blackwell Publishing with offices at 350 Main St., Malden, MA 02148 USA; 9600 Garsington Road, Oxford, OX4 2ZG UK; and 600 North Bridge Rd, #05-01 Parkview Square, 18878 Singapore.

Information for subscribers: For new orders, renewals, sample copy requests, claims, changes of address and all other subscription correspondence please contact the Journals Department at your nearest Blackwell office (address details listed above). UK office phone: +44 (0)1865 778315, fax +44 (0)1865 471775; US office phone: 1-800-835-6770 (toll free US) or 1-781-388-8599; fax: 1-781-388-8232; Asia office phone: +65 6511 8000, fax; +44 (0)1865 471775, Email: customerservices@blackwellpublishing.com

Subscription rates:
Institutional Premium The Americas: $4043 Rest of World: £2246
The Premium institutional price also includes online access to full-text articles from 1997 to present, where available. For other pricing options or more information about online access to Blackwell Publishing journals, including access information and terms and conditions, please visit www.blackwellpublishing. com/nyas
*Customers in Canada should add 6% GST or provide evidence of entitlement to exemption.
**Customer in the UK or EU: add the appropriate rate for VAT EC for non-registered customers in countries where this is applicable. If you are registered for VAT please supply your registration number.

Mailing: The *Annals of the New York Academy of Sciences* is mailed Standard Rate. Mailing to rest of world by DHL Smart & Global Mail. Canadian mail is sent by Canadian publications mail agreement number 40573520. **Postmaster:** Send all address changes to *Annals of the New York Academy of Sciences*, Blackwell Publishing Inc., Journals Subscription Department, 350 Main St., Malden, MA 02148-5020.

Membership information: Members may order copies of *Annals* volumes directly from the Academy by visiting www.nyas.org/annals, emailing membership@nyas.org, faxing 212-298-3650, or calling 800-843-6927 (US only), or 212-298-8640 (International). For more information on becoming a member of the New York Academy of Sciences, please visit www.nyas.org/membership. Claims and inquiries on member orders should be directed to the Academy at email: membership@nyas.org or Tel: 212-298-8640 (International) or 800-843-6927 (US only).

Annals are available to subscribers online at the New York Academy of Sciences and also at Blackwell Synergy. Visit www.blackwell-synergy.com or www.annalsnyas.org to search the articles and register for table of contents e-mail alerts. Access to full text and PDF downloads of *Annals* articles are available to nonmembers and subscribers on a pay-per-view basis at www.blackwell-synergy.com and www.annalsnyas.org.

The paper used in this publication meets the minimum requirements of the National Standard for Information Sciences Permanence of Paper for Printed Library Materials, ANSI Z39.48_1984.

ISSN: 0077-8923 (print); 1749-6632 (online)
ISBN-10: 1-57331-661-X (paper); ISBN-13: 978-1-57331-661-3 (paper)

A catalogue record for this title is available from the British Library.

ANNALS OF THE NEW YORK ACADEMY OF SCIENCES

Volume 1098
March 2007

ORAL-BASED DIAGNOSTICS

Editors
DANIEL MALAMUD AND R. SAM NIEDBALA

This volume is the result of a conference entitled **Oral-based Diagnostics** held on October 10–13, 2006 at Emerald Point Resort, Lake Lanier Islands, Georgia.

CONTENTS

Part III. Meeting Unmet Needs in Diagnostics

Part IV. Oral Samples/Oral Diseases

Part V. Molecular and Protein Markers of Disease

Part VI. Horizons in Oral Diagnostics

Part VII. Short Papers

Financial assistance was received from:

- Dräger Safety
- Dräger USA
- International Diagnostic Systems Corp
- K Street Associates, LLC
- Lehigh University
- National Institute of Dental and Craniofacial Research – NIH
- New York University College of Dentistry
- OraSure Technologies, Inc.
- Salimetrics LLC
- StatSure Diagnostic Systems (SDS)

Dedication to Irwin Mandel

This volume is dedicated to Irwin Mandel, Professor Emeritus of Dentistry at Columbia University. Without question, he is the father of the field of salivary research. We recognize him as the major general of the "Salivation Army," and a mighty good general too. His dedication, his discoveries, and the large group of his students and colleagues that have been influenced by his teachings are immense. The fields of salivary research and oral-based diagnostics highlighted in this meeting and in this volume of the *Annals of the New York Academy of Sciences* are dominated by his ideas. As his students, colleagues, and friends, we thank you, Irwin, for your continued enthusiasm and commitment to the development of salivary diagnostics. We salute your role in creating this field, and cherish your support and your friendship.

—DANIEL MALAMUD
New York University College of Dentistry
New York, New York

Ann. N.Y. Acad. Sci. 1098: xiii (2007). © 2007 New York Academy of Sciences.
doi: 10.1196/annals.1384.045

Introduction

In 1992 Drs. Daniel Malamud and Larry Tabak organized the first New York Academy of Sciences meeting on Saliva as a Diagnostic Fluid, which was attended by approximately 100 scientists from industry, academia, and government agencies. Over the course of the meeting, many promising applications and uses for saliva-based diagnostics were discussed, and attendees left the meeting with the feeling that the future would be broad and bright for this emerging field.

Now, almost 14 years later a similar conference was organized by Drs. Daniel Malamud and Sam Niedbala. The meeting was expanded to accommodate more than 150 persons who once again came from diverse backgrounds and from around the world, working in industry, academia, and government. The meeting was held at Lake Lanier, Georgia; formal presentations were scheduled in the mornings and evening, while the afternoons were left free for viewing poster presentations, engaging in informal discussions and debates, and developing new research collaborations and friendships.

The breath and scope of topics at the 2006 meeting was immense. The topic was broadened from saliva to oral-based diagnostics. When the first meeting was held, little was known about nucleic acids resident in the mouth. Given routine techniques for amplification and detection of disease-specific nucleic acid sequences, the oral cavity has become a viable medium for genetic testing. Also, a detailed description of the total human salivary proteome is nearing completion. This information will permit the discovery of protein "signatures" that are necessary for multiparameter detection of oral and systemic diseases.

During the first meeting, many possibilities for routine testing with saliva were discussed. These included therapeutic drug monitoring, detection of salivary antibodies, monitoring of oral disease, and detection of drugs of abuse. At that time, very few oral fluid tests were performed in commercial laboratories, whereas today millions of oral samples are tested each year. The two largest areas of commercial success have been diagnostic antibody tests for HIV and panels of tests for drugs of abuse. An additional benefit of the achievement of these commercial tests has been the acknowledgment by government agencies that an oral fluid test can be as reliable as tests using blood and/or urine.

The end result of more than a decade of research has been the successful demonstration of the use of oral fluids and other oral-derived samples such as buccal cells, plaque, and volatiles for research and commercial purposes. There are many areas of diagnostic testing that have yet to benefit from the use of oral diagnostics. For example, diabetes and cardiovascular disease still lack viable

Ann. N.Y. Acad. Sci. 1098: xiv–xv (2007).
doi: 10.1196/annals.1384.046

oral cavity–derived biomarkers. It is anticipated that with the development of new tools, major advances will be made to meet these challenges. The topics covered at this meeting, which are reflected in the articles in this volume, outline new strategies for future research. A host of new microfluidic platforms for testing oral fluids at point-of-care and lab-based settings are presented. New methods to amplify target analytes in minute concentrations and new ways to collect and process an oral sample are on the horizon.

Reflected in Dr. Tabak's keynote speech was the concept of using the oral cavity as a window to the body. This classic point of view, espoused by Dr. Irwin Mandel, is now becoming a reality, and in the future remote sensing may provide real-time data for general health profiling. Many exciting technologies and success stories have been written since the first New York Academy of Sciences meeting in 1992. We should realize that along with these successes were some failures and dead-ends. This is the nature of science and serves to refocus research directions and lead us to future discoveries. Oral samples are relatively easy to collect and analyze, and it is hoped that in the future oral-based diagnostics will identify oral and systemic diseases long before the appearance of clinical symptoms. Ultimately this capability would support a lowering in the cost of health care while increasing the quality and length of life.

DANIEL MALAMUD
New York University College of Dentistry
New York, New York, USA

R. SAM NIEDBALA
Lehigh University
Bethlehem, Pennsylvania, USA

Implications for Diagnostics in the Biochemistry and Physiology of Saliva

ARIE v. NIEUW AMERONGEN, ANTOON J. M. LIGTENBERG, AND ENNO C. I. VEERMAN

Department of Oral Biochemistry, Academic Centre for Dentistry Amsterdam, ACTA, Vrije Universiteit and Universiteit van Amsterdam, the Netherlands

ABSTRACT: Oral fluid mainly consists of a mixture of glandular salivas. In addition, it is contaminated by some crevicular fluid, containing serum constituents. The contribution of the various salivary glands shows a continuous variation, resulting in wide ranges of concentrations for all constituents of oral fluid. As a consequence, the collection of oral fluid for diagnostic purposes should be standardized. Oral fluid can be used to detect a number of diseases and recent use of illicit drugs. It can also be used to monitor therapeutic drug concentrations. The development of microchips for salivary components offers great possibilities to use oral fluid for point-of-care testing.

KEYWORDS: diagnostics; saliva; salivary proteins; standardization; variables

INTRODUCTION

As a noninvasive method salivary research for diagnostics is highly attractive.[1,2] However, in this volume a number of pitfalls and opportunities will be demonstrated concerning how to handle saliva reliable for diagnostic purposes.[3,4] The fluid present in the oral cavity is in general designated "saliva." However, per definition, saliva is the fluid originating from the salivary glands. And indeed, the larger part of the oral fluid originates from the salivary glands. Oral fluid largely originates from the major salivary glands: gl. parotideae, gl. submandibulares, and gl. sublinguales (FIG. 1). Moreover, numerous minor salivary glands in the lip, cheek, tongue, and palate contribute small volumes of glandular saliva.

But in addition, some oral fluid comes from the serum either via the crevicular fluid or via mucosal damage and by leakage. When our oral tissues are in a state of perfect health, the contribution from serum to the oral fluid is

Address for correspondence: Dr. Arie V. Nieuw Amerongen, Department of Oral Biochemistry, ACTA, Vrije Universiteit, Medical Faculty, Van der Boechorststraat 7, 1081 BT Amsterdam, the Netherlands. Voice: 0031-20-444-86-75; fax: 0031-20- 444-86-85.
a.vannieuwamerongen@VUmc.NL

**Ann. N.Y. Acad. Sci. 1098: 1–6 (2007). © 2007 New York Academy of Sciences.
doi: 10.1196/annals.1384.033**

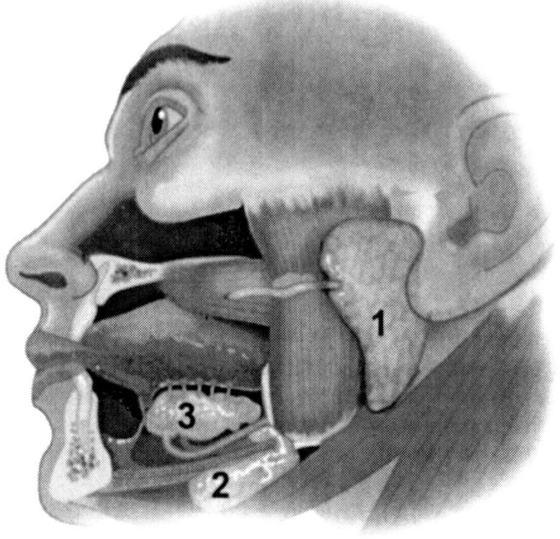

FIGURE 1. The anatomical location of the three pair large salivary glands: parotid glands (1), submandibular glands (2), and sublingual glands (3). (from Nieuw Amerongen: Salivary Glands and Oral Health, 2004, Bohn Stafleu Van Loghum).

minimal. However, under pathological conditions the serum constituents can make a substantial contribution to oral fluid and can even be used, for example, as markers for mucosal or epithelial inflammation. Because the term *saliva* is generally been used to designate the oral fluid, this term will be used here also.

In other words, with respect to salivary diagnostics it should be borne in mind that serum components can also be determined and be used for diagnosing systemic diseases, but that they are not real salivary constituents. If, on the other hand, only one of the salivary glands has a disorder, it is preferable to use glandular saliva for analysis.

BIOCHEMISTRY AND PHYSIOLOGY OF SALIVA

Saliva is derived from several types of salivary glands. Each type of salivary gland secretes saliva with characteristic composition and properties.[5] Because the contribution of each of these sources can vary, depending on, for example, the activity of the sympathetic and parasympathetic branches of the nervous system the salivary composition may exhibit considerable variations over time. As a consequence it is difficult to indicate normal values for salivary parameters. Only ranges in concentration of a salivary component can be given. For example, the concentration of the inorganic ion sodium can vary from 5 mM under resting conditions to as high as 60 mM in highly stimulated parotid saliva. To a lesser degree this also holds true for the organic composition of saliva. This can be exemplified for the mucins. These glycoproteins are absent from parotid saliva. Under resting conditions hardly any contribution by the parotid glands occurs, implying that the concentration of mucins is high in resting saliva, which under these conditions mainly originates from the (sero)mucous glands. On the other hand, after strong stimulation of the parotid glands, the concentration of the mucins in saliva will be decreased drastically. In other words, some increase or decrease in concentration of salivary constituents has hardly any diagnostic value. On the other hand, it is the challenge for saliva diagnostics to identify typical biomarkers for a systemic or a glandular disease.

FUNCTIONS OF SALIVA

The main function of saliva is to protect all oral tissues against decay, damage, and microbial inflammation. Therefore, a great number of proteins are present in saliva (FIG. 1). Particularly the mucins, which are secreted from virtually all seromucous salivary glands, are highly essential for nearly all functions of saliva.[6] They make the largest contribution to the rheological properties of saliva, such as viscosity, elasticity, and stickiness. In saliva two types of mucins are present, designated MUC5B and MUC7. The MUC5B mucin, formerly named MG1, is the high molecular weight mucin, which is particularly involved in protection of the tooth enamel surface against wear and encapsulation of a number of microorganisms. MUC5B lends to saliva its viscous and elastic properties. In addition, MUC7, formerly named MG2, makes saliva sticky to all surfaces, including mucosa and microorganisms. These characteristic rheological properties may, however, hamper saliva analysis. To lower the viscosity and to remove cellular contaminations, saliva is routinely vortexed and centrifuged, thus making it more accessible to analysis. The mucins remain scientifically highly intriguing glycoproteins because their diversity in the oligosaccharide chains is great. Particularly, the terminal sugars depend on blood group and secretor status and, in addition, on pathological conditions. In other words, the study and analysis of salivary mucins is essential for understanding the properties of an individual's saliva, but remains a challenge for highly trained researchers.

PATHOLOGIC CHANGES IN SALIVARY PROTEINS

Saliva contains a considerable number of (glyco)proteins and peptides in the concentration range from 10 μg/mL to more than about 500 μg/mL (FIG. 2).

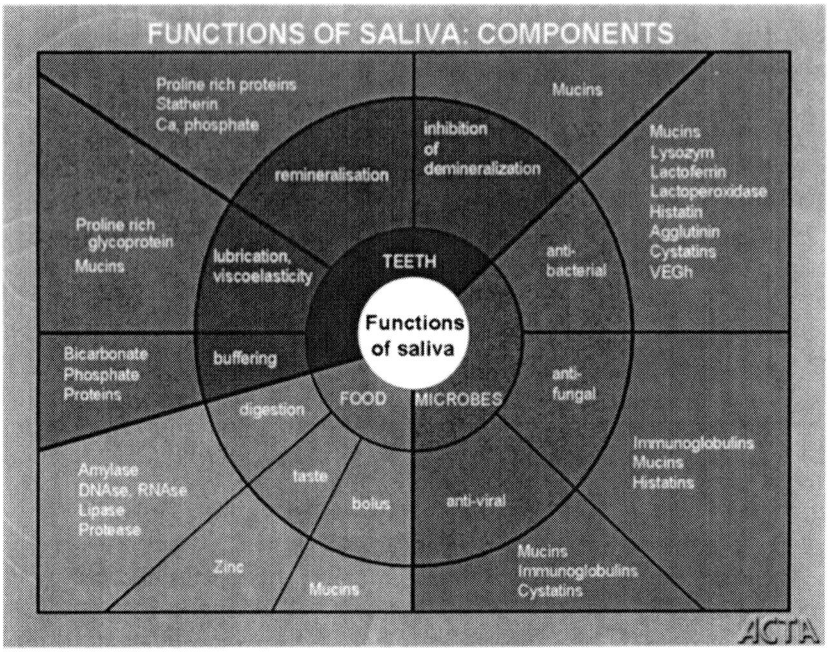

FIGURE 2. Overview of the relationship between the various functions of saliva and the salivary constituents involved. A number of salivary proteins participate in more than one function.

In addition, hundreds of trace proteins have been detected using sophisticated mass spectrometric techniques. As a consequence, proteomics of saliva collected under strictly standardized conditions may give a great impetus to the field of oral-based diagnostics.

In a number of oral diseases the protein composition will change,[7] for example, in periodontitis[8,9] and systemic diseases, such as epilepsy.[10] Particularly, screening of minor proteins may be valuable for early diagnosis of a number of diseases, such as malignancies.[11] Among others, salivary levels of the soluble form of c-erbB-2, a tumor marker for breast tumors, may have potential use in the initial detection and/or follow-up screening for the recurrence of breast cancer in women. Another possible application is in the detection of inflammation markers in saliva, for example, C-reactive protein (CRP) and cytokines for early diagnosis and monitoring of Sjögren syndrome and periodontal diseases.

IMPLICATIONS FOR DIAGNOSTICS

It will be clear that, although saliva, because of its easy, noninvasive mode of collection, is potentially an attractive diagnostic fluid, standardization of the collection procedure is of ultimate importance in order to obtain reproducible results.[12–14]

Therefore, it is very important to know both the limitations and the possibilities of salivary diagnostics.[15,16] Standardization of the conditions for collection of saliva is strictly essential for achieving reliability and interpretation of the data. This is particularly important when the exact concentrations of intrinsic salivary constituents are used as diagnostic criteria. In cases in which the mere presence of an analyte is indicative for a certain disease or condition, standardized collection is somewhat less critical. Examples of the latter are: identification of HIV-positive individuals, based on the detection of anti-HIV antibodies in saliva, monitoring of smoking habits, or for the surveillance of drug or alcohol abuse.[17–20] In addition, by novel sophisticated analysis techniques (MS, MALDI-TOF), hundreds of trace proteins have been detected in saliva. Undoubtedly, the availability of such highly sensitive analytical tools will give a great impetus to saliva-based diagnostics.[21] It thus seems appropriate to end this contribution by paraphrasing the title of a guest editorial by Daniel Malamud in the Journal of the American Dental Association in 2006: "Salivary diagnostics. The future is now."[22]

REFERENCES

1. HOFMAN, L.F. 2001. Human saliva as a diagnostic specimen. J. Nutr. **131:** 1621S–1625S.
2. KALK, W.W.I., A. VISSINK, *et al.* 2002. Sialometry and sialochemistry. A non-invasive approach for diagnosing Sjögren's syndrome. Ann. Rheum. Dis. **61:** 137–144.
3. SREEBNY, L.M. 2000. Saliva in health and disease: an appraisal and update. Int. Dent. J. **50:** 140–161.
4. KAUFMAN, E. & I.B. LAMSTER. 2002. The diagnostic applications of saliva—A review. Crit. Rev. Oral Biol. Med. **13:** 197–212.
5. VEERMAN, E.C.I., P.A.M. VAN DEN KEIJBUS, A. VISSINK & A.V. NIEUW AMERONGEN. 1996. Human glandular salivas: their separate collection and analysis. Eur. J. Oral Sci. **104:** 346–352.
6. NIEUW AMERONGEN, A.V., J.G.M. BOLSCHER & E.C.I. VEERMAN. 1995. Salivary mucins: protective functions in relation to their diversity. Glycobiology **5:** 733–740.
7. COHEN-BROWN, G. & J.A. SHIP. 2004. Diagnosis and treatment of salivary gland disorders. Quintess. Int. **35:** 108–123.
8. HENSKENS, Y.M.C., P.A.M. VAN DEN KEIJBUS, E.C.I. VEERMAN, *et al.* 1996. Protein composition of whole and parotid saliva in healthy and periodontitis subjects. J. Periodont. Res. **31:** 57–65.
9. MILLER, C.S., C.P. KING, *et al.* 2006. Salivary biomarkers of existing periodontal disease. A cross-sectional study. JADA **137:** 322–329.
10. NIEUW AMERONGEN, A.V., H. STROOKER, C.H. ODERKERK, *et al.* 1992. Changes in saliva of epileptic patients. J. Oral Pathol. Med. **21:** 203–208.

11. Li, Y., M.A.R. St. John, *et al.* 2004. Salivary transcriptome diagnostics for oral cancer detection. Clin. Cancer Res. **10:** 8442–8450.
12. George, J.R. & J.H. Fitchen. 1997. Future applications of oral fluid specimen technology. Am. J. Med. **102:** 21–25.
13. Streckfus, C.F. & L.R. Bigler. 2002. Saliva as a diagnostic fluid. Oral Dis. **8:** 69–76.
14. Forde, M.D., S. Koka, *et al.* 2006. Systemic assessments utilizing saliva: Part 1. General considerations and current assessments. Int. J. Prosthodont. **19:** 43–52.
15. Nagler, R.M., O. Hershkovich, *et al.* 2002. Saliva analysis in the clinical setting: revisiting an underused diagnostic tool. J. Invest. Med. **50:** 214–225.
16. Kadehjian, L. 2005. Legal issues in oral fluid testing. Forensic Sci. Int. **150:** 151–160.
17. Samyn, N., G. De Boeck & A.G. Verstraete. 2002. The use of oral fluid and sweat wipes for the detection of drugs of abuse in drivers. J. Forensic Sci. **47:** 1–8.
18. Walsh, J.M., R. Flegel, *et al.* 2003. An evaluation of rapid point-of-collection oral fluid drug-testing devices. J. Anal. Toxic. **27:** 429–439.
19. Lo Muzio, L., S. Falaschini, *et al.* 2005. Saliva as a diagnostic matrix for drug abuse. Int. J. Immunopathol. Pharmacol. **18:** 567–573.
20. Verstraete, A.G. 2005. Oral fluid testing for driving under the influence of drugs: history, recent progress and remaining challenges. Forensic Sci. Int. **150:** 143–150.
21. Wong, D.T. 2006. Salivary diagnostics powered by nanotechnologies, proteomics and genomics. JADA **137:** 313–321.
22. Malamud, D. 2006. Salivary diagnostics. The future is now. Guest editorial. JADA **137:** 284–285.

Point-of-Care Diagnostics Enter the Mouth

LAWRENCE A. TABAK

National Institute of Dental and Craniofacial Research, National Institutes of Health, Department of Health and Human Services, Bethesda, Maryland 20892-2290, USA

ABSTRACT: In this succinct review, I delineate a case supporting point-of-care (POC) diagnostics to provide a brief outline of why oral fluid/saliva–based POC offer several advantages over more traditional blood-based tests and conclude with a focused overview of the ethical, legal, and social implications of more widespread access to oral fluid/saliva–based POC diagnostics.

KEYWORDS: diagnostic; saliva; point-of-care

WHY POINT-OF-CARE (POC) DIAGNOSTICS?

> FALSTAFF: Sirrah, you giant, what says the doctor to my water?
>
> PAGE: He said, sir, the water itself was a good healthy water, but, for the party that owed it, he might have more diseases than he knew for.
>
> *Henry IV, Part 2*
> WILLIAM SHAKESPEARE

POC diagnostics are not new. There are numerous references to "uroscopy" by the time of Galen (second century) to aid diagnosis. Both woodcuttings (e.g., *Feldtbuch der Wundtartzney* by Hans Von Gersdorff, 1517) and paintings (e.g., *A Surgeon and a Line of Patients*, 1482, by an unknown artist[a] and *The Village Doctor*, 17th century, by David Teniers the Younger) depict physicians analyzing patients' urine in their presence. In addition to examining the urine's color for evidence of blood and clarity for evidence of microbial infection, diagnosticians tasted the specimen to determine whether the sugar content was excessive.

[a]JAMA **294:** 2277 (2005).

Address for correspondence: Lawrence A. Tabak, National Institute of Dental and Craniofacial Research, National Institutes of Health, Department of Health and Human Services, 31 Center Drive, Rm. 2C39, MSC 2290, Bethesda, MD 20892-2290. Voice: 301-496-3571; fax: 301-402-2185.
tabakl@mail.nih.gov

Ann. N.Y. Acad. Sci. 1098: 7–14 (2007).
doi: 10.1196/annals.1384.043

The goal of POC diagnostics to provide results rapidly that accelerate clinical decision making and, in more modern times, lower the cost of health care has remained constant throughout history.[1] As outlined by Dr. Elias Zerhouni, Director of the National Institutes of Health (NIH), there is urgent need to transform the practice of medicine in the 21st century. Intervention must occur at the earliest possible moment of a disease process. To ensure this, the intimate details of the preclinical events must be well understood, thereby allowing patients most at risk to be identified. The net result will be a health-care system that is more cost-effective.[2] POC diagnostics can play an important role in this transformation.

POC diagnostics are most often used to measure biological or physical characteristics termed "biomarkers," which are indicative of a specific underlying physiologic state. Thus, biomarkers can be used to ascertain disease risk, assess disease severity, and guide treatment by identifying patients who will respond to specific drugs as well as individuals who will suffer untoward consequences. Blood pressure and blood sugar level are perhaps the two most well-known biomarkers. Other examples include cholesterol, an indicator of cardiovascular health, and the presence of antibodies directed against HIV as a marker of HIV infection. Under the auspices of the Foundation for the National Institutes of Health (FNIH), a public–private partnership was recently launched among the Food and Drug Administration (FDA), the National Institutes of Health (NIH), and the Pharmaceutical Research and Manufacturers of America (PhRMA). For additional details, refer to the FNIH website (http://www.fnih.org/).

From the patient's health perspective, the effectiveness of POC is only as good as action taken following testing.[3] While there are a number of examples of POC diagnostics improving patient outcomes, perhaps the most dramatic demonstration was in-home monitoring of blood glucose levels for diabetics in the Diabetes Control and Complications Trial.[4,5] Patients either adjusted their insulin levels daily by monitoring at least four times per day using a POC blood glucose monitor or kept their insulin levels largely constant, rarely adjusting the daily dosage of insulin. The former group had statistically significant reductions in diabetic complications. While this trial was not designed to test the efficacy of the POC testing *per se*, it is clear that the ready availability of blood glucose levels and patients' willingness to adjust their insulin dosages led to the significant benefits.

A more recent example is "rapid" diagnostic tests for sexually transmitted diseases that can be used within the community and ultimately in the privacy of one's home. This category of POC tests has found increasing use particularly in the developing world.[6] These tests speed appropriate treatment of infections at the earliest and typically asymptomatic stages of the disease, reducing further transmission.[7] Rapid results provide the additional advantage of "instant feedback." In 2000, over 30% of Americans tested for HIV were lost to follow-up because they did not return for their test results.[8] Coupled with this advantage comes the responsibility of providing access to appropriate follow-up counseling. The more generalized goal of using POC diagnostics for resource-poor settings is slowly being realized, but the need is increasingly urgent.[9,10] Infectious diseases account for almost 15 million deaths annually worldwide. In the poorest nations, they account for half of the deaths compared to less than 5% of the fatalities in the wealthiest countries.[11] The current "gold standard" test for malaria, which kills over 1 million people annually worldwide, is smear microscopy, which requires highly trained individuals and quality control that is often difficult to achieve in field settings.[11] The organization Doctors without Borders is currently validating a new test, which emerged from their sponsored program, designed to develop a POC diagnostic for malaria optimized for field use in the hot and humid conditions found in Africa.[12]

As disease patterns shift from acute conditions of the past to more chronic conditions, there has been an increased appreciation of the complex interplay among genes, environment, diet and nutrition, infectious agents, and societal influences as underlying causes of "complex diseases."[14–16] Continued advances in bioengineering using microfluidics and nanotechnologies will yield POC diagnostics that have the potential to monitor each of these variables simultaneously and rapidly in a community setting.[17–20] Ultimately, implantable sensors designed for real-time monitoring will emerge, affording the opportunity to perform health surveillance.[21] As outlined in the next section, the use of oral fluids/saliva as a diagnostic platform would markedly enhance the feasibility of such a goal.

WHY ORAL FLUID/SALIVA–BASED POC DIAGNOSTICS?

TABLE 1 summarizes potential advantages of using oral fluids/saliva for POC testing. The underlying rationale and pitfalls have been well documented in numerous reviews on the subject and is one of the subjects of this volume.[22–25]

The use of oral fluids/saliva as a diagnostic medium markedly enhances both the potential value and abuse of POC tests for a wide range of analytes

TABLE 1. Advantages of oral fluids/saliva as a diagnostic medium

1. Collection of saliva is "safe"
Considered "acceptable and noninvasive" by patients
Reduces transmission of infectious disease by eliminating the potential for accidental needle sticks
2. Self-collection is possible
Allows for community- or home-based or special population (e.g., confined, remote, hemophiliac, morbidly obese, pediatric) sampling
3. It is economical
Eliminates the need for a health care intermediary (e.g., phlebotomist)
4. Salivary levels of many drugs reflect the unbound (and hence available) fraction in plasma

TABLE 2. Examples of oral fluids/saliva as a diagnostic medium

Analytes	Example	Recent references
Antibodies	HIV	Yapijakis *et al.*[26]
	Hepatitis C	Champion *et al.*[27]
Hormones	Androgen	Azurmendi *et al.*[28]
	Cortisol	Gutteling *et al.*[29]
	Estradiol and progesterone	Chatterton *et al.*[30]; Mylonas *et al.*[31]
	Testosterone	Loney *et al.*[32]
Nucleic acids	Human genomic	Etter *et al.*[33]; Quinque *et al.*[34]; Ng *et al.*[35]
	mRNA	Hu *et al.*[36]
	Mitochondrial	Jiang *et al.*[37]
	Microbial	
	Bacterial (*Helicobacter*)	Tiwari *et al.*[38]
	Viral (HIV; HPV)	Yapijakis *et al.*[39]; Zhao *et al.*[40]
Pharmaceuticals		
	Alcohol	Degutis *et al.*[41]
	Cotinine	Maziak *et al.*[42]
	Drugs of abuse	Tonnes *et al.*[43]; Lennox *et al.*[44]
	Phenytoin	Yager *et al.*[45]

and an equally broad set of purposes (TABLE 2). No attempt has been made to catalog the many recent references in this area—rather, recent examples have been selected to demonstrate the breadth of this growing field. These examples underscore the main point—the option of using whole mouth fluids/saliva as the basis of POC diagnostic tests has expanded greatly the possibility of home- and community-based testing.

ETHICAL, LEGAL, AND SOCIAL IMPLICATIONS OF ORAL FLUID/SALIVA–BASED POC DIAGNOSTICS

The advent of oral fluid/saliva–based diagnostics has already yielded perplexing ethical, legal, and social implications. A simple Google search of the internet using the search term "home drug testing" yielded over 950,000 potential sites. A review of internet-based home drug testing products revealed only a single site with an explicit statement regarding testing a minor against his/her will.[46]

As oral fluids contain sufficient quantities of host DNA to decode genotype, one must look at an oral fluid/saliva sample no differently than a blood sample.[47–49] Genetic testing has raised many privacy concerns, particularly the potential to discriminate against individuals for insurability and/or employability.[50,51] Principles of population screening as applied to genetic susceptibility to disease have been articulated, although there remain many gray areas.[52]

Advances in biomedical engineering will make it possible for the current "outer body" experience of salivary diagnostics to evolve into real-time monitoring of analytes through placement of in-dwelling biosensors in the mouth. Increasing use of salivary diagnostics will help catalyze a shift from disease diagnosis to health surveillance. However, with advances in this technology comes the additional obligation to ensure the privacy and rights of patients.[53] Indeed, the health surveillance of the general population is proving to be a contentious issue. For example, the plan to monitor hemoglobin A1 C tests as a means of curtailing the dramatic increases in diabetes in New York City has drawn increasing concern over patient's privacy rights.[54] In particular, critics have argued that diabetes poses no public danger—it is neither communicable nor does it put others at risk.

ACKNOWLEDGMENTS

This short review is dedicated to Dr. Irwin D. Mandel, who pioneered the field of Salivary Diagnostics.

I thank Ms. Carol Lowe for her help with the preparation of this manuscript.

REFERENCES

1. PRICE, C. 2001. Point of care testing. Br. Med. J. **322:** 1285–1288.
2. CULLITON, B. 2006. Extracting knowledge from science: a conversation with Elias Zerhouni. Health Aff. **25:** 94–103.
3. PRICE, C. 2001. Point of care testing. Br. Med. J. **322:** 1285–1288.
4. GUTIERRES, S.L. & T.E. WELTY. 2004. Point-of-care testing: an introduction. Ann. Pharmacother. **38:** 119–125.
5. WRITING TEAM FOR THE DIABETES CONTROL AND COMPLICATIONS TRIAL/EPIDEMIOLOGY OF DIABETES INTERVENTIONS AND COMPLICATIONS RESEARCH GROUP. 2003. Sustained effect of intensive treatment of type 1 diabetes mellitus on development and progression of diabetic nephropathy: the epidemiology of diabetes interventions and complications (EDIC) study. J. Am. Med. Assoc. **290:** 2159–2167.
6. YAGER, P., T. EDWARDS, E. FU, *et al.* 2006. Microfluidic diagnostic technologies for global public health. Nature **442:** 412–418.
7. MABEY, D., R.W. PEELING & M.D. PERKINS. 2001. Rapid and simple point of care diagnostics for STIs. Sex. Transm. Infect. **77:** 397–401.
8. WRIGHT, A.A. & I.T. KATZ. 2006. Home testing for HIV. N. Engl. J. Med. **354:** 437–440.
9. USDIN, M., M. GUILLERM & P. CHIRAC. 2006. Neglected tests for neglected patients. Nature **441:** 283–284.
10. YAGER, P., T. EDWARDS, E. FU, *et al.* 2006. Microfluidic diagnostic technologies for global public health. Nature **442:** 412–418.
11. YAGER, P., T. EDWARDS, E. FU, *et al.* 2006. Microfluidic diagnostic technologies for global public health. Nature **442:** 412–418.

12. Carter, R. & K. Mendis. 2002. Evolutionary and historical aspects of the burden of malaria. Clin. Microbiol. Rev. **15:** 564–594.
13. Usdin, M., M. Guillerm & P. Chirac. 2006. Neglected tests for neglected patients. Nature **441:** 283–284.
14. Yach, D., C. Hawkes, C.L. Gould & K.J. Hofman. 2004. The global burden of chronic diseases: overcoming impediments to prevention and control. J. Am. Med. Assoc. **291:** 2616–2622.
15. Zerhouni, E. 2005. U.S. biomedical research: basic, translational, and clinical sciences. J. Am. Med. Assoc. **294:** 1352–1358.
16. Kiberstis, P. & L. Roberts. 2002. It's not just the genes. Science **296:** 685.
17. Yager, P., T. Edwards, E. Fu, *et al.* 2006. Microfluidic diagnostic technologies for global public health. Nature **442:** 412–418.
18. Dittrich, P. & A. Manz. 2006. Lab-on-a-chip: microfluidics in drug discovery. Nature **5:** 210–218.
19. Cheng, M. M., G. Cuda, Y. Bunimovich, *et al.* 2005. Nanotechnologies for biomolecular detection and medical diagnostics. Curr. Opin. Chem. Biol. **10:** 1–9.
20. Friedrich, M.J. 2005. Nanoscale biosensors show promise. J. Am. Med. Assoc. **293:** 1965.
21. Frost, M.C. & M.E. Meyerhoff. 2002. Implantable chemical sensors for real-time clinical monitoring: progress and challenges. Curr. Opin. Chem. Biol. **6:** 633–641.
22. Malamud, D. & L.A. Tabak, Eds. 1993. Saliva as a Diagnostic Fluid. Ann. N.Y. Acad. Sci. Vol. 694.
23. Tabak, L.A. 2001. A revolution in biomedical assessment: the development of salivary diagnostics. J. Dent. Educ. **65:** 1335–1339.
24. Mukhopadhyay, R. 2006. Devices to drool for: miniaturized analytical techniques are now sensitive enough to detect traces of biomarkers in saliva. Is saliva ready for point-of-care diagnostic devices. Anal. Chem. **78:** 4255–4259.
25. Wong, D.T. 2006. Salivary diagnostics powered by nanotechnologies, proteomics and genomics. J. Am. Dent. Assoc. **137:** 313–321.
26. Yapijakis, C. P., V. Panis, N. Koufaliotis, *et al.* 2006. Immunological and molecular detection of human immunodeficiency virus in saliva and comparison with blood testing. Eur. J. Oral Sci. **114:** 175–179.
27. Champion, J.K., A. Taylor, S. Hutchinson, *et al.* 2004. Incidence of hepatitis C virus infection and associated risk factors among Scottish prison inmates: a cohort study. Am. J. Epidemiol. **159:** 514–519.
28. Azurmendi, A., F. Braza, G. Ainhoa, *et al.* 2006. Aggression, dominance, and affiliation: their relationships with androgen levels and intelligence in 5-year-old children. Horm. Behav. **50:** 132–140.
29. Gutteling, B.M., C. de Weerth, S.H.N. Willemsen-Swinkels, *et al.* 2005. The effects of prenatal stress on temperament and problem behavior of 27-month-old toddlers. Eur. Child Adolesc. Psychiatry **14:** 41–51.
30. Chatterton, R.T. Jr., E.T. Mateo, N. Hou, *et al.* 2005. Characteristics of salivary profiles of oestradiol and progesterone in premenopausal women. J. Endocrinol. **186:** 77–84.
31. Mylonas, P.G., M. Makri, N.A. Georgopoulos, *et al.* 2006. Adequacy of saliva 17-hydroxprogesterone determination using various collection methods. Steroids **71:** 273–276.

32. Loney, B. R., M.A. Butler, E.N. Lima, *et al.* 2006. The relation between salivary cortisol, callous-unemotional traits, and conduct problems, in an adolescent non-referred sample. J. Child Psychol. Psychiatry **47:** 30–36.
33. Etter, J.-F., E. Neidhart, S. Bertrand, *et al.* 2005. Collecting saliva by mail for genetic and cotinine analyses in participants recruited through the internet. Eur. J. Epidemiol. **20:** 833–838.
34. Quinque, D., R. Kittler, M. Kayser, *et al.* 2006. Evaluation of saliva as a source of human DNA for population and association studies. Anal. Biochem. **353:** 272–277.
35. Ng, D.P.K., D. Koh, S. Choo & C. Kee-Seng. 2006. Saliva as a viable alternative source of human genomic DNA in genetic epidemiology. Int. J. Clin. Chem. **367:** 81–85.
36. Hu, S., Y. Li, J. Wang, *et al.* 2006. Human saliva proteome and transcriptome. J. Dent. Res. **85:** 1129–1133.
37. Jiang, W.-W., B. Masayesva, M. Zahurak, *et al.* 2005. Increased mitochondrial DNA content in saliva associated with head and neck cancer. Clin. Cancer Res. **11:** 2486–2491.
38. Tiwari, S.K., A.A. Khan, K.S. Ahmed, *et al.* 2005. Rapid diagnosis of *Helicobacter pylori* infection in dyspeptic patients using salivary secretion: a non-invasive approach. Singapore Med. J. **46:** 224–228.
39. Yapijakis, C.P., V. Panis, N. Koufaliotis, *et al.* 2006. Immunological and molecular detection of human immunodeficiency virus in saliva and comparison with blood testing. Eur. J. Oral Sci. **114:** 175–179.
40. Zhao, M., E. Rosenbaum, A.L. Carvalho, *et al.* 2005. Feasibility of quantitative PCR-based salivary rinse screening of HPV for head and neck cancer. Int. J. Cancer **117:** 605–610.
41. Degutis, L.C., R. Rabinovici, A. Sabbaj. 2004. The saliva strip test is an accurate method to determine blood alcohol level in trauma patients. Acad. Emerg. Med. **11:** 885–887.
42. Maziak, W., K.D. Ward & T. Eissenberg. 2006. Measuring exposure to environmental tobacco smoke (ETS): a developing country's perspective. Prev. Med. **42:** 409–414.
43. Toennes, S.W., S. Steinmeyer, H.-J. Maurer, *et al.* 2005. Screening for drugs of abuse in oral fluid-correlation of analysis results with serum in forensic cases. J. Anal. Toxicol. **29:** 22–28.
44. Lennox, R., M.L. Dennis, C.K. Scott & R. Funk. 2006. Combining psychometric and biometric measures of substance use. Drug Alcohol Depend. **83:** 95–103.
45. Yager, P., T. Edwards, E. Fu, *et al.* 2006. Microfluidic diagnostic technologies for global public health. Nature **442:** 412–418.
46. Levy, S., S. Van Hook & J. Knight. 2004. A review of internet-based home drug-testing products for parents. Pediatrics **113:** 720–726.
47. Etter, J.-F., E. Neidhart, S. Bertrand, *et al.* 2005. Collecting saliva by mail for genetic and cotinine analyses in participants recruited through the internet. Eur. J. Epidemiol. **20:** 833–838.
48. Quinque, D., R. Kittler, M. Kayser, *et al.* 2006. Evaluation of saliva as a source of human DNA for population and association studies. Anal. Biochem. **353:** 272–277.
49. Ng, D.P.K., D. Koh, S. Choo & C. Kee-Seng. 2006. Saliva as a viable alternative source of human genomic DNA in genetic epidemiology. Int. J. Clin. Chem. **367:** 81–85.

50. KHOURY, M.J., L.L. MCCABE & E.R.B. MCCABE. 2003. Population screening in the age of genomic medicine. N. Engl. J. Med. **348:** 50–58.
51. BRANDT-RAUF, P.W. & S. BRANDT-RAUL. 2004. Genetic testing in the workplace: ethical, legal, and social implications. Annu. Rev. Public Health **25:** 139–153.
52. KHOURY, M.J., L.L. MCCABE & E.R.B. MCCABE. 2003. Population screening in the age of genomic medicine. N. Engl. J. Med. **348:** 50–58.
53. TABAK, L.A. 2001. A revolution in biomedical assessment: the development of salivary diagnostics. Journal of Dental Education **65:** 1335–1339.
54. FAIRCHILD, A.L. 2006. Diabetes and disease surveillance. Science **313:** 175–176.

Autoimmune Diseases and Sjögren's Syndrome

An Autoimmune Exocrinopathy

PHILIP C. FOX[a,b]

[a]*Sjögren's Syndrome Foundation, Bethesda, Maryland 20814, USA*

[b]*Department of Oral Medicine, Carolinas Medical Center, Charlotte, North Carolina 28203, USA*

ABSTRACT: Autoimmune diseases include a diverse group of over 80 conditions. Sjögren's syndrome is the second most common autoimmune rheumatic disease, with an estimated prevalence in the United States of 2–4 million persons. There are prominent and consistent oral and dental findings in Sjögren's syndrome related to the autoimmune-mediated loss of normal salivary function. Additionally, nonoral clinical manifestations of Sjögren's syndrome include: dry eyes (with specific ocular surface changes termed keratoconjunctivitis sicca); other xeroses, such as dryness of the nose, throat, skin, and vagina; peripheral (and less frequently central) neuropathies; myalgias and arthralgias; thyroid disorders (particularly autoimmune thyroiditis); pulmonary disorders; renal disorders; and lymphoma. There is a significant (20- to 40-fold) increase in the incidence of malignant lymphoma, particularly in primary Sjögren's syndrome. Establishing the diagnosis of Sjögren's syndrome has been difficult in the light of its nonspecific symptoms (dry eyes and mouth), disagreement on diagnostic criteria, and a lack of both sensitive and specific laboratory markers. Many serum and salivary biomarkers for Sjögren's syndrome have been proposed although, to date, none has proven to be sufficiently specific for diagnostic purposes or has been well correlated with disease activity measures. Investigators have recently begun to apply modern genomic and proteomic approaches to identify candidate biomarkers in Sjögren's syndrome. The results of these investigations promise to provide a wealth of information on candidate biomarkers and possible etiopathological mechanisms underlying this disorder. Further, this information will improve clinical outcomes by fostering the design of new rational therapeutics and assisting in the monitoring of clinical disease.

Address for correspondence: Philip C. Fox, DDS, PC Fox Consulting, LLC, 6509 Seven Locks Road, Cabin John, MD 20818. Voice: 301-320-8200; fax: 301-320-3884.
pcfox@comcast.net

Ann. N.Y. Acad. Sci. 1098: 15–21 (2007). © 2007 New York Academy of Sciences.
doi: 10.1196/annals.1384.003

KEYWORDS: salivary glands; biomarkers; xerostomia; diagnosis and therapy of Sjögren's syndrome

Autoimmune diseases are a diverse group comprising over 80 conditions with an estimated prevalence in the United States of between 5% and 8% of the population, a total of 14–22 million persons.[1] This places the cumulative number of affected individuals in the same category as cancer (estimated 9 million affected) and heart disease (22 million). While manifestations of the different autoimmune conditions vary widely, these disorders share underlying defects in the immune response leading the body to attack its own organs and tissues. Many autoimmune conditions are difficult to diagnose and, for virtually all, the cause is not known or incompletely understood. Most of these diseases disproportionately affect women, or carry a substantial gender predilection.

Sjögren's syndrome is the second most common autoimmune rheumatic disease, with a prevalence in the United States estimated at 2–4 million persons.[2] Sjögren's syndrome has been termed an autoimmune exocrinopathy, characterized by dryness of the mouth and eyes resulting from a chronic, progressive loss of secretory function of the salivary and lacrimal glands. Sjögren's syndrome has a female-to-male ratio of 9:1, the greatest of any autoimmune disorder. While the condition is found in children and men, it is most commonly seen in peri- or postmenopausal women.

Although exocrine involvement is a defining feature of Sjögren's syndrome, it is a systemic disorder. Sjögren's syndrome is classified as primary or secondary, the major distinguishing factor being that secondary Sjögren's syndrome occurs in conjunction with another connective tissue disease. Associated connective tissue diseases include rheumatoid arthritis, systemic lupus erythematosus, scleroderma, and primary biliary cirrhosis. There is an approximately even incidence of primary and secondary Sjögren's syndrome.

There are prominent and consistent oral and dental findings in Sjögren's syndrome.[3] They are related to the autoimmune-mediated loss of normal salivary function and are similar to the findings in other conditions with decreased salivary gland output, such as radiation- or drug-induced salivary gland hypofunction. The clinical features of Sjögren's syndrome include an increased caries rate, mucosal dryness, pain and atrophy, increased infections (both fungal and bacterial), altered rheological properties of secretions (thicker, opaque, or viscous secretions), and enlargement of the salivary glands. Other clinical symptoms include difficulties in chewing, swallowing, and speaking, altered or diminished taste acuity, compromised nutrition, and a markedly diminished quality of life. Salivary gland hypofunction has a negative impact on communication and alimentation abilities, critical aspects for satisfactory quality of life.

Histopathological examination of affected labial minor salivary glands, a common diagnostic technique in Sjögren's syndrome, reveals characteristic changes consisting of a focal, periductal mononuclear cell infiltrate, a loss of acinar cells, and the relative preservation of ductal cells.[4] The infiltrate is composed of T and B cells (80%:20%), which produce a plethora of immunologically active products.[5] The end results of the pathologic processes in the salivary glands in Sjögren's syndrome are a reduction in functional acinar tissue, a loss of secretory output, and symptoms of oral dryness.

Other, nonoral clinical manifestations of Sjögren's syndrome include dry eyes (with specific ocular surface changes termed keratoconjunctivitis sicca); other xeroses, such as dryness of the nose, throat, skin, and vagina; peripheral (and less frequently central) neuropathies; myalgias and arthralgias; thyroid disorders (particularly autoimmune thyroiditis); pulmonary disorders; renal disorders; and lymphoma. There is a significant (20- to 40-fold) increase in the incidence of malignant lymphoma, particularly in primary Sjögren's syndrome.[6] Often these tumors involve the salivary glands. In virtually all cases, they are B cell mucosally associated lymphoid tissue tumors (MALTomas).

Patients with both primary and secondary Sjögren's syndrome demonstrate numerous serological markers of autoimmune reactivity.[2] These include a marked hypergammaglobulinemia (IgG>IgA>IgM), elevated total protein and sedimentation rate, persistent rheumatoid factors, a decreased WBC count, and the presence of autoantibodies directed against the ribonuclear proteins SS-A/Ro and SS-B/La. These changes are found in a large percentage of patients but are nonspecific, being seen in many autoimmune inflammatory conditions.

Management of Sjögren's syndrome is still primarily palliative, although current research is targeting the underlying systemic autoimmunity.[7] Palliative measures are aimed at diminishing symptoms through increased hydration. Patients are encouraged to sip water frequently, use available rinses, gels, sprays, and mouthwashes, avoid or minimize alcohol, caffeine, and intense flavorings, and increase humidity in the local environment, particularly at night.

Preventive measures help to minimize the effects of salivary hypofunction and include supplemental fluoride applications, use of remineralizing solutions and xylitol-containing products, meticulous oral hygiene, and adoption of a noncariogenic diet.

A number of approaches have been proposed to stimulate secretion and thus relieve dryness symptoms. Sugar-free gums or mints provide masticatory and gustatory stimulation for the salivary glands and produce transient increases in salivation. Acupuncture has been proposed as a treatment modality, although results have been mixed.[8] Electrical stimulation of secretion has been tested with minimal success, although recent trials with newer devices show promise.[9] The most widely used and successful therapy is salivary stimulation with the systemic secretogogues cevimeline (Evoxac) or pilocarpine (Salagen).[7] These agents are FDA-approved for relief of oral dryness in Sjögren's syndrome and also provide significant transient increases in salivary output for up to 3–4 h. Although they have a substantial incidence of side effects, most adverse events are mild and tolerable and these agents are thus considered both safe and effective. However, they afford only temporary relief of the dryness symptoms and have not been shown to preserve or improve long-term salivary function.

Recent research interest has focused on biologic agents that are directed against the underlying autoimmune inflammatory process. Initial enthusiasm over the anti-TNF-α agents proved to be premature and these have not been found to be effective in Sjögren's syndrome.[10,11] Ongoing clinical studies examining the effects of the anti-B-cell agents rituximab (anti-CD-20)[12] and epratuzumab (anti-CD-22)[13] show promise and may prove to be viable therapeutics to address this condition.

Establishing the diagnosis of Sjögren's syndrome has been difficult in the light of its nonspecific symptoms (dry eyes and mouth) and lack of both sensitive and specific laboratory markers. Further, there has been a lack of agreement on diagnostic criteria. The recent American–European Consensus Classification Criteria have improved the situation, particularly for clinical trials.[14] However, clinical diagnosis remains problematic and there is a great need for a means of monitoring disease activity. Establishment of valid diagnostic and disease severity markers will have a tremendous impact on clinical trials as well as clinical practice.

Toward that end, a host of serum biomarkers for Sjögren's syndrome have been proposed. These include numerous cytokines (IL-6, TNF-α, BAFF); lysosomal enzymes; functional autoantibodies (anti-fodrin, anti-M3 receptor); soluble CD40 ligand; soluble E-cadherin; various B cell markers; and β2 microglobulin.[15–24] None has proven to be sufficiently specific for diagnostic purposes or has been correlated well with disease activity measures.

There have also been a number of salivary biomarkers suggested, as it was felt that examination of the product of an affected organ might provide greater specificity. Proposed biomarkers include cytokines (IL-6, TNF-α); lactoferrin; lysozyme; β2 microglobulin; and various B cell markers.[18,25–28] However, these too have not demonstrated sufficient specificity or sensitivity to serve as clinically useful markers.

Investigators have recently begun to apply modern proteomic approaches to identify candidate biomarkers in Sjögren's syndrome. A study by Ryu and colleagues using SELDI-TOF-MS of parotid saliva identified a number of potential biomarkers for Sjögren's syndrome.[29] Both increased and decreased potential protein biomarkers were found. Those increased were lactoferrin, β2 microglobulin, Igκ light chain, polymeric Ig receptor, lysozyme C, and cystatin C. Decreased proteins were amylase, carbonic anhydrase VI, and two proline-rich proteins. Interestingly, these findings mirrored earlier studies (see above) using conventional clinical chemical and immunohistochemical techniques. Appropriate validation remains to be done on these findings and the issues of specificity must be addressed. Also, as all studies are done on subjects with established disease, it remains an open question whether changes identified are a cause or an effect of the disorder.

Similar proteomic analyses were recently presented in abstract form[30] at the IXth International Sjögren's Syndrome Symposium by Baldini *et al.* and Zoukhri *et al.* Baldini *et al.* reported that whole saliva of Sjögren's syndrome patients had decreased amylase, carbonate dehydratase VI precursor, cystatin and prolactin-inducible protein, and numerous increased unidentified proteins. Zoukhri *et al.* conducted MALDI-o-TOF-MS analyses of submandibular saliva (and tears) and identified three candidate biomarkers in preliminary modeling. All these results require validation and amplification of the findings, but the commonality of findings in these early proteomic investigations using several different gland secretions (parotid, submandibular and whole saliva, and tears) of Sjögren's syndrome patients gives increased confidence in the results. Extending these sorts of proteomic approaches has great promise for identifying specific and sensitive biomarkers that will aid in the diagnosis and monitoring of Sjögren's syndrome.

Moser and colleagues at the University of Minnesota have used a genomic screening approach to identify candidate biomarkers in Sjögren's syndrome.[31,32] They have conducted gene expression profiling of peripheral blood of Sjögren's syndrome patients and controls using commercially available microarrays. Preliminary results reported at the IXth International Sjögren's Syndrome Symposium identified a signature of dysregulation of the type I interferon pathway—a feature that Sjogren's syndrome shares with systemic lupus erythematosus, dermatomyositis, and psoriasis.[33] Other groups are conducting similar studies,[34] some of which will be reported in this volume. The results of these investigations promise to provide a wealth of information on candidate biomarkers and possible etiopathological mechanisms.[33] This should aid in the design of rationale therapeutics and assist in monitoring of clinical disease.

REFERENCES

1. The Autoimmune Diseases Coordinating Committee. 2005. Progress in Autoimmune Diseases Research: A Report to Congress. U.S. Department Of Health and Human Services, National Institutes of Health, National Institute of Allergy and Infectious Diseases. (Access at www.niaid.nih.gov/publications/pdf/ADCCFinal.pdf)
2. Kassan, S.S. & H.M. Moutsopoulos. 2004. Clinical manifestations and early diagnosis of Sjögren's syndrome. Arch. Intern. Med. **64:** 1275–1284.
3. Daniels, T.E. & P.C. Fox. 1992. Salivary and oral components of Sjogren's syndrome. Rheum. Dis. Clin. North Am. **18:** 571–589.
4. Daniels, T.E. 1984. Labial salivary gland biopsy in Sjogren's syndrome. Assessment as a diagnostic criterion in 362 suspected cases. Arthritis Rheum. **27:** 147–156.
5. Yamamoto, K. 2003. Pathogenesis of Sjogren's syndrome. Autoimmun. Rev. **2:** 13–18.

6. THEANDER, E. *et al.* 2006. Lymphoma and other malignancies in primary Sjogren's syndrome: a cohort study on cancer incidence and lymphoma predictors. Ann. Rheum. Dis. **65:** 796–803.
7. FOX, P.C. 2004. Salivary enhancement therapies. Caries Res. **38:** 241–246.
8. JEDEL, E. 2005. Acupuncture in xerostomia—a systematic review. J. Oral Rehabil. **32:** 392–396.
9. STRIETZEL, F.P. *et al.* 2006. Electrostimulating device in the management of xerostomia. Oral Dis. In press.
10. MARIETTE, X. *et al.* 2004. Inefficacy of infliximab in primary Sjogren's syndrome: results of the randomized, controlled trial of Remicade in primary Sjogren's syndrome (TRIPSS). Arthritis Rheum. **50:** 1270–1276.
11. SANKAR, V. *et al.* 2004. Etanercept in Sjogren's syndrome: a twelve-week randomized, double-blind, placebo-controlled pilot clinical trial. Arthritis Rheum. **50:** 2240–2245.
12. PIJPE, J. *et al.* 2005. Rituximab treatment in patients with primary Sjogren's syndrome: an open-label phase II study. Arthritis Rheum. **52:** 2740–2750.
13. STEINFELD, S.D. & P. YOUINOU. 2006. Epratuzumab (humanised anti-CD22 antibody) in autoimmune diseases. Expert Opin. Biol. Ther. **6:** 943–949.
14. VITALI, C. *et al.* 2002. Classification criteria for Sjögren's syndrome: a revised version of the European criteria proposed by the American-European Consensus Group. Ann. Rheum. Dis. **61:** 554–558.
15. OXHOLM, P. 1992. Primary Sjogren's syndrome—clinical and laboratory markers of disease activity. Semin. Arthritis Rheum. **22:** 114–126.
16. SOLIOTIS, F.C. & H.M. MOUTSOPOULOS. 2004. Sjögren's syndrome. Autoimmunity **37:** 305–307.
17. LYONS, R. *et al.* 2005. Effective use of autoantibody tests in the diagnosis of systemic autoimmune disease. Ann. NY Acad. Sci. **1050:** 217–228.
18. GRISIUS, M.M., D.K. BERMUDEZ & P.C. FOX. 1997. Salivary and serum interleukin 6 in primary Sjogren's syndrome. J. Rheumatol. **24:** 1089–1091.
19. SZODORAY, P. *et al.* 2004. Circulating cytokines in primary Sjogren's syndrome determined by a multiplex cytokine array system. Scand. J. Immunol. **59:** 592–599.
20. GOULES, A. *et al.* 2006. Elevated levels of soluble CD40 ligand (sCD40L) in serum of patients with systemic autoimmune diseases. J. Autoimmun. **26:** 165–171.
21. JONSSON, M.V. *et al.* 2005. Elevated serum levels of soluble E-cadherin in patients with primary Sjogren's syndrome. Scand. J. Immunol. **62:** 552–559.
22. SOHAR, N., I. SOHAR & H. HAMMER. 2005. Lysosomal enzyme activities: new potential markers for Sjogren's syndrome. Clin. Biochem. **38:** 1120–1126.
23. SZODORAY, P. *et al.* 2005. Distinct profiles of Sjogren's syndrome patients with ectopic salivary gland germinal centers revealed by serum cytokines and BAFF. Clin. Immunol. **117:** 168–176.
24. GOTTENBERG, J.E. *et al.* 2005. Correlation of serum B lymphocyte stimulator and beta2 microglobulin with autoantibody secretion and systemic involvement in primary Sjogren's syndrome. Ann. Rheum. Dis. **64:** 1050–1055.
25. TURKCAPAR, N. *et al.* 2005. Vasculitis and expression of vascular cell adhesion molecule-1, intercellular adhesion molecule-1, and E-selectin in salivary glands of patients with Sjogren's syndrome. J. Rheumatol. **32:** 1063–1070.
26. ZIGON, P. *et al.* 2005. Are autoantibodies against a 25-mer synthetic peptide of M3 muscarinic acetylcholine receptor a new diagnostic marker for Sjogren's syndrome? Ann. Rheum. Dis. **64:** 1247.

27. Miyazaki, K. *et al.* 2005. Analysis of in vivo role of alpha-fodrin autoantigen in primary Sjogren's syndrome. Am. J. Pathol. **167:** 1051–1059.
28. Tishler, M. *et al.* 1999. Increased salivary interleukin-6 levels in patients with primary Sjogren's syndrome. Rheumatol. Int. **18:** 125–127.
29. Ryu, O.H. *et al.* 2006. Rheumatology **45:** 1077–1086.
30. Abstracts accessed at www.sjogrens.org/research/ISSS/poster_abstracts.html.
31. Baechler, E.C. *et al.* 2006. Gene expression profiling in human autoimmunity. Immunol. Rev. **210:** 120–137.
32. Forabosco, P. *et al.* 2006. Meta-analysis of genome-wide linkage studies of systemic lupus erythematosus. Genes Immun. Sep 14 [Epub ahead of print].
33. Centola, M. *et al.* 2006. Genome-scale assessment of molecular pathology in systemic autoimmune diseases using microarray technology: a potential breakthrough diagnostic and individualized therapy-design tool. Scand. J. Immunol. **64:** 236–242.
34. Feng, Y. *et al.* 2004. Parallel detection of autoantibodies with microarrays in rheumatoid diseases. Clin. Chem. **50:** 416–422.

Salivary Proteome and Its Genetic Polymorphisms

FRANK G. OPPENHEIM, ERDJAN SALIH, WALTER L. SIQUEIRA, WEIMIN ZHANG, AND EVA J. HELMERHORST

Department of Periodontology and Oral Biology, Boston University, Goldman School of Dental Medicine, Boston, Massachusetts 02118, USA

ABSTRACT: Salivary diagnostics for oral as well as systemic diseases is dependent on the identification of biomolecules reflecting a characteristic change in presence, absence, composition, or structure of saliva components found under healthy conditions. Most of the biomarkers suitable for diagnostics comprise proteins and peptides. The usefulness of salivary proteins for diagnostics requires the recognition of typical features, which make saliva as a body fluid unique. Salivary secretions reflect a degree of redundancy displayed by extensive polymorphisms forming families for each of the major salivary proteins. The structural differences among these polymorphic isoforms range from distinct to subtle, which may in some cases not even affect the mass of different family members. To facilitate the use of modern state-of-the-art proteomics and the development of nanotechnology-based analytical approaches in the field of diagnostics, the salient features of the major salivary protein families are reviewed at the molecular level. Knowledge of the structure and function of salivary gland–derived proteins/peptides has a critical impact on the rapid and correct identification of biomarkers, whether they originate from exocrine or non-exocrine sources.

KEYWORDS: saliva; protein; polymorphism; isoforms; genetic; proteomics; oral diagnostics

INTRODUCTION

Salivary Polymorphisms

The complex field of salivary biochemistry has enjoyed considerable progress over the last decades. Early efforts in this area using classical electrophoretic and chromatographic techniques have been successful in obtaining pure proteins and have made it feasible to gain insights into the structural

Address for correspondence: Frank G. Oppenheim, D.M.D., Ph.D., Department of Periodontology and Oral Biology, 700 Albany Street, CABR W-201, Boston, MA 02118. Voice: 617-638-4756; fax: 617-638-4765.
fropp@bu.edu

Ann. N.Y. Acad. Sci. 1098: 22–50 (2007).
doi: 10.1196/annals.1384.030

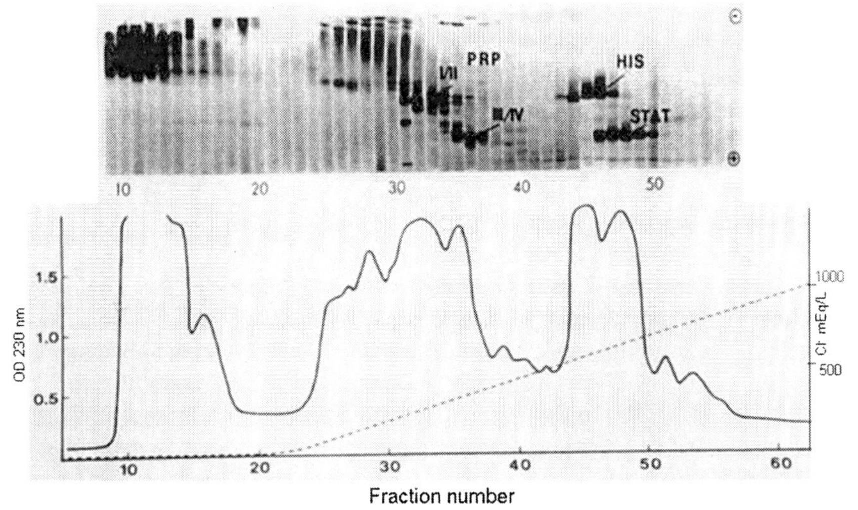

FIGURE 1. Anion-exchange chromatogram of 1 g of parotid saliva protein. *Top panel* shows the electrophoretic mobility of proteins present in individual collected fractions. Data from Oppenheim *et al.*, 1982.[1]

and functional characteristics of the major salivary proteins. To demonstrate the multitude of proteins present in human parotid secretion, FIGURE 1 shows an anion-exchange chromatogram in combination with gel electrophoresis of individual fractions. This method demonstrates the presence of at least 40 major proteins ranging in size and net charge and this separation has been amenable to generate fractions for the final purification of individual components.[1] Such investigations have resulted in the discovery and characterization of the major salivary proteins comprising amylases,[2–5] acidic, basic, and glycosylated proline-rich proteins (PRPs),[6–18] statherins,[19,20] histatins,[21–23] peroxidases,[24] cystatins,[25–33] and mucins.[34–37] A unique feature of many of these major proteins is that they display genetic polymorphisms and thus give rise to several families of structurally and functionally closely related molecules. The biosynthetic basis for these polymorphisms has been linked to gene duplication, differential mRNA splicing, and posttranslational modifications, such as proteolysis, phosphorylation, and glycosylation. In some cases, polymorphic forms differ significantly in size. In other instances, polymorphic forms show only minor amino acid rearrangements that do not affect the mass of such polymorphic molecules. The fact that genetic polymorphisms are so commonly found among the major salivary proteins indicates that these proteins have been subject to evolutionary pressures, which may reflect nature's selection for improved function.

In light of the recent developments to use salivary parameters for diagnostic purposes, the abundance of salivary polymorphisms has added significant complexity to diagnostic exploitation. Particularly, isoforms that differ minimally in structure create obstacles for the employment of assays with inadequate power of discrimination. To facilitate the development of oral salivary diagnostic tests it is important to recognize the salient structural features of the major salivary protein families.

Histatins

Histatins are one of the smallest molecular weight proteins in human salivary secretions. They are secreted by both the parotid and submandibular glands and are unusual in their high content of histidine,[23] an amino acid whose occurrence is one of the lowest in all mammalian proteins. The major histatins in glandular secretions are histatins 1, 3, and 5, which account for about 80% of all histatins present in glandular secretions.[21,22] Twelve minor histatins have been isolated from parotid secretion by chromatography and sequenced by Edman degradation,[38] and additional smaller fragments may be found with mass spectrometry.[39] Two histatin genes, *HIS1* and *HIS2*, are responsible for the biosynthesis of histatin 1 and histatin 3, respectively. One distinct difference between these two products is that histatin 1 is phosphorylated at Ser in position 2, whereas histatin 3 is not. The lack of phosphorylation of histatin 3 is likely related to the substitution of Glu for Ala at position 4, thereby altering the phosphokinase recognition site. For histatin 5, no cDNA or gene has been found, and it is well accepted that this protein derives posttranslationally from a chymotryptic-like cleavage event carboxy-terminally to Tyr24 in histatin 3 (TABLE 1). Allelic variation has been reported in *HIS2* in subjects of African descent. The *HIS2(2)* allele encodes for a histatin 3 isomer, designated histatin 3-2, in which Arg22 is replaced by Gln22. Furthermore, because of the introduction of a stop codon, the histatin 3-2 is 27 residues instead of 32 residues in length.[40] Because of the loss of the proteolytic cleavage site after Arg22 in histatin 3-2, the histatin degradation pattern in subjects homozygous for this mutation is distinctly different, but without apparent pathological consequences.

TABLE 1. Polymorphic forms of histatins

Gene[a]	Protein	Amino acid sequence[b]
HIS1	Histatin 1	D**S**HEKRHHGYRRKFHEKHHSHREFPFYGDYGSNYLYDN (38)
HIS2	Histatin 3	DSHAKRHHGYKRKFHEKHHSHR-----G-YRSNYLYDN (32)
HIS2[2]	Histatin 3-2	DSHAKRHHGYKRKFHEKHHSHG-----G-YRSN----- (27)
-	Histatin 5	DSHAKRHHGYKRKFHEKHHSHR-----G-Y-------- (24)
		* * ****** ********** * *

[a]Data from references: 21, 22; Swiss-Prot accession numbers P15515 and P15516 for histatin 1 and 3, respectively.

[b]Bolded **S** = phosphorylated serine residue; * indicates amino acid homology.

Saliva is the only body fluid in which histatins have been found.[41] The concentrations of histatins in submandibular/sublingual secretion (SMSL), parotid secretion (PS), and whole saliva (WS) are shown in TABLE 7. The concentrations of histatins in WS are markedly lower than in pure glandular secretions. These concentration differences are likely due to the high proteolytic activity of WS causing a rapid degradation of histatins upon their release into the oral cavity.[23,42,43] In addition, complex formation and binding to oral hard and soft tissues may reduce free histatin levels in WS.[44,45] Multiple functions have been postulated for histatins, based on their antifungal and antibacterial activities studied extensively *in vitro,*[21,43,46–48] lipopolysaccharide, and tannin-binding properties[49,50] and their capacity to inhibit a variety of host and bacteria-derived proteases and toxins.[51–54] Overall, the functional activities of histatins in the oral cavity clearly point toward an important role for these proteins in the maintenance of a balanced oral microbial ecology.

Statherin

Statherin is also a small molecular weight salivary protein synthesized and secreted by the parotid and submandibular glands. Statherin contains two vicinal phosphoserine residues in positions 2 and 3 and is rich in tyrosine and proline.[19] For a long time statherin was considered to be present as a single protein until shorter statherin variants were discovered.[20] The major statherin comprises 43 residues, whereas statherin variant SV2 is missing an internal 10-residue segment (TABLE 2). SV1 and SV3 are lacking the carboxyl-terminal phenylalanine residue, but are otherwise identical to statherin and SV2, respectively. Close inspection of the gene structure of statherin shows that the SV2 variant is generated by an alternate mRNA splicing event omitting the transcription of exon 4 that codes for the internal 10 amino acid residues missing from this variant. The removal of the carboxyl-terminal phenylalanine in SV1 and SV3, however, is believed to represent posttranslational proteolysis steps.[20,55] Statherin levels in pure glandular secretions are shown in TABLE 7. Notably, as for histatins, statherin levels are substantially lower in WS. Besides proteolytic degradation, statherins may disappear from WS on account of their

TABLE 2. Polymorphic forms of statherins

Gene[a]	Protein	Amino acid sequence[b]	
STATH	Statherin	D**SS**EEKFLRRIGRFGYGYGPYQPVPEQPLYPQPYQPQYQQYTF	(43)
-	SV1	D**SS**EEKFLRRIGRFGYGYGPYQPVPEQPLYPQPYQPQYQQYT-	(42)
STATH	SV2	D**SS**EE----------YGYGPYQPVPEQPLYPQPYQPQYQQYTF	(33)
-	SV3	D**SS**EE----------YGYGPYQPVPEQPLYPQPYQPQYQQYT-	(32)
		***** ***************************	

[a]Data from references: 19, 20; Swiss-Prot accession number P02808.
[b]Bolded **S** = phosphorylated serine residue; * indicates amino acid homology.

strong affinity for tooth surfaces. Statherin is the only salivary protein that inhibits both primary calcium phosphate precipitation (spontaneous precipitation) and secondary calcium phosphate precipitation (crystal growth)[19,56–58] and thus is believed to be a principal player in maintaining oral fluid to be supersaturated with respect to calcium phosphate salts. This function is critical for the remineralization capacity of human saliva.

Cystatins

Cystatins comprise four superfamilies of cysteine-containing proteins that are widespread among tissues and body fluids of mammalian origin. The major cystatins occurring in salivary glandular secretion belong to the "secretory type 2" cystatins.[59] This family comprises cystatins S, SA, SN, C, and D, which are 120 to 122 residues in length.[27,59,60] The five genes encoding for cystatins SN, SA, C, S, and D have been characterized and named *CST1, 2, 3, 4,* and *5*, respectively.[26,28,29,61–64] There are, however, as many as 11 different cystatin isoforms that have been characterized to date. While the *CST1* gene encodes only for cystatin SN, the *CST2* gene comprises two alleles, generating cystatins SA1 and SA2. The *CST3* gene produces either cystatins C or a disease-associated variant containing a Leu to Gln substitution at position 68.[29] The *CST4* gene is represented by four alleles, encoding for unphosphorylated cystatin S, and three phosphorylated isomers that have been designated cystatins S1, S2, and SAIII. These phosphorylated cystatin S isomers differ only with respect to the number and the position of the phosphate groups.[27,31,65] The *CST5* gene is responsible for the expression of cystatin D. Two forms of cystatin D are known, one of which contains a fifth cysteine residue at position 26.[30] In all cystatin isoforms belonging to the "secretory type 2" class, the four cysteine residues located in the carboxyl-terminal half of the polypeptide chain are highly conserved, and confer the specific secondary structure to these proteins through intramolecular disulfide bridging.

On the transcriptional level it is interesting to note that *CST2* and *CST5* are uniquely expressed in submandibular and parotid glands. *CST1* and *CST4* are expressed not only in salivary glands, but also in lacrimal and tracheal glands, and in the epithelial lining of the gallbladder and seminal vesicles, whereas *CST3* is ubiquitously expressed in a wide variety of tissues.[66,67] The total amount of secreted cystatins is much more pronounced in SMSL than in PS.[68,69] Cystatins S, SA, and SN are abundant and can be readily detected in SMSL secretions. Their concentrations in PS, SMSL, and WS are shown in TABLE 7.[66] The data show that there are significant amounts of cystatins present in WS. This is surprising since many salivary proteins undergo rapid proteolytic degradation in the oral environment. Cystatins apparently resist in part such enzymatic breakdown, which is consistent with their protease-inhibiting functions described below.

Cystatins are potent inhibitors of cysteine proteases (CP) and as such have been suggested to play a vital role in controlling the proteolytic activity of lysosomal CP, such as cathepsins.[66] While intact cystatins do not inhibit serine proteases, specific domains of the cystatin polypeptide chain are homologous to known serine protease inhibitors.[70] This raises the possibility that cystatin fragmentation could lead to the generation of inhibitors of this class of enzymes, provided proper cleavage occurs in the oral cavity. The phosphorylated cystatins are major components of the *in vivo* formed acquired enamel pellicle[59] and these cystatins have also been shown to inhibit secondary calcium phosphate precipitation.[71] These latter features are similar to those found for other salivary phosphoproteins and contribute to the mineral homeostasis of tooth surfaces.[70–73]

Amylases

Alpha-amylase (α-1,4-α-D-glucan 4-glucanohydrolase) is the most predominant protein in human saliva and is mainly synthesized in the parotid salivary gland. Amylase is not only produced in the salivary glands, but is also synthesized by the pancreas, providing a second level of digestion of dietary starch.[74,75] The amylase genes encoding for salivary and pancreatic amylase are *AMY1* and *AMY2.* Some cancer tissues produce an amylase encoded by the *AMY2B* gene, formerly known as *AMY3* (TABLE 4). The predicted polypeptide chain product of the salivary amylase gene contains 496 amino acid residues and shows only a 3% amino acid sequence variation when compared to pancreatic amylase and 2% when compared to carcinoid amylase.

Salivary amylases consist of two families of isoenzymes, designated family A and family B. Family A enzymes are about 63 kDa in size and are N-glycosylated, most likely at Asn in the only consensus sequence consisting of Asn-Gly-Ser at residues 412–414.[2] Family A consists of three isoenzymes, isoenzyme 1, 3, and 5, of which isoenzyme 1 is the primary gene product, isoenzyme 3 originates from isoenzyme 1 after the incorporation of a sialic acid, and isoenzyme 5 is the deamidated form of isoenzyme 3. Deamidation of Asn residues has been suggested to occur in the regions comprising Glu-Asn-Gly-Lys-Asp at position 349–353 and Asn-Gly-Asn-Cys at position 459–462.[3] Family B enzymes exhibit a molecular weight of about 59 kDa, are nonglycosylated, and comprise isoenzymes 2, 4, and 6.[76–78] Isoenzyme 2 is the primary gene product, and isoenzymes 4 and 6 are generated after deamidation.[2,3,79] While the mechanism and regulation of deamidation are not fully understood, this posttranslational modification leads to polymorphic isoforms showing significant charge differences but only minor changes in size, resulting in the typical streaking patterns observed in two-dimensional gel electrophoretograms.[80]

TABLE 3. Polymorphic forms of cystatins

Locus/Gene[a]	Protein	Amino acid sequence[b]
CST1	SN	WSPKEEDRIIPGGIYNADLNDEWVQRALHFAISEYNKAT-KDDYYRRPLRVLRARQQTVG (59)
*CST2*1*	SA1	WSPQEEDRIIEGGIYDADLNDERVQRALHFVISEYNKAT-EDEYYRRLLRVLRAREQIVG (59)
*CST2*2*	SA2	WSPQEEDRIIEGGIYDADLNDERVQRALHFVISEYNKAT-EDEYYRRLLRVLRAREQIVD (59)
CST3	C	-SSPGKPPRLVGGPMDASVEEEGVRRALDFAVGEYNKAS-NDMYHSRALQVVRARKQIVA (58)
CST3	L68Q/C	-SSPGKPPRLVGGPMDASVEEEGVRRALDFAVGEYNKAS-NDMYHSRALQVVRARKQIVA (58)
CST4	S	SSSKEENRIIPGGIYDADLNDEWVQRALHFAISEYNKAT-EDEYYRRPLQVLRAREQTFG (59)
CST4	S1	SS**S**KEENRIIPGGIYDADLNDEWVQRALHFAISEYNKAT-EDEYYRRPLQVLRAREQTFG (59)
CST4	S2	**SSS**KEENRIIPGGIYDADLNDEWVQRALHFAISEYNKAT-EDEYYRRPLQVLRAREQTFG (59)
CST4	SAIII	SS**S**KEENRIIPGGIYDADLNDEWVQRALHFAISEYNKAT-EDEYYRRPLQVLRAREQTFG (59)
CST5	D	GSASAQSRTLAGGIHATDLNDKSVQCALDFAISEYNKVINKDEYYSRPLQVMAAYQQIVG (60)
CST5	R^{26}/D	GSASAQSRTLAGGIHATDLNDKSVQRALDFAISEYNKVINKDEYYSRPLQVMAAYQQIVG (60)
		* ** * ** * **** * * * * * * *
CST1	SN	GVNYFFDVEVGRTICTKSQPNLDTCAFHEQPELQKKQLCSFEIYEVPWENRRSLVKSRCQ (119)
*CST2*1*	SA1	GVNYFFDIEVGRTICTKSQPNLDTCAFHEQPELQKKQLCSFQIYEVPWEDRMSLVNSRCQ (119)
*CST2*2*	SA2	GVNYFFDIEVGRTICTKSQPNLDTCAFHEQPELQKKQLCSFQIYEVPWEDRMSLVNSRCQ (119)
CST3	C	GVNYFLDVELGRTTCTKTQPNLDNCPFHDQPHLKRKAFCSFQIYAVPWQGTMTLSKSTCQ (118)

Continued.

TABLE 3. Continued

Locus/Gene[a]	Protein	Amino acid sequence[b]
CST3	C/L68Q	GVNYFLDVEQGRTTCTKTQPNLDNCPFHDQPHLKRKAFCSFQIYAVPWQGTMTLSKSTCQ (118)
CST4	S	GVNYFFDVEVGRTICTKSQPNLDTCAFHEQPELQKKQLCSFEIYEVPWEDRMSLVNSRCQ (119)
CST4	S1	GVNYFFDVEVGRTICTKSQPNLDTCAFHEQPELQKKQLCSFEIYEVPWEDRMSLVNSRCQ (119)
CST4	S2	GVNYFFDVEVGRTICTKSQPNLDTCAFHEQPELQKKQLCSFEIYEVPWEDRMSLVNSRCQ (119)
CST4	SAIII	GVNYFFDVEVGRTICTKSQPNLDTCAFHEQPELQKKQLC**S**FEIYEVPWEDRM**S**LVN**S**RCQ (119)
CST5	D	GVNYYFNVKFGRTTCTKSQPNLDNCPFNDQPKLKEEEFCSFQINEVPWEDKISILNYKCR (120)
CST5	R^{26}/D	GVNYYFNVKFGRTTCTKSQPNLDNCPFNDQPKLKEEEFCSFQINEVPWEDKISILNYKCR (120)
		**** *** *** ***** * * ** * *** * *** *
CST1	SN	ES (121)
*CST2*1*	SA1	EA (121)
*CST2*2*	SA2	DA (121)
CST3	C	DA (120)
CST3	C/L68Q	DA (120)
CST4	S	EA (121)
CST4	S1	EA (121)
CST4	S2	EA (121)
CST4	SAIII	EA (121)
CST5	D	KV (122)
CST5	R^{26}/D	KV (122)

[a]Data from references:27–33; Swiss-Prot accession numbers for cystatin SN, SA, C, S, and D are: P01037, P09228, P01034, P01036, and P28325.

[b]Bolded **S** = phosphorylated serine residue; * indicates amino acid homology.

TABLE 4. Polymorphic forms of amylases

Locus/Gene[a]	Protein	Amino acid sequence
AMY1	Salivary amylase	QYSSNTQQGRTSIVHLFEWRWVDIALECERYLAPKGFGGVQVSPPNENVAIHNPFRPWWE (60)
AMY2	Pancreatic amylase	QYSPNTQQGRTSIVHLFEWRWVDIALECERYLAPKGFGGVQVSPPNENVAIYNPFRPWWE (60)
AMY2B	Carcinoid amylase	QYSPNTQQGRTSIVHLFEWRWVDIALECERYLAPKGFGGVQVSPPNENVAIHNPFRPWWE (60)
		*** *** ********
AMY1	Salivary amylase	RYQPVSYKLCTRSGNEDEFRNMVTRCNNVGVRIYVDAVINHMCGNAVSAGTSSTCGSYFN (120)
AMY2	Pancreatic amylase	RYQPVSYKLCTRSGNEDEFRNMVTRCNNVGVRIYVDAVINHMCGNAVSAGTSSTCGSYFN (120)
AMY2B	Carcinoid amylase	RYQPVSYKLCTRSGNEDEFRNMVTRCNNVGVRIYVDAVINHMSGNAVSAGTSSTCGSYFN (120)
		** *****************
AMY1	Salivary amylase	PGSRDFPAVPYSGWDFNDGKCKTGSGDIENYNDATQVRDCRLSGLLDLALGKDYVRSKIA (180)
AMY2	Pancreatic amylase	PGSRDFPAVPYSGWDFNDGKCKTGSGDIENYNDATQVRDCRLTGLLDLALEKDYVRSKIA (180)
AMY2B	Carcinoid amylase	PGSRDFPAVPYSGWDFNDGKCKTGSGDIENYNDATQVRDCRLVGLLDLALEKDYVRSKIA (180)
		** ******* *********
AMY1	Salivary amylase	EYMNHLIDIGVAGFRIDASKHMWPGDIKAILDKLHNLNSNWFPEGSKPFIYQEVIDLGGE (240)
AMY2	Pancreatic amylase	EYMNHLIDIGVAGFRLDASKHMWPGDIKAILDKLHNLNSNWFPAGSKPFIYQEVIDLGGE (240)
AMY2B	Carcinoid amylase	EYMNHLIDIGVAGFRLDASKHMWPGDIKAILDKLHNLNSNWFPAGSKPFIYQEVIDLGGE (240)
		*************** *************************** ****************
AMY1	Salivary amylase	PIKSSDYFGNGRVTEFKYGAKLGTVIRKWNGEKMSYLKNWGEGWGFMPSDRALVFVDNHD (300)
AMY2	Pancreatic amylase	PIKSSDYFGNGRVTEFKYGAKLGTVIRKWNGEKMSYLKNWGEGWGFVPSDRALVFVDNHD (300)
AMY2B	Carcinoid amylase	PIKSSDYFGNGRVTEFKYGAKLGTVIRKWNGEKMSYLKNWGEGWGFMPSDRALVFVDNHD (300)
		** *************

Continued.

TABLE 4. Continued

Locus/Gene[a]	Protein	Amino acid sequence
AMY1	Salivary amylase	NQRGHGAGGASILTFWDARLYKMAVGFMLAHPYGFTRVMSSYRWPRYFENGKDVNDWVGP (360)
AMY2	Pancreatic amylase	NQRGHGAGGASILTFWDARLYKMAVGFMLAHPYGFTRVMSSYRWPRQFQNGNDVNDWVGP (360)
AMY2B	Carcinoid amylase	NQRGHGAGGASILTFWDARLYKMAVGFMLAHPYGFTRVMSSYRWPRQFQNGNDVNDWVGP (360)
		** * ** ********
AMY1	Salivary amylase	PNDNGVTKEVTINPDTTCGNDWVCEHRWRQIRNMVNFRNVVDGQPFTNWYDNGSNQVAFG (420)
AMY2	Pancreatic amylase	PNNNGVIKEVTINPDTTCGNDWVCEHRWRQIRNMVIFRNVVDGQPFTNWYDNGSNQVAFG (420)
AMY2B	Carcinoid amylase	PNNNGVIKEVTINPDTTCGNDWVCEHRWRQIRNMVNFRNVVDGQPFTNWYDNGSNQVAFG (420)
		** *** **************************** ************************
AMY1	Salivary amylase	RGNRGFIVFNNDDWTFSLTLQTGLPAGTYCDVISGDKINGNCTGIKIYVSDDGKAHFSIS (480)
AMY2	Pancreatic amylase	RGNRGFIVFNNDDWSFSLTLQTGLPAGTYCDVISGDKINGNCTGIKIYVSDDGKAHFSIS (480)
AMY2B	Carcinoid amylase	RGNRGFIVFNNDDWTFSLTLQTGLPAGTYCDVISGDKINGNCTGIKIYVSDDGKAHFSIS (480)
		************** ***
AMY1	Salivary amylase	NSAEDPFIAIHAESKL (496)
AMY2	Pancreatic amylase	NSAEDPFIAIHAESKL (496)
AMY2B	Carcinoid amylase	NSAEDPFIAIHAESKL (496)
		****************

[a]Data from references: 4, 5; Swiss-Prot accession numbers P04547, P04746, and P19961 for AMY1, AMY2, and AMY2B, respectively.
*indicates amino acid homology; underlined: putative deamidated residues; boxed: putative glycosylated regions.

Salivary amylase is most prominent in PS and virtually absent from SMSL secretion (TABLE 7). Functionally, α–amylase plays a key role in the partial digestion of dietary starch and glycogen initiated in the oral cavity by hydrolyzing α-1,4-glycosidic bonds. Salivary amylase also binds with high affinity to several species of oral streptococci,[81–84] suggesting an additional role of this protein in bacterial clearance.[85] Because of its high abundance, it is not surprising that amylase has been detected in the acquired enamel pellicle and in dental plaque,[72,80,86–88] but a clear functional role for this protein on the tooth surface has not been described.

Proline-Rich Proteins (PRPs)

Probably the most extensive polymorphism at the protein level is seen in the PRP families. PRPs also constitute a large fraction of the proteins of human parotid and submandibular secretions and comprise basic, acidic as well as glycosylated PRP.[89–91] Their unusually high content of proline, ranging between 25 and 40% of all amino acids, has only been detected in collagen-type proteins belonging to the connective tissue class and is therefore unique among such highly soluble proteins. Furthermore, their high proline, glycine, and glutamine content in the absence of any hydroxyproline and hydroxylysine are characteristics only found in these salivary proteins. PRPs have been characterized with respect to their physicochemical properties, amino acid sequences, posttranslational modifications, and protein polymorphisms.[6,7,10,55,91–94] PRPs constitute a large fraction of the PS and SMSL protein pool, representing approximately 30% of their protein content (TABLE 7).

The basic PRP (bPRP) family is encoded by genes in four separate loci comprising *PRB1*, *PRB2*, *PRB3*, and *PRB4*.[9,89,95–97] The *PRB1* and *PRB2* genes give rise to the nonglycosylated bPRP, and *PRB3* and *PRB4* generate glycosylated bPRP (TABLE 5). Each of the *PRB* genes is associated with several alleles giving rise to an extensive and complex pattern of polymorphism. The variety of these bPRP polymorphisms is dictated by individual amino acid deletions/insertions, different tandem repeat sequences, glycosylation, and proteolytic posttranslational modifications. The heterogeneity of these proteins both at the level of primary amino acid sequence, size and posttranslational modifications suggests significant functional diversity. The glycosylated bPRPs have been proposed to act as lubricants,[98] whereas nonglycosylated bPRPs have abilities to precipitate tannin and hence are believed to prevent absorption of this potential toxin from the alimentary canal.[99] The proteolytic cleavage of bPRPs as posttranslational modification may act as modulator of their biological function, such as in their interactions with bacteria and ability to precipitate tannin.

Acidic PRPs (aPRPs) are encoded for by two gene loci named *PRH1* and *PRH2*.[90] The *PRH1* locus contains alleles Db, PIF, and Pa, encoding for aPRPs

TABLE 5. Polymorphic forms of basic PRPs

Locus/gene[a]	Protein	Amino acid sequence[b]
PRB1	bPRP1L	---QNLNEDVSQEESPSLIAGNPQGPSPQGGNKPQGPPPPGKPQGPPPQGGNKPQGPPP (57)
PRB1	bPRP1M	---------------------NPQGPSPQGGNKPQGPPPPGKPQGPPPQGGNKPQGPPP (39)
PRB1	bPRP1S	---------------------NPQGPSPQGGNKPQGPPPPGKPQGPPPQGGNKPQGPPP (39)
PRB2	bPRP2L	NPQGAPPQGGNKPQGPPSPPGKPQGPPPQGGNQPQGPPPPGKPQGPPPQGGNKPQGPPP (60)
PRB3	bPRP3L	---QSLNEDVSQEESPSVISGKPEGRPPQGGNQSQGPPPRPGKPEGPPPQGGNQSQGPPP (57)
PRB3	bPRP3M	---------------------KPEGRRPQGGNQPQRTPPPPGKPEGRPPQGGNQSQGPPP (39)
PRB4	bPRP4L	---------------------KPQGRRPQGGNQPQRPPPPPGKPQGPPPQGGNQSQGPPP (39)
PRB4	bPRP4M	--------------------------------------PPPGKPQGRRPQGGNQPQRPPP (22)
PRB4	bPRP4S	---ESSSEDVSQEESLFL (15)
		↓
PRB1	bPRP1L	-PGKPQGPPPQG-DKSRSPRSPPGKPQGPPPQGGNQPQGPPPPPGKPQGPPPQGGNKPQG (115)
PRB1	bPRP1M	-PGKPQGPPPQG-DKSRSPRSPPGKPQGPPPQGGNQPQGPPPPPGKPQGPPPQGGNKPQG (97)
PRB1	bPRP1S	-PGKPQGPPPQG-DKSRSPRSPPGKPQGPPPQGGNQPQGPPPPPGKPQGPPPQGGN---- (93)
PRB2	bPRP2L	-PGKPQGPPPQG-DKSRSPRSPPGKPQGPPPQGGNQPQGPPPPPGKPQGPPPQGGNKPQG (118)
PRB3	bPRP3L	RPGKPEGQPPQGGNQSQGPPPRPGKPEGPPPQGGNQSQGPPPRPGEPEGPPPQGGNQSQG (117)
PRB3	bPRP3M	RPGKPEGPPPQGGNQSQGPPPRPGKPEGQPPQGGNQSQGPPPRPGKPEGPPPQGGNQSQG (99)
PRB4	bPRP4L	PPGKPEGRPPQGGNQSQGPPPHPGKPERPPPQGGNQSQGPPPHPGKPESRPPQGGHQSQG (99)
PRB4	bPRP4M	PPGKPQGPPPQGGNQSQGPPPHPGKPERPPPQGGNQSQGPPPHPGKPERPPPQGGNQSQG (82)
PRB4	bPRP4S	ISGKPEGRRPQGGNQPQRPPPPPGKPQGPPPQGGNQSQGPPPPPGKPEGRPPQGGNQSQG (75)
		↓
PRB1	bPRP1L	PPPPGKPQGPPPQGD-KSQSPRSPPGKPQGPPPQGGNQPQGPPPPPGKPQGPPPQGGNKP (174)
PRB1	bPRP1M	PPPPGKPQGPPP-- (109)
PRB1	bPRP1S	------------------↓--- (93)
PRB2	bPRP2L	PPPPGKPQGPPPQGDNKSRSSRSPPGKPQGPPPQGGNQPQGPPPPPGKPQGPPPQGGNKP (178)
PRB3	bPRP3L	PPPH-- (121)

Continued.

TABLE 5. Continued

Locus/gene[a]	Protein	Amino acid sequence[b]
PRB3	bPRP3M	PPPR-- (103)
PRB4	bPRP4L	PPPT-- (103)
PRB4	bPRP4M	PPPT-- (86)
PRB4	bPRP4S	PPPH-- (79)
		↓
PRB1	bPRP1L	QGPPPPGKPQGPPPQGD-KSQSPRSPPGKPQGPPPQGGNQPQGPPPPPGKPQGPPQQGGN (233)
PRB1	bPRP1M	--------------QGD-KSQSPRSPPGKPQGPPPQGGNQPQGPPPPPGKPQGPPQQGGN (154)
PRB1	bPRP1S	-- (93)
PRB2	bPRP2L	QGPPPPGKPQGPPPQGDNKSQSARSPPGKPQGPPPQGGNQPQGPPPPPGKPQGPPPQGGN (238)
PRB3	bPRP3L	---------------------------PGKPEGPPPQGGNQSQGPPPRPGKPEGPPPQGGN (155)
PRB3	bPRP3M	---------------------------PGKPEGPPPQGGNQSQGPPPHPGKPEGPPPQGGN (137)
PRB4	bPRP4L	---------------------------PGKPEGPPPQGGNQSQGTPPPPGKPEGRPPQGGN (137)
PRB4	bPRP4M	---------------------------PGKPEGPPPQGGNQ------------------- (100)
PRB4	bPRP4S	---------------------------PGKPERPPPQGGNQ------------------- (93)
PRB1	bPRP1L	RPQGPPP-PGKPQGPPPQG-DKSRSPQSPPGKPQGPPPQGGNQPQGPPPPPGKPQGPPPQ (291)
PRB1	bPRP1M	RPQGPPP-PGKPQGPPPQG-DKSRSPQSPPGKPQGPPPQGGNQPQGPPPPPGKPQGPPPQ (212)
PRB1	bPRP1S	RPQGPPP-PGKPQGPPPQG-DKSRSPRSPPGKPQGPPPQGGNQPQGPPPPPGKPQGPPPQ (151)
PRB2	bPRP2L	KSQGPPP-PGKPQGPPPQGGSKSRSSRSPPGKPQGPPPQGGNQPQGPPPPPGKPQGPPPQ (297)
PRB3	bPRP3L	QSQGPPPRPGKPEGPPPQGGNQSQGPPPRPGKPEGPPPQGGNQSQGPPPRPGKPEGSPSQ (215)
PRB3	bPRP3M	QSQGPPPHPGKPEGPPPQGGNQSQGPPPRPGKPEGPPPQGGNQSQGPPPRPGKPEGSPSQ (197)

Continued.

TABLE 5. Continued

Locus/gene[a]	Protein	Amino acid sequence[b]
PRB4	bPRP4L	QSQGPPPHPGKPERPPPQGGNQSHRPPPPPGKPERPPPQGGNQSQGPPPHPGKPEGPPPQ (197)
PRB4	bPRP4M	-SQGPPPHPGKPERPPPQGGNQSHRPPPPPGKPERPPPQGGNQSQGPPPHPGKPEGPPPQ (159)
PRB4	bPRP4S	-SQGTPPPPGKPERPPPQGGNQSHRPPPPPGKPERPPPQGGNQSQGPPPHPGKPEGPPPQ (152)
PRB1	bPRP1L	GGNKPQGP-PPPGKPQGPPAQGGSKSQSARAPPGKPQGPPQQEGNNPQGPPPPAGGNPQQ (350)
PRB1	bPRP1M	GGNKPQGP-PPPGKPQGPPAQGGSKSQSARAPPGKPQGPPQQEGNNPQGPPPPAGGNPQQ (271)
PRB1	bPRP1S	GGNKPQGP-PPPGKPQGPPAQGGSKSQSARSPPGKPQGPPQQEGNNPQGPPPPAGGNPQQ (210)
PRB2	bPRP2L	GGNKPQGP-PPPGKPQGPPPQGGSKSRSARSPPGKPQGPPQQEGNNPQGPPPPAGGNPQQ (356)
PRB3	bPRP3L	GGNKPRGPPPHPGKPQGPPPQEGNK----------PQRPPPP-RR-PQGPPPP-GGNPQQ (262)
PRB3	bPRP3M	GGNKPQGPPPHPGKPQGPPPQEGNK----------PQRPPPPPGR-PQGPPPP-GGNPQQ (245)
PRB4	bPRP4L	EGNKSRSARSPPGKPQGPPQQEGNK----------PQGPPPP-GK-PQGPPPP-GGNPQQ (244)
PRB4	bPRP4M	EGNKSRSARSPPGKPQGPPQQEGNK----------PQGPPPP-GK-PQGPPPP-GGNPQQ (206)
PRB4	bPRP4S	EGNKSRSARSPPGKPQGPPQQEGNK----------PQGPPPP-GK-PQGPPPA-GGNPQQ (199)
PRB1	bPRP1L	PQAPPAGQPQGPPRPPQGGRPSRPPQ------ (376)
PRB1	bPRP1M	PQAPPAGQPQGPPRPPQGGRPSRPPQ------ (297)
PRB1	bPRP1S	PQAPPAGQPQGPPRPPQGGRPSRPPQ------ (236)
PRB2	bPRP2L	PQAPPAGQPQGPPRPPQGGRPSRPPQ------ (382)
PRB3	bPRP3L	PLPPPAGKPQGPPPPPQGGRPHRPPQGQ-PPQ (293)
PRB3	bPRP3M	PLPPPAGKPQGPPPPPQGGRPHRPPQGQ-PPQ (276)
PRB4	bPRP4L	PQAPPAGKPQGPPPPPQGGRPPRPAQGQQPPQ (276)
PRB4	bPRP4M	PQAPPAGKPQGPPPPPQGGRPPRPAQGQQPPQ (238)
PRB4	bPRP4S	PQDPPAGKPQGPPPPPQGGRPPRPAQGQQPPQ (231)

[a]Data from references: 11–16, 89; Swiss-Prot accession numbers for bPRB1L, bPRB2L, bPRB3L, bPRB4L, bPRB4M, and bPRB4S are P04280, P02812, Q04118, P10162, P10161, P10163, respectively.

[b]Arrows point to proteolytic posttranslational cleavage sites, and underlined residues represent potential glycosylation sites.

Db-s, PIF-s and Pa, respectively. The *PRH2* locus contains the alleles *PR1* and *PR2*, encoding for PRP1 and PRP2, respectively. PIF-s, Pa, PRP1, and PRP2 are 150 residues in length whereas Db-s contains 171 residues on account of a 21-residue insert at position 82–102. Posttranslational proteolysis of PIF-s, PRP1, PRP2, and Db-s occurs through a tryptic cleavage event at Arg106 in PIF-s, PRP1, and PRP2 and at Arg127 in Db-s leading to the formation of PIF-f, PRP3, PRP4, and Db-f, respectively.[8,18] The structural similarities between the nine acidic PRPs are depicted in TABLE 6. The minor structural differences between the aPRP at positions 4 (Asn or Asp), 50 (Asn or Asp), and 26 (Ile or Leu) are apparent, as is the unique substitution of an Arg for a Cys at residue 124 in Pa. Because of this substitution, Pa is the only aPRP that appears in a monomeric as well as in a dimeric, disulfide-linked, form. Despite the minor structural differences among the aPRPs, they have all been resolved by anion exchange chromatography using shallow sodium chloride gradients for their elution.[90]

Acidic PRPs show a high affinity for hydroxyapatite and are potent inhibitors of secondary calcium phosphate precipitation, which is in large part due to their two phosphate groups linked covalently to Ser residues in positions 8 and 22.[100] Therefore, their predominant role in the oral cavity is believed to be related to mineral homeostasis and the maintenance of tooth integrity.

Mucins

Mucins are high molecular weight glycoproteins that cover the epithelial surfaces of the gastrointestinal, respiratory, and reproductive tracts.[101] SMSL has been shown to contain a large, gel-forming mucous glycoprotein of 20–40 million Da, referred to as MG1, and a much smaller mucin of 130–180 kDa, designated MG2.[34,65,102] MG1 represents a heterogeneous family of proteins comprising *MUC5B*, *MUC4,* and *MUC19* gene products.[36,103–105] Gel-forming mucins are typically large and characterized by a high number of oligosaccharide chains. The linear polypeptide backbone, called apomucin, contains alternating glycosylated and nonglycosylated domains. The highly glycosylated regions display an extended structure caused by charge repulsion between neighboring oligosaccharide groups. Biochemical analysis of gel-forming mucins has shown that they comprise approximately 15% protein, 78% carbohydrate, and 7% sulfate.[102,106] Individual mucin monomers consist of a large region located in the middle of the polypeptide chain containing variable numbers of tandem repeated sequences ("VNTR"), which are enriched in serine, threonine, alanine, glycine, and proline residues.[106] Mucins are primarily O-glycosylated at serine and threonine residues, which are clustered in the central tandem repeat regions.[107] The N- and C-terminal regions of MG1 are rich in cysteines, which are involved in intermolecular disulfide bridging.[108] Some mucins exhibit a significant degree of genetic polymorphism in the length of the genomic DNA sequence encoding for the VNTR regions.[106,109] Similar to MG1-like mucins, MG2, the second salivary mucin, contains threonine-rich, glycosylated tandem repeat regions and some degree of polymorphism in the number of repeats of these regions.[110] It differs from MG1 mucins in that it is approximately 200× smaller in size, contains 68% carbohydrate compared to 80% in MG1 mucins, and is represented only by one gene product designated MUC7.[35,111]

In the oral cavity, mucins are secreted by submandibular and sublingual glands, and various minor salivary glands, but not by the parotid glands.[112] Mucins constitute the third most abundant group of proteins, in saliva (TABLE 7) and form various complexes with other salivary proteins, thereby modulating their activities. The main function of mucins is related to the protection of the epithelium from desiccation, mechanical injury, and microbial attack. Mucins add to the viscoelastic and rheological properties of saliva, facilitating mastication, swallowing, speech, and protecting the teeth from abrasive forces. The large degree of microheterogeneity in the glycosylated regions of MG1 and MG2, which is typical for such large glycosylated molecules, combined with polymorphism in the number of VNTR and the high viscosity of these components, poses a considerable challenge in the analysis of mucin molecules for diagnostic purposes. The exploitation of mucin molecules for diagnostics is gaining increasing interest in a variety of disease conditions.[113] While there have not been adequate studies performed to fully explore the diagnostic value of mucin isoforms in saliva, the potential of these molecules to reflect health and various disease conditions cannot be ignored and will likely be subject to further exploration.

Proteomics in Saliva Biochemistry

The completion of the Human Genome Project and advances made in state-of-the-art mass spectrometric techniques have led to the emergence of a variety of studies encompassing the large field of biological/biomedical sciences to analyze complex biological samples at the protein level. One of the areas of interest in salivary proteomics is the field of salivary diagnostics since oral fluid has the potential to reflect both oral and systemic health conditions.[114–117] The most widely used proteomics approaches to characterize complex mixtures of proteins/peptides consist of the popular so-called "bottom-up" mass spectrometric identification of proteins based on peptides generated with trypsinization. Application of such an approach using high-throughput and sensitive MS technology to perform rapid peptide profiling and sequencing by MS/MS makes it feasible to define global differential protein expressions.[117–122] However, the small structural differences found among polymorphic forms of salivary proteins render unequivocal identification of such proteins in mixtures a challenging task. This is further complicated by some annotation characteristics, which are based predominantly on genetic information and do not

TABLE 6. Polymorphic forms of acidic PRPs

Locus/Gene[a]	Protein	Amino acid sequence[b]
PRH2/PR1	PRP1	QDLDEDV**S**QEDVPLVISDGGD**S**EQFIDEERQGPPLGGQQSQPSAGDGNQNDGPQQGPPQQ (60)
-	PRP3	QDLDEDV**S**QEDVPLVISDGGD**S**EQFIDEERQGPPLGGQQSQPSAGDGNQNDGPQQGPPQQ (60)
PRH2/PR2	PRP2	QDLDEDV**S**QEDVPLVISDGGD**S**EQFIDEERQGPPLGGQQSQPSAGDGNQDDGPQQGPPQQ (60)
-	PRP4	QDLDEDV**S**QEDVPLVISDGGD**S**EQFIDEERQGPPLGGQQSQPSAGDGNQDDGPQQGPPQQ (60)
PRH1/Pa	Pa	QDLNEDV**S**QEDVPLVISDGGD**S**EQFLDEERQGPPLGGQQSQPSAGDGNQDDGPQQGPPQQ (60)
PRH1/PIF	PIF-s	QDLNEDV**S**QEDVPLVISDGGD**S**EQFIDEERQGPPLGGQQSQPSAGDGNQDDGPQQGPPQQ (60)
-	PIF-f	QDLNEDV**S**QEDVPLVISDGGD**S**EQFIDEERQGPPLGGQQSQPSAGDGNQDDGPQQGPPQQ (60)
PRH1/Db	Db-s	QDLNEDV**S**QEDVPLVISDGGD**S**EQFLDEERQGPPLGGQQSQPSAGDGNQDDGPQQGPPQQ (60)
-	Db-f	QDLNEDV**S**QEDVPLVISDGGD**S**EQFLDEERQGPPLGGQQSQPSAGDGNQDDGPQQGPPQQ (60)
		*** ******************** ************************ **********
PRH2/PR1	PRP1	GGQQQQGPPPPQGKPQGPPQQGGHPPPPQGRPQGPPQQGGHPRPPRGRPQGPPQQGGHQQ (120)
-	PRP3	GGQQQQGPPPPQGKPQGPPQQGGHPPPPQGRPQGPPQQGGHPRPPR-------------- (106)
PRH2/PR2	PRP2	GGQQQQGPPPPQGKPQGPPQQGGHPPPPQGRPQGPPQQGGHPRPPRGRPQGPPQQGGHQQ (120)
-	PRP4	GGQQQQGPPPPQGKPQGPPQQGGHPPPPQGRPQGPPQQGGHPRPPR-------------- (106)
PRH1/Pa	Pa	GGQQQQGPPPPQGKPQGPPQQGGHPPPPQGRPQGPPQQGGHPCPPRGRPQGPPQQGGHQQ (120)
PRH1/PIF	PIF-s	GGQQQQGPPPPQGKPQGPPQQGGHPPPPQGRPQGPPQQGGHPRPPRGRPQGPPQQGGHQQ (120)
-	PIF-f	GGQQQQGPPPPQGKPQGPPQQGGHPPPPQGRPQGPPQQGGHPRPPR-------------- (106)
PRH1/Db	Db-s	GGQQQQGPPPPQGKPQGPPQQGGQ------QQQGPP-------PPQGKPQGPPQQGGH-- (105)
-	Db-f	GGQQQQGPPPPQGKPQGPPQQGGQ------QQQGPP-------PPQGKPQGPPQQGGH-- (105)
		*********************** ****** **

Continued.

TABLE 6. Continued

Locus/Gene[a]	Protein	Amino acid sequence[b]
PRH2/PR1	PRP1	GPPPPPPGKPQGPPPQGGRPQGPPQGQSPQ------------------------------ (150)
-	PRP3	-- (106)
PRH2/PR2	PRP2	GPPPPPPGKPQGPPPQGGRPQGPPQGQSPQ------------------------------ (150)
-	PRP4	-- (106)
PRH1/Pa	Pa	GPPPPPPGKPQGPPPQGGRPQGPPQGQSPQ------------------------------ (150)
PRH1/PIF	PIF-s	GPPPPPPGKPQGPPPQGGRPQGPPQGQSPQ------------------------------ (150)
-	PIF-f	-- (106)
PRH1/Db	Db-s	--PPPPQGRPQGPPQQGGHPR-PPRGRPQGPPQQGGHQQGPPPPPPGKPQGPPPQGGRPQ (163)
-	Db-f	--PPPPQGRPQGPPQQGGHPR-PPR----------------------------------- (127)
PRH2/PR1	PRP1	--------- (150)
-	PRP3	--------- (106)
PRH2/PR2	PRP2	--------- (150)
-	PRP4	--------- (106)
PRH1/Pa	Pa	--------- (150)
PRH1/PIF	PIF-s	--------- (150)
-	PIF-f	--------- (106)
PRH1/Db	Db-s	GPPQGQSPQ (172)
-	Db-f	--------- (127)

[a]Data from references: 8, 18; Swiss-Prot accession number P02810.

[b]Bolded **S** = phosphorylated serine residue; * indicates amino acid homology.

TABLE 7. Concentrations (μg/mL) of salivary proteins in glandular salivary secretions and WS[a]

	PS	*SMSL*	*WS*
Amylase	650–800	–	380–500
MG1	–	80–560	80–500
MG2	–	21–230	10–200
Histatin	30–55	13–70	2–8
Statherin	18–209	20–150	2–12
aPRP	190–800	100–400	90–180
Cystatin	2–4	267–280	240–280
Total protein	1,000–2,000	1,000–3,000	2,000–4,000

[a]Values represent best estimates of data compiled from References 57, and 125–131. The values represent concentrations comprising all isoforms of a particular protein family.

fully take into account the structural data obtained from the characterization of isolated proteins.

Modern MS analysis of peptides involves a variety of proteomics instruments and technologies. The use of liquid chromatography electrospray ionization mass spectrometry (LC-ESI-MS) in conjunction with peptide sequence analysis through collision-induced dissociation (CID) and sophisticated search algorithms has enabled large-scale protein identification, with minimal requirement for protein purification.[123,124] In general, in the bottom-up approach, the protein samples are digested, and the smaller constituent peptide fragments are the basis for subsequent protein identification. A prerequisite for definitive protein identifications is that at least one of the tryptic peptides that is generated is unique for a particular protein. From the high sequence similarity of the salivary protein isoforms shown in TABLES 1 to 6, it is evident that many tryptic peptides are not unique for a particular isoform within the same salivary protein family. To illustrate this, FIGURE 2A shows the mass spectrum obtained with an LTQ LC-mass spectrometer (Thermo-Finnigan) of a mixture of tryptic peptides generated from purified PRP1. The tandem MS spectrum of one of these peptides is shown in FIGURE 2B and its sequence was identified to represent residues K.PQGPPQQGGHPPPPQGR.P. Since this peptide is common to all aPRPs, a GenBank database search would not have been able to specify that this peptide belongs to PRP1. Indeed, neither would a manual approach have allowed the assignment of the specific parent aPRP isoform since the sequence is present in residues 75–91 in PRP1-4, Pa, PIF-s and PIF-f, and residues 109–125 in Db-s and Db-f. This example illustrates the general problems associated with the identification of a member of a polymorphic protein family based on common peptides.

WS, representing a mixture of salivary proteins and their proteolytic degradation patterns, has classically been considered more valuable for diagnostic purposes than glandular secretions for a variety of reasons. When the bottom-up approach was applied to a sample of WS, a PRP-derived peptide was

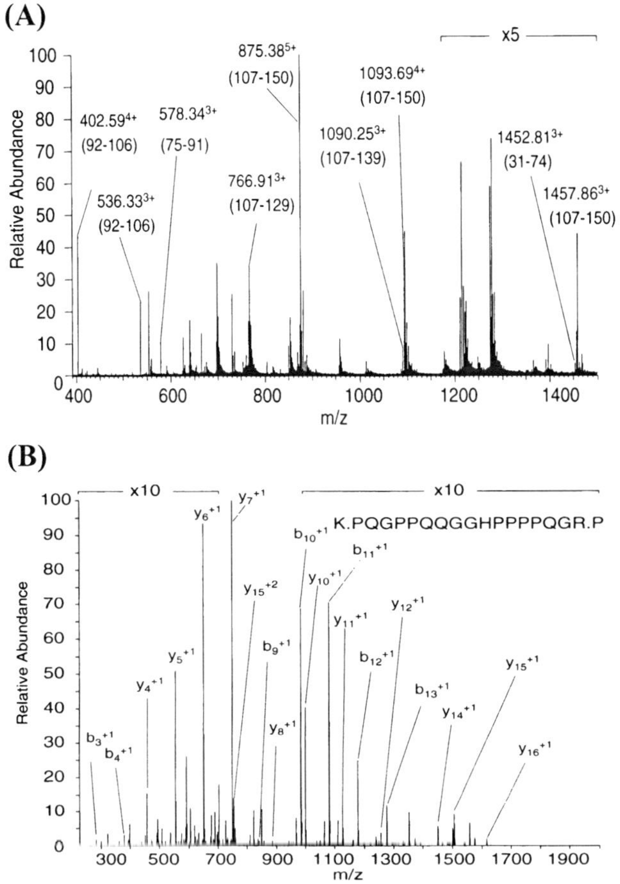

FIGURE 2. MS and MS/MS sequence analysis of the tryptic digest of purified aPRP1. (**A**) Peptide fingerprint of the purified aPRP1 tryptic digest by MS analysis. Approximately 300 femtomoles of the PRP1 tryptic digest was analyzed, which provided almost complete coverage of the tryptic peptides generated, each of which was identified from the MS/MS sequence data. The specific tryptic peptide regions corresponding to the aPRP1 amino acid sequences are indicated by the m/z and the residue numbers. (**B**) MS/MS spectrum of one of the aPRP1 tryptic peptides from **A** above. The peptide with m/z = 578.34^{3+} corresponding to the aPRP1 peptide with sequence K.PQGPPQQGGHPPPPQGR.P was identified and the corresponding "b" and "y" ion series are indicated.

identified with the sequence R.PQGPPQQGGHPRPPR.G (FIG. 3A). The sequence of this peptide did not allow unequivocal determination from which of the PRP isoforms it was derived. In fact, there is only one theoretical aPRP-derived peptide that would allow proper assignment, and that is a peptide containing the cysteine residue that is unique to the Pa isoform (TABLE 6). The challenges associated with the identification of salivary proteins based

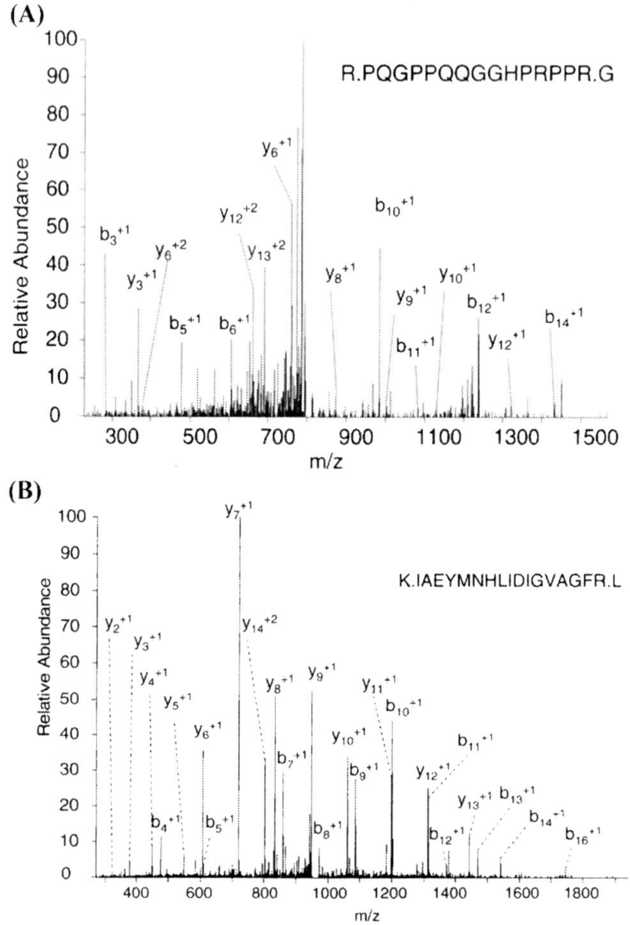

FIGURE 3. MS/MS sequence analysis of a WS tryptic digest. (**A**) MS/MS spectrum of one of the aPRP tryptic peptides identified in WS. The "b" and "y" fragment ions are as indicated. The derived sequence corresponded to peptide R.PQGPPQQGGHPRPPR.G, representing residues 92–106 in aPRP1, aPRP2, aPRP3, aPRP4, PIF-s, and PIF-f and residues 113–127 in Db-s and Db-f. (**B**) MS/MS spectrum of one of the α-amylase tryptic peptides identified in WS. The fragment corresponded to peptide K.IAEYMNHLIDIGVAGFR.L., representing residues 179–195 in various amylase isoforms.

on proteolytic digests of highly homologous groups of salivary proteins and MS/MS have made it virtually impossible to discriminate between individuals carrying a particular isoform pattern. Furthermore, comparing isoforms of salivary proteins shown in TABLES 1–6 reveals that problems of differentiation between phenotypes are not only relevant for the presence but also for the absence of a particular isoform. For example, the absence of PRP1 could remain entirely unnoticed in a tryptic digest of salivary proteins. Another challenge to overcome is associated with the database-dependent assignment of a tryptic peptide to a protein. Databases are biased toward one or the other protein promoted by "greedy" search algorithms selecting proteins depending on the order of their listing in the database. An example of this is given with peptide K.IAEYMNHLIDIGVAGFR.L, which was identified in WS (FIG. 3B). This peptide is present in both pancreatic and salivary amylase, but was assigned by the database search to pancreatic amylase. The probability, however, that this peptide was in fact derived from pancreatic amylase in the oral fluid sample is extremely low.

Regardless of the protein source, differences between health and disease may be manifested by only minor structural differences in proteins. These structural differences may be caused, for example, by point mutations, or subtle changes in posttranslational modifications, such as phosphorylation and glycosylation. In order to be successful in using human body fluids for diagnostic purposes, technologies must be available that can discriminate between such small structural differences among proteins. This is particularly true if individual proteins from a polymorphic protein family, such as typically found in salivary secretions, are affected by or are causative of a particular disease state. Advances in mass spectrometry paralleled by the rapid progress in microfluidics and microsensor array technologies hold great promise to bring oral fluid diagnostics to a level of discrimination adequate for widespread use in clinical medicine.

ACKNOWLEDGMENTS

This work was supported by NIH/NIDCR Grants DE05672, DE07652, and DE14950.

REFERENCES

1. OPPENHEIM, F.G., G.D. OFFNER & R.F. TROXLER. 1982. Phosphoproteins in the parotid saliva from the subhuman primate *Macaca fascicularis*. Isolation and characterization of a proline-rich phosphoglycoprotein and the complete covalent structure of a proline-rich phosphopeptide. J. Biol. Chem. **257:** 9271–9282.
2. KELLER, P.J. *et al.* 1971. Further studies on the structural differences between the isoenzymes of human parotid α-amylase. Biochemistry **10:** 4867–4874.
3. BANK, R.A. *et al.* 1991. Electrophoretic characterization of posttranslational modifications of human parotid salivary alpha-amylase. Electrophoresis **12:** 74–79.
4. NISHIDE, T. *et al.* 1986. Primary structure of human salivary alpha-amylase gene. Gene **41:** 299–304.
5. NISHIDE, T. *et al.* 1984. Corrected sequences of cDNAs for human salivary and pancreatic alpha-amylases [corrected]. Gene **28:** 263–270.

6. OPPENHEIM, F.G., D.I. HAY & C. FRANZBLAU. 1971. Proline-rich proteins from human parotid saliva. I. Isolation and partial characterization. Biochemistry **10:** 4233–4238.
7. BENNICK, A. & G.E. CONNELL. 1971. Purification and partial characterization of four proteins from human parotid saliva. Biochem. J. **123:** 455–464.
8. HAY, D.I. *et al.* 1988. The primary structures of six human salivary acidic proline-rich proteins (PRP-1, PRP-2, PRP-3, PRP-4, PIF-s and PIF-f). Biochem. J. **255:** 15–21.
9. LYONS, K.M. *et al.* 1988. Many protein products from a few loci: assignment of human salivary proline-rich proteins to specific loci. Genetics **120:** 255–265.
10. WONG, R.S. & A. BENNICK. 1980. The primary structure of a salivary calcium-binding proline-rich phosphoprotein (protein C), a possible precursor of a related salivary protein A. J. Biol. Chem. **255:** 5943–5948.
11. ISEMURA, S., E. SAITOH & K. SANADA. 1979. Isolation and amino acid sequences of proline-rich peptides of human whole saliva. J. Biochem. (Tokyo). **86:** 79–86.
12. ISEMURA, S., E. SAITOH & K. SANADA. 1980. The amino acid sequence of a salivary proline-rich peptide, P-C, and its relation to a salivary proline-rich phosphoprotein, protein C. J. Biochem. (Tokyo) **87:** 1071–1077.
13. ISEMURA, S., E. SAITOH & K. SANADA. 1982. Fractionation and characterization of basic proline-rich peptides of human parotid saliva and the amino acid sequence of proline-rich peptide P-E. J. Biochem. (Tokyo) **91:** 2067–2075.
14. SAITOH, E., S. ISEMURA & K. SANADA. 1983. Complete amino acid sequence of a basic proline-rich peptide, P-D, from human parotid saliva. J. Biochem. (Tokyo) **93:** 495–502.
15. SAITOH, E., S. ISEMURA & K. SANADA. 1983. Complete amino acid sequence of a basic proline-rich peptide, P-F, from human parotid saliva. J. Biochem. (Tokyo) **93:** 883–888.
16. SAITOH, E., S. ISEMURA & K. SANADA. 1983. Further fractionation of basic proline-rich peptides from human parotid saliva and complete amino acid sequence of basic proline-rich peptide P-H. J. Biochem. (Tokyo) **94:** 1991–1999.
17. AZEN, E.A. *et al.* 1990. Alleles at the PRB3 locus coding for a disulfide-bonded human salivary proline-rich glycoprotein (Gl 8) and a null in an Ashkenazi Jew. Am. J. Hum. Genet. **47:** 686–697.
18. AZEN, E.A. *et al.* 1987. Alleles at the PRH1 locus coding for the human salivary-acidic proline-rich proteins Pa, Db, and PIF. Am. J. Hum. Genet. **41:** 1035–1047.
19. SCHLESINGER, D.H. & D.I. HAY. 1977. Complete covalent structure of statherin, a tyrosine-rich acidic peptide which inhibits calcium phosphate precipitation from human parotid saliva. J. Biol. Chem. **252:** 1689–1695.
20. JENSEN, J.L. *et al.* 1991. Multiple forms of statherin in human salivary secretions. Arch. Oral Biol. **36:** 529–534.
21. OPPENHEIM, F.G. *et al.* 1988. Histatins, a novel family of histidine-rich proteins in human parotid secretion. Isolation, characterization, primary structure, and fungistatic effects on *Candida albicans*. J. Biol. Chem. **263:** 7472–7477.
22. OPPENHEIM, F.G. *et al.* 1986. The primary structure and functional characterization of the neutral histidine-rich polypeptide from human parotid secretion. J. Biol. Chem. **261:** 1177–1182.
23. BAUM, B.J. *et al.* 1976. Studies on histidine-rich polypeptides from human parotid saliva. Arch. Biochem. Biophys. **177:** 427–436.

24. MANSSON-RAHEMTULLA, B., F. RAHEMTULLA & M.G. HUMPHREYS-BEHER. 1990. Human salivary peroxidase and bovine lactoperoxidase are cross-reactive. J. Dent. Res. **69:** 1839–1846.
25. ISEMURA, S., E. SAITOH & K. SANADA. 1986. Characterization of a new cysteine proteinase inhibitor of human saliva, cystatin SN, which is immunologically related to cystatin S. FEBS Lett. **198:** 145–149.
26. THIESSE, M., S.J. MILLAR & D.P. DICKINSON. 1994. The human type 2 cystatin gene family consists of eight to nine members, with at least seven genes clustered at a single locus on human chromosome 20. DNA Cell Biol. **13:** 97–116.
27. LAMKIN, M.S. *et al.* 1991. Salivary cystatin SA-III, a potential precursor of the acquired enamel pellicle, is phosphorylated at both its amino- and carboxyl-terminal regions. Arch. Biochem. Biophys. **288:** 664–670.
28. FREIJE, J.P. *et al.* 1991. Structure and expression of the gene encoding cystatin D, a novel human cysteine proteinase inhibitor. J. Biol. Chem. **266:** 20538–20543.
29. ABRAHAMSON, M. *et al.* 1992. Hereditary cystatin C amyloid angiopathy: identification of the disease-causing mutation and specific diagnosis by polymerase chain reaction based analysis. Hum. Genet. **89:** 377–380.
30. BALBIN, M. *et al.* 1993. A sequence variation in the human cystatin D gene resulting in an amino acid (Cys/Arg) polymorphism at the protein level. Hum. Genet. **90:** 668–669.
31. ISEMURA, S. *et al.* 1991. Identification of full-sized forms of salivary (S-type) cystatins (cystatin SN, cystatin SA, cystatin S, and two phosphorylated forms of cystatin S) in human whole saliva and determination of phosphorylation sites of cystatin S. J. Biochem. (Tokyo) **110:** 648–654.
32. SHINTANI, M. *et al.* 1994. Genetic polymorphisms of the CST2 locus coding for cystatin SA. Hum. Genet. **94:** 45–49.
33. GRUBB, A. & H. LOFBERG. 1982. Human gamma-trace, a basic microprotein: amino acid sequence and presence in the adenohypophysis. Proc. Natl. Acad. Sci. USA **79:** 3024–3027.
34. LOOMIS, R.E. *et al.* 1987. Biochemical and biophysical comparison of two mucins from human submandibular-sublingual saliva. Arch. Biochem. Biophys. **258:** 452–464.
35. BOBEK, L.A. *et al.* 1993. Molecular cloning, sequence, and specificity of expression of the gene encoding the low molecular weight human salivary mucin (MUC7). J. Biol. Chem. **268:** 20563–20569.
36. TROXLER, R.F. *et al.* 1997. Molecular characterization of a major high molecular weight mucin from human sublingual gland. Glycobiology **7:** 965–973.
37. TROXLER, R.F. *et al.* 1995. Molecular cloning of a novel high molecular weight mucin (MG1) from human sublingual gland. Biochem. Biophys. Res. Commun. **217:** 1112–1119.
38. TROXLER, R.F. *et al.* 1990. Structural relationship between human salivary histatins. J. Dent. Res. **69:** 2–6.
39. HARDT, M. *et al.* 2005. Toward defining the human parotid gland salivary proteome and peptidome: identification and characterization using 2D SDS-PAGE, ultrafiltration, HPLC, and mass spectrometry. Biochemistry **44:** 2885–2899.
40. SABATINI, L.M. & E.A. AZEN. 1994. Two coding change mutations in the HIS2(2) allele characterize the salivary histatin 3–2 protein variant. Hum. Mutat. **4:** 12–19.

41. SCHENKELS, L.C., E.C. VEERMAN & A.V. NIEUW AMERONGEN. 1995. Biochemical composition of human saliva in relation to other mucosal fluids. Crit. Rev. Oral Biol. Med. **6:** 161–175.
42. HELMERHORST, E.J. *et al.* 2006. Oral fluid proteolytic effects on histatin 5 structure and function. Arch. Oral Biol. **51:** 1061–1070.
43. PAYNE, J.B. *et al.* 1991. Selective effects of histidine-rich polypeptides on the aggregation and viability of *Streptococcus mutans* and *Streptococcus sanguis*. Oral Microbiol. Immunol. **6:** 169–176.
44. IONTCHEVA, I. *et al.* 2000. Molecular mapping of statherin- and histatin-binding domains in human salivary mucin MG1 (MUC5B) by the yeast two-hybrid system. J. Dent. Res. **79:** 732–739.
45. BRUNO, L.S. *et al.* 2005. Two-hybrid analysis of human salivary mucin MUC7 interactions. Biochim. Biophys Acta **1746:** 65–72.
46. POLLOCK, J.J. *et al.* 1984. Fungistatic and fungicidal activity of human parotid salivary histidine-rich polypeptides on *Candida albicans*. Infect. Immun. **44:** 702–707.
47. EDGERTON, M. *et al.* 1998. Candidacidal activity of salivary histatins. Identification of a histatin 5-binding protein on *Candida albicans*. J. Biol. Chem. **273:** 20438–20447.
48. HELMERHORST, E.J. *et al.* 1997. Synthetic histatin analogues with broad-spectrum antimicrobial activity. Biochem. J. **326**(Pt 1): 39–45.
49. YAN, Q. & A. BENNICK. 1995. Identification of histatins as tannin-binding proteins in human saliva. Biochem. J. **311**(Pt 1): 341–347.
50. SUGIYAMA, K. 1993. Anti-lipopolysaccharide activity of histatins, peptides from human saliva. Experientia **49:** 1095–1097.
51. NISHIKATA, M. *et al.* 1991. Salivary histatin as an inhibitor of a protease produced by the oral bacterium *Bacteroides gingivalis*. Biochem. Biophys. Res. Commun. **174:** 625–630.
52. BASAK, A. *et al.* 1997. Histidine-rich human salivary peptides are inhibitors of proprotein convertases furin and PC7 but act as substrates for PC1. J. Pept. Res. **49:** 596–603.
53. GUSMAN, H. *et al.* 2001. Salivary histatin 5 is a potent competitive inhibitor of the cysteine proteinase clostripain. FEBS Lett. **489:** 97–100.
54. MURAKAMI, Y. *et al.* 2002. Inhibitory effect of synthetic histatin 5 on leukotoxin from *Actinobacillus actinomycetemcomitans*. Oral Microbiol. Immunol. **17:** 143–149.
55. CASTAGNOLA, M. *et al.* 2003. Determination of the post-translational modifications of salivary acidic proline-rich proteins. Eur. J. Morphol. **41:** 93–98.
56. HAY, D.I., S.K. SCHLUCKEBIER & E.C. MORENO. 1982. Equilibrium dialysis and ultrafiltration studies of calcium and phosphate binding by human salivary proteins. Implications for salivary supersaturation with respect to calcium phosphate salts. Calcif. Tissue Int. **34:** 531–538.
57. HAY, D.I. *et al.* 1984. Relationship between concentration of human salivary statherin and inhibition of calcium phosphate precipitation in stimulated human parotid saliva. J. Dent. Res. **63:** 857–863.
58. MORENO, E.C., K. VARUGHESE & D.I. HAY. 1979. Effect of human salivary proteins on the precipitation kinetics of calcium phosphate. Calcif. Tissue Int. **28:** 7–16.
59. BOBEK, L.A. & M.J. LEVINE. 1992. Cystatins—inhibitors of cysteine proteinases. Crit. Rev. Oral Biol. Med. **3:** 307–332.

60. Isemura, S., E. Saitoh & K. Sanada. 1987. Characterization and amino acid sequence of a new acidic cysteine proteinase inhibitor (cystatin SA) structurally closely related to cystatin S, from human whole saliva. J. Biochem. (Tokyo) **102:** 693–704.
61. Saitoh, E. *et al.* 1992. Characterization of two members (CST4 and CST5) of the cystatin gene family and molecular evolution of cystatin genes. Agents Actions Suppl. **38**(Pt 1): 340–348.
62. Saitoh, E. *et al.* 1987. Human cysteine-proteinase inhibitors: nucleotide sequence analysis of three members of the cystatin gene family. Gene **61:** 329–338.
63. Saitoh, E., K. Minaguchi & O. Ishibashi. 1998. Production and characterization of two variants of human cystatin SA encoded by two alleles at the CST2 locus of the type 2 cystatin gene family. Arch. Biochem. Biophys. **352:** 199–206.
64. Dickinson, D.P. *et al.* 1993. Genomic cloning, physical mapping, and expression of human type 2 cystatin genes. Crit. Rev. Oral Biol. Med. **4:** 573–580.
65. Ramasubbu, N. *et al.* 1991. Large-scale purification and characterization of the major phosphoproteins and mucins of human submandibular-sublingual saliva. Biochem. J. **280**(Pt 2): 341–352.
66. Dickinson, D.P. 2002. Cysteine peptidases of mammals: their biological roles and potential effects in the oral cavity and other tissues in health and disease. Crit. Rev. Oral Biol. Med. **13:** 238–275.
67. Dickinson, D.P., M. Thiesse & M.J. Hicks. 2002. Expression of type 2 cystatin genes CST1-CST5 in adult human tissues and the developing submandibular gland. DNA Cell Biol. **21:** 47–65.
68. Henskens, Y.M. *et al.* 1996. Protein composition of whole and parotid saliva in healthy and periodontitis subjects. Determination of cystatins, albumin, amylase and IgA. J. Periodontal Res. **31:** 57–65.
69. Baron, A.C. *et al.* 1999. Cysteine protease inhibitory activity and levels of salivary cystatins in whole saliva of periodontally diseased patients. J. Periodontal Res. **34:** 437–444.
70. Lamkin, M.S. & F.G. Oppenheim. 1993. Structural features of salivary function. Crit. Rev. Oral Biol. Med. **4:** 251–259.
71. Johnsson, M. *et al.* 1991. The effects of human salivary cystatins and statherin on hydroxyapatite crystallization. Arch. Oral Biol. **36:** 631–636.
72. Al-Hashimi, I. & M.J. Levine. 1989. Characterization of in vivo salivary-derived enamel pellicle. Arch. Oral Biol. **34:** 289–295.
73. Baron, A., A. DeCarlo & J. Featherstone. 1999. Functional aspects of the human salivary cystatins in the oral environment. Oral Dis. **5:** 234–240.
74. Merritt, A.D. *et al.* 1973. Salivary and pancreatic amylase: electrophoretic characterizations and genetic studies. Am. J. Hum. Genet. **25:** 510–522.
75. Tomita, N. *et al.* 1989. A novel type of human alpha-amylase produced in lung carcinoid tumor. Gene **76:** 11–18.
76. Yamashita, K. *et al.* 1980. Structural studies of the sugar chains of human parotid alpha-amylase. J. Biol. Chem. **255:** 5635–5642.
77. Stiefel, D.J. & P.J. Keller. 1973. Preparation and some properties of human pancreatic amylase including a comparison with human parotid amylase. Biochim. Biophys. Acta **302:** 345–361.
78. Mayo, J.W. & D.M. Carlson. 1974. Isolation and properties of four alpha-amylase isozymes from human submandibular saliva. Arch. Biochem. Biophys. **163:** 498–506.

79. KAUFFMAN, D.L. *et al.* 1970. The isoenzymes of human parotid amylase. Arch. Biochem. Biophys. **137:** 325–339.
80. YAO, Y. *et al.* 2003. Identification of protein components in human acquired enamel pellicle and whole saliva using novel proteomics approaches. J. Biol. Chem. **278:** 5300–5308.
81. SCANNAPIECO, F.A. *et al.* 1989. Characterization of salivary alpha-amylase binding to *Streptococcus sanguis*. Infect. Immun. **57:** 2853–2863.
82. SCANNAPIECO, F.A. *et al.* 1990. Structural relationship between the enzymatic and streptococcal binding sites of human salivary alpha-amylase. Biochem. Biophys. Res. Commun. **173:** 1109–1115.
83. DOUGLAS, C.W., A.A. PEASE & R.A. WHILEY. 1990. Amylase-binding as a discriminator among oral streptococci. FEMS Microbiol. Lett. **54:** 193–0197.
84. BERGMANN, J.E. & H.J. GULZOW. 1995. Detection of binding of denatured salivary alpha-amylase to *Streptococcus sanguis*. Arch. Oral Biol. **40:** 973–974.
85. SCANNAPIECO, F.A., G. TORRES & M.J. LEVINE. 1993. Salivary alpha-amylase: role in dental plaque and caries formation. Crit. Rev. Oral Biol. Med. **4:** 301–307.
86. ORSTAVIK, D. & F.W. KRAUS. 1973. The acquired pellicle: immunofluorescent demonstration of specific proteins. J. Oral Pathol. **2:** 68–76.
87. ORSTAVIK, D. & F.W. KRAUS. 1974. The acquired pellicle: enzyme and antibody activities. Scand J. Dent. Res. **82:** 202–205.
88. DIPAOLA, C., M.S. HERRERA & I.D. MANDEL. 1984. Host proteins in dental plaques of caries-resistant versus caries-susceptible human groups. Arch. Oral Biol. **29:** 353–355.
89. STUBBS, M. *et al.* 1998. Encoding of human basic and glycosylated proline-rich proteins by the PRB gene complex and proteolytic processing of their precursor proteins. Arch. Oral Biol. **43:** 753–770.
90. HAY, D.I. *et al.* 1994. Human salivary acidic proline-rich protein polymorphisms and biosynthesis studied by high-performance liquid chromatography. J. Dent. Res. **73:** 1717–1726.
91. AZEN, E.A. & F.G. OPPENHEIM. 1973. Genetic polymorphism of proline-rich human salivary proteins. Science **180:** 1067–1069.
92. AZEN, E.A. & C.L. DENNISTON. 1974. Genetic polymorphism of human salivary proline-rich proteins: further genetic analysis. Biochem. Genet. **12:** 109–120.
93. AZEN, E.A. & C. DENNISTON. 1981. Genetic polymorphism of PIF (parotid isoelectric focusing variant) proteins with linkage to the PPP (parotid proline-rich protein) gene complex. Biochem. Genet. **19:** 475–485.
94. SCHLESINGER, D.H. & D.I. HAY. 1986. Complete covalent structure of a proline-rich phosphoprotein, PRP-2, an inhibitor of calcium phosphate crystal growth from human parotid saliva. Int. J. Pept. Protein Res. **27:** 373–379.
95. AZEN, E.A. & P.L. YU. 1984. Genetic polymorphisms of Pe and Po salivary proteins with probable linkage of their genes to the salivary protein gene complex (SPC). Biochem. Genet. **22:** 1065–1080.
96. MAEDA, N. *et al.* 1985. Differential RNA splicing and post-translational cleavages in the human salivary proline-rich protein gene system. J. Biol. Chem. **260:** 11123–11130.
97. ROBINSON, R. *et al.* 1989. Primary structure and possible origin of the non-glycosylated basic proline-rich protein of human submandibular/sublingual saliva. Biochem. J. **263:** 497–503.

98. Hatton, M.N. *et al.* 1985. Masticatory lubrication. The role of carbohydrate in the lubricating property of a salivary glycoprotein-albumin complex. Biochem. J. **230:** 817–820.
99. Lu, Y. & A. Bennick. 1998. Interaction of tannin with human salivary proline-rich proteins. Arch. Oral Biol. **43:** 717–728.
100. Hay, D.I. *et al.* 1987. Inhibition of calcium phosphate precipitation by human salivary acidic proline-rich proteins: structure-activity relationships. Calcif. Tissue Int. **40:** 126–132.
101. Offner, G.D. *et al.* 1998. The amino-terminal sequence of MUC5B contains conserved multifunctional D domains: implications for tissue-specific mucin functions. Biochem. Biophys. Res. Commun. **251:** 350–355.
102. Levine, M.J. *et al.* 1987. Structural aspects of salivary glycoproteins. J. Dent. Res. **66:** 436–441.
103. Nielsen, P.A. *et al.* 1997. Identification of a major human high molecular weight salivary mucin (MG1) as tracheobronchial mucin MUC5B. Glycobiology **7:** 413–419.
104. Liu, B. *et al.* 1998. MUC4 is a major component of salivary mucin MG1 secreted by the human submandibular gland. Biochem. Biophys. Res. Commun. **250:** 757–761.
105. Chen, Y. *et al.* 2004. Genome-wide search and identification of a novel gel-forming mucin MUC19/Muc19 in glandular tissues. Am. J. Respir. Cell Mol. Biol. **30:** 155–165.
106. Strous, G.J. & J. Dekker. 1992. Mucin-type glycoproteins. Crit. Rev. Biochem. Mol. Biol. **27:** 57–92.
107. Van Klinken, B.J. *et al.* 1998. Strategic biochemical analysis of mucins. Anal. Biochem. **265:** 103–116.
108. Desseyn, J.L. *et al.* 1997. Genomic organization of the 3′ region of the human mucin gene MUC5B. J. Biol. Chem. **272:** 16873–16883.
109. Debailleul, V. *et al.* 1998. Human mucin genes MUC2, MUC3, MUC4, MUC5AC, MUC5B, and MUC6 express stable and extremely large mRNAs and exhibit a variable length polymorphism. An improved method to analyze large mRNAs. J. Biol. Chem. **273:** 881–890.
110. Bobek, L.A. *et al.* 1996. Structure and chromosomal localization of the human salivary mucin gene, MUC7. Genomics **31:** 277–282.
111. Reddy, M.S. *et al.* 1992. Structural features of the low-molecular-mass human salivary mucin. Biochem. J. **287**(Pt 2): 639–643.
112. Tabak, L.A. 1995. In defense of the oral cavity: structure, biosynthesis, and function of salivary mucins. Annu. Rev. Physiol. **57:** 547–564.
113. Yin, B.W., A. Dnistrian & K.O. Lloyd. 2002. Ovarian cancer antigen CA125 is encoded by the MUC16 mucin gene. Int. J. Cancer **98:** 737–740.
114. Malamud, D. 2006. Salivary diagnostics: the future is now. J. Am. Dent. Assoc. **137:** 284–286.
115. Slavkin, H.C. 1998. Toward molecularly based diagnostics for the oral cavity. J. Am. Dent. Assoc. **129:** 1138–1143.
116. Tabak, L.A. 2001. A revolution in biomedical assessment: the development of salivary diagnostics. J. Dent. Educ. **65:** 1335–1339.
117. Hu, S. *et al.* 2005. Large-scale identification of proteins in human salivary proteome by liquid chromatography/mass spectrometry and two-dimensional gel electrophoresis-mass spectrometry. Proteomics **5:** 1714–1728.

118. MANN, M. & M. WILM. 1994. Error-tolerant identification of peptides in sequence databases by peptide sequence tags. Anal. Chem. **66:** 4390–4399.
119. LIN, Z. *et al.* 2003. High-throughput analysis of protein/peptide complexes by immunoprecipitation and automated LC-MS/MS. J. Biomol. Tech. **14:** 149–155.
120. AEBERSOLD, R. & M. MANN. 2003. Mass spectrometry-based proteomics. Nature **422:** 198–207.
121. YATES, J.R., III. 2004. Mass spectrometry as an emerging tool for systems biology. Biotechniques **36:** 917–919.
122. HU, S., J.A. LOO & D.T. WONG. 2006. Human body fluid proteome analysis. Proteomics **6:** 6326–6353.
123. ENG, J.K., A.L. MCMORMACK & J.R. YATES. 1994. An approach to correlate tandem mass spectral data of peptide with aminoacid sequences in a protein database. J. Am. Soc. Mass Spectrom. **5:** 976–989.
124. PERKINS, D.N. *et al.* 1999. Probability-based protein identification by searching sequence databases using mass spectrometry data. Electrophoresis **20:** 3551–3567.
125. JENKINS, G.N. 1978. The Physiology and Biochemistry of the Mouth. Blackwell Scientific Oxford, UK.
126. KOUSVELARI, E.E. *et al.* 1980. Immunochemical identification and determination of proline-rich proteins in salivary secretions, enamel pellicle, and glandular tissue specimens. J. Dent. Res. **59:** 1430–1438.
127. BAUM, B.J., E.E. KOUSVELARI & F.G. OPPENHEIM. 1982. Exocrine protein secretion from human parotid glands during aging: stable release of the acidic proline-rich proteins. J. Gerontol. **37:** 392–395.
128. JOHNSON, D.A., C.K. YEH & M.W. DODDS. 2000. Effect of donor age on the concentrations of histatins in human parotid and submandibular/sublingual saliva. Arch. Oral Biol. **45:** 731–740.
129. RAYMENT, S.A. *et al.* 2000. Salivary mucin: a factor in the lower prevalence of gastroesophageal reflux disease in African-Americans? Am. J. Gastroenterol. **95:** 3064–3070.
130. LI, J. *et al.* 2004. Statherin is an in vivo pellicle constituent: identification and immuno-quantification. Arch. Oral Biol. **49:** 379–385.
131. AGUIRRE, A. *et al.* 1987. Immunochemical quantitation of alpha-amylase and secretory IgA in parotid saliva from people of various ages. Arch. Oral Biol. **32:** 297–301.

Interpretation of Oral Fluid Tests for Drugs of Abuse

EDWARD J. CONE[a] AND MARILYN A. HUESTIS[b]

[a]*Johns Hopkins School of Medicine, Baltimore, Maryland, USA*

[b]*Chemistry and Drug Metabolism Section, IRP, NIDA, NIH, 5500 Nathan Shock Drive, Baltimore, Maryland, USA*

ABSTRACT: Oral fluid testing for drugs of abuse offers significant advantages over urine as a test matrix. Collection can be performed under direct observation with reduced risk of adulteration and substitution. Drugs generally appear in oral fluid by passive diffusion from blood, but also may be deposited in the oral cavity during oral, smoked, and intranasal administration. Drug metabolites also can be detected in oral fluid. Unlike urine testing, there may be a close correspondence between drug and metabolite concentrations in oral fluid and in blood. Interpretation of oral fluid results for drugs of abuse should be an iterative process whereby one considers the test results in the context of program requirements and a broad scientific knowledge of the many factors involved in determining test outcome. This review delineates many of the chemical and metabolic processes involved in the disposition of drugs and metabolites in oral fluid that are important to the appropriate interpretation of oral fluid tests. Chemical, metabolic, kinetic, and analytic parameters are summarized for selected drugs of abuse, and general guidelines are offered for understanding the significance of oral fluid tests.

KEYWORDS: oral fluid; saliva; interpretation; testing; advantages; limitations

INTRODUCTION

Testing for drugs of abuse in biological fluids and tissues is an international phenomenon, practiced in many different settings for a variety of reasons. The majority of testing in the United States occurs in workplace programs, but a significant volume of testing also is performed in the United States and abroad in drug treatment centers, hospital emergency rooms, pain treatment clinics, sports organizations, and by courts and other legal authorities. Historically, urine has been the matrix of choice for testing, but recent advances in technology and the introduction of commercial oral fluid assays have effectively

Address for correspondence: Edward J. Cone, ConeChem Research, LLC, 441 Fairtree Drive, Severna Park, MD 21146. Voice: 410-315-8643; fax: 410-315-9067.
edward.cone@comcast.net

Ann. N.Y. Acad. Sci. 1098: 51–103 (2007). © 2007 New York Academy of Sciences.
doi: 10.1196/annals.1384.037

established oral fluid as a viable alternative. The popularity of oral fluid as a test matrix has arisen because of its distinct advantages over urine. Oral fluid collection can be performed in almost any location, with less embarrassment and under directly observed conditions. The noninvasiveness of collection and reduced opportunity for adulteration and substitution, in contrast to urine testing, are key factors to oral fluid's success. Some programs now use oral fluid as the test matrix of choice, especially in the private-sector workplace, drug treatment, and legal settings. The obvious advantages of oral fluid and concomitant advances in technology led the Department of Health and Human Services (DHHS) to propose oral fluid analysis for inclusion in the federal workplace drug-testing program.[1]

There is now a significant body of scientific literature on oral fluid testing. Several reviews document aspects of oral fluid testing including drug disposition,[2–6] detection times,[7] diagnostics,[8] legal issues[9] and application of state-of-the-art technologies.[10–12] This review addresses chemical, physiological, and pharmacological factors that determine the outcome of oral fluid results and offers general guidelines for interpretation of positive tests and discussion of the limitations of oral fluid testing.

FACTORS INVOLVED IN INTERPRETATION OF ORAL FLUID DRUG TESTS

Interpretation of oral fluid drug results is dependent in part on the purpose of testing. Workplace drug testing is primarily performed as a means of identification of individuals, frequently in safety-sensitive positions, who are abusing drugs and as a deterrent to drug misuse by the general workforce. Drug treatment specialists perform drug testing to foster drug abstinence and compliance with program requirements. Testing in legal settings documents program violations, compliance, and provides evidence of impairment.

There are numerous considerations in drug test interpretation. FIGURE 1 illustrates the flow of the test process and factors important to accurate interpretation. Questions posed during the interpretational process depend upon the nature of the drug-testing program, and may be limited in scope, complex, and sometime seek answers that go beyond reasonable scientific certainty. Metabolic disposition patterns should be understood for each drug class. The body of evidence, including what is known with certainty and what is not known, must be brought to bear during the interpretation process.

Oral fluid is a composite tissue consisting primarily of saliva, mixed with gingival crevicular fluid, buccal and mucosal transudates, cellular debris, bacteria, and residues of ingested products. Salivary glands are highly perfused, allowing rapid transference of drug in blood to salivary glands. Within minutes of parenteral drug administration, drug can appear in oral fluid.[13] Transfer of drug and metabolites from blood to saliva occurs primarily by passive diffusion

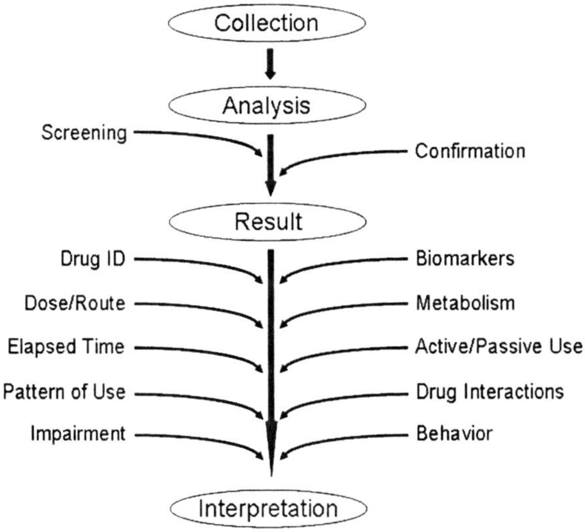

FIGURE 1. Schematic of oral fluid drug testing and interpretation of results.

and is dependent upon numerous factors, including chemical properties of the drug, salivary pH, concentration of un-ionized drug (ionized drug does not passively diffuse across cellular membranes), drug–protein binding (only the free fraction can diffuse), and membrane characteristics. Because of the relative acidity of saliva compared to plasma, basic drugs (e.g., amphetamine and cocaine) are frequently found in higher concentrations than in plasma, yielding saliva/plasma ratios (s/p) greater than unity.[13] Although the terms "saliva" and "oral fluid" are used somewhat interchangeably in scientific literature, *oral fluid* more accurately describes the biological characteristics of this matrix. Thus, oral fluid is the term used predominantly in this review.

Interpretation of oral fluid tests requires an understanding of the unique features of this biological matrix, chemical and physiological factors that affect drug transfer into oral fluid, analytical factors, kinetics of drug disposition, drug metabolic patterns, and potential risks of oral contamination and passive exposure. Clearly, oral fluid tests are most useful in detection of recent drug use. Drugs and metabolites can generally be detected for a period of several hours to several days following drug exposure.[7] Drug concentrations in oral fluid are generally related to content in blood, but also may be present as residual drug in the oral cavity. The frequently observed high correspondence of drug concentrations in oral fluid to blood makes oral fluid an attractive matrix for use in detection of recent drug use and in interpretation of possible drug-induced behavioral effects.

A table is included for each drug that summarizes key chemical, metabolic, kinetic and analytic parameters for each drug. Following each table, general guidelines are offered, although they should not be considered absolute criteria.

A list defining each parameter and the usefulness of factors influencing oral fluid test outcome follows:

- Drug/metabolite(s) in oral fluid and urine: predominant drug or metabolite(s) most often detected in oral fluid and urine following drug exposure.
- Fb: free drug fraction in blood not bound to plasma protein. Only free un-ionized drug can passively diffuse across biological membranes; hence, highly protein-bound drugs generally will be present in lower concentration in oral fluid than in plasma.
- Plasma/oral fluid t1/2: period of time required for drug concentration in plasma or oral fluid to decrease by 50%. Drugs with longer half-lives generally can be detected for longer periods of time.
- Plasma: oral fluid correlation: relationship between drug concentrations in plasma and oral fluid. A high plasma: oral fluid correlation reflects the close relationship between plasma and oral fluid.
- s/p and s/b: saliva (oral fluid) to plasma or blood ratio. Saliva to plasma ratios greater than unity (>1) generally are observed for drugs whose molecular structures contain a basic nitrogen moiety and exhibit low plasma–protein binding. Basic drugs tend to accumulate in oral fluid in higher concentrations than in plasma because of the lower pH of saliva. Values of s/p or s/b reported in this review are actual measured values of drug in saliva (oral fluid), plasma, and whole blood.
- P and log P: partition coefficient or logarithm of the partition coefficient of a drug. These parameters express the relative distribution of drug between oil and water under specified conditions, for example, octanol/water at 37°C and pH 7.4. Drugs with higher P or log P are more lipophilic and generally distribute more rapidly and to a greater degree into bodily tissues and fluids.
- pKa: negative logarithm of equilibrium coefficient of neutral and charged forms of a compound. pKa is a constant for each drug. For basic compounds, those with higher pKa constants exhibit greater ionization in plasma (pH 7.4) than drugs with lower pKa constants. For acidic drugs, those with lower pKa constants will exhibit greater ionization in plasma (pH 7.4) than drugs with higher pKa constants. A drug's pKa affects the transfer of basic drugs from plasma to saliva in two ways. Only un-ionized drug will transfer across epithelial membranes, and the acidic nature of saliva allows accumulation of greater concentrations of basic drugs than neutral or acidic drugs.
- Screening test: initial laboratory test commonly employed to rapidly eliminate negative specimens and identify presumptively positive specimens. Most screening procedures are commercial immunoassay-based tests; administrative cutoff concentrations are utilized to distinguish between positive and negative specimens; drugs that do not react sufficiently in

these assays may, by necessity, be tested by chromatographic methods.

- Confirmation test: laboratory test utilized to confirm the presence of a specific drug analyte(s) at or above an administrative cutoff concentration(s).
- Detection time (cutoff, ng/mL): typical duration of time a drug can be detected in a biological specimen at a specified ng/mL cutoff concentration. This parameter serves as a useful guideline to establish the probable time that drug use occurred prior to specimen collection. This parameter is frequently determined following administration of a single drug dose and may underestimate detection times following multiple drug administrations or chronic drug usage. Detection times also are highly influenced by cutoff concentrations employed in the analytical test.
- Interpretation resources: cites key clinical studies and reports that were used in development of interpretational guidelines for oral fluid.
- Possible drug/metabolite sources: potential sources of drug, or drug-associated metabolite(s), which might result in a positive oral fluid or urine test.
- Biomarkers: drug-related metabolites and analytes that predict or signal, with a high degree of reliability, additional information for interpretation, for example, identification of anhydroecognine methyl ester indicates that cocaine has been administered by the smoking route. It should be noted that biomarkers are only useful, however, if it can be established that they were not present originally as contaminants of the administered drug or produced as artifacts during specimen storage and analysis. Of additional note, many of the biomarkers proposed in this review are based on urine studies and have not been established as biomarkers in oral fluid tests. Further, when biomarkers are present, they offer additional assurances of drug use, but their absence does not lessen the value of a confirmed test.

DRUG INTERPRETATION

Amphetamines

Amphetamine

Overview: Amphetamine is a sympathomimetic amine with central nervous stimulant activity (TABLE 1). The d-isomer (dextro-amphetamine) is three to four times more potent than the l-isomer (levo-amphetamine) as a stimulant. It is prescribed for treatment of attention-deficit hyperactivity disorder (ADHD) and narcolepsy. Amphetamine is available in immediate-release and sustained-release formulations in doses ranging from 3 to 30 mg. Some formulations, such as Adderall®, contain combinations of d- and l-isomers.

Amphetamine is excreted intact in urine and also as metabolites. Amphetamine metabolites are produced primarily by oxidative enzymes, leading

TABLE 1. Summary of chemical, metabolic, kinetic, and analytic factors important in interpretation of oral fluid tests for amphetamine

Factor	Amphetamine information
Drug/metabolite(s) in oral fluid	Amphetamine[16]
Drug/metabolite(s) in urine	Amphetamine; *p*-hydroxy-amphetamine; norephedrine; *p*-hydroxynorephedrine[20]
Amphetamine Fb	16% (similar for d- and l-isomers)[16]
Amphetamine plasma t1/2	d-Amphetamine (acidic urine): 6.8 h d-amphetamine (alkaline urine) 17.0 h l-amphetamine (acid urine): 7.7 h l-amphetamine (alkaline urine) 23.7 h[16]
Amphetamine oral fluid t1/2	d-Amphetamine (acidic urine): 6.8 h d-amphetamine (alkaline urine) 17.2 h l-amphetamine (acidic urine): 8.5 h l-amphetamine (alkaline urine) 24.6 h[16]
Amphetamine plasma: oral fluid correlation	Proportional to plasma in postabsorptive phase[16]
Amphetamine s/p	2.76 (postabsorptive phase)[16]
Amphetamine log P (HPLC)	0.82[127]
Amphetamine pKa	10[16]
Amphetamine screening test	Amphetamines
Amphetamine confirmation test	Amphetamine
Amphetamine detection time in oral fluid [cutoff, ng/mL]	Amphetamine GC, [10]: 20–50 h[16]
Amphetamine detection time in urine [cutoff, ng/mL]	Single dose: EIA [1000] 24–72 h[7,128]; chronic: EIA [300]: 9 days; EIA [1000]: 9 days[129]
Amphetamine oral fluid interpretation resources[16]	–

HPLC=high-performance liquid chromatography; GC=gas chromatography; EIA=enzyme immunoassay.

to mostly inactive products. The amount of amphetamine excreted in urine is highly dependent upon urinary pH. Under acid conditions, as much as 57–66% of the dose may be excreted unchanged, whereas under alkaline conditions, the amount may decrease to 1–5%.[14] Analytical methods are available for determining the enantiomeric forms of amphetamine and methamphetamine.[15] The mean elimination half-life of d-amphetamine is slightly shorter than that of l-amphetamine.[16] Amphetamine also is produced as a metabolite of methamphetamine and from a variety of pharmaceutical products (see *Interpretation*).

Amphetamine appears rapidly in oral fluid following administration and parallels plasma drug concentrations. Wan *et al.*[16] evaluated the influence of systemic administration of sodium bicarbonate and ammonium chloride on the disposition of amphetamine in plasma and oral fluid. Although salivary pH was relatively constant with both treatments, the half-lives of d- and l-amphetamine were reduced considerably under acidic conditions.

Interpretation of Oral Fluid Amphetamine Tests:

- Positive test for amphetamine (no methamphetamine)
  - Interpretation: amphetamine use; must rule out possibility that amphetamine presence is from metabolism of another drug. Determination of d/l-isomer ratio will assist interpretation.[17] L-amphetamine only (no l-methamphetamine) would not occur from use of nasal inhaler.
- Possible sources of amphetamine
  - Prescription products containing amphetamine
  - Illicit amphetamine
  - Drugs that are metabolized to amphetamine (and methamphetamine) include: amfetaminil (amphetamine only); benzphetamine (d-isomers of amphetamine and methamphetamine); clobenzorex (d-amphetamine only); dimethylamphetamine (d-isomers of amphetamine and methamphetamine); ethylamphetamine (amphetamine only); famprofazone (d/l-isomers of amphetamine and methamphetamine); fencamine (d/l-isomers of amphetamine and methamphetamine); fenethylline (d/l-amphetamine only); fenproporex (d/l-amphetamine only); furfenorex (amphetamine and methamphetamine); mefenorex (amphetamine only); mesocarb (amphetamine only); prenylamine (d/l-amphetamine only); and selegiline (l-deprenyl; l-isomers of amphetamine and methamphetamine).[17–19]
  - At present, although no oral fluid or urine studies have been reported on passive amphetamine smoke exposure, the risk of a positive test from passive exposure does not appear likely.
- Biomarkers
  - The confirmed presence of oxidative metabolites of amphetamine such as *p*-hydroxyamphetamine, norephedrine, and *p*-hydroxynorephedrine in oral fluid would be useful to substantiate use.[20]

Methamphetamine

Overview: Methamphetamine is a sympathomimetic amine with central nervous stimulant activity similar to amphetamine (TABLE 2). The d-isomer (dextro-methamphetamine) is prescribed for treatment of ADHD and for exogenous obesity. It is available as an immediate-release formulation containing 5 mg of methamphetamine hydrochloride. The usual effective dose is 20 to 25 mg daily. The l-isomer (levmetamfetamine; l-desoxyephedrine), marketed as a nasal inhaler decongestant product, is sold over the counter (OTC). The nasal inhaler typically contains 50 mg of levmetamfetamine. Illicit methamphetamine is widely available and abused. Methamphetamine is easily manufactured in clandestine laboratories and is the most prevalent synthetic drug manufactured in the United States. The synthetic route determines which isomeric form of methamphetamine is produced. Currently, the most common

TABLE 2. Summary of chemical, metabolic, kinetic, and analytic factors important in interpretation of oral fluid tests for methamphetamine

Factor	Methamphetamine information
Drug/metabolite(s) in oral fluid	Methamphetamine; amphetamine[24]
Drug/metabolite(s) in urine	Methamphetamine; amphetamine, *p*-hydroxy-methamphetamine[25]
Methamphetamine Fb	10–20%[130]
Methamphetamine plasma t1/2	9.3–11.1 h[24]; 8.3–18.2 h[23]; 10.1 h[131]
Methamphetamine oral fluid t1/2	8.1–11.1 h[24]
Methamphetamine plasma: oral fluid correlation	$r^2 = 0.22$[24]; $r^2 = 0.50$[131]
Methamphetamine s/p	Oral: 2[24]; 7.76[131]; IV: 6[23]
Methamphetamine log P (HPLC)	0.86[127]
Methamphetamine pKa	10.1[24]
Methamphetamine screening test	Amphetamines EIA
Methamphetamine confirmation test	Methamphetamine; amphetamine
Methamphetamine/metabolite detection time in oral fluid [cutoff, ng/mL]	Single dose X4, methamphetamine: GC-MS [2.5]: 6–76 h; methamphetamine/amphetamine: GC-MS [50/2.5]: 24 h[24]
Methamphetamine detection time in urine [cutoff, ng/mL]	Single dose X4, methamphetamine: GC-MS [250]: 31–96 h; methamphetamine/ amphetamine: GC-MS [250/100]: 25–96 h; methamphetamine/amphetamine: GC-MS [500/200]: 22–92 h[132]
Methamphetamine oral fluid interpretation resources[23,24,131]	

HPLC=high-performance liquid chromatography; GC-MS=gas chromatography-mass spectrometry.

route yields the d-isomer, but older methods primarily yielded a racemic mixture.

Methamphetamine is abused by numerous routes including smoking, snorting, injection, and oral administration. The l-isomer of methamphetamine, levmetamfetamine contained in nasal inhalers, also can produce positive results for methamphetamine in urine.[21] Methamphetamine is excreted intact in urine and also as metabolites following enzymatic oxidation and conjugation. Like amphetamine, the amount of methamphetamine excreted is highly dependent upon urinary pH. Amphetamine, a metabolite of methamphetamine, is excreted in urine in variable amounts. Kim *et al.*[22] reported that the ratio of amphetamine to methamphetamine in urine increased over time from 0.13 to 0.36 following oral doses of 10 or 20 mg of d-methamphetamine hydrochloride. Cook *et al.*[23] reported a similar finding for subjects administered intravenous and smoked d-methamphetamine.

Methamphetamine and amphetamine appear rapidly in plasma and oral fluid following administration. The relative amounts of amphetamine to methamphetamine in plasma and oral fluid were reported to be approximately 21% and

24%, respectively.[24] Oral fluid methamphetamine concentrations were higher than plasma concentrations by approximately fourfold.

Interpretation of Oral Fluid Methamphetamine Tests:

- Positive test for methamphetamine (no amphetamine)
  - Interpretation: methamphetamine use; must rule out possibility that methamphetamine presence is from metabolism of another drug or from use of over-the-counter nasal inhaler. Determination of d/l-isomer ratio will assist interpretation[17]
- Positive test for methamphetamine and amphetamine (methamphetamine > amphetamine)
  - Interpretation: methamphetamine use; must rule out possibility that methamphetamine and amphetamine presence is from metabolism of another drug or from use of OTC nasal inhaler. Determination of d/l-isomer ratio will assist interpretation[17]
- Positive test for methamphetamine and amphetamine (methamphetamine < amphetamine)
  - Interpretation: possible combined use of methamphetamine and amphetamine; must rule out possibility that presence is due to metabolism of other drugs. OTC nasal inhaler use would not account for high abundance of amphetamine. Determination of d/l-isomer ratio will assist interpretation[17]
- Possible sources of methamphetamine
  - Prescription products containing methamphetamine
  - Illicit methamphetamine
  - Drugs that may be metabolized to methamphetamine (and amphetamine): include: benzphetamine (d-isomers); dimethylamphetamine (d-isomers); famprofazone (d/l-isomers); fencamine (d/l-isomers); furfenorex; and selegiline (l-deprenyl; l-isomers)[17–19]
  - Nasal inhaler (l-methamphetamine)
  - At present, although no oral fluid or urine studies have been reported on passive methamphetamine smoke exposure, the risk of a positive test from passive exposure does not appear likely
- Biomarkers
  - The confirmed presence of oxidative metabolites of methamphetamine such as *p*-hydroxymethamphetamine in oral fluid would be useful to substantiate use[25]
  - The confirmed presence of amphetamine in oral fluid would be useful to substantiate methamphetamine use. DHHS has proposed that a positive confirmed test for methamphetamine be accompanied by the confirmed presence of amphetamine at the assay's limit of detection.[1]

3,4-Methylenedioxymethamphetamine (MDMA)

Overview: MDMA is a synthetic, ring-substituted amphetamine derivative that has become popular as an illicit recreational drug (TABLE 3). MDMA is one of numerous "designer" drugs, often referred to as "club drugs," because of induction of feelings of euphoria, energy, and a desire to socialize. MDMA street names include "ecstasy," "XTC," "E," and "Adam." "Ecstasy" may also be used to denote the group of ring-substituted illicit amphetamines. MDMA exerts multiple effects on neurotransmitter systems and may produce acute toxic reactions such as tachycardia, hypertension, arrhythmia, panic attack, and psychosis.[26] Recreational doses are generally in the range of 75 to 150 mg. Illicit synthetic routes generally produce a racemic mixture of R- and S-isomers. The S(+) isomer of MDMA is considered to be responsible for psychostimulant and empathic effects and the R(−) isomer for its hallucinogenic properties.[27]

MDMA is typically administered by the oral route and reaches maximal blood concentrations in approximately 2 h.[27] It is metabolized by multiple pathways, primarily involving *N*-demethylation and *O*-demethylenation. *N*-demethylation of MDMA yields 3,4-methylenedioxyamphetamine (MDA), an active metabolite exhibiting similar pharmacological properties as

TABLE 3. Summary of chemical, metabolic, kinetic, and analytic factors important in interpretation of oral fluid tests for MDMA

Factor	MDMA information
Drug/metabolite(s) in oral fluid	MDMA; MDA; HMMA[30]
Drug/metabolite(s) in urine	MDMA; MDA; HHMA; HMMA; HMA[27]
MDMA Fb	34–40%[133]
MDMA plasma t1/2	8–9 h[27]; 7.2 h[30]
MDMA oral fluid t1/2	5.6 h[30]
MDMA plasma: oral fluid correlation	0.81[30]
MDMA s/p	6.4–18.1[30]; 0.8–22.4[35]
MDMA log P (octanol/water, 37°C, pH 7.4)	1.68[134]
MDMA pKa	10.4[133]
MDMA screening test	Methamphetamine/MDMA
MDMA confirmation test	MDMA; MDA
MDMA/metabolite detection time in oral fluid [cutoff, ng/mL]	MDMA: GC-MS [LOD = 5.7]: 24 h; MDA: GC-MS [LOD = 1]: 24 h[30]
MDMA detection time in urine [cutoff, ng/mL]	MDMA/HMMA, GC-MS [20 ng/mL]: 48 h[7]
MDMA oral fluid interpretation resources[30,31,35]	–

MDMA=3,4-methylenedioxymethamphetamine; MDA=3,4-methylenedioxyamphetamine; HHMA=3,4-dihydroxymethamphetamine; HMMA=4-hydroxy-3-methoxymethamphetamine; HMA=4-hydroxy-3-methoxyamphetamine; GC-MS=gas chromatography-mass spectrometry; LOD = limit of detection.

the parent drug. *O*-demethylenation of MDMA and MDA produces 3,4-dihydroxymethamphetamine (HHMA) and 3,4-dihydroxyamphetamine (HHA), respectively. Additional metabolites are formed by *O*-methylation of HHMA to 4-hydroxy-3-methoxymethamphetamine (HMMA) and of HHA to 4-hydroxy-3-methoxyamphetamine (HMA), deamination, and conjugation. Nonlinear pharmacokinetics of MDMA was observed in humans; small increases in recreational doses gave rise to disproportionate increases in MDMA plasma concentrations.[28] Fallon *et al.*[29] reported that the plasma half-life in humans of (R)-MDMA (5.8 $\pm$ 2.2 h) was significantly longer than that of (S)-MDMA (3.6 $\pm$ 0.9 h). Approximately 15% of the dose of MDMA is excreted in urine in 24 h as intact MDMA, 1.5% as MDA, 17.7% as HHMA, 22.7% as HMMA, and 1.3% as HMA.[27]

Oral administration of 100 mg of MDMA to humans yielded peak oral fluid and plasma concentrations of MDMA at 1.5 h after drug intake.[30] Oral fluid concentrations of MDMA were an order of magnitude higher than in plasma. HMMA, the major metabolite of MDMA, was detected in oral fluid in trace amounts, and MDA was present at approximately 4–5% of MDMA. Oral fluid concentrations of MDMA were highly correlated with plasma MDMA. The high concentrations of MDMA in oral fluid relative to plasma were attributed to the high pKa of MDMA and low plasma–protein binding.

Interpretation of Oral Fluid MDMA Tests:

- Positive test for MDMA (no MDA)
  - Interpretation: illicit MDMA use
- Positive test for MDMA and MDA
  - MDMA >> MDA
    - Interpretation: illicit MDMA use; presence of MDA likely due to metabolism of MDMA to MDA
  - MDA $\geq$ MDMA
    - Interpretation: combined use of illicit MDMA and illicit MDA, or illicit MDMA contaminated with MDA
- Possible sources of MDMA
  - Illicit MDMA
- Biomarkers
  - The confirmed presence of MDA or HMMA in oral fluid would be useful to substantiate MDMA use.[30] It should be noted that MDA could be present either from metabolism of MDMA or use of MDA.

3,4-Methylenedioxyamphetamine (MDA)

Overview: MDA is a synthetic, ring-substituted amphetamine derivative available as an illicit recreational drug (TABLE 4). It appears to produce similar pharmacological effects as MDMA. MDA street names include "love drug" and "love pill." Recreational doses of MDA are similar to MDMA and generally in the range of 75 to 150 mg. Like MDMA, MDA has a chiral center (R- and S-isomers), and illicit synthesis leads to production of the racemic mixture. MDA also serves as a precursor in the synthesis of MDMA and 3,4-methylenedioxy-*N*-ethylamphetamine (MDEA).

Although MDMA has been extensively studied in humans, there are few metabolic data on MDA. MDA is an active metabolite of MDMA and is produced by *O*-demethyleneation (see MDMA). HHA and HMA are metabolites of MDA. A number of studies have characterized the kinetics of MDA (as a metabolite of MDMA),[18,28,31] but there appears to be little information on MDA administered under controlled clinical settings.

MDA has been reported to appear in oral fluid following the administration of MDMA in concentrations representing approximately 4–5% of MDMA.[32]

Interpretation of Oral Fluid MDA Tests:

- Positive test for MDA (no MDMA)
  - Interpretation: illicit MDA use
- Possible sources of MDA
  - Illicit MDA
  - Metabolite of illicit MDMA
  - Metabolite of illicit MDEA

TABLE 4. Summary of chemical, metabolic, kinetic, and analytic factors important in interpretation of oral fluid tests for MDA

Factor	MDA information
Drug/metabolite(s) in oral fluid	MDA[30]
Drug/metabolite(s) in urine	MDA; HHA; HMA[27]
MDA Fb	34–40%[133]
MDA plasma t1/2	24.9 h (metabolite of MDMA)[27]
MDA oral fluid t1/2	MDA (MDMA dosed): 5.6–37.3 h[28]; MDA (MDEA administration): 3.4 h[33]
MDA plasma: oral fluid correlation	Not available
MDA s/p	Not available
MDA log P (octanol/water, 37°C, pH 7.4)	1.38[134]
MDA pKa	10.0[133]
MDA screening test	Amphetamines/MDA
MDA confirmation test	MDA
MDA detection time in oral fluid [cutoff, ng/mL]	Not available
MDA detection time in urine [cutoff, ng/mL]	Not available
MDA oral fluid interpretation resources[30]	–

MDA=3,4-methylenedioxyamphetamine; HMA=4-hydroxy-3-methoxyamphetamine; HHA=3,4-dihydroxyamphetamine.

TABLE 5. Summary of chemical, metabolic, kinetic, and analytic factors important in interpretation of oral fluid tests for MDEA

Factor	MDEA information
Drug/metabolite(s) in oral fluid	MDEA; MDA
Drug/metabolite(s) in urine	MDEA; MDA; HME[34]
MDEA Fb	Not available
MDEA plasma t1/2	(R)-MDEA: 7.5 h; (S)-MDEA: 4.2 h[135]
MDEA oral fluid t1/2	Not available
MDEA plasma: oral fluid correlation	Not available
MDEA s/p	Not available
MDEA log P (octanol/water, 37°C, pH 7.4)	Not available
MDEA pKa	Not available
MDEA screening test	Amphetamine/MDEA
MDEA confirmation test	MDEA; MDA
MDEA detection time in oral fluid [cut-off, ng/mL]	Not available
MDEA detection time in urine [cut-off, ng/mL]	MDEA, FPIA [300]: 33–62 h; MDEA, GC-MS [5]: 33–62 h; MDA, GC-MS [5]: 32–36 h; HME, GC-MS [10]: 7–8 days[136]
MDEA oral fluid interpretation resources	–

MDEA=3,4-methylenedioxyethylamphetamine; MDA=3,4-methylenedioxyamphetamine; HME=*N*-ethyl-4-hydroxy-3-methoxyamphetamine; FPIA=fluorescent polarization immunoassay; GC-MS=gas chromatography-mass spectrometry.

- Biomarkers
  - The confirmed presence of HHA and/or HMA in oral fluid would be useful to substantiate MDA use.[27]

3,4-Methylenedioxyethylamphetamine (MDEA)

Overview: When an ethyl group is substituted for the methyl group of MDMA, the synthetic analogue MDEA, is formed (TABLE 5). Like MDMA and MDA, MDEA has a chiral center and exists as two isomers. The S(+)-isomer appears to be primarily responsible for its psychotropic effect (higher affinity for presynaptic 5-HT transporters), whereas the R(−)-isomer appears to mediate the hallucinogenic effects.[33] MDEA is referred to by street names such as "eve" and "intellect." Recreational doses of MDEA are generally in the range of 60 mg to 175 mg.

MDEA is metabolized by *O*-demethylenation and by *N*-dealkylation of the ethyl-group. The major metabolite is formed by *O*-demethylenation to yield *N*-ethyl-4-hydroxy-3-methoxyamphetamine (HME); *N*-dealkylation leads to formation of the active metabolite MDA.[34]

MDEA has been reported in oral fluid of recreational drug users in concentrations ranging from 332 to 3320 ng/mL[35] and suspected users in concentrations >100 ng/mL.[32]

Interpretation of Oral Fluid MDEA Tests:

- Positive test for MDEA (no MDA)
  - Interpretation: illicit MDEA use
- Positive test for MDEA and MDA
  - MDEA >> MDA
    - Interpretation: Illicit MDEA use, presence of MDA likely due to metabolism of MDEA to MDA
  - MDA ≥ MDEA
    - Interpretation: combined use of illicit MDEA and illicit MDA, or illicit MDEA contaminated with MDA
- Possible sources of MDEA
  - Illicit MDEA
- Biomarkers
  - The confirmed presence of MDA or HME in oral fluid would be useful to substantiate MDEA use.[34] It should be noted that MDA could be present either from use of MDA or metabolism of MDEA.

Barbiturates

Overview

Barbiturates are sedatives used for seizure disorders, induction of anesthesia, and management of increased intracranial pressure. Barbiturate actions range from the short-acting, fast-starting pentobarbital and secobarbital to the long-acting, slow-starting phenobarbital, amobarbital, and butabarbital. Depending on the dose, frequency, and duration of use, one can rapidly develop tolerance, physical dependence, and psychological dependence on barbiturates. With the development of tolerance, the margin of safety between the effective dose and the lethal dose becomes narrow. Because of the risks associated with barbiturate abuse, and because new and safer drugs such as the benzodiazepines are now available, barbiturates are less frequently prescribed than in the past. Nearly all barbiturates are structurally related to barbituric acid. Although over 2,000 derivatives of barbituric acid have been synthesized, only about a dozen are currently used. Barbiturates available in the United States include amobarbital, aprobarbital, butabarbital, mephobarbital, pentobarbital, phenobarbital, and secobarbital. One preparation contains a combination of amobarbital and secobarbital. Barbiturate abusers appear to prefer the short-acting (e.g., pentobarbital, secobarbital) and intermediate drugs (e.g., amobarbital), but all barbiturates have significant abuse liability properties.

When used as sedative/hypnotics, barbiturates are typically administered orally. Barbiturates are efficiently absorbed from the gastrointestinal tract and exhibit high bioavailability. Generally, barbiturates are metabolized in the liver by oxidation and conjugation prior to renal excretion.

Studies of barbiturate disposition in oral fluid have been somewhat limited. Because amobarbital is one of the more commonly abused barbiturates and characterized most thoroughly, it is selected as the model compound for interpretation (TABLE 6). The disposition of amobarbital in oral fluid and serum of human subjects has been reported following ingestion of 120 mg of amobarbital.[36,37] Peak concentrations of amobarbital were observed simultaneously in oral fluid and saliva at approximately 1 h and exhibited similar elimination curves. Concentrations were lower in oral fluid than serum with an average ratio of 0.35.[36] Amobarbital was detectable in oral fluid for approximately 50 h.

Studies on the detection of other barbiturates in oral fluid also have been reported including pentobarbital,[37] hexobarbital,[38,39] and phenobarbital.[40–45]

Interpretation of Oral Fluid Amobarbital Tests

- Positive test for amobarbital
  - Interpretation: amobarbital use
- Possible sources of amobarbital
  - Prescription amobarbital

TABLE 6. Summary of chemical, metabolic, kinetic, and analytic factors important in interpretation of oral fluid tests for amobarbital

Factor	Amobarbital information
Drug/metabolite(s) in oral fluid	Amobarbital[36]
Drug/metabolite(s) in urine	Amobarbital; 3′-hydroxy-amobarbital; *N*-glucosyl amobarbital; 3′-carboxyamobarbital[130]
Amobarbital Fb	58%[36]
Amobarbital serum t1/2	20.6 h[36]
Amobarbital oral fluid t1/2	20.0 h[36]
Amobarbital plasma: oral fluid correlation	$r = 0.96$[36]
Amobarbital s/p	0.35[36]
Amobarbital log P (octanol/water, 37°C, pH 7.4)	2.07*
Amobarbital pKa	7.95[36]
Amobarbital screening test	Barbiturates
Amobarbital confirmation test	Amobarbital
Amobarbital detection time in oral fluid [cutoff, ng/mL]	Amobarbital: GC [100]: 50 h[36]
Amobarbital detection time in urine [cutoff, ng/mL]	3′-Hydroxy-amobarbital: GC [NR]: 5–6 days[137]
Amobarbital oral fluid interpretation resources[36,37]	–

GC=gas chromatography; NR=not reported.
*Source: Internet: Amobarbital: http://chrom.tutms.tut.ac.jp/JINNO/DRUGDATA/22amobarbital.html (accessed September 29, 2006).

- Biomarkers
  - The confirmed presence of oxidative metabolites of amobarbital in oral fluid would be useful to substantiate use.

Benzodiazepines

Overview

Benzodiazepines are a large class of structurally related compounds with anxiolytic, sedative, hypnotic, antipsychotic, and antiepileptic effects. Over 40 different drugs have been marketed worldwide and approximately 14 different benzodiazepines are marketed in the United States. The benzodiazepines, as a class, are generally lipophilic drugs with low solubility in water and exhibit efficient absorption from the gastrointestinal tract, first-pass metabolism, and high plasma–protein binding (70–99%). Potencies vary widely over this class of drugs; for example, the usual dose of triazolam is 0.125 mg, whereas 50–100 mg of chlordiazepoxide is usually taken for relief of symptoms of acute alcoholism.

A number of bezodiazepines share sufficient structural similarity such that metabolic conversions from one compound to another may occur. For example, diazepam may be metabolically converted in the human body to temazepam (3-hydroxydiazepam), nordiazepam, and oxazepam. In addition to this complexity, medazepam and ketazolam may be metabolically converted to diazepam. Thus, interpretation of benzodiazepine tests requires a thorough knowledge of the metabolic profiles of benzodiazepines. A number of benzodiazepines contain unique structural substitutions and use distinct metabolic pathways.

Studies of benzodiazepine disposition in oral fluid have been somewhat limited. Because diazepam has been studied most thoroughly, it is selected as the model compound for interpretation (TABLE 7). The disposition of diazepam and its metabolites in oral fluid has been reported following a single oral dose,[46–51] multiple daily doses,[48,51] and chronic dosing.[49,52] Following oral diazepam, concentrations in plasma and oral fluid peak at approximately 0.75 h.[47] Metabolites identified in oral fluid include 3-hydroxydiazepam, nordiazepam, and oxazepam.[51] At steady state, a significant correlation was observed between diazepam and nordiazepam in oral fluid.[52] The metabolite, nordiazepam, typically is found in higher concentration than diazepam in oral fluid at steady state with a ratio of 1.58 (nordiazepam/diazepam).[52] A significant correlation also was reported for diazepam concentration in oral fluid and CSF.[49]

Studies on the detection of other benzodiazepines in oral fluid also have been reported including chlordiazepoxide,[53] clorazepate,[49] flunitrazepam,[54] midazolam,[55] nitrazepam,[56,57] and oxazepam.[51]

Interpretation of Oral Fluid Diazepam Tests

- Positive test for diazepam (other metabolites of diazepam may be present including nordiazepam, oxazepam, and 3-hydroxydiazepam)

TABLE 7. Summary of chemical, metabolic, kinetic, and analytic factors important in interpretation of oral fluid tests for diazepam

Factor	Diazepam information
Drug/metabolite(s) in oral fluid	Diazepam; 3-hydroxydiazepam (temazepam); nordiazepam; oxazepam[51]
Drug/metabolite(s) in urine	Diazepam; nordiazepam; oxazepam conjugates[51]
Diazepam Fb	96–98%[47]
Diazepam plasma t1/2	21–37 h[138]
Diazepam oral fluid t1/2	Not available
Diazepam plasma: oral fluid correlation	$r = 0.97$[47,48]
Diazepam s/p	0.035[47]
Diazepam log P (octanol/water)	2.99[a]
Diazepam pKa	3.3[52]
Diazepam screening test	Benzodiazepines
Diazepam confirmation test	Diazepam; nordiazepam; oxazepam
Diazepam detection time in oral fluid [cutoff, ng/mL]	Diazepam GC [NR]: <5–50 h[3]
Diazepam detection time in urine [cutoff, ng/mL]	Diazepam, EMIT [500]: 2–7 days[139]
Diazepam oral fluid interpretation resources[46–52,140,141]	–

GC=gas chromatography; NR=not reported; EMIT=enzyme multiplied immunoassay technique.
[a]Source: National Toxicology Program, http://ntp.niehs.nih.gov/index.cfm?objectid=7182FF48-BDB7-CEBA-F8980E5DD01A1E2D (accessed September 27, 2006).

  - Interpretation: diazepam use, but also possible combined use of diazepam with nordiazepam, oxazepam, or 3-hydroxydiazepam
- Possible sources of diazepam
  - Diazepam
  - Metabolite of medazepam
  - Metabolite of ketazolam
- Biomarkers
  - The confirmed presence of oxidative metabolites of diazepam such as 3-hydroxydiazepam (temazepam), nordiazepam, oxazepam[51] in oral fluid would be useful to substantiate use, but it should be noted that these drug/metabolites may arise from use of other benzodiazepines.

Cannabis (Marijuana, THC)

Overview

Delta-9-tetrahydrocannabinol (THC) is a naturally occurring psychoactive constituent of *Cannabis sativa* L. (marijuana or cannabis; TABLE 8). Cannabis is the most widely used illegal substance in the world. THC also is found in pharmaceutical preparations, for example, dronabinol, a light yellow resinous oil insoluble in water and formulated in sesame oil. Dronabinol capsules are supplied as round, soft gelatin capsules containing either 2.5, 5, or 10 mg dronabinol for oral administration for the treatment of anorexia associated with weight loss in patients with AIDS and for treatment of nausea and vomiting associated with cancer chemotherapy. THC also is found in small amounts in cannabis products sold commercially such as hemp seeds and hemp oil and may produce positive urine tests for cannabinoid metabolite.[58]

The primary route of administration of cannabis is by smoking, but ingestion of cannabis products as foodstuffs is not uncommon. THC appears rapidly in plasma following the smoking of cannabis products.[59] Oral ingestion generally produces lower blood concentrations and delays in time-to-peak effects.[60,61] The highly lipophilic nature of THC allows rapid tissue uptake with concomitant decreases in plasma. THC appears to be released slowly from tissue resulting in a prolonged half-life of THC and metabolites. THC is metabolized by hydroxylation to an active metabolite, 11-hydroxy-THC, which in

TABLE 8. Summary of chemical, metabolic, kinetic, and analytic factors important in interpretation of oral fluid tests for THC

Factor	THC information
Drug/metabolite(s) in oral fluid	THC; THCCOOH (minor)[62–65]
Drug/metabolite(s) in urine	THCCOOH; multiple minor metabolites; conjugates[142]
THC Fb	97%[143]
THC plasma t1/2	THC: 0.75 h[65]; THC: (multiphasic): >20 h [144,145]; THCCOOH: 5–6 days[146]
THC oral fluid t1/2	0.8 h[65]
THC plasma: oral fluid correlation	High correlation[65]
THC s/p	1.18 (range 0.5–2.2) h[65]
THC log P (octanol/water)	6.97[a]
THC pKa	10.6[143]
Cannabis screening test	Cannabinoids
Cannabis confirmation test	THC
THC detection time in oral fluid [cutoff, ng/mL]	Single doses, THC: RIA [1]: 2.0–24.0 h[65]
THCCOOH detection time in urine [cutoff, ng/mL]	Single dose: cannabinoid metabolites: EMIT [20]: 9.3–78.4 h[65]; cannabinoid metabolites: EMIT [50]: 6.4–54.0 h;THCCOOH: GC-MS [15]: 8.0–122.3 h[147]; chronic users: cannabinoid metabolites: EMIT [20]: up to 67 days[148]
THC oral fluid interpretation resources[62–65,149,150]	–

THC=delta-9-tetrahydrocannabinol; THCCOH=11-nor-9-carboxy-delta-9-tetrahydrocannabinol; RIA=radioimmunoassay; EMIT=enzyme multiplied immunoassay technique; GC-MS=gas chromatography-mass spectrometry.

[a]Source: National Toxicology Program, http://ntp.niehs.nih.gov/index.cfm?objectid=7182FF48-BDB7-CEBA-F8980E5DD01A1E2D (accessed September 27, 2006).

turn, is oxidized to 11-nor-9-carboxy-Δ9-tetrahydrocannabinol (THCCOOH). THCCOOH is excreted in urine as the water-soluble glucuronic acid conjugate.

THC is found in oral fluid following smoked[62,63] and oral ingestion[64] of cannabis. THCCOOH is also found in oral fluid at very low concentrations. Extremely high (typically >200 ng/mL), but declining levels of THC, have been reported immediately after smoking cannabis.[62,63,65] Thereafter, THC appeared to decline in similar fashion to plasma concentrations.[65] On the basis of evidence to date, it appears that THC is present in oral fluid primarily as a result of deposition in the oral cavity, rather than from transfer from blood.[65] Following ingestion of hemp oil liquid containing THC and capsules of dronabinol, positive oral fluid tests for THC did not occur.[66] Early studies on passive inhalation of cannabis smoke indicated a potential risk for detection of low concentrations of THC in oral fluid for up to 30 min following exposure;[63] however, a more recent study in which methods were taken to eliminate contamination during specimen collection indicated that passive inhalation did not produce positive oral fluid tests.[62]

Interpretation of Oral Fluid THC Tests

- Positive test for THC
    - Interpretation: cannabis use, but must rule out possibility that presence is due to use of pharmaceutical THC (Sativex®). Ingestion of Marinol® and hemp seed oil does not produce positive oral fluid tests for THC.[66]
- Possible sources of THC
    - Illicit cannabis products
    - Hempseed products (does not give positive THC test)
    - Sativex
    - Marinol (does not give positive THC oral fluid test)
    - A positive THC test from passive cannabis smoke exposure does not appear to be feasible on the basis of recent studies[62]
- Biomarkers
    - The confirmed presence of oxidative metabolites of THC such as 11-hydroxy-THC and THCCOOH in oral fluid would be useful to substantiate use.
    - The confirmed presence of conjugates of THC and 11-hydroxy-THC in oral fluid would be useful to substantiate use.[67]
    - The use of 11-nor-delta-9-tetrahydrocannabivarin-9-carboxylic acid (THCV-COOH) has been proposed as a biomarker in urine to distinguish the use of synthetic THC (Marinol) from cannabis use.[68] THCV-COOH is an oxidative metabolite of delta-9-tetrahydrocannabivarin found naturally in cannabis, but not in synthetic THC. The confirmed presence of THCV-COOH in oral fluid would be useful to differentiate cannabis use from synthetic THC.

Cocaine

Overview

Cocaine is a local anesthetic and vasoconstrictor found in abundance in leaves of the coca plant (TABLE 9). Pharmaceutical preparations are used primarily for topical anesthesia of the upper respiratory tract. Illicit cocaine is broadly available in illicit markets in two main forms, cocaine hydrochloride (for intravenous and intranasal administration) and free-base cocaine (for smoking). Self-administration of cocaine produces stimulation and short-lived euphoria and is frequently followed by a desire for more drug.

Cocaine has a short half-life (approximately 1 h) and is rapidly hydrolyzed by hepatic esterases to benzoylecgonine (BZE) and ecognine methyl ester (EME). A variety of other metabolites of cocaine also have been identified.[69] When cocaine is administered in concert with ethanol, a transesterification product, cocaethylene, is formed in minor amounts and is metabolized in a similar manner as cocaine. Cocaine and its metabolites appear rapidly in oral fluid following all routes of administration.[70] High concentrations of cocaine and BZE in oral fluid are observed shortly after intranasal and smoked administration. Cocaine concentrations decrease rapidly within approximately 1 h; thereafter, oral fluid concentrations appear to decline in parallel with blood.[70]

Interpretation of Oral Fluid Cocaine Tests

- Positive test for cocaine (no BZE)
  - Interpretation: very recent cocaine use (e.g., within 8 h)
- Positive tests for cocaine and BZE
  - Cocaine concentration > BZE concentration
    - Interpretation: cocaine use likely within 2–8 h
  - Cocaine concentration < BZE concentration
    - Interpretation: cocaine use likely within 12 h for occasional users
    - Interpretation: cocaine use likely within 48 h for daily users
      - Positive test for BZE (no cocaine)
    - Interpretation: cocaine use likely within 48 h for occasional users
    - Interpretation: cocaine use likely within 48–96 days for daily users
      - Possible sources of cocaine
- Pharmaceutical cocaine, for example, use of cocaine as an anesthetic in surgery
- Illicit cocaine
- Coca tea
- At present, no studies on oral fluid testing have been reported with passive

TABLE 9. Summary of chemical, metabolic, kinetic, and analytic factors important in interpretation of oral fluid tests for cocaine

Factor	Cocaine information
Drug/metabolite(s) in oral fluid	Cocaine; BZE; EME[70]
Drug/metabolite(s) in urine	BZE; EME; norcocaine; benzoylnorecgonine; hydroxylated metabolites[69]
Cocaine Fb	Cocaine: negligible binding; BZE: negligible binding[151]
Cocaine Plasma t1/2	Cocaine: 1.4–1.8 h; BZE: 5.4–7.6 h; EME: 2.7–5.6 h[152]
Oral fluid t1/2	Cocaine: 1–2 h; BZE: 4.8–9.2 h; EME: 3.9–5.4 h[152]
Cocaine plasma:oral fluid correlation	$r = 0.89$[153]
Cocaine s/p	0.4–9.7[152]
Cocaine log P (octanol/water, 37°C, pH 7.4)	0.55[154]
Cocaine log P (HPLC)	3.35[127]
BZE log P (HPLC)	0.44[127]
EME log P (HPLC)	0.72[127]
Cocaine pKa	Cocaine: 8.81; BZE: 2.1; 11.75[151]
Cocaine screening test	Cocaine metabolite
Cocaine confirmation test	Cocaine; BZE
Detection times in oral fluid [cutoff, ng/mL]	Single dose: cocaine, GC-MS [1]: up to 12 h; GC-MS [1]: BZE: >12 h[70]; Chronic: cocaine, GC-MS [1]: 8.0–48.0 h; BZE, GC-MS [1]: 36.0–72.0 h[155]
Cocaine detection time in urine [cutoff, ng/mL]	Single dose: BZE, EIA [300]: 14–59 h; BZE, GC-MS [150]: 21–59 h[69]; chronic: BZE, GC-MS [10]: 123–218 h[152]; BZE, EIA [300]: 11–147 h[156]
Cocaine oral fluid interpretation resources[70,125,152,153,155]	–

BZE=benzoylecgonine; EME=ecognine methyl ester; GC-MS=gas chromatography-mass spectrometry; EIA=enzyme immunoassay.

cocaine smoke exposure; however, the risk of a positive test from passive exposure does not appear likely on the basis of urine studies.[71]

- Biomarkers
  - The confirmed presence of metabolites of oxidative metabolites of cocaine such as norcocaine, benzoylnorecgonine, and other hydroxy-metabolites in oral fluid would be useful to substantiate use.
  - The presence of cocaethylene in oral fluid may indicate combined cocaine and ethanol use; however, cocaethylene also may be present as a contaminant of both pharmaceutical and illicit cocaine.
  - The presence of anhydroecgonine methyl ester[70] or ecgonidine[72] would be indicative of smoked administration of cocaine.

Nicotine/Cotinine

Overview

Nicotine is the principal alkaloid that accounts for the widespread human use of tobacco products throughout the world. Tobacco products come in different forms including cigarettes, cigars, pipe tobacco, chewing tobacco, and smokeless tobacco (snuff). Almost 30% of the population of the United States are current users of tobacco products, the majority being cigarette smokers. Passive smoking also delivers nicotine to nonsmokers. Passive smoking is exposure to tobacco smoke that occurs when a nonsmoker is exposed to the sidestream smoke of a cigarette. Nicotine in high doses can be toxic, but serious direct toxicity is rare. The major health effect of nicotine is by mediating tobacco use, which results in millions of premature deaths yearly. A number of nicotine replacement products have been developed as smoking cessation medications including nicotine gum, nicotine patches, and nicotine sprays.

Nicotine is efficiently absorbed during use of tobacco products and nicotine replacement medications. It has a half-life in blood of approximately 2 h and is readily distributed to tissues, metabolized, and excreted in urine as nicotine and metabolites. Quantitatively, the most important metabolites of nicotine are cotinine and 3′-hydroxycotinine. The metabolite, cotinine, exhibits a considerably longer half-life of approximately 17 h.[73]

Nicotine appears in oral fluid at peak concentrations within 2–5 min following nicotine infusion in humans.[74] Oral fluid/plasma ratios were >1 for 60–120 min after nicotine administration. Cotinine also appears rapidly in oral fluid and is higher in concentration, but parallel to serum concentrations.[75] The higher concentration and longer half-life of cotinine makes it the preferred marker in plasma, oral fluid, or urine for monitoring nicotine intake after passive and active smoking (TABLE 10).[76] Passive smokers usually have cotinine concentrations in oral fluid <5 ng/mL, but heavy passive exposure can result in concentrations ≥10 ng/mL.[77] Cotinine concentrations in oral fluid between 10–100 ng/mL are seen in infrequent smokers and concentrations >100 ng/mL are generally associated with regular active smoking.[77] Trans-3′-hydroxycotinine concentrations in oral fluid of light- and heavy-smoking pregnant women have been reported to range from 14.4–117.7 ng/mL and 38.3–184.4 ng/mL, respectively.[78] A ratio of 0.41 for trans-3′-hydroxycotinine/cotinine was most effective in differentiating light from heavy tobacco use.

Interpretation of Oral Fluid Cotinine Tests

- Positive test for cotinine
  - Cutoff concentrations for cotinine in oral fluid to distinguish environmental smoke exposure from smoking have been suggested, but overlap between groups occurs.[77]

TABLE 10. Summary of chemical, metabolic, kinetic, and analytic factors important in interpretation of oral fluid tests for nicotine/cotinine

Factor	Nicotine/cotinine information
Drug/metabolite(s) in oral fluid	Nicotine; cotinine; trans-3′-hydroxycotinine; norcotinine[78]
Drug/metabolite(s) in urine	Nicotine; cotinine; trans-3′-hydroxycotinine[78]
Nicotine Fb	5–20%[157]
Cotinine Fb	2.6%[158]
Nicotine plasma t1/2	2 h[159]
Cotinine plasma t1/2	16.6 h[160]
Nicotine oral fluid t1/2	Not available
Cotinine oral fluid t1/2	17.4 h[160]
Nicotine plasma:oral fluid correlation	None[74]
Cotinine plasma:oral fluid correlation	$r = 0.99$[159]
Nicotine s/p	>1[74]
Cotinine s/p	1.2; 1.3–1.8[159,161]
Nicotine log P	0.93[161]
Cotinine log P	0.04[161]
Nicotine pKa	9.13[161]
Cotinine pKa	4.72[161]
Nicotine/cotinine screening test	Cotinine
Nicotine/cotinine confirmation test	Cotinine; trans-3′-hydroxycotinine; nicotine
Nicotine detection time in oral fluid [cutoff, ng/mL]	Not available
Cotinine detection time in oral fluid [cutoff, ng/mL]	light smokers: RIA [10]: 1–5 days[162] oral ingestion of 28 mg of nicotine/day: GC [<10]: 4 days[160]
Nicotine detection time in urine [cutoff, ng/mL]	Nicotine, GC [NR]: 16 h[163]
Cotinine detection time in urine [cutoff, ng/mL]	Cotinine, daily smokers: GC-MS [5]: > 5 days[164]; oral ingestion of 28 mg of nicotine per day: GC [<50]: 4 days[160]
Nicotine/cotinine oral fluid interpretation resources[73–78,159,162,165–167]	–

GC=gas chromatography; RIA=radioimmunoassay; NR=not reported.

- ■ Cotinine (0–10 ng/mL)
  - □ Interpretation: passive tobacco smoke exposure
- ■ Cotinine (>10–100 ng/mL)
  - □ Interpretation: passive tobacco smoke exposure or light smoking
- ■ Cotinine (>100 ng/mL)
  - □ Interpretation: light to heavy smoking

- Possible sources of cotinine
  - o Metabolite of nicotine-containing replacement medications, for example, nicotine gum, nicotine patch, nicotine nasal spray
  - o Metabolite of nicotine in tobacco products

  - Metabolite of nicotine from environmental tobacco smoke exposure
  - Metabolite of nicotine present in some foods (insignificant amounts compared with environmental tobacco smoke exposure)[79]
- Biomarkers
  - The confirmed presence of oxidative metabolites of cotinine such as trans-3′-hydroxycotinine and norcotinine in oral fluid would be useful to substantiate use.[78]

Opioids

Heroin

Overview: Heroin is a diacetyl derivative of morphine prepared from opium for the illicit drug market (TABLE 11). Illicit heroin contains minor amounts of other alkaloids including codeine and acetylcodeine. Pharmaceutical grade heroin is prepared from pure morphine and is generally free from impurities. Pharmaceutical heroin is used in some European countries for therapeutic treatment of heroin addiction.[80]

Heroin is most commonly administered by intravenous and other parenteral routes and also may be smoked. The greater lipophilicity of heroin and 6-acetylmorphine, as compared to morphine, allows rapid entry into the central nervous system, resulting in fast onset of euphoria and other pharmacological effects. Heroin has an extremely short half-life (minutes) and is rapidly converted to 6-acetlymorphine and morphine.[80]

Heroin and 6-acetylmorphine appear in oral fluid within 2 min following administration.[81] Drug and metabolite concentrations in oral fluid generally are similar to blood concentrations following intravenous administration, but may be substantially higher than blood following smoking. The elevated drug and metabolite concentrations following smoking are presumably due to residual drug deposited in the oral cavity. Thirty to sixty minutes after smoked heroin, concentrations in oral fluid diminish considerably and begin to reflect blood concentrations.

Interpretation of Oral Fluid Heroin Tests:

- Positive test for heroin, 6-acetylmorphine and morphine
  - Interpretation: heroin use
- Positive test for 6-acetylmorphine and morphine
  - Interpretation: heroin use
- Positive test for 6-acetylmorphine (only)
  - Interpretation: heroin use; consistent with very recent use by smoked or snorted route
- Positive test for morphine (only)
  - Interpretation: heroin or morphine use; possible poppy seed ingestion within last hour

TABLE 11. Summary of chemical, metabolic, kinetic, and analytic factors important in interpretation of oral fluid tests for heroin

Factor	Heroin information
Drug/metabolite(s) in oral fluid	Heroin, morphine; 6-AM[81]
Drug/metabolite(s) in urine	Morphine; morphine-3-glucuronide; morphine-6-glucuronide; 6-AM; normorphine (minor); heroin (minor)[168]
Heroin Fb	Heroin: 20–39%[169]; 6-AM: not available; Morphine: 20–37.5%[170,171]
Heroin plasma t1/2	Heroin: 3–4 min; 6-AM: 22–26 min; morphine: 177–184 min[80]
Heroin oral fluid t1/2	Heroin (IV): 1–67 min; heroin (SM): 26–294 min[81]
Heroin plasma:oral fluid correlation	Not available
Heroin s/b	Heroin s/b (IV): 0–1.9; heroin s/b (SM): 0–784; 6-AM (IV): 0–7.2; 6-AM (SM): 0–333; morphine (IV): 0–1.8; morphine (SM): 0–29[81]
Heroin log P (octanol/water, 37°C, pH 7.4)	Heroin: 1.076; 6-AM: 0.791; morphine: 0.083[154]
Heroin log P (HPLC)	6-AM: 1.48; morphine: 0.5[127]
Heroin pKa	Heroin: 7.6[80]; 6-AM: 8.19[172]; morphine: 7.9; 9.6[173]
Heroin screening test	Opiates
Heroin confirmation test	Total morphine; 6-AM; morphine
Heroin detection time in oral fluid [cutoff, ng/mL]	Heroin, GC-MS [1]: 2–24 h; 6-AM, GC-MS [1]: 0.5–8 h; morphine, GC-MS [1]: 2–12 h[81]
Heroin detection time in urine [cutoff, ng/mL]	Single dose: opiates, EMIT [300]: 7–54 h[174]; chronic: total morphine, GC-MS [25]: up to 270 h; 6-AM, GC-MS [10]: up to 34.5 h[175]
Oral fluid interpretation resources[81]	—

6-AM=6-acetylmorphine; IV=intravenous; SM=smoked; GC-MS=gas chromatography-mass spectrometry; EMIT=enzyme multiplied immunoassay technique.

- Positive test for morphine and codeine
  - Codeine concentration >> morphine concentration
    - Interpretation: codeine use
  - Morphine concentration ≥ codeine concentration
    - Interpretation:
      - Possible heroin or morphine use. Codeine presence may arise as impurity of heroin or secondary use in combination with morphine, but generally in low concentration.
      - Possible recent poppy seed ingestion within hour
- Possible sources of heroin
  - Pharmaceutical heroin
  - Illicit heroin

- Possible sources of 6-AM
  - Pharmaceutical heroin
  - Illicit heroin
- Possible sources of morphine
  - Pharmaceutical heroin
  - Illicit heroin
  - Morphine
  - Poppy seeds
  - Metabolism of codeine to morphine
- At present, although no oral fluid or urine studies have been reported on passive heroin smoke exposure, the risk of a positive test from passive exposure does not appear likely.
- Biomarkers
  - The confirmed presence of 6-AM in oral fluid is indicative of heroin use.
  - The confirmed presence of 6-acetylcodeine in oral fluid would be useful to differentiate illicit heroin use from pharmaceutical grade heroin.[82]

Morphine

Overview: Morphine, like most therapeutic opioids, produces effects through interaction with μ opioid receptors (TABLE 12).[83] The pharmacological actions of morphine encompass a wide range of effects including analgesia, mood alteration, and drug-seeking behavior. Historically, opioids have been the mainstay of pain treatment and continue to be widely used for this purpose. Numerous morphine prescription medications for treatment of acute and chronic pain are available in immediate-release formulations for oral and parenteral administration. Controlled-release formulations are available for oral administration.

Pharmaceutical morphine preparations have significant abuse liability. Drug abusers prefer immediate-release formulations that can be administered by the oral, intranasal, parenteral, and smoked routes. Controlled-release formulations generally contain higher doses, but may be more difficult to administer by alternate routes. Instructions may be found on the Internet for tampering with morphine formulations to allow faster administration, higher doses, and conversion of morphine to heroin.[84]

Following parenteral morphine administration, morphine appears rapidly in saliva. Cone[13] reported an approximate 45-min delay in equilibration of morphine concentrations in saliva compared to plasma following intramuscular administration of 10- and 20-mg doses; thereafter, saliva concentrations paralleled plasma concentrations. Kopecky *et al.*[85] measured morphine in

TABLE 12. Summary of chemical, metabolic, kinetic, and analytic factors important in interpretation of oral fluid tests for morphine

Factor	Morphine information
Drug/metabolite(s) in oral fluid	Morphine[13]
Drug/metabolite(s) in urine	Morphine; morphine-3-glucuronide; morphine-6-glucuronide; normorphine (minor); hydromorphone (minor)[92]
Morphine Fb	20–37.5%[170,171]
Morphine plasma t1/2 (h)	2.9-4.5 h[176]
Morphine oral fluid t1/2 (h)	Similar to plasma[13]
Morphine plasma:oral fluid correlation	High correlation[13]; no correlation[85]
Morphine s/p	<1[13]
Morphine log P (octanol/water, 37°C, pH 7.4)	0.083[154]
Morphine log P (HPLC)	0.5[127]
Morphine pKa	7.9; 9.6[173]
Morphine screening test	Opiates
Morphine confirmation test	Total morphine
Morphine detection time in oral fluid [cutoff, ng/mL]	Morphine, GC-MS [0.6]: 24 h[93]
Morphine detection time in urine [cutoff, ng/mL]	Total morphine, GC-MS [300] 36–48 h[88]
Morphine oral fluid interpretation[13,93] resources	–

HPLC=high performance liquid chromatography; GC-MS=gas chromatography-mass spectrometry.

saliva and plasma of pediatric patients following parenteral administration, but failed to demonstrate significant correlation. Although morphine is rapidly metabolized by conjugation, only free morphine has been reported in saliva. Morphine has been detected in oral fluid following intravenous and smoked heroin administration[81] and poppy seed ingestion,[86] but has not been detected following codeine administration.[87]

Morphine is a metabolite of heroin,[88] codeine[88] and a natural component of poppy seeds.[86] Pholcodine, a synthetic derivative of morphine with codeine-like effects, also has been reported to be metabolized to morphine in minor amounts.[89–91] Interpretation of oral fluid morphine results must take into account these multiple sources of morphine (e.g., heroin, codeine, or poppy seeds).

Interpretation of Oral Fluid Morphine Tests:

- Positive test for morphine (no other opioids present)
  - Interpretation: possible use of heroin or morphine or ingestion of poppy seed

        - Although 6-AM is frequently detected in oral fluid, it has a shorter half-life, and may not be present.
        - Poppy seed ingestion may result in low concentrations of morphine for up to 1 h after consumption.[86]
        - Codeine use has not been reported to produce detectable levels of morphine in oral fluid.[87]
        - Pholcodine use has been reported to result in low concentrations of morphine in urine,[89,90] but not likely to be found in oral fluid in absence of high concentration of phlocodine.
- Positive tests for morphine and 6-AM
    - Interpretation: heroin use
- Positive tests for morphine and codeine
    - Interpretation: codeine concentration >> morphine concentration
        - Codeine use
    - Interpretation: morphine concentration > codeine concentration
        - Heroin or combined morphine/codeine use
        - Codeine presence may arise as impurity of heroin or secondary use in combination with morphine
- Possible sources of morphine
    - Pharmaceutical heroin
    - Illicit heroin
    - Morphine
    - Poppy seeds[86]
    - Codeine
    - Pholcodine[89,90]
- Biomarkers
    - The confirmed presence of oxidative metabolites such as normorphine in oral fluid would be useful to substantiate use.[92]

Codeine

Overview: Codeine is an analgesic and cough suppressant usually marketed as an oral preparation, frequently in combination with additional active ingredients such as acetaminophen, aspirin, caffeine, guaifenesin, and ibuprofen (TABLE 13). Prescription doses range from 15 to 60 mg. Recreational drug users prefer higher doses and may attempt various purification methods available on the Internet, for example, "cold water extraction," to eliminate unwanted active components and excipients present in over-the-counter and prescription formulations.[84] Codeine appears to be abused primarily by the oral route.

Following oral administration of 60 and 120 mg, codeine appeared in oral fluid within an hour and reached maximum concentration in approximately 1.6–1.7 h.[87] Concentrations in oral fluid correlated significantly with plasma concentration and were three to four times higher in oral fluid than plasma.

Codeine could be detected in oral fluid for approximately 21 and 7 h at cutoff concentrations of 2.5 and 40 ng/mL, respectively.[87] Following intramuscular codeine of 60 and 120 mg, codeine appeared rapidly in oral fluid and reached maximal concentrations in 0.5–0.75 h.[93]

Codeine is metabolized by oxidation to morphine and norcodeine and by conjugation. Codeine is not a metabolite of morphine.[88] Only free codeine and norcodeine have been detected in oral fluid; morphine was not detected.[87] Codeine is known to be present in minor amounts in illicit heroin and poppy seeds. 6-Acetylcodeine, a derivative of codeine produced during heroin manufacture, was proposed as a possible urinary marker of illicit heroin use,[82] and has been identified in oral fluid of opioid-dependent women.[94] Interpretation of oral fluid codeine results must take into account possible sources of codeine (e.g., heroin or poppy seeds).

Interpretation of Oral Fluid Codeine Tests:

- Positive test for codeine (no other opioid present)
  - Interpretation: codeine use
- Positive tests for codeine and morphine
  - Codeine concentration >> morphine concentration

TABLE 13. Summary of chemical, metabolic, kinetic, and analytic factors important in interpretation of oral fluid tests for codeine

Factor	Codeine information
Drug/metabolite(s) in oral fluid	Codeine; norcodeine[87]
Drug/metabolite(s) in urine	Codeine; codeine-6-glucuronide; morphine-6-glucuronide; norcodeine[177]; hydrocodone (minor)[101]
Codeine Fb	14.2–28.5%[171]
Codeine plasma t1/2	1.4–3.5 h[87]
Codeine oral fluid t1/2	1.0–3.8 h[87]
Codeine plasma:oral fluid correlation	$r = 0.22$[87]
Codeine s/p (range)	4.1 (1–17.2)[87]
Codeine log P (octanol/water, 37°C, pH 7.4)	0.54[154]
Codeine pKa	8.2[87]
Codeine screening	Opiates
Codeine confirmation	Total codeine
Codeine detection time in oral fluid [cutoff, ng/mL]	Codeine, GC-MS [2.5]:7–21 h[87]
Codeine detection time in urine [cutoff, ng/mL]	Codeine, GC-MS [300]: 24–48 h[88]
Codeine oral fluid interpretation resources[87,93,178]	–

GC-MS=gas chromatography-mass spectrometry.

            - Interpretation: codeine use
        - Morphine concentration > codeine concentration
            - Interpretation
- Possible heroin or morphine use. Codeine presence may arise as impurity of heroin or secondary codeine use in combination with morphine.
- Possible recent poppy seed ingestion within an hour
- Possible sources of codeine
    - Codeine
    - Illicit heroin
    - Poppy seeds
    - Codeine is not a metabolite of morphine
- Biomarkers
    - The confirmed presence of oxidative metabolites such as norcodeine in oral fluid would be useful to substantiate use.[87]

Hydromorphone

Overview: Hydromorphone (Dilaudid®) is a hydrogenated ketone derivative of morphine sold as an opioid analgesic for relief of pain (TABLE 14). It is available commercially in various immediate-release forms such as injectable, oral liquid, tablets, and suppositories. Doses range from 1 to 10 mg. Hydromorphone is approximately seven times more potent than morphine when injected intravenously.[95]

Following oral administration, hydromorphone reaches maximal plasma levels in approximately 1 h, demonstrates an elimination half-life of approximately 4.1 h, and has an absolute bioavailability of 50–60%.[95,96] Hydromorphone is metabolized primarily by conjugation and to a lesser extent by 6-keto reduction to α- and β-isomers of hydromorphol.[97] Evidence for the metabolic transformation of morphine to hydromorphone in small amounts in subjects chronically treated with high-dose morphine has been reported.[98]

Hydromorphone was reported to appear in saliva rapidly following intravenous administration.[96] Initial saliva/plasma (s/p) ratios were lower during the distribution phase (up to approximately 40 min post drug administration) than the elimination phase.

Interpretation of Oral Fluid Hydromorphone Tests:

- Positive test for hydromorphone (no hydrocodone/morphine)
    - Interpretation: hydromorphone use
- Positive test for hydromorphone and morphine
    - Hydromorphone > morphine
        - Interpretation: hydromorphone and morphine use
    - Hydromorphone << morphine

TABLE 14. Summary of chemical, metabolic, kinetic, and analytic factors important in interpretation of oral fluid tests for hydromorphone

Factor	Hydromorphone information
Drug/metabolite(s) in oral fluid	Hydromorphone[96]
Drug/metabolite(s) in urine	Hydromorphone; hydromorphone-3-glucuronide; hydromorphol (α/β-isomers)[97]
Hydromorphone Fb	7%[96]
Hydromorphone plasma t1/2	2.5–2.6 h[95]; 2.4–4.1 h[96]
Hydromorphone oral fluid t1/2	2.12 h[96]
Hydromorphone plasma:oral fluid correlation	Not available
Hydromorphone s/p	0.25–2.32[96]
Hydromorphone log P (octanol/water, 37°C, pH 7.4)	0.161[154]
Hydromorphone log P (octanol/phosphate buffer, pH 7.4)	0.308[96]
Hydromorphone pKa	8.08; 9.47[96]
Hydromorphone screening test	Opiates
Hydromorphone confirmation test	Total hydromorphone
Hydromorphone detection time in oral fluid [cutoff, ng/mL]	Single doses: hydromorphone, RIA [1]: 6 h[96]
Hydromorphone detection time in urine [cutoff, ng/mL]	Single doses: hydromorphone: EMIT [300]: 6.3–11.3 h; hydromorphone: GC-MS [300]: 12–24 h[103]
Hydromorphone oral fluid interpretation resources[96]	–

RIA=radioimmunoassay; EMIT=enzyme multiplied immunoassay technique; GC-MS=gas chromatography-mass spectrometry.

    - Interpretation: possible hydromorphone use; possible occurrence of hydromorphone as a minor metabolite of morphine[98]
- Possible sources of hydromorphone
    - Hydromorphone
    - Hydrocodone metabolism to hydromorphone in minor amounts
    - Chronic morphine administration with minor metabolic conversion of morphine to hydromorphone[98]
- Biomarkers
    - The confirmed presence of reduced metabolites such as hydromorphol in oral fluid would be useful to substantiate use.[97]

Hydrocodone

Overview: Hydrocodone is a semisynthetic opioid analgesic and antitussive with multiple actions qualitatively similar to those of codeine (TABLE 15). Misuse of hydrocodone appears to be primarily by the oral route, likely because most preparations are compounded with other active ingredients. Other

TABLE 15. Summary of chemical, metabolic, kinetic, and analytic factors important in interpretation of oral fluid tests for hydrocodone

Factor	Hydrocodone information
Drug/metabolite(s) in oral fluid	Hydrocodone
Drug/metabolite(s) in urine	Hydrocodone; norhydrocodone; hydrocodol (α/β-isomers); hydromorphone; hydromorphol (α/β-isomers); conjugated metabolites[100]
Hydrocodone Fb	Not available
Hydrocodone plasma t1/2	Extensive metabolizers: 4.2 h; poor metabolizers: 6.2 h[99]
Hydrocodone oral fluid t1/2	Not available
Hydrocodone plasma:oral fluid correlation	Not available
Hydrocodone s/p	Not available
Hydrocodone log P (octanol/water, 37°C, pH 7.4)	0.143[154]
Hydrocodone pKa	8.9[130]
Hydrocodone screening test	Opiates
Hydrocodone confirmation test	Hydrocodone
Hydrocodone detection time in oral fluid [cutoff, ng/mL]	Expected to be similar to codeine
Hydrocodone detection time in urine [cutoff, ng/mL]	Single doses: hydrocodone: EMIT [300]: 12–36 h; hydrocodone: GC-MS [300]: 11–24 h; hydromorphone: GC-MS [300]: 0–11.8 h[103]
Hydrocodone oral fluid interpretation resources	–

EMIT=enzyme multiplied immunoassay technique; GC-MS=gas chromatography-mass spectrometry.

routes of administration include intranasal, smoked, and rectal administration. Pharmaceutical hydrocodone is supplied as tablets, syrup, suspension, and oral solution, and as an extended-release formulation. Most formulations are immediate-release and contain hydrocodone in doses ranging from 2.5 to 10 mg.

Following oral administration, hydrocodone reaches maximal serum concentrations in approximately 1 h and has an elimination half-life of approximately 4–6 h. Hydrocodone is metabolized by *O*-demethylation to hydromorphone by CYP2D6, a genetically polymorphic enzyme.[99] Hydromorphone is a potent analgesic and may be primarily responsible for hydrocodone's analgesic effects. Additional hydrocodone metabolic pathways include *N*-demethylation, keto-reduction, and conjugation.[100] Evidence for human metabolic transformation in small amounts of codeine[101] and dihydrocodeine[102] to hydrocodone has been reported.

Although commercial testing services are available in the United States for hydrocodone in oral fluid, there appears to be little information on the time course of appearance and disappearance of this drug in oral fluid.

Interpretation of Oral Fluid Hydrocodone Tests:

- Positive test for hydrocodone (no hydromorphone/codeine)
  - Interpretation: hydrocodone use
- Positive test for hydrocodone and hydromorphone
  - Hydrocodone $>$ hydromorphone
    - Interpretation: hydrocodone use; presence of hydromorphone likely due to metabolism of hydrocodone to hydromorphone
  - Hydrocodone $<$ hydromorphone
    - Interpretation: likely combined use of hydrocodone and hydromorphone, but could occur at low concentrations toward end of excretion following hydrocodone use[103]
- Positive test for hydrocodone and codeine
  - Hydrocodone $\geq$ codeine
    - Interpretation: likely combined use of hydrocodone and codeine
  - Hydrocodone $<<$ codeine
    - Interpretation: possible hydrocodone use; possible occurrence of hydrocodone as a minor metabolite of codeine[101]
- Possible sources of hydrocodone
  - Hydrocodone
  - Metabolite of codeine in minor amounts; but would occur only when codeine was present in very high concentration[101]
  - Metabolite of dihydrocodeine in minor amounts; but would occur only when dihydrocodeine was present in very high concentration[102]
  - Hydrocodone is not a metabolite of hydromorphone
- Biomarkers
  - The confirmed presence of oxidative metabolites such as norhydrocodone and hydromorphone in oral fluid would be useful to substantiate use.[100]
  - The confirmed presence of reduced metabolites such as hydrocodol in oral fluid would be useful to substantiate use.[100]

Oxymorphone

Overview: Oxymorphone is a potent semisynthetic opioid substitute for morphine (TABLE 16). The approximate potency of oxymorphone is some 10 times that of morphine following parenteral administration.[104] Oxymorphone is available as an injection and as a suppository for relief of moderate-to-severe pain. Oxymorphone is available as an injection in concentrations of 1–1.5 mg/mL and as a 5 mg suppository.

Oral oxymorphone is approximately one-sixth as potent as injectable oxymorphone.[105] It is metabolized primarily by conjugation and to a lesser extent by 6-keto-reduction to oxymorphol. Oxymorphol has been shown in animal studies to have analgesic activity.[106] Following a 10-mg oral dose, approximately 46% of the dose was reported to be excreted in urine as free and conjugated oxymorphone.[107] 6-Keto-reduced metabolites accounted for an additional 2.7% of the dose.

Oral fluid testing for oxymorphone is quite feasible, but there appears to be no information currently available on the time course of appearance and disappearance of this drug in oral fluid.

Interpretation of Oral Fluid Oxymorphone Tests:

- Positive test for oxymorphone (no oxycodone)
  - Interpretation: oxymorphone use
- Possible sources of oxymorphone
  - Oxymorphone
  - Oxycodone metabolism to oxymorphone in minor amounts
- Biomarkers
  - The confirmed presence of oxidative metabolites such as noroxymorphone in oral fluid would be useful to substantiate use.

TABLE 16. Summary of chemical, metabolic, kinetic, and analytic factors important in interpretation of oral fluid tests for oxymorphone

Factor	Oxymorphone information
Drug/metabolite(s) in oral fluid	Oxymorphone
Drug/metabolite(s) in urine	Oxymorphone; conjugated oxymorphone; oxymorphol (α/β-isomers)[107]
Oxymorphone Fb	10–12%[a]
Oxymorphone plasma t1/2	Parenteral: 1.3 h; oral: 7.3–9.4 h[179]
Oxymorphone oral fluid t1/2	Not available
Oxymorphone plasma:oral fluid correlation	Not available
Oxymorphone s/p	Not available
Oxymorphone log P (octanol/water, 37°C, pH 7.4)	–0.105[154]
Oxymorphone pKa	8.5, 9.3[130]
Oxymorphone screening test	Oxycodone
Oxymorphone confirmation test	Total oxymorphone
Oxymorphone detection time in oral fluid [cutoff, ng/mL]	Not available
Oxymorphone detection time in urine [cutoff, ng/mL]	Single doses: oxymorphone: opiate EMIT [300]: 0 h; oxymorphone: GC-MS [300]: 11.3–36 h[103]
Oxymorphone oral fluid interpretation resources[107]	–

EMIT=enzyme multiplied immunoassay technique; GC-MS=gas chromatography-mass spectrometry.

[a]Source: http://www.opana.com/pdfs/Opana_IR_PI.pdf#search=%22oxymorphone%20plasma%20binding%22 (accessed September 29, 2006).

TABLE 17. Summary of chemical, metabolic, kinetic, and analytic factors important in interpretation of oral fluid tests for oxycodone

Factor	Oxycodone information
Drug/metabolite(s) in oral fluid	Oxycodone
Drug/metabolite(s) in urine	Oxycodone; oxymorphone; conjugated oxymorphone; oxycodol (α/β-isomers); oyxmorphol (α/β-isomers)[109]
Oxycodone Fb	0.45[130]
Plasma t1/2	Oxycodone: 2.25 h[104]; oxycodone 2.1–5.1 h[180]; oxymorphone: 2.40 h[104]; noroxycodone: 4.15 h[104]
Oxycodone oral fluid t1/2	Not available
Oxycodone plasma:oral fluid correlation	Not available
Oxycodone s/p	Not available
Oxycodone log P (octanol/water, 37°C, pH 7.4)	0.243[154]
Oxycodone pKa	8.53[181]
Oxycodone screening test	Oxycodone
Oxycodone confirmation test	Oxycodone
Oxycodone detection time in oral fluid [cutoff, ng/mL]	Not available
Oxycodone detection time in urine [cutoff, ng/mL]	Single doses: oxycodone: opiate EMIT [300]: 0–6.6 h; oxycodone: GC-MS [300]: 12–36 h; oxymorphone: GC-MS [300]: 12–48 h[103]
Oxycodone oral fluid interpretation resources[182]	–

EMIT=enzyme multiplied immunoassay technique; GC-MS=gas chromatography-mass spectrometry.

- The confirmed presence of reduced metabolites such as oxymorphol in oral fluid would be useful to substantiate use.[107]

Oxycodone

Overview: Oxycodone is an opioid analgesic supplied as tablets, capsules, and oral solution (Table 17). It is available primarily as the hydrochloride salt with some formulations containing a combination of oxycodone hydrochloride and oxycodone terephthalate. Some formulations also contain aspirin or acetaminophen. Immediate-release formulations contain oxycodone doses ranging from 2.5 to 30 mg; a controlled-release formulation contains from 10 to 160 mg.

Oxycodone is effective orally and has a bioavailability of 50 to 90%.[108] It is metabolized by *O*-demethylation to oxymorphone, by *N*-demethylation to noroxycodone, and to minor metabolites by 6-keto-reduction.[109] CYP2D6 is the enzyme responsible for conversion of oxycodone to oxymorphone, a metabolite that is more potent than the parent drug. Although the analgesic potency of oxymorphone is higher, it is unclear how much activity is contributed to the actions of oxycodone by this metabolite. The analgesic potency of noroxycodone appears to be lower than that of oxycodone.[104]

Oral fluid testing for oxycodone is quite feasible, but little information is currently available on the time course of appearance and disappearance of this drug in oral fluid.

Interpretation of Oral Fluid Oxycodone Tests:

- Positive test for oxycodone (no oxymorphone)
  - Interpretation: oxycodone use
- Positive test for oxycodone and oxymorphone
  - Oxycodone > oxymorphone
    - Interpretation: oxycodone use; presence of oxymorphone likely due to metabolism of oxycodone to oxymorphone
  - Oxycodone < oxymorphone
    - Interpretation: likely combined use of oxycodone and oxymorphone, but could occur at low concentrations toward end of excretion following oxycodone use[103]
- Possible sources of oxycodone
  - Oxycodone
  - Oxycodone is not a metabolite of oxymorphone
- Biomarkers
  - The confirmed presence of oxidative metabolites such as noroxycodone and oxymorphone in oral fluid would be useful to substantiate use.[109]
  - The confirmed presence of reduced metabolites such as oxycodol in oral fluid would be useful to substantiate use.[109]

Methadone

Overview: Methadone is a long-acting opioid μ-receptor agonist with pharmacological properties similar to those of morphine (TABLE 18). Methadone is available as an oral concentrate and dispensable tablets for relief of chronic pain, treatment of opioid abstinence syndromes, and treatment of heroin dependence.

Methadone undergoes extensive metabolism in the liver to form cyclic metabolites, 2-ethylidene-1,5-dimethyl-3,3-diphenylpyrrolidine (EDDP) and 2-ethyl-5-methyl-3,3-diphenylpyrrolidine (EMDP), and other minor metabolites. The amount of methadone excreted in urine is increased when urine is acidified.[110] EDDP is excreted in urine in approximately equal amounts to methadone, but with apparent less variability. Thus, there may be advantages in testing for EDDP when monitoring patient compliance.[111]

Methadone and EDDP appear rapidly in oral fluid and correlate with plasma concentrations.[112] Oral fluid pH appears to be a factor in determining concentration of methadone in oral fluid, but is less important for EDDP detection.[112]

Interpretation of Oral Fluid Methadone Tests:

- Positive test for methadone (no EDDP)
  - Interpretation: methadone use within 24–48 h or possible methadone doping
- Positive tests for methadone and EDDP
  - Interpretation: methadone use within 24–48 h
- Possible sources of methadone
  - Methadone
- Biomarkers
  - The confirmed presence of oxidative metabolites such as EDDP and EMDP in oral fluid would be useful to substantiate use.[112]

TABLE 18. Summary of chemical, metabolic, kinetic, and analytic factors important in interpretation of oral fluid tests for methadone

Factor	Methadone information
Drug/metabolite(s) in oral fluid	Methadone; EDDP[112]
Drug/metabolite(s) in urine	Methadone; EDDP; EMDP; minor metabolites[183]
Methadone Fb	88.1–89.1%[184]
Methadone plasma t1/2	15–55 h[185]; 23–27 h[186]; 33–46 h[185]
Methadone oral fluid t1/2	Not available
Methadone plasma:oral fluid correlation	$r = 0.81$[187]; $r = 0.54$[112]
Methadone s/p	0.51[188]; 0.6–7.2[112]; 1.3[187]
Methadone P (chloroform/water, 25°C, pH 7.4)	Methadone: 14.56; EDDP: 2.38[110]
Methadone pKa	Methadone 8.62[110];9.2[186]; 8.2[112]; EDDP: 10.42[110]
Methadone screening tests	Methadone; EDDP
Methadone confirmation test	Methadone; EDDP
Methadone detection time in oral fluid [cutoff, ng/mL]	Methadone, GC-MS [20]: 24 h[188]
Methadone detection time in urine [cutoff, ng/mL]	Methadone, EIA [300]: 24–96 h[189]
Methadone oral fluid interpretation resources[112,187,188]	–

EDDP=2-ethylidene-1,5-dimethyl-3,3-diphenylpyrrolidine; EMDP=2-ethyl-5-methyl-3,3-diphenylpyrrolidine; EIA=enzyme immunoassay; GC-MS=gas chromatography-mass spectrometry.

TABLE 19. Summary of chemical, metabolic, kinetic, and analytic factors important in interpretation of oral fluid tests for buprenorphine

Factor	Buprenorphine information
Drug/metabolite(s) in oral fluid	Buprenorphine; norbuprenorphine[13]
Drug/metabolite(s) in urine	Buprenorphine; conjugated buprenorphine; norbuprenorphine; conjugated norbuprenorphine[115]
Buprenorphine Fb	95–98%[190]
Plasma t1/2	Buprenorphine: 40.7 h; norbuprenorphine: 73.3 h[118]
Buprenorphine oral fluid t1/2	Not available
Buprenorphine plasma:oral fluid correlation	High correlation[13]
Buprenorphine s/p	Intramuscular: 0.05–0.41; sublingual: >1[13]
Buprenorphine log P (octanol/water, 37°C, pH 7.4)	4.98[172]
Buprenorphine pKa	8.24, 10[190]
Buprenorphine screening test	Buprenorphine
Buprenorphine confirmation test	Buprenorphine; norbuprenorphine; conjugates
Buprenorphine detection time in oral fluid [cutoff, ng/mL]	Buprenorphine, RIA [1]: up to 5 days[13]
Buprenorphine detection time in urine [cutoff, ng/mL]	total buprenorphine/norbuprenorphine, GC [10/5]: 4–8 days[191]; total buprenorphine/norbuprenorphine, GC-MS [0.25/0.2]: 4–8 days[192]
Buprenorphine oral fluid interpretation resources[13]	–

RIA=radioimmunoassay; GC=gas chromatography; GC-MS=gas chromatography-mass spectrometry.

Buprenorphine

Overview: Buprenorphine is a thebaine-derived synthetic opioid whose actions are limited by a ceiling effect (TABLE 19). It is available in injectable form as an analgesic (0.3 mg) and as sublingual tablets (2 mg and 8 mg) for treatment of opioid dependence. One type of sublingual tablet contains only buprenorphine hydrochloride and the second contains buprenorphine hydrochloride in combination with naloxone hydrochloride in a ratio of 4:1 buprenorphine:naloxone (ratio of free bases).

Following sublingual administration, buprenorphine reaches maximal plasma concentrations in 1.3–1.6 h.[113] The bioavailability of sublingual buprenorphine (ethanol solution) ranges from 28 to 36%. Sublingual bioavailability of buprenorphine tablets (relative to ethanolic solution) is 49% ± 25%.[114] Buprenorphine is metabolized by conjugation and *N*-demethylation, and is excreted in urine primarily as conjugated metabolites.[115]

Cone reported measurement of buprenorphine in saliva following intramuscular and sublingual administration of single doses of buprenorphine.[13] Drug concentrations in saliva were substantially lower than plasma at all times following intramuscular administration and were substantially higher following sublingual administration. The low s/p ratio following intramuscular administration is likely due to the high fraction of drug that is protein-bound in plasma. The high s/p observed following sublingual administration was attributed to an oral mucosal drug depot.[116] Close correspondence in saliva and plasma buprenorphine concentrations was observed for subjects administered sublingual buprenorphine daily or on an every-other-day basis.[13]

Interpretation of Oral Fluid Buprenorphine Tests:

- Positive test for buprenorphine
  - Interpretation: buprenorphine use
- Positive tests for buprenorphine and norbuprenorphine
  - Buprenorphine >> norbuprenorphine
    - Interpretation: acute buprenorphine use (suggested from plasma data)[117]
  - Buprenorphine ≤ norbuprenorphine
    - Interpretation: chronic buprenorphine use (suggested from plasma data[118]
- Possible sources of buprenorphine
  - Buprenorphine
- Biomarkers
  - The confirmed presence of oxidative metabolites such as norbuprenorphine in oral fluid would be useful to substantiate use.[115]

Phencyclidine (PCP)

Overview

PCP is a dissociative anesthetic reported in the 1950s to be effective in surgery without respiratory depression, but also produced unpleasant side effects leading to its discontinuation of clinical use (TABLE 20). It was used in veterinary medicine, but withdrawn in 1978, leaving only illicit synthesis as a source of the drug. PCP is illicitly marketed under a number of street names including Angel Dust, Supergrass, Killer Weed, Embalming Fluid, and Rocket Fuel. Among PCP's least desirable side effects are delirium, visual disturbances, and hallucinations and, occasionally, violence.

PCP is well absorbed and readily penetrates the central nervous system after intravenous, smoked, intranasal, oral, and percutaneous administration, and it can be passively absorbed from the environment.[119] The plasma elimination half-life is 7–50 h (mean: 17.6 h) in normal volunteers.[120] PCP is excreted in urine in small amounts (approximately 10% of dose) along with

TABLE 20. Summary of chemical, metabolic, kinetic, and analytic factors important in interpretation of oral fluid tests for PCP

Factor	PCP information
Drug/metabolite(s) in oral fluid	PCP[124]
Drug/metabolite(s) in urine	PCP; PPC; PCHP[120]
PCP Fb	64%[123]; 68–69%[120]
PCP plasma t1/2	7–50 h[120]
PCP oral fluid t1/2	Not available
PCP plasma:oral fluid correlation	$r = 0.87$[120]
PCP serum:oral fluid correlation	$r = 0.92$[124]; $r = 0.4$[123]
PCP s/p	1.5–3[193]
PCP log P (octanol/water, 37°C, pH 7.4):	1.96[154]
PCP pKa	8.5[123]
PCP screening test	PCP
PCP confirmation test	PCP
PCP detection time in oral fluid [cutoff, ng/mL]	Not available
PCP detection time in urine [cutoff, ng/mL]	Chronic: PCP; GC [5]: up to 30 days[194]
PCP oral fluid interpretation resources[120,121,123,124]	–

PCP=phencyclidine; PPC=4-(1-piperidinyl)-cyclohexanol; PCHP=1-(1-phenylcyclohexyl)-4-hydroxypiperidine; GC=gas chromatography.

a number of polar metabolites.[120] Acidification of urine by ingestion of ammonium chloride[121] or by infusion of 0.1 N HCl[122] increased PCP excretion in urine. Hydroxy-metabolites of PCP excreted in urine include 4-phenyl-4-(1-piperidinyl)-cyclohexanol (PPC) and 1-(1-phenylcyclohexyl)-4-hydroxypiperidine (PCHP).[120]

PCP has been measured by radioimmunoassay in oral fluid and serum specimens of emergency room patients suspected of PCP intoxication.[123] Of 74 patients with positive oral fluid tests, 73 were accompanied by positive serum result. Only two specimens were positive for serum and negative for oral fluid. Concentrations of PCP in oral fluid and plasma ranged from 2–600 ng/mL and 5–400 ng/mL, respectively. PCP also has been measured in oral fluid of humans following administration of small doses of radiolabeled drug.[120,121,124] The radioactivity in oral fluid was principally (>90%) parent drug. Acidification or alkalinization of urine, respectively, increased and decreased urinary excretion of PCP but did not significantly alter concentration in oral fluid.

Interpretation of Oral Fluid PCP Tests

- Positive test for PCP
  - Interpretation: PCP use

- Possible sources of PCP
  - Pharmaceutical PCP
  - Illicit PCP
- Biomarkers
  - The confirmed presence of oxidative metabolites such as PPC and PCHP in oral fluid would be useful to substantiate use.[120]

LIMITATIONS

Interpretation of biological tests for drugs of abuse will always be limited by available scientific evidence. At present, much of the available knowledge on oral fluid and urine tests has been generated from single- or multiple-dosing studies, but there is limited information in chronic users. Significant ethical issues exist in the study of many licit and illicit drugs that preclude their study under conditions that simulate "real-world use," and relevant information may never be available. There also are significant cost and resource limitations for clinical studies that limit the number of drug studies that can be performed. Despite a substantial number of clinical studies on drug disposition in oral fluid, many psychoactive drugs have not been studied. Many benzodiazepines and barbiturates and some opioid products have received limited or no evaluation in oral fluid; controlled dosing studies of hallucinogens in humans are virtually nonexistent.

Collection of oral fluid is an important process that deserves comment in the context of limitations to interpretation. The dynamic nature of oral fluid, especially pH, can substantially affect drug concentrations of basic drugs. For example, Kato *et al.*[125] showed that oral fluid collected from cocaine users under nonstimulated conditions produced substantially higher drug concentrations in oral fluid than under stimulated conditions. A similar finding has been reported for cotinine.[126] Collection of oral fluid with an absorptive device further introduces issues of drug/metabolite recovery. If drug/metabolite is not fully recoverable from the collection device, concentrations in oral fluid will be lower relative to neat oral fluid. If so, cutoff concentrations for oral fluid may, by necessity, be lower than for neat oral fluid. Obviously, comparison of oral fluid studies across different collection conditions is problematic if drug recovery is not equivalent.

Aside from collection, many other factors that may affect oral fluid test outcomes and interpretation have not been studied. Do adulterants exist that can be safely placed in the mouth and negate screening tests, thus producing false negative results? Observed collection, coupled with a waiting period, would likely be an effective procedure to overcome the risk of adulteration, but this remains to be clearly established. Does passive exposure to drug smoke result in positive oral fluid tests? Studies on passive inhalation of cannabis smoke indicate that the risk of a positive from exposure is equivalent to or less than that for urine testing.[62] However, there is virtually no information on the risk of passive inhalation with heroin, methamphetamine, and PCP. Would passive exposure to these drugs result in positive tests in oral fluid? Experience in passive exposure studies with cannabis and cocaine suggests that passive exposure to other drugs is highly unlikely to produce positive oral fluid tests. Also, it should be noted that the limitation in information on the risk of passive inhalation for many drugs extends to urine testing as well.

Overall, it is clear that interpretation of oral fluid tests should be limited to available scientific knowledge. How oral fluid tests perform under many different conditions is not available, nor is it available for other biological matrices, for example, urine, sweat, and hair. It can be confidently predicted that scientific studies can never address all questions that may arise regarding how drug exposure affects test outcome. As always, science is an evolving process building upon prior knowledge that sometimes strikes out in new directions. Use of oral fluid as a test matrix is a relatively new science showing exceptional promise for detection of recent drug use. Informed interpretation of results requires a broad understanding of the unique characteristics of oral fluid.

ACKNOWLEDGMENT

E. J. Cone is a consultant to OraSure Diagnostics, a company that manufactures oral fluid diagnostic products.

REFERENCES

1. DHHS. 2004. Proposed revisions to mandatory guidelines for federal workplace drug testing programs. Fed. Regist. **69:** 19673–19732.
2. Caddy, B. 1984. Saliva as a specimen for drug analysis. *In* Advances in Analytical Toxicology. R.C. Baselt, Ed.: 198–254. Biomedical Publications, Foster City, CA.
3. Schramm, W., R.H. Smith, P.A. Craig & D.A. Kidwell. 1992. Drugs of abuse in saliva: a review. J. Anal. Toxicol. **16:** 1–9.
4. Idowu, O.R. & B. Caddy. 1982. A review of the use of saliva in the forensic detection of drugs and other chemicals. J. Forensic Sci. Soc. **22:** 123–135.
5. Mucklow, J.C., M.R. Bending, G.C. Kahn & C.T. Dollery. 1978. Drug concentration in saliva. Clin. Pharmacol. Ther. **24:** 563–570.
6. Wylie, F.M., H. Torrance, A. Seymour, *et al.* 2005. Drugs in oral fluid. Part II. Investigation of drugs in drivers. Forensic Sci. Int. **150:** 199–204.
7. Verstraete, A.G. 2004. Detection times of drugs of abuse in blood, urine, and oral fluid. Ther. Drug Monit. **26:** 200–205.

8. Choo, R.E. & M.A. Huestis. 2004. Oral fluid as a diagnostic tool. Clin. Chem. Lab Med. **42:** 1273–1287.
9. Kadehjian, L. 2005. Legal issues in oral fluid testing. Forensic Sci. Int. **150:** 151–160.
10. Maurer, H.H. 2005. Advances in analytical toxicology: the current role of liquid chromatography-mass spectrometry in drug quantification in blood and oral fluid. Anal. Bioanal. Chem. **381:** 110–118.
11. Rivier, L. 2000. Techniques for analytical testing of unconventional samples. Baillieres Best Pract. Res. Clin. Endocrinol. Metab. **14:** 147–165.
12. Samyn, N., A. Verstraete, C. van Haeren & P. Kintz. 1999. Analysis of drugs of abuse in saliva. Forensic Sci. Rev. **11:** 1–19.
13. Cone, E.J. 1993. Saliva testing for drugs of abuse. Ann. N. Y. Acad. Sci. **694:** 91–127.
14. Beckett, A.H. & M. Rowland. 1965. Urinary excretion kinetics of amphetamine in man. J. Pharm. Pharmacol. **17:** 628–638.
15. Cody, J.T. 1992. Determination of methamphetamine enantiomer ratios in urine by gas chromatography-mass spectrometry. J. Chromatogr. **580:** 77–95.
16. Wan, S.H., S.B. Matin & D.L. Azarnoff. 1978. Kinetics, salivary excretion of amphetamine isomers, and effect of urinary pH. Clin. Pharmacol. Ther. **23:** 585–590.
17. Cody, J.T. 2002. Precursor medications as a source of methamphetamine and/or amphetamine positive drug testing results. J. Occup. Environ. Med. **44:** 435–450.
18. Kraemer, T. & H.H. Maurer. 2002. Toxicokinetics of amphetamines: metabolism and toxicokinetic data of designer drugs, amphetamine, methamphetamine, and their N-alkyl derivatives. Ther. Drug Monit. **24:** 277–289.
19. Torre, R. de la, M. Farre, M. Navarro, *et al.* 2004. Clinical pharmacokinetics of amfetamine and related substances: monitoring in conventional and non-conventional matrices. Clin. Pharmacokinet. **43:** 157–185.
20. Cho, A.K. & J. Wright. 1978. Minireview: pathways of metabolism of amphetamine. Life Sci. **22:** 363–372.
21. Poklis, A. & K.A. Moore. 1995. Response of EMIT amphetamine immunoassays to urinary desoxyephedrine following Vicks inhaler use. Ther. Drug Monit. **17:** 89–94.
22. Kim, I., J.M. Oyler, E.T. Moolchan, *et al.* 2004. Urinary pharmacokinetics of methamphetamine and its metabolite, amphetamine following controlled oral administration to humans. Ther. Drug Monit. **26:** 664–672.
23. Cook, C.E., A.R. Jeffcoat, J.M. Hill, *et al.* 1993. Pharmacokinetics of methamphetamine self-administered to human subjects by smoking S-(+)-methamphetamine hydrochloride. Drug Metab. Dispos. **21:** 717–723.
24. Schepers, R.J., J.M. Oyler, R.E. Joseph, Jr., *et al.* 2003. Methamphetamine and amphetamine pharmacokinetics in oral fluid and plasma after controlled oral methamphetamine administration to human volunteers. Clin. Chem. **49:** 121–132.
25. Caldwell, J., L.G. Dring & R.T. Williams. 1972. Metabolism of [14C] methamphetamine in man, the guinea pig and the rat. Biochem. J. **129:** 11–22.
26. Mas, M., M. Farre, R. de. la. Torre, *et al.* 1999. Cardiovascular and neuroendocrine effects and pharmacokinetics of 3, 4-methylenedioxymethamphetamine in humans. J. Pharmacol. Exp. Ther. **290:** 136–145.

27. Torre, R. de la, M. Farre, P.N. Roset, *et al.* 2004. Human pharmacology of MDMA: pharmacokinetics, metabolism, and disposition. Ther. Drug Monit. **26:** 137–144.
28. Torre, R. de la, M. Farre, J. Ortuno, *et al.* 2000. Non-linear pharmacokinetics of MDMA ('ecstasy') in humans. Br. J. Clin. Pharmacol. **49:** 104–109.
29. Fallon, J.K., A.T. Kicman, J.A. Henry, *et al.* 1999. Stereospecific analysis and enantiomeric disposition of 3, 4-methylenedioxymethamphetamine (Ecstasy) in humans. Clin. Chem. **45:** 1058–1069.
30. Navarro, M., S. Pichini, M. Farre, *et al.* 2001. Usefulness of saliva for measurement of 3,4-methylenedioxymethamphetamine and its metabolites: correlation with plasma drug concentrations and effect of salivary pH. Clin. Chem. **47:** 1788–1795.
31. Laloup, M., G. Tilman, V. Maes, *et al.* 2005. Validation of an ELISA-based screening assay for the detection of amphetamine, MDMA and MDA in blood and oral fluid. Forensic Sci. Int. **153:** 29–37.
32. Samyn, N. & C. van Haeren. 2000. On-site testing of saliva and sweat with Drugwipe and determination of concentrations of drugs of abuse in saliva, plasma and urine of suspected users. Int. J. Legal Med. **113:** 150–154.
33. Freudenmann, R.W. & M. Spitzer. 2004. The neuropsychopharmacology and toxicology of 3,4-methylenedioxy-N-ethyl-amphetamine (MDEA). CNS. Drug Rev. **10:** 89–116.
34. Ensslin, H.K., H.H. Maurer, E. Gouzoulis, *et al.* 1996. Metabolism of racemic 3,4-methylenedioxyethylamphetamine in humans. Isolation, identification, quantification, and synthesis of urinary metabolites. Drug Metab. Dispos. **24:** 813–820.
35. Samyn, N., G. De Boeck, M. Wood, *et al.* 2002. Plasma, oral fluid and sweat wipe ecstasy concentrations in controlled and real life conditions. Forensic Sci. Int. **128:** 90–97.
36. Inaba, T. & W. Kalow. 1975. Salivary excretion of amobarbital in man. Clin. Pharmacol. Ther. **18:** 558–562.
37. Dilli, S. & D. Pillai. 1980. Analysis of trace amounts of barbiturates in saliva. J. Chromatogr. **190:** 113–118.
38. van der Graaff, M., N.P.E. Vermeulen, P. Heij, *et al.* 1986. Pharmacokinetics of orally administered hexobarbital in plasma and saliva of healthy subjects. Biopharm. Drug Dispos. **7:** 265–272.
39. Tjaden, U.R., J.C. Kraak & J.F.K. Huber. 1977. Rapid trace analysis of barbiturates in blood and saliva by high pressure liquid chromatography. J. Chromatogr. **143:** 183–194.
40. Nishihara, K., K. Uchino, Y. Saitoh, *et al.* 1979. Estimation of plasma unbound phenobarbital concentration by using mixed saliva. Epilepsia **20:** 37–45.
41. Mucklow, J.C., C.J. Bacon, A.M. Hierons, *et al.* 1981. Monitoring of phenobarbitone and phenytoin therapy in small children by salivary samples. Ther. Drug Monit. **3:** 275–277.
42. Cook, C.E., E. Amerson, W.K. Poole, *et al.* 1975. Phenytoin and phenobarbital concentrations in saliva and plasma measured by radioimmunoassay. Clin. Pharmacol. Ther. **18:** 742–747.
43. Schmidt, D. & H.J. Kupferberg. 1975. Diphenylhydantoin, phenobarbital, and primidone in saliva, plasma and cerebrospinal fluid. Epilepsia **16:** 735–741.
44. Troupin, A.S. & P. Friel. 1975. Anti-convulsant level in saliva, serum, and cerebrospinal fluid. Epilepsia **16:** 223–227.

45. GOLDSMITH, R.F. & R.A. OUVRIER. 1981. Salivary anticonvulsant levels in children: a comparison of methods. Ther. Drug Monit. **3:** 151–157.
46. GILES, H.G., D.H. ZILM, R.C. FRECKER, *et al.* 1977. Saliva and plasma concentrations of diazepam after a single oral dose. Br. J. Clin. Pharmacol. **4:** 711–712.
47. DI-GREGORIO, G.J., A.J. PIRAINO & E. RUCH. 1978. Diazepam concentrations in parotid saliva, mixed saliva, and plasma. Clin. Pharmacol. Ther. **24:** 720–725.
48. DE GIER, J.J., B. J. 'T HART, P.F. WILDERINK & F.A. NELEMANS. 1980. Comparison of plasma and saliva levels of diazepam. J. Clin. Pharmacol. **10:** 151–155.
49. HALLSTROM, C., M.H. LADER & S.H. CURRY. 1980. Diazepam and N-desmethyldiazepam concentrations in saliva, plasma and CSF. Br. J. Clin. Pharmacol. **9:** 333–339.
50. KAPAS, I.K. & L. VERECZKEY. 1982. A rapid capillary gas chromatographic method for the determination of diazepam in human plasma and saliva. Acta Pharmaceutics Hungarica **52:** 246–252.
51. TJADEN, U.R., M.T.H.A. MEELES, C.P. THYS & M. VAN DER KAAY. 1980. Determination of some benzodiazepines and metabolites in serum, urine and saliva by high-performance liquid chromatography. J. Chromatogr. **181:** 227–241.
52. GILES, H.G., R. MILLER, S.M. MACLEOD & E.M. SELLERS. 1980. Diazepam and N-desmethyldiazepam in saliva of hospital inpatients. J. Clin. Pharmacol. **20:** 71–76.
53. LUCEK, R. & R. DIXON. 1980. Chlordiazepoxide concentrations in saliva and plasma measured by radioimmunoassay. Res. Commun. Chem. Path. Pharmacol. **27:** 397–400.
54. SAMYN, N., G. DE BOECK, V. CIRIMELE, *et al.* 2002. Detection of flunitrazepam and 7-aminoflunitrazepam in oral fluid after controlled administration of rohypnol. J. Anal. Toxicol. **26:** 211–215.
55. QUINTELA, O., A. CRUZ, M. CONCHEIRO, *et al.* 2004. A sensitive, rapid and specific determination of midazolam in human plasma and saliva by liquid chromatography/electrospray mass spectrometry. Rapid Commun. Mass Spectrom. **18:** 2976–2982.
56. KANGAS, L., H. ALLONEN, R. LAMMINTAUSTA, *et al.* 1979. Pharmacokinetics of nitrazepam in saliva and serum after a single oral dose. Acta Pharmacol. Toxicol. **45:** 20–24.
57. HART, B.J., J. WILTING & J.J. DE GIER. 1987. Complications in correlation studies between serum, free serum and saliva concentrations of nitrazepam. Meth. Find. Exp. Clin. Pharmacol. **9:** 127–131.
58. COSTANTINO, A., R.H. SCHWARTZ & P. KAPLAN. 1997. Hemp oil ingestion causes positive urine tests for delta9-tetrahydrocanabinol carboxylic acid. J. Anal. Toxicol. **21:** 482–485.
59. HUESTIS, M.A., A.H. SAMPSON, B.J. HOLICKY, *et al.* 1992. Characterization of the absorption phase of marijuana smoking. Clin. Pharmacol. Ther. **52:** 31–41.
60. ALT, A. & G. REINHARDT. 1998. Positive cannabis results in urine and blood samples after consumption of hemp food products. J. Anal. Toxicol. **22:** 80–81.
61. CONE, E.J., R.E. JOHNSON, B.D. PAUL, *et al.* 1988. Marijuana-laced brownies: behavioral effects, physiologic effects, and urinalysis in humans following ingestion. J. Anal. Toxicol. **12:** 169–175.

62. NIEDBALA, R.S., K.W. KARDOS, E.F. FRITCH, *et al.* 2005. Passive cannabis smoke exposure and oral fluid testing. II. Two studies of extreme cannabis smoke exposure in a motor vehicle. J. Anal. Toxicol. **29:** 607–615.
63. NIEDBALA, S., K. KARDOS, S. SALAMONE, *et al.* 2004. Passive cannabis smoke exposure and oral fluid testing. J. Anal. Toxicol. **28:** 546–552.
64. NIEDBALA, R.S., K.W. KARDOS, D.F. FRITCH, *et al.* 2001. Detection of marijuana use by oral fluid and urine analysis following single-dose administration of smoked and oral marijuana. J. Anal. Toxicol. **25:** 289–303.
65. HUESTIS, M.A. & E.J. CONE. 2004. Relationship of delta-9-tetrahydrocannabinol in oral fluid to plasma after controlled administration of smoked cannabis. J. Anal. Toxicol. **28:** 394–399.
66. DARWIN, W.D., R.A. GUSTAFSON, J. SKLEROV, *et al.* 2004. Cannabinoid analysis of oral fluid collected from clinical subjects following oral tetrahydrocannabinol (THC) administration (hemp oil and dronabinol) [abstract]. 2004 Joint SOFT/TIAFT Meeting, Washington, DC August 30–September 3, 2004: 300.
67. KEMP, P.M., I.K. ABUKHALAF, J.E. MANNO, *et al.* 1995. Cannabinoids in humans I. Analysis of delta-9-tetrahydrocannabinol and six metabolites in plasma and urine using GC-MS. J. Anal. Toxicol. **19:** 285–291.
68. ELSOHLY, M.A., H. DEWIT, S.R. WACHTEL, *et al.* 2001. Delta9-tetrahydrocannabivarin as a marker for the ingestion of marijuana versus Marinol: results of a clinical study. J. Anal. Toxicol. **25:** 565–571.
69. CONE, E.J., A.H. SAMPSON-CONE, W.D. DARWIN, *et al.* 2003. Urine testing for cocaine abuse: metabolic and excretion patterns following different routes of administration and methods for detection of false-negative results. J. Anal. Toxicol. **27:** 386–401.
70. CONE, E.J., J. OYLER & W.D. DARWIN. 1997. Cocaine disposition in saliva following intravenous, intranasal, and smoked administration. J. Anal. Toxicol. **21:** 465–475.
71. CONE, E.J., D. YOUSEFNEJAD, M.J. HILLSGROVE, *et al.* 1995. Passive inhalation of cocaine. J. Anal. Toxicol. **19:** 399–411.
72. PAUL, B.D., S. LALANI, T. BOSY, *et al.* 2005. Concentration profiles of cocaine, pyrolytic methyl ecgonidine and thirteen metabolites in human blood and urine: determination by gas chromatography-mass spectrometry. Biomed. Chromatogr. **19:** 677–688.
73. WALL, M.A., J. JOHNSON, P. JACOB & N.L. BENOWITZ. 1988. Cotinine in the serum, saliva, and urine of nonsmokers, passive smokers, and active smokers. AJPH **78:** 699–701.
74. JENKINS, A.J., J.E. HENNINGFIELD & E.J. CONE. 2005. Relationship between plasma and oral fluid nicotine concentrations in humans: a pilot study. Ther. Drug Monit. **27:** 345–348.
75. JARCZYK, L., G. SCHERER & F. ADLKOFER. 1989. Serum and saliva concentrations of cotinine in smokers and passive smokers. J. Clin. Chem. Clin. Biochem. **27:** 230–231.
76. JARVIS, M.J., H. TUNSTALL-PEDOE, C. FEYERABEND, *et al.* 1987. Comparison of tests used to distinguish smokers from nonsmokers. AJPH **77:** 1435–1438.
77. ETZEL, R.A. 1990. A review of the use of saliva cotinine as a marker of tobacco smoke exposure. Prev. Med. **19:** 190–197.
78. KIM, I., A. WTSADIK, R.E. CHOO, *et al.* 2005. Usefulness of salivary trans-3′-hydroxycotinine concentration and trans-3′-hydroxycotinine/cotinine ratio as

biomarkers of cigarette smoke in pregnant women. J. Anal. Toxicol. **29:** 689–695.
79. Benowitz, N.L. 1996. Cotinine as a biomarker of environmental tobacco smoke exposure. Epidemiol. Rev. **18:** 188–204.
80. Rook, E.J., J.M. van Ree, B.W. van den, *et al.* 2006. Pharmacokinetics and pharmacodynamics of high doses of pharmaceutically prepared heroin, by intravenous or by inhalation route in opioid-dependent patients. Basic Clin. Pharmacol. Toxicol. **98:** 86–96.
81. Jenkins, A.J., J.M. Oyler & E.J. Cone. 1995. Comparison of heroin and cocaine concentrations in saliva with concentrations in blood and plasma. J. Anal. Toxicol. **19:** 359–374.
82. O'Neal, C.L. & A. Poklis. 1998. The detection of acetylcodeine and 6-acetylmorphine in opiate positive urines. Forensic Sci. Int. **95:** 1–10.
83. Chen, Z.R., R.J. Irvine, A.A. Somogyi & F. Bochner. 1991. Mu receptor binding of some commonly used opioids and their metabolites. Life Sci. **48:** 2165–2171.
84. Cone, E.J. 2006. Ephemeral profiles of prescription drug and formulation tampering: evolving pseudoscience on the Internet. Drug Alcohol Depend. **83** (Suppl 1): S31–S39.
85. Kopecky, E.A., S. Jacobson, J. Klein, *et al.* 1997. Correlation of morphine sulfate in blood plasma and saliva in pediatric patients. Ther. Drug Monit. **19:** 530–534.
86. Rohrig, T.P. & C. Moore. 2003. The determination of morphine in urine and oral fluid following ingestion of poppy seeds. J. Anal. Toxicol. **27:** 449–452.
87. Kim, I., A.J. Barnes, J.M. Oyler, *et al.* 2002. Plasma and oral fluid pharmacokinetics and pharmacodynamics after oral codeine administration. Clin. Chem. **48:** 1486–1496.
88. Cone, E.J., S. Dickerson, B.D. Paul & J.M. Mitchell. 1993. Forensic drug testing for opiates V. Urine testing for heroin, morphine, and codeine with commercial opiate immunoassays. J. Anal. Toxicol. **17:** 156–164.
89. Johansen, M., K.E. Rasmussen & A.S. Christophersen. 1990. Determination of pholcodine and its metabolites in urine by capillary gas chromatography. J. Chromatogr. **532:** 277–284.
90. Johansen, M., K.E. Rasmussen, A.S. Christophersen & B. Skuterud. 1991. Metabolic study of pholcodine in urine using enzyme multiplied immunoassay technique (EMIT) and capillary gas chromatography. Acta Pharm. Nord. **3:** 91–94.
91. Meadway, C., S. George & R. Braithwaite. 2002. Interpretation of GC-MS opiate results in the presence of pholcodine. Forensic Sci. Int. **127:** 131–135.
92. Yeh, S.Y., C.W. Gorodetzky & H.A. Krebs. 1977. Isolation and identification of morphine 3-and 6-glucuronides, morphine 3,6-diglucuronide, morphine 3-ethereal sulfate, normorphine, and normorphine 6-glucuronide as morphine metabolites in humans. J. Pharm. Sci. **66:** 1288–1293.
93. Cone, E.J. 1990. Testing human hair for drugs of abuse. I. Individual dose and time profiles of morphine and codeine in plasma, saliva, urine, and beard compared to drug-induced effects on pupils and behavior. J. Anal. Toxicol. **14:** 1–7.
94. Dams, R., R.E. Choo, W.E. Lambert, *et al.* 2007. Oral fluid as an alternative matrix to monitor opiate and cocaine use in substance abuse treatment patients. Drug Alcohol Depend. **87:** 258–267.

95. Vallner, J.J., J.T. Stewart, J.A. Kotzan, *et al.* 1981. Pharmacokinetics and bioavailability of hydromorphone following intravenous and oral administration to human subjects. J. Clin. Pharmacol. **21:** 152–156.
96. Ritschel, W., P.V. Parab, D.D. Denson, *et al.* 1987. Absolute bioavailability of hydromorphone after peroral and rectal administration in humans: saliva/plasma ratio and clinical effects. J. Clin. Pharmacol. **27:** 647–653.
97. Cone, E.J., B.A. Phelps & C.W. Gorodetzky. 1977. Urinary excretion of hydromorphone and metabolites in humans, rats, dogs, guinea pigs, and rabbits. J. Pharm. Sci. **66:** 1709–1713.
98. Cone, E.J., H.A. Heit, Y.H. Caplan & D. Gourlay. 2006. Evidence of morphine metabolism to hydromorphone in pain patients chronically treated with morphine. J. Anal. Toxicol. **30:** 1–5.
99. Otton, S.V., M. Schadel, S.W. Cheung, *et al.* 1993. CYP2D6 phenotype determines the metabolic conversion of hydrocodone to hydromorphone. Clin. Pharmacol. Ther. **54:** 463–472.
100. Cone, E.J., W.D. Darwin, C.W. Gorodetzky & T. Tan. 1978. Comparative metabolism of hydrocodone in man, rat, guinea pig, rabbit, and dog. Drug Metab. Dispos. **6:** 488–493.
101. Oyler, J.M., E.J. Cone, R.E. Joseph, Jr. & M.A. Huestis. 2000. Identification of hydrocodone in human urine following controlled codeine administration. J. Anal. Toxicol. **24:** 530–535.
102. Balikova, M., V. Maresova & V. Habrdova. 2001. Evaluation of urinary dihydrocodeine excretion in human by gas chromatography-mass spectrometry. J. Chromatogr. B Biomed. Sci. Appl. **752:** 179–186.
103. Smith, M.L., R.O. Hughes, B. Levine, *et al.* 1995. Forensic drug testing for opiates. VI. Urine testing for hydromorphone, hydrocodone, oxymorphone, and oxycodone with commercial opiate immunoassays and gas chromatography-mass spectrometry. J. Anal. Toxicol. **19:** 18–26.
104. Heiskanen, T., K.T. Olkkola & E. Kalso. 1998. Effects of blocking CYP2D6 on the pharmacokinetics and pharmacodynamics of oxycodone. Clin. Pharmacol. Ther. **64:** 603–611.
105. Beaver, W.T., S.L. Wallenstein, R.W. Houde & A. Rogers. 1977. Comparisons of the analgesic effects of oral and intramuscular oxymorphone and of intramuscular oxymorphone and morphine in patients with cancer. J. Clin. Pharmacol. **17:** 186–198.
106. Adams, M.P. & H. Ahdieh. 2004. Pharmacokinetics and dose-proportionality of oxymorphone extended release and its metabolites: results of a randomized crossover study. Pharmacotherapy **24:** 468–476.
107. Cone, E.J., W.D. Darwin, W.F. Buchwald & C.W. Gorodetzky. 1983. Oxymorphone metabolism and urinary excretion in human, rat, guinea pig, rabbit, and dog. Drug Metab. Dispos. **11:** 446–450.
108. Takala, A., V. Kaasalainen, T. Seppala, *et al.* 1997. Pharmacokinetic comparison of intravenous and intranasal administration of oxycodone. Acta Anaesthesiol. Scand. **41:** 309–312.
109. Baldacci, A. & W. Thormann. 2005. Analysis of oxycodol and noroxycodol stereoisomers in biological samples by capillary electrophoresis. Electrophoresis **26:** 1969–1977.
110. Baselt, R.C. & L.J. Casarett. 1971. Urinary excretion of methadone in man. Clin. Pharmacol. Ther. **13:** 64–70.

111. PRESTON, K.L., D.H. EPSTEIN, D. DAVOUDZADEH & M.A. HUESTIS. 2003. Methadone and metabolite urine concentrations in patients maintained on methadone. J. Anal. Toxicol. **27:** 332–341.
112. BERMEJO, A.M., A.C.S. LUCAS & M.J. TABERNERO. 2000. Saliva/plasma ratio of methadone and EDDP. J. Anal. Toxicol. **24:** 70–72.
113. MENDELSON, J., R.A. UPTON, E.T. EVERHART, *et al.* 1997. Bioavailability of sublingual buprenorphine. J. Clin. Pharmacol. **37:** 31–37.
114. NATH, R.P., R.A. UPTON, E.T. EVERHART, *et al.* 1999. Buprenorphine pharmacokinetics: relative bioavailability of sublingual tablet and liquid formulations. J. Clin. Pharmacol. **39:** 619–623.
115. CONE, E.J., C.W. GORODETZKY, D. YOUSEFNEJAD, *et al.* 1984. The metabolism and excretion of buprenorphine in humans. Drug Metab. Dispos. **12:** 577–581.
116. CONE, E.J., S.L. DICKERSON, W.D. DARWIN, *et al.* 1991. Elevated drug saliva levels suggest a "depot-like" effect in subjects treated with sublingual buprenorphine. NIDA Res. Monogr. **105:** 569.
117. KUHLMAN, J.J., S. LALANI, J. MAGLUILO, *et al.* 1996. Human pharmacokinetics of intravenous, sublingual, and buccal buprenorphine. J. Anal. Toxicol. **20:** 369–378.
118. KUHLMAN, J.J., B. LEVINE, R.E. JOHNSON, *et al.* 1998. Relationship of plasma buprenorphine and norbuprenorphine to withdrawal symptoms during dose induction, maintenance and withdrawal from sublingual buprenorphine. Addiction **93:** 549–559.
119. PITTS, F.N., JR., R.E. ALLEN, O. ANILINE & L.S. YAGO. 1981. Occupational intoxication and long-term persistence of phencyclidine (PCP) in law enforcement personnel. Clin. Toxicol. **18:** 1015–1020.
120. COOK, C.E., M. PEREZ-REYES, A.R. JEFFCOAT & D.R. BRINE. 1983. Phencyclidine disposition in humans after small doses of radiolabeled drug. Fed. Proc. **42:** 2566–2569.
121. PEREZ-REYES, M., S. DI GUISEPPI, D.R. BRINE, *et al.* 1982. Urine pH and phencyclidine excretion. Clin. Pharmacol. Ther. **32:** 635–641.
122. DOMINO, E.F. & A.E. WILSON. 1977. Effects of urine acidification on plasma and urine phencyclidine levels in overdosage. Clin. Pharmacol. Ther. **22:** 421–424.
123. MCCARRON, M.M., C.B. WALBERG, J.R. SOARES, *et al.* 1984. Detection of phencyclidine usage by radioimmunoassay of saliva. J. Anal. Toxicol. **8:** 197–201.
124. COOK, C.E., D.R. BRINE, A.R. JEFFCOAT, *et al.* 1982. Phencyclidine disposition after intravenous and oral doses. Clin. Pharmacol. Ther. **31:** 625–634.
125. KATO, K., M. HILLSGROVE, L. WEINHOLD, *et al.* 1993. Cocaine and metabolite excretion in saliva under stimulated and nonstimulated conditions. J. Anal. Toxicol. **17:** 338–341.
126. BINNIE, V., S. MCHUGH, L. MACPHERSON, *et al.* 2004. The validation of self-reported smoking status by analysing cotinine levels in stimulated and unstimulated saliva, serum and urine. Oral Dis. **10:** 287–293.
127. NAKAHARA, Y., K. TAKAHASHI & R. KIKURA. 1995. Hair analysis for drugs of abuse X. Effect of physiochemical properties of drugs on the incorporation rates into hair. Biol. Pharm. Bull. **18:** 1223–1227.
128. BRAITHWAITE, R.A., D.R. JARVIE, P.S.B. MINTY, *et al.* 1995. Screening for drugs of abuse. I: Opiates, amphetamines and cocaine. Ann. Clin. Biochem. **32:** 123–153.

129. SMITH-KIELLAND, A., B. SKUTERUD & J. MORLAND. 1997. Urinary excretion of amphetamine after termination of drug abuse. J. Anal. Toxicol. **21:** 325–329.
130. BASELT, R.C. 2000. Disposition of Toxic Drugs and Chemicals in Man. Chemical Toxicology Institute, Foster City, CA.
131. COOK, C.E., A.R. JEFFCOAT, B.M. SADLER, *et al.* 1992. Pharmacokinetics of oral methamphetamine and effects of repeated daily dosing in humans. Drug Metab. Dispos. **20:** 856–862.
132. OYLER, J.M., E.J. CONE, R.E. JOSEPH, JR., *et al.* 2002. Duration of detectable methamphetamine and amphetamine excretion in urine after controlled oral administration of methamphetamine to humans. Clin. Chem. **48:** 1703–1714.
133. GARRETT, E.R., K. SEYDA & P. MARROUM. 1991. High performance liquid chromatographic assays of the illicit designer drug "ecstasy" a modified amphetamine, with applications to stability, partitioning and plasma protein binding. Acta Pharm. Nordica **3:** 9–14.
134. BARFKNECHT, C.F. & D.E. NICHOLS. 1975. Correlation of psychotomimetic activity of phenethylamines and amphetamines with 1-octanol-water partition coefficients. J. Med. Chem. **18:** 208–210.
135. SPITZER, M., B. FRANKE, H. WALTER, *et al.* 2001. Enantio-selective cognitive and brain activation effects of N-ethyl-3,4-methylenedioxyamphetamine in humans. Neuropharmacology **41:** 263–271.
136. ENSSLIN, H.K., K.A. KOVAR & H.H. MAURER. 1996. Toxicological detection of the designer drug 3,4-methylenedioxyethylamphetamine (MDE, "Eve") and its metabolites in urine by gas chromatography-mass spectrometry and fluorescence polarization immunoassay. J. Chromatogr. Biomed. Appl. **683:** 189–197.
137. GROVE, J. & P.A. TOSELAND. 1971. The excretion of hydroxyamylobarbitone in man after oral administration of amylobarbitone and hydroxyamylobarbitone. J. Pharm. Pharmacol. **23:** 936–940.
138. KAPLAN, S.A., M.L. JACK, K. ALEXANDER & R.E. WEINFELD. 1973. Pharmacokinetic profile of diazepam in man following single intravenous and oral and chronic oral administration. J. Pharm. Sci. **62:** 1789–1796.
139. VEREBEY, K., D. JUKOFSKY & S.J. MULE. 1982. Confirmation of EMIT benzodiazepine assay with GLC/NPD. J. Anal. Toxicol. **6:** 305–308.
140. DIXON, R. & T. CREWS. 1978. Diazepam: determination in micro samples of blood, plasma, and saliva by radioimmunoassay. J. Anal. Toxicol. **2:** 210–213.
141. SHARP, M.E., S.M. WALLACE, K.W. HINDMARSH & H.W. PEEL. 1983. Monitoring saliva concentrations of methaqualone, codeine, secobarbital, diphenhydramine and diazepam after single oral doses. J. Anal. Toxicol. **7:** 11–14.
142. ETZEL, R.A., S. CARLSSON, S.L. KANTER, *et al.* 1982. Urinary metabolites of delta-9-tetrahydrocannabinol in man. Arzneimittelforchung **32:** 764–768.
143. GARRETT, E.R. & C.A. HUNT. 1974. Physicochemical properties, solubility, and protein binding of delta-9-tetrahydrocannabinol. J. Pharm. Sci. **63:** 1056–1064.
144. OHLSSON, A., J.E. LINDGREN, A. WAHLEN, *et al.* 1982. Single dose kinetics of deuterium labelled delta-1-tetrahydrocannabinol in heavy and light cannabis users. Biomed. Environ. Mass Spectrom. **9:** 6–10.
145. GROTENHERMEN, F. 2003. Pharmacokinetics and pharmacodynamics of cannabinoids. Clin. Pharmacokinet. **42:** 327–360.
146. KELLY, P. & R.T. JONES. 1992. Metabolism of tetrahydrocannabinol in frequent and infrequent marijuana users. J. Anal. Toxicol. **16:** 228–235.

147. HUESTIS, M.A., J.M. MITCHELL & E.J. CONE. 1995. Detection times of marijuana metabolites in urine by immunoassay and GC-MS. J. Anal. Toxicol. **19:** 443–449.
148. ELLIS, G.M., M.A. MANN, B.A. JUDSON, *et al.* 1985. Excretion patterns of cannabinoid metabolites after last use in a group of chronic users. Clin. Pharmacol. Ther. **38:** 572–578.
149. THOMPSON, L.K. & E.J. CONE. 1987. Determination of delta-9-tetrahydrocannabinol in human blood and saliva by high-performance liquid chromatography with amperometric detection. J. Chromatogr. **421:** 91–97.
150. KINTZ, P., V. CIRIMELE & B. LUDES. 2000. Detection of cannabis in oral fluid (saliva) and forehead wipes (sweat) from impaired drivers. J. Anal. Toxicol. **24:** 557–561.
151. GARRETT, E.R. & K. SEYDA. 1983. Prediction of stability in pharmaceutical preparations. XX: Stability evaluation and bioanalysis of cocaine and benzoylecgonine by high-performance liquid chromatography. J. Pharm. Sci. **72:** 258–271.
152. JUFER, R.A., A. WSTADIK, S.L. WALSH, *et al.* 2000. Elimination of cocaine and metabolites in plasma, saliva, and urine following repeated oral administration to human volunteers. J. Anal. Toxicol. **24:** 467–477.
153. CONE, E.J., K. KUMOR, L.K. THOMPSON & M. SHERER. 1988. Correlation of saliva cocaine levels with plasma levels and with pharmacologic effects after intravenous cocaine administration in human subjects. J. Anal. Toxicol. **12:** 200–206.
154. CONE, E. J. Personal communication, 2006.
155. JUFER, R., S.L. WALSH, E.J. CONE & A.H. SAMPSON-CONE. 2006. Effect of repeated cocaine administration on detection times in oral fluid and urine. J. Anal. Toxicol. **30:** 458–462.
156. PRESTON, K.L., D.H. EPSTEIN, E.J. CONE, *et al.* 2002. Urinary elimination of cocaine metabolites in chronic cocaine users during cessation. J. Anal. Toxicol. **26:** 393–400.
157. SVENSSON, C.K. 1987. Clinical pharmacokinetics of nicotine. Clin. Pharmacokinet. **12:** 30–40.
158. BEHERA, D., R. UPPAL & S. MAJUMDAR. 2003. Urinary levels of nicotine & cotinine in tobacco users. Indian J. Med. Res. **118:** 129–133.
159. CURVALL, M. & C.R. ENZELL. 1986. Monitoring absorption by means of determination of nicotine and cotinine. Arch. Toxicol. **9:** 88–102.
160. JARVIS, M.J., M.A.H. RUSSELL, N.L. BENOWITZ & C. FEYERABEND. 1988. Elimination of cotinine from body fluids: implications for noninvasive measurement of tobacco smoke exposure. AJPH **78:** 696–698.
161. 1999. Analytical Determination of Nicotine and Related Compounds and Their Metabolites. J.W. Gorrod & P. Jacob, III, Eds. Elsevier, Amsterdam.
162. CAREY, K.B. & D.B. ABRAMS. 1988. Properties of saliva cotinine in young adult light smokers. AJPH **78:** 842–843.
163. BECKETT, A.H., J.W. GORROD & P. JENNER. 1972. A possible relation between pKa, and lipid solubility and the amounts excreted in urine of some tobacco alkaloids given to man. J. Pharm. Pharmacol. **24:** 115–120.
164. BUCHHALTER, A.R., M.C. ACOSTA, S.E. EVANS, *et al.* 2005. Tobacco abstinence symptom suppression: the role played by the smoking-related stimuli that are delivered by denicotinized cigarettes. Addiction **100:** 550–559.

165. MURRAY, D.M., C. MCBRIDE, R. LINDQUIST & J.D. BELCHER. 1991. Sensitivity and specificity of saliva thiocyanate and cotinine for cigarette smoking: a comparison of two collection methods. Addict. Behav. **16:** 161–166.
166. MCNEILL, A.D., M.J. JARVIS, R. WEST, *et al.* 1987. Saliva cotinine as an indicator of cigarette smoking in adolescents. Br. J. Addict. **82:** 1355–1360.
167. GALEAZZI, R.L., P. DAENENS & M. GUGGER. 1985. Steady-state concentration of cotinine as a measure of nicotine-intake by smokers. Eur. J. Clin. Pharmacol. **28:** 301–304.
168. CONE, E.J., R. JUFER & W.D. DARWIN. 1996. Forensic drug testing for opiates. VII. Urinary excretion profile of intranasal (snorted) heroin. J. Anal. Toxicol. **20:** 379–391.
169. COHN, G.L., J.A. CRAMER, W. MCBRIDE, *et al.* 1974. Heroin and morphine binding with human serum proteins and red blood cells. Proc. Soc. Exp. Biol. Med. **147:** 664–666.
170. OLSEN, G.D. 1975. Morphine binding to human plasma proteins. Clin. Pharmacol. Ther. **17:** 31–35.
171. JUDIS, J. 1977. Binding of codeine, morphine, and methadone to human serum proteins. J. Pharm. Sci. **66:** 802–806.
172. AVDEEF, A., D.A. BARRETT, P.N. SHAW, *et al.* 1996. Octanol-, chloroform-, and propylene glycol dipelargonat-water partitioning of morphine-6-glucuronide and other related opiates. J. Med. Chem. **39:** 4377–4381.
173. DAHLSTROM, B. 1985. Pharmacokinetics of morphine in relation to the analgesic effect. *In* Pharmacokinetics and Pharmacodynamics of Psychoactive Drugs. G. Barnett & C.N. Chiang, Eds.: 185–208. Biomedical Publications, Foster City.
174. SMITH, M.L., E.T. SHIMOMURA, J. SUMMERS, *et al.* 2000. Detection times and analytical performance of commercial urine opiate immunoassays following heroin administration. J. Anal. Toxicol. **24:** 522–529.
175. REITER, A., J. HAKE, C. MEISSNER, *et al.* 2001. Time of drug elimination in chronic drug abusers. Case study of 52 patients in a "low-step" detoxification ward. Forensic Sci. Int. **119:** 248–253.
176. STANSKI, D.R., D.J. GREENBLATT & E. LOWENSTEIN. 1978. Kinetics of intravenous and intramuscular morphine. Clin. Pharmacol. Ther. **24:** 52–59.
177. CONE, E.J., P. WELCH, B.D. PAUL & J.M. MITCHELL. 1991. Forensic drug testing for opiates. III. Urinary excretion rates of morphine and codeine following codeine administration. J. Anal. Toxicol. **15:** 161–166.
178. KINTZ, P., V. CIRIMELE & B. LUDES. 1998. Codeine testing in sweat and saliva with the Drugwipe. Int. J. Leg. Med. **111:** 82–84.
179. ADAMS, M.P. & H. AHDIEH. 2005. Single- and multiple-dose pharmacokinetic and dose-proportionality study of oxymorphone immediate-release tablets. Drugs R. D. **6:** 91–99.
180. LEOW, K.P., T. CRAMOND & M.T. SMITH. 1995. Pharmacokinetics and pharmacodynamics of oxycodone when given intravenously and rectally to adult patients with cancer pain. Anesth. Analg. **80:** 296–302.
181. TIEN, J.H. 1991. Transdermal-controlled administration of oxycodone. J. Pharm. Sci. **80:** 741–743.
182. JONES, J., K. TOMLINSON & C. MOORE. 2002. The simultaneous determination of codeine, morphine, hydrocodone, hydromorphone, 6-acetylmorphine, and oxycodone in hair and oral fluid. J. Anal. Toxicol. **26:** 171–175.
183. SULLIVAN, H.R. & S.L. DUE. 1973. Urinary metabolites of dl-methadone in maintenance subjects. J. Med. Chem. **16:** 909–913.

184. WILKINS, J.N., A. ASHOFTEH, D. SETODA, *et al.* 1997. Ultrafiltration using the Amicon MPS-1 for assessing methadone plasma protein binding. Ther. Drug Monit. **19:** 83–87.
185. WOLFF, K., A. ROSTAMI-HODJEGAN, S. SHIRES, *et al.* 1997. The pharmacokinetics of methadone in healthy subjects and opiate users. Br. J. Clin. Pharmacol. **44:** 325–334.
186. GARRIDO, M.J. & I.F. TROCONIZ. 1999. Methadone: a review of its pharmacokinetic/pharmacodynamic properties. J Pharmacol. Toxicol. **42:** 61–66.
187. WOLFF, K., A. HAY & D. RAISTRICK. 1991. Methadone in saliva. Clin. Chem. **37:** 1297–1298.
188. KANG, G.I. & F.S. ABBOTT. 1982. Analysis of methadone and metabolites in biological fluids with gas chromatography-mass spectrometry. J. Chromatogr. **231:** 311–319.
189. SIMPSON, D., R.A. BRAITHWAITE, D.R. JARVIE, *et al.* 1997. Screening for drugs of abuse (II): cannabinoids, lysergic acid diethylamide, buprenorphine, methadone, barbiturates, benzodiazepines and other drugs. Ann. Clin. Biochem. **34:** 460–510.
190. GARRETT, E.R. & V.R. CHANDRAN. 1985. Pharmacokinetics of morphine and its surrogates VI: bioanalysis, solvolysis kinetics, solubility, pKa values, and protein binding of buprenorphine. J. Pharm. Sci. **74:** 515–524.
191. CONE, E.J., C.W. GORODETZKY, D. YOUSEFNEJAD & W.D. DARWIN. 1985. Ni electron-capture gas-chromatographic assay for buprenorphine and metabolites in human urine and feces. J. Chromatogr. **337:** 291–300.
192. VINCENT, F., J. BESSARD, J. VACHERON, *et al.* 1999. Determination of buprenorphine and norbuprenorphine in urine and hair by gas chromatography-mass spectrometry. J. Anal. Toxicol. **23:** 270–279.
193. WALL, M.E., D.R. BRINE, A.R. JEFFCOAT, *et al.* 1981. Phencyclidine metabolism and disposition in man following a 100 ug intravenous dose. Res. Commun. Subst. Abuse **2:** 161–172.
194. SIMPSON, G.M., A.M. KHAJAWALL, E. ALATORRE & F.R. STAPLES. 1982. Urinary phencyclidine excretion in chronic abusers. J. Toxicol. Clin. Toxicol. **19:** 1051–1059.

Methamphetamine Disposition in Oral Fluid, Plasma, and Urine

MARILYN A. HUESTIS[a] AND EDWARD J. CONE[b]

[a]*Chemistry and Drug Metabolism Section, IRP, NIDA, NIH, 5500 Nathan Shock Drive, Baltimore, Maryland, USA*

[b]*Johns Hopkins School of Medicine, Baltimore, Maryland, USA*

ABSTRACT: This review of the disposition of methamphetamine in oral fluid, plasma, and urine is based on a comprehensive controlled dosing study involving five healthy, drug-free research volunteers who resided on a closed clinical ward for 12 weeks. Subjects were administered four low (10 mg) and high (20 mg) daily oral doses of methamphetamine in two separate sessions. Near-simultaneous collections of oral fluid and plasma were performed on the first day of each low- and high-dose session. Thereafter, oral fluid was provided on each day of dosing by different oral fluid collection methods. All urine specimens were collected on an *ad libitum* basis throughout the study. Specimens were analyzed by gas-chromatography mass spectrometry for methamphetamine and the metabolite, amphetamine, with a limit of quantification of 2.5 ng/mL for each analyte. Methamphetamine and metabolite concentrations in oral fluid appeared to follow a similar time course in oral fluid as in plasma and were dose-proportional, but oral fluid concentrations exceeded plasma concentrations. Urine drug concentrations were substantially higher than those in oral fluid. Some drug accumulation was noted with daily dosing, but generally did not markedly influence detection times or detection rates of oral fluid tests. Detection times and detection rates for oral fluid and urine were determined at cessation of 4 days of dosing. Generally, detection times and rates for urine were longer than those observed for oral fluid at conventional cutoff concentrations. When contemplating selection of oral fluid as a test matrix, the advantages of oral fluid collection should be weighed against its shorter time of detection compared to that of urine.

KEYWORDS: methamphetamine; collection; oral fluid; saliva; urine; plasma

INTRODUCTION

Dextromethamphetamine is a powerful central nervous stimulant with a high potential for abuse. Currently, there are only a few accepted medical conditions

Address for correspondence: Marilyn A. Huestis, Ph.D., Chemistry and Drug Metabolism Section, IRP, NIDA, NIH, 5500 Nathan Shock Drive, Baltimore, MD. Voice: 410-550-2711; fax: 410-550-2971. MHUESTIS@intra.nida.nih.gov

Ann. N.Y. Acad. Sci. 1098: 104–121 (2007). © 2007 New York Academy of Sciences.
doi: 10.1196/annals.1384.038

(attention deficit disorder and short-term treatment of obesity) for pharmaceutically derived d-methamphetamine. Illicitly derived methamphetamine is widely available at relatively low cost because of its ease of manufacture in clandestine laboratories. Most synthetic procedures for illicit manufacture generally lead to production of relatively pure d-methamphetamine although some methods result in d/l-mixtures. The l-isomer (levmetamfetamine; l-desoxyephedrine) is considerably less potent and is sold over-the-counter as a nasal inhaler decongestant product. Abuse of illicit methamphetamine has increased substantially over the past decade and dominates drug use trends in many parts of the world. In 2004 there were a reported 583,000 current methamphetamine users in the United States.[1]

The effects of d-methamphetamine last from 6 to 12 h. After the initial "rush," many individuals remain hyperactive, restless, and irritable. Some users are reported to be prone to aggression and violence.[2] Illicit methamphetamine is used by a variety of routes including smoking and intravenous, intranasal ("snorted"), and oral administration. Methamphetamine is rapidly absorbed by all routes, but time to onset of effects is fastest by intravenous and smoked routes. The primary site of metabolism is in the liver by aromatic hydroxylation, *N*-demethylation (to form the metabolite amphetamine), and deamination. The basic nitrogen moiety in methamphetamine's chemical structure (pKa = 10.1), its relatively high lipophilicity, and low plasma–protein binding (10–20%) influences distribution and excretion processes. Acidic urine enhances methamphetamine excretion and shortens its half-life in the body, whereas basic urine slows excretion and prolongs residence time.[3] Transfer across membranes and excretion in saliva is also facilitated by methamphetamine's lipid nature, low protein-binding, and ion-trapping in saliva. Typically, salivary fluid is more acidic than is plasma, thus producing higher concentrations of methamphetamine in saliva compared to plasma.[4]

Oral fluid testing for methamphetamine and other drugs of abuse has emerged as an important alternative to urine testing in workplace and drug treatment programs. The primary advantage of oral fluid testing is its ease of collection. Oral fluid can be noninvasively collected under direct observation, thereby reducing or eliminating problems such as adulteration and substitution that are often associated with urine collection. Disadvantages include shorter drug detection windows for oral fluid than for urine and a need for greater assay sensitivity.

Understanding methamphetamine disposition in oral fluid, blood, and urine is essential to the development and use of oral fluid tests. How closely does oral fluid parallel plasma drug concentrations? What are the influences of different oral fluid collection devices on concentration? Is there a dose–concentration relationship for drug in oral fluid similar to that found for plasma? What is the effect of repeated dosing on oral fluid concentrations? What are the detection times and detection rates for drug in oral fluid compared to plasma and urine? These are fundamental questions of vital importance to the use and interpretation of oral fluid tests. This review addresses these issues by providing a detailed summary of methamphetamine disposition in oral fluid, with comparisons to its disposition in plasma, and urine. Data for this review are derived from recent controlled dosing studies performed by Huestis and coworkers.[4–6]

METHAMPHETAMINE STUDY PROTOCOL

Methamphetamine was administered to five drug-free, stimulant-experienced subjects who resided on a closed research ward for the duration of the study. The protocol was approved by an Institutional Review Board, and each subject provided written informed consent. Each subject participated in two dosing sessions. The first consisted of administration of single 10-mg oral doses of methamphetamine (10 mg METH) daily for 4 days. Generally, dosing was administered on four consecutive days, but occasionally, dosing days were separated by 48–72 h. Following a 3-week washout period, each subject participated in a second dosing session and received single 20-mg oral doses of methamphetamine (20 mg METH) on four subsequent days. Blood specimens were collected at timed intervals on the first day of each dose administration, but only limited sampling was conducted on subsequent days. Oral fluid specimens were collected at near-simultaneous times as plasma specimens on the first day of dosing. Subsequent oral fluid collections on days 2–4 were made with the same sampling schedule as day 1. Different oral fluid collection methods were employed for assessment of the effect of different devices on drug concentration. Oral fluid was collected on days 1 and 2 (low and high doses) by stimulation with a citric acid sourball candy (Brach's Confections Inc., Dallas, TX, USA), designated as "sour candy," on day 3 of each session with a cotton swab treated with 20 mg of citric acid (Salivette® cotton swab citric acid preparation; Sarstedt, Newton, NC, USA) designated as "CAC 3," and on day 4 of each session with cotton swabs without citric acid (Salivette cotton swab without preparation; Sarstedt), designated as "SAL 4." Complete urine collections (*ad libitum*) were conducted throughout the study. No attempt was made to control water intake or urine pH during dosing and subsequent specimen collections. All specimens were analyzed for methamphetamine and amphetamine by gas-chromatography mass spectrometry (GC-MS) with a limit of quantification (LOQ) of 2.5 ng/mL for each analyte. All data reported in this review represent drug concentrations determined by GC-MS. Details of the study design and pharmacokinetic analyses have been reported.[4,5]

Disposition of Methamphetamine and Amphetamine in Oral Fluid, Plasma, and Urine after Administration of Single Oral Doses of Methamphetamine

Following first-day administration of single oral doses of 10 and 20 mg of methamphetamine, drug was detected in concentrations >2.5 ng/mL in oral

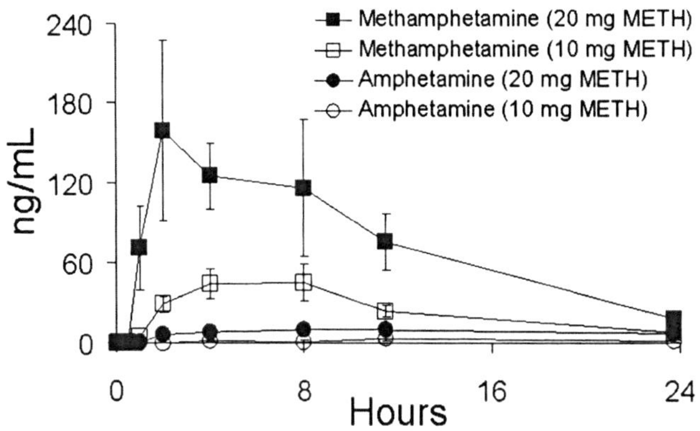

FIGURE 1. Mean oral fluid concentrations of methamphetamine (METH) and amphetamine in five subjects following single-dose oral administration of 10 and 20 mg methamphetamine. Oral fluid was collected by stimulation with sour candy. Error bars represent standard error of the mean.

fluid and plasma within 0.5 to 2 h and reached maximal concentration from 2–11.5 h. FIGURES 1 and 2 illustrate mean data for methamphetamine and amphetamine in oral fluid and plasma, respectively, on day 1 of the dosing sessions. First detection of amphetamine occurred in the range of 1.0–23.8 h and was slightly delayed relative to methamphetamine in oral fluid and plasma. Maximal concentrations occurred from 2.0–23.8 h. A summary of C_{max}, T_{max}, and time to first positive methamphetamine and metabolite (amphetamine) specimens in oral fluid, plasma, and urine is shown in TABLE 1. Approximate proportional increases in drug concentration were observed between the

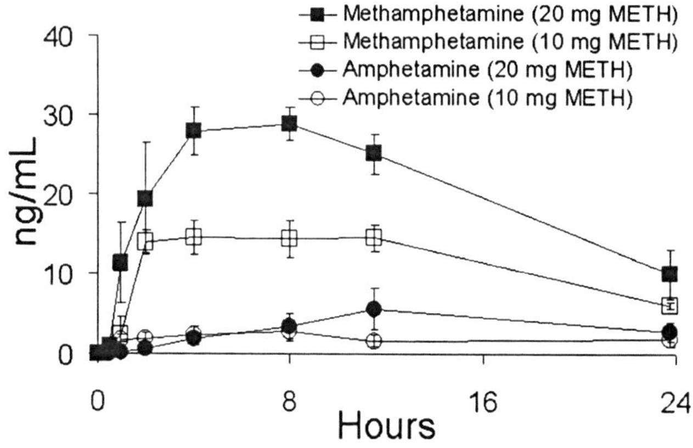

FIGURE 2. Mean plasma concentrations of methamphetamine (METH) and amphetamine in five subjects following single-dose oral administration of 10 and 20 mg of methamphetamine. Error bars represent standard error of the mean.

TABLE 1. Comparison of methamphetamine and amphetamine C_{max}, T_{max}, and time to first positive specimen at the LOQ (2.5 ng/mL) in oral fluid, plasma, and urine following a single dose (first day, mean data) of 10 and 20 mg of methamphetamine administered to five subjects. Oral fluid was collected by stimulation with sour candy

Meth dose (mg)	Analyte	Specimen	*N*	C_{max} ± SEM, ng/mL (range)	T_{max} ± SEM, h (range)	Time to first positive (LOQ) ± SEM, h (range)
10	Meth	Oral fluid	5	57.1 ± 12.9 (24.7–90.2)	5.6 ± 1.0 (4.0–8.0)	1.6 ± 0.2 (1.0–2.0)
	Amp	Oral fluid	3	6.5 ± 1.5 (4.6–9.4)	10.3 ± 1.2 (8.0–11.5)	6.5 ± 2.5 (4.0–11.5)
20	Meth	Oral fluid	5	192.2 ± 54.0 (75.3–321.7)	4.7 ± 1.8 (2.0–11.5)	0.8 ± 0.1 (0.5–1.0)
	Amp	Oral fluid	5	14.3 ± 3.0 (2.8–20.2)	8.2 ± 1.7 (2.0–11.5)	4.1 ± 1.9 (1.0–11.5)
10	Meth	Plasma	5	17.4 ± 1.1 (14.5–20.3)	5.2 ± 1.2 (2.0–8.0)	1.7 ± 0.3 (0.5–2.0)
	Amp	Plasma	4	4.2 ± 1.5 (2.0–8.4)	15.9 ± 4.5 (8.0–23.8)	10.1 ± 5.1 (1.0–23.8)
20	Meth	Plasma	5	32.4 ± 3.5 (26.2–44.3)	7.5 ± 1.5 (2.0–11.5)	1.1 ± 0.2 (0.5–2.0)
	Amp	Plasma	5	5.6 ± 1.4 (2.9–10.6)	13.9 ± 2.5 (11.5–23.8)	7.4 ± 1.9 (2.0–11.5)
10	Meth	Urine	5	2969.7 ± 607.6 (1573.7–4844.6)	15.0 ± 1.6 (11.4–19.6)	5.0 ± 2.2 (0.7–11.3)
	Amp	Urine	5	354.5 ± 33.8 (258.7–435.3)	15.0 ± 1.6 (11.4–19.6)	5.8 ± 1.9) (1.4–11.3)
20	Meth	Urine	5	4038.4 ± 700.9 (1871.4–6003.5)	14.0 ± 3.2 (4.8–19.7)	3.3 ± 1.4 (0.9–8.8)
	Amp	Urine	5	737.7 ± 226.3 (266.7–1556.1)	17.0 ± 1.4 (13.1–19.7)	4.3 ± 1.3 (1.2–8.8)

C_{max} = maximum observed drug concentration; T_{max} = time to maximum drug concentration; Meth = methamphetamine; Amp = amphetamine.

low- and high-dose conditions for both oral fluid and plasma. Concentrations of methamphetamine in oral fluid ranged from 1.4 to 8.2 times higher than in plasma. FIGURE 3 illustrates mean oral fluid/plasma ratios over the first-day dosing sessions.

Reproducibility of methamphetamine concentrations on days 1 and 2 of the low- and high-dose session are illustrated in FIGURES 4 and 5 for two subjects (subjects AA and BB). Oral fluid specimen collections were made on these days by the same collection method (stimulation with sour candy). This afforded the opportunity to compare methamphetamine concentrations within the same subjects over the course of two sequential days. Day 2 concentrations for both subjects were slightly higher than day 1 as a result of carryover from the first doses. Given this consideration, responses on day 2 were relatively similar to those of day 1. Of greater note, in FIGURES 4 and 5 is the large intersubject

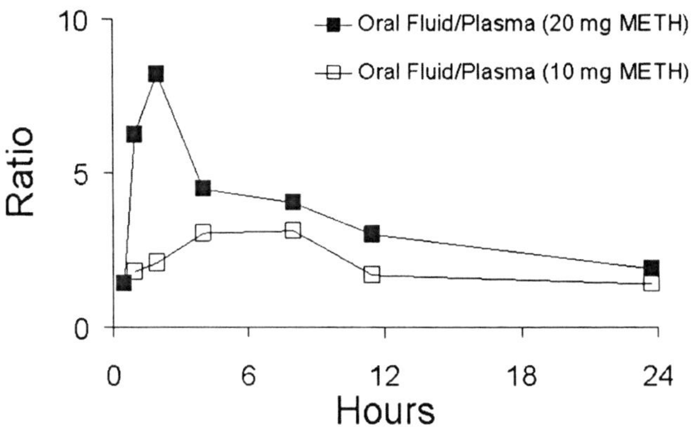

FIGURE 3. Mean oral fluid/plasma ratios of methamphetamine (METH) in five subjects following single-dose oral administration of 10 and 20 mg of methamphetamine. Oral fluid was collected by use of sour candy.

variability in drug concentrations. Methamphetamine concentrations in oral fluid specimens from subject BB were approximately threefold higher than for subject AA.

Appearance of the metabolite, amphetamine, in oral fluid, plasma, and urine was highly variable and occurred in the range of 1–23.5 h in considerably lower concentrations compared to the parent drug, methamphetamine. FIGURE 6 illustrates first-day urine concentrations of methamphetamine and amphetamine for a typical subject (subject AA).

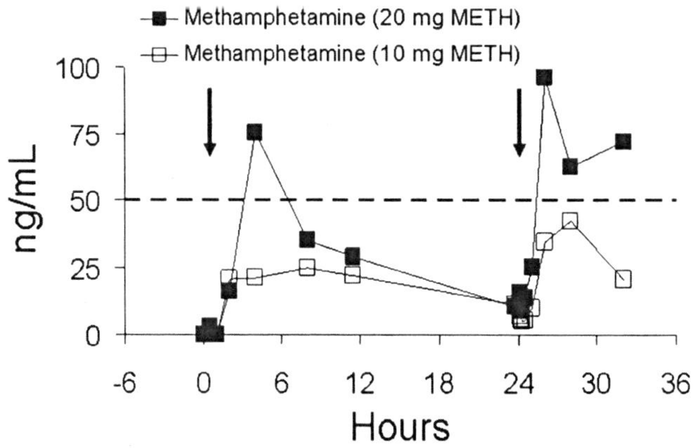

FIGURE 4. Oral fluid concentrations of methamphetamine (METH) for subject AA following single-dose oral administration of 10 and 20 mg of methamphetamine on two sequential days. Oral fluid was collected by use of sour candy. The *arrows* represent time of methamphetamine administration, and the *dotted line* represents the recommended DHHS cutoff concentration of 50 ng/mL for methamphetamine in oral fluid.

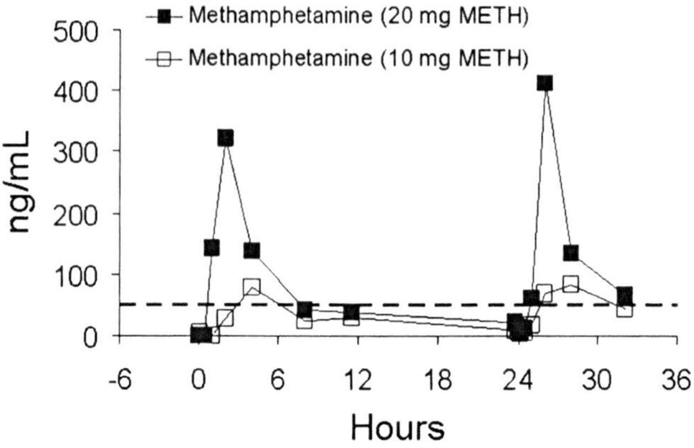

FIGURE 5. Oral fluid concentrations of methamphetamine (METH) for subject BB following single-dose oral administration of 10 and 20 mg of methamphetamine on two sequential days. Oral fluid was collected by use of sour candy. The *arrows* represent time of methamphetamine administration and the *dotted line* represents the recommended DHHS cutoff concentration of 50 ng/mL for methamphetamine in oral fluid.

Detection of Methamphetamine and Amphetamine in Oral Fluid and Urine at Administrative Cutoff Concentrations following First-Day Administration of Single Oral Doses of Methamphetamine

Detection rates for methamphetamine in oral fluid (10 specimens collected after dosing at timed intervals of 0.08, 0.17, 0.25, 0.5, 1, 2, 4, 8, 11.5, and 23.5 h) at the proposed Department of Health and Human Services (DHHS) administrative cutoff of 50 ng/mL (accompanied by amphetamine metabolite at assay limit of detection [LOD])[7] were low to moderate following 10- and 20-mg

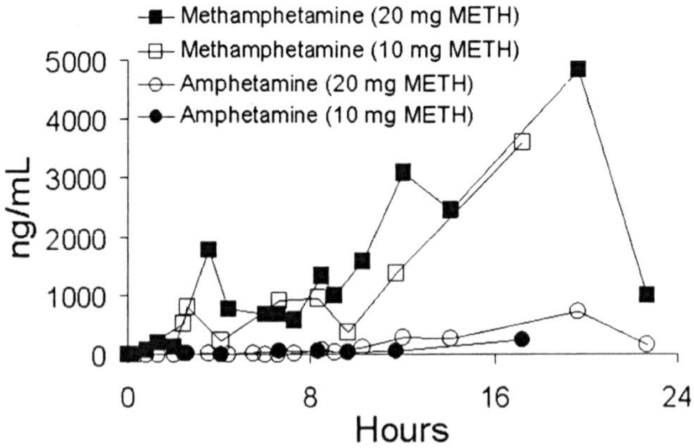

FIGURE 6. Urine concentrations of methamphetamine (METH) and amphetamine for subject AA following single-dose oral administration of 10 and 20 mg of methamphetamine.

doses of methamphetamine. Following the low dose, two subjects produced a single positive result at the DHHS cutoff concentrations for specimens collected at 4 h (4% detection rate over 24 h). A total of 25 specimens were positive for methamphetamine at 2.5 ng/mL (LOQ). After the high dose, one subject failed to produce a positive specimen at the DHHS cutoff concentration (on account of the absence of amphetamine) and four subjects produced a total of 15 positives (28.6% detection rate over 24 h). A total of 49 specimens was positive for methamphetamine at 2.5 ng/mL (LOQ). Positive specimens were generally produced over the 1–24 h period post dose. Because oral dosing was utilized in this study, the delay in absorption and metabolism of methamphetamine generally led to negative specimens collected over the first hour of specimen collection. It should be noted that oral fluid collections made during the first hour were a factor in reducing the detection rates.

Following administration of the first dose (day 1), methamphetamine was first detected in urine at the current DHHS administration confirmation cutoff concentration of 500 ng/mL[8] at mean times of 5.5 and 4.3 h, respectively, for the low and high doses of methamphetamine. Appearance of the metabolite, amphetamine, at the required cutoff concentration of 200 ng/mL was delayed relative to methamphetamine and occurred at mean times of 14.5 and 14.2 h, respectively. At the proposed lower DHHS confirmation cutoff concentrations of 250 and 100 for methamphetamine and amphetamine, first detection of methamphetamine was approximately equivalent to that observed for the higher cutoff concentration; however, amphetamine was first detected at mean times of 9.9 and 9.1 h, respectively. The data for detection of the first positive specimen are summarized in TABLE 2. It should be noted that DHHS criteria for a "positive" methamphetamine require that methamphetamine be accompanied by amphetamine at 200 ng/mL. It is evident from detection time data (TABLE 2) that many methamphetamine positives did not meet DHHS criteria

TABLE 2. Time to detection of the first positive specimen of methamphetamine and amphetamine in urine at different DHHS confirmation cutoff concentrations following a single dose (first day, mean data) of 10 and 20 mg of methamphetamine administered to five subjects

Meth dose (mg)	Analyte	*N*	Cutoff concentration (ng/mL)	Time to first positive ± SEM, h (range)
10	Meth	5	500	5.5 ± 2.0 (1.4–11.3)
	Amp	5	200	14.5 ± 1.9 (9.3–19.6)
20	Meth	5	500	4.3 ± 1.3 (1.2–8.8)
	Amp	5	200	14.2 ± 2.1 (7.9–19.6)
10	Meth	5	250	5.5 ± 2.0 (1.4–11.3)
	Amp	5	100	9.9 ± 2.2 (4.2–17.2)
20	Meth	5	250	3.9 ± 1.3 (1.2–8.8)
	Amp	5	100	9.1 ± 1.6 (3.3–13.1)

Meth = methamphetamine; Amp = amphetamine.

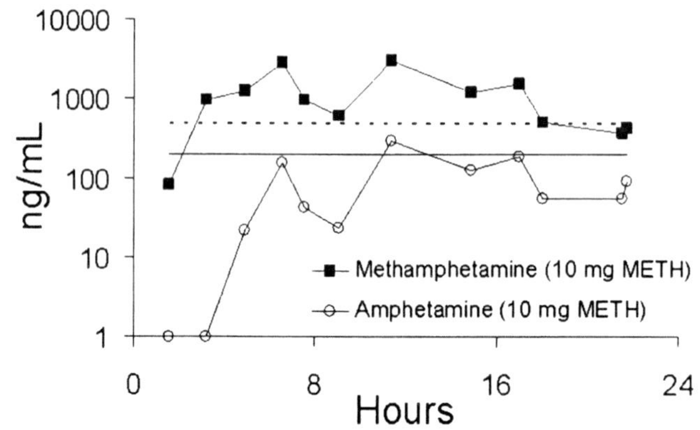

FIGURE 7. Methamphetamine (METH) and amphetamine excretion in urine of one subject (subject BB) following administration of a single dose (first day) of 10 mg methamphetamine. The *dotted line* illustrates the 500 ng/mL DHHS confirmation cutoff concentration for methamphetamine and the *solid line* illustrates the 200 ng/mL DHHS confirmation cutoff concentration for amphetamine.

as "positive" because of low amphetamine concentrations. This is illustrated in FIGURE 7 for one subject (subject BB) following administration of a 10-mg dose of methamphetamine. Although methamphetamine exceeded the 500 ng/mL confirmation cutoff concentration for the majority of time over the first 24 h, amphetamine concentrations were below the required concentration of 200 ng/mL in all but one specimen.

Detection rates for methamphetamine in urine (collected *ad libitum*) over the first 24 h (postdosing, day 1) at the current DHHS administrative confirmation cutoff concentration of 500 ng/mL (accompanied by 200 ng/mL of amphetamine metabolite) for the low and high doses were 31.0% (range = 7.7–66.7%) and 39.4% (range = 9.1–75.0%), respectively. Lowering the cutoff concentration requirement increased detection rates. Detection rates at the proposed confirmation cutoff concentration of 250 ng/mL (accompanied by 100 ng/mL of amphetamine) were 52.6% (range = 11.1–100.0%) and 63.5% (range = 27.3–100.0%), respectively. As noted earlier, with both cutoff concentration requirements, approximately 75% or more of false negative tests were produced as a result of amphetamine concentrations not meeting the required concentration and 25% or less were produced by both methamphetamine and amphetamine not meeting the required concentrations. Detection rates for methamphetamine at 2.5 ng/mL (LOQ) were >96% for both low- and high-dosing conditions.

Effect of Repeated Dosing on Methamphetamine Disposition

Although much of the information on drug disposition in oral fluid, plasma, and urine have been based on single-dose studies, recreational misuse and abuse frequently involves repeated daily self-administration. The design of the current study offered a unique opportunity to evaluate drug disposition in oral fluid and urine following repeated sequential and nonsequential daily dosing. Four doses of methamphetamine were administered in two sessions (low- and high-dose conditions). Oral fluid and urine collections were made throughout the study; however, plasma collections were primarily made only on the first day of each session because of blood-volume limitations. Oral fluid collections in both low- and high-dosing sessions were made by stimulation with sour candy on days 1 and 2, on day 3 with CAC 3, and on day 4 with SAL 4. Hence, it should be noted that oral fluid collections were conducted differently on days 3 and 4 and may have influenced concentrations. Drug was generally administered on a sequential daily basis, and thus, for most subjects, sessions were completed over a 4-day period. However, administration of some doses was delayed from 1 to 2 days. In particular, day 4 dosing was delayed for 2 days for four subjects (subjects S, Y, AA, low dose; subject W, high dose) because of the intervention of a weekend during which time dosing did not occur. A few other sessions had delays of 1 day. Any delays in dosing would be expected to produce less drug accumulation.

It is instructive to evaluate the effect of different daily dosing patterns. The relatively long half-life of methamphetamine[4] (approximately 9–11 h) suggests that some drug accumulation should result from sequential daily dosing. FIGURE 8 illustrates oral fluid methamphetamine concentrations during

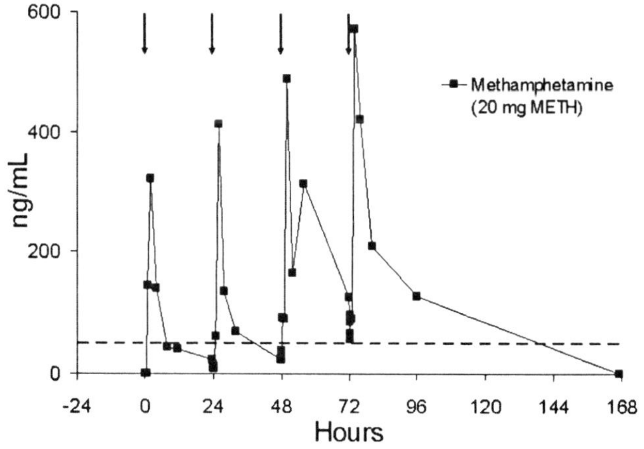

FIGURE 8. Oral fluid concentrations of methamphetamine (METH) in subject BB following sequential daily single-dose oral administration of 20 mg methamphetamine on different days. Oral fluid was collected by use of sour candy on days 1 and 2, Salivette with citric acid (CAC 3) on day 3, and Salivette without citric acid (SAL 4) on day 4. The *arrows* represent time of methamphetamine administration and the *dotted line* represents the recommended DHHS cutoff concentration of 50 ng/mL for methamphetamine in oral fluid.

sequential daily dosing without delay over 4 days for one subject (subject BB). There is clear indication of accumulation of methamphetamine in oral fluid from day-to-day. Note that by day 4, substantial methamphetamine accumulation occurred, and that oral fluid specimens collected prior to dosing were positive at the 50 ng/mL confirmation cutoff. Specimens for this subject continued to test positive for approximately 24 h following the fourth dose.

FIGURE 9 illustrates methamphetamine oral fluid concentrations during sequential daily dosing over 4 days for subject Y. Some accumulation is apparent from day-to-day, but not sufficient for the subject to test positive at the beginning of dosing on day 4. Specimens continued to test positive for approximately 24 h following the fourth dose.

FIGURE 10 illustrates methamphetamine oral fluid concentrations during nonsequential daily dosing over 4 days for another subject (subject W), in which two delays in dosing occurred during the high-dose methamphetamine session. There was a 1-day delay in daily dosing between days 1 and 2, and a 2-day delay in daily dosing between days 3 and 4. No accumulation was apparent on day 4 as a result of administration of the three earlier doses. In this example, the subject tested positive for approximately 8 h following the fourth dose.

The effect of multiple dosing across four sequential days on methamphetamine urine concentrations is illustrated in FIGURE 11 for one subject (subject BB). There was indication of accumulation and carryover to the next

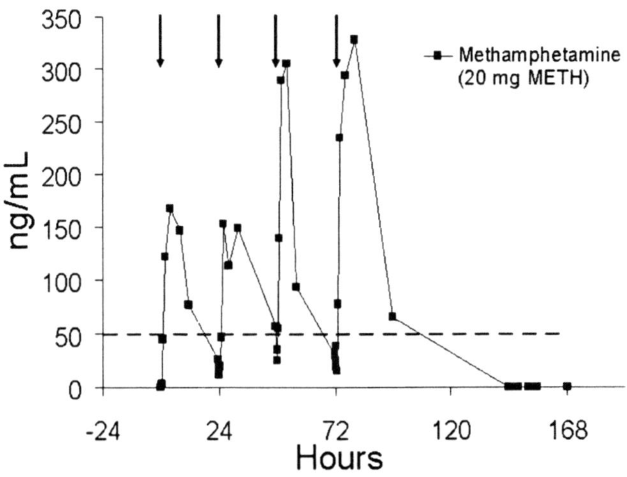

FIGURE 9. Oral fluid concentrations of methamphetamine (METH) in subject Y following single-dose oral administration of 20 mg methamphetamine on different days. Oral fluid was collected by use of sour candy on days 1 and 2, Salivette with citric acid (CAC 3) on day 3, and Salivette without citric acid (SAL 4) on day 4. The *arrows* represent time of methamphetamine administration and the *dotted line* represents the recommended DHHS cutoff concentration of 50 ng/mL for methamphetamine in oral fluid.

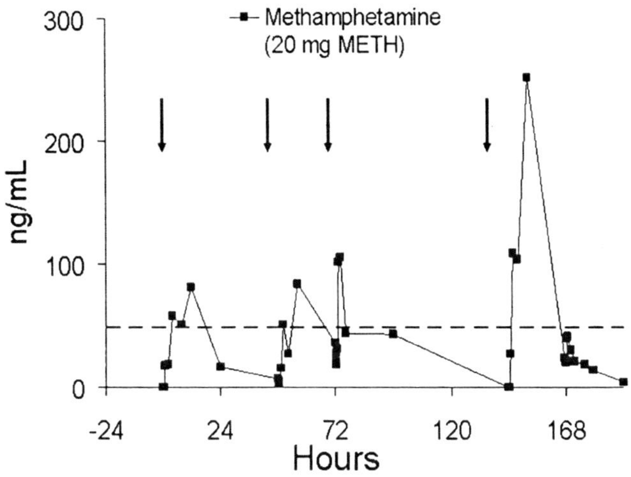

FIGURE 10. Oral fluid concentrations of methamphetamine (METH) in subject W following single-dose oral administration of 20 mg methamphetamine on different days. Oral fluid was collected by use of sour candy on days 1 and 2, Salivette with citric acid (CAC 3) on day 3, and Salivette without citric acid (SAL 4) on day 4. The *arrows* represent time of methamphetamine administration and the *dotted line* represents the recommended DHHS cutoff concentration of 50 ng/mL for methamphetamine in oral fluid.

dose on a day-to-day basis. Considerably higher methamphetamine concentrations occurred on days 3 and 4 of dosing. As a result of drug accumulation, most urine specimens for this subject were positive at the DHHS confirmation cutoff concentration throughout the 4 days of dosing and for approximately 46 h following the last dose.

FIGURE 12 illustrates methamphetamine urine concentrations during non-sequential daily dosing over 4 days for another subject (subject S), in which two delays in dosing occurred during the high-dose methamphetamine session. There was a 1-day delay in daily dosing between day 1 and day 2 and a 2-day delay in daily dosing between days 3 and 4. Urine concentrations fell below the DHHS confirmation cutoff during the 1-day delay following the first dose and during the 2-day delay in dosing following the third dose. Following the fourth dose, positive specimens were intermixed with negative specimens over the remaining collection period.

Detection Times and Rates in Oral Fluid and Urine Following Repeated Dosing

Detection times or "windows of detection" are terms often used to describe how long after drug administration a test will be positive. Many factors go into determining detection times including dosage, pattern of usage, specimen

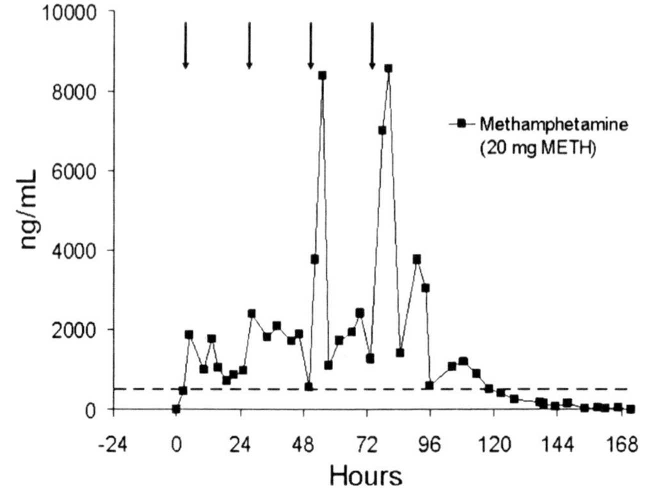

FIGURE 11. Urine concentrations of methamphetamine (METH) in subject BB following sequential daily single-dose oral administration of 20 mg methamphetamine on different days. The *arrows* represent time of methamphetamine administration and the *dotted line* represents the DHHS cutoff concentration of 500 ng/mL for methamphetamine in urine.

type, choice of analyte(s), and assay characteristics (sensitivity, specificity, cutoff concentration). Percent detection rates are a measure of the "efficiency" of the test over a specified period (defined as the total number of positive specimens divided by the total number of specimens tested and multiplied by 100).

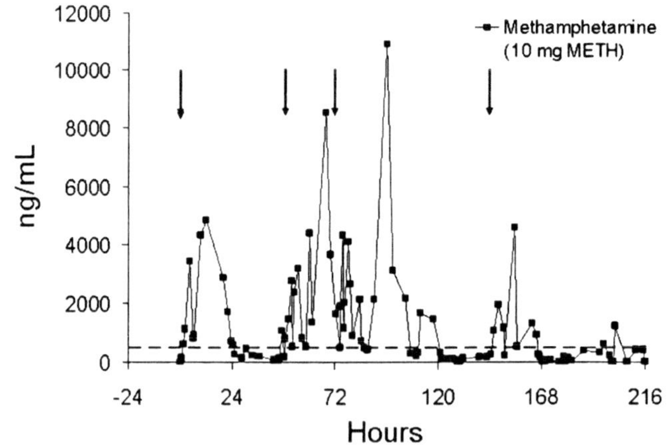

FIGURE 12. Urine concentrations of methamphetamine (METH) in subject S following nonsequential daily single-dose oral administration of 10 mg methamphetamine on different days. The *arrows* represent time of methamphetamine administration and the *dotted line* represents the DHHS cutoff concentration of 500 ng/mL for methamphetamine in urine.

For example, if a methamphetamine test has an average detection window of 48 h, one might find that the detection rate over the first 24 h was 100%, that is, all specimens test positive, but over the entire period of collection, the detection rate could be considerably less than 100%. Thus, both detection times and detection rates are useful in interpretation of drug tests. Detection times in urine are often also complicated by the occurrence of negative specimens interspersed with positive specimens. As drug concentration in the body diminishes as a result of metabolism and excretion, sequential collections may result in positive tests followed by negative tests and occasional reappearance of positive specimens. In addition, excess water intake can dilute urine and drug concentration and produce negative tests, whereas dehydration can concentrate urine. Although methamphetamine concentrations in oral fluid should be less influenced by these factors, drug concentration is influenced by saliva pH, degree of stimulation, and device characteristics. Thus, two approaches were adopted in this review for the analysis of detection times. Detection times are reported in terms of: (1) detection times for sequentially positive specimens (how long specimens remained positive) and (2) detection time to the last positive (how long a positive specimen could be detected).

In the current study, detection times for methamphetamine in oral fluid and urine were determined on the basis of specimens collected immediately following administration of the fourth dose in the low- and high-dose sessions (day 4). Detection times are summarized in TABLE 3. For oral fluid, mean

TABLE 3. Methamphetamine detection times to last consecutive positive and time to last positive in oral fluid and urine at different confirmation cutoff concentrations following a series of four single daily doses (fourth day, mean data) of 10 and 20 mg of methamphetamine administered to five subjects. Detection times are based on time of administration of fourth dose. Oral fluid was collected on day 4 with SAL 4

Meth dose (mg)	Specimen	Cutoff concentration for methamphetamine/ amphetamine (ng/mL)	Time to last consecutive positive ± SEM h (range)	Time to last positive ± SEM, h (range)
10	OF	50/2.5	0.02 ± 0.02 (0.0–0.1)	4.0 ± 1.8 (0.0–8.0)
10	OF	2.5(methamphetamine)	5.6 ± 4.6 (0.0–23.8)	24.6 ± 9.9 (4.0–56.0)
20	OF	50/2.5	4.8 ± 4.8 (0.0–23.8)	20.6 ± 3.2 (8.0–23.8)
20	OF	2.5(methamphetamine)	29.5 ± 12.5 (0.0–76.0)	74.1 ± 17.0 (23.8–126.9)
10[a]	Urine	500/200	10.8 ± 9.5 (0.0–48.7)	43.6 ± 7.5 (24.6–59.5)
10[a]	Urine	250/100	21.0 ± 10.9 (0.8–48.7)	59.5 ± 5.7 (44.3–73.0)
20	Urine	500/200	32.9 ± 11.0 (6.1–59.5)	66.9 ± 7.5 (45.9–91.7)
20	Urine	250/100	57.0 ± 7.7 (31.6–76.8)	79.7 ± 8.8 (50.8–95.5)

Meth = methamphetamine; OF = oral fluid.

[a]Urine detection times for the 10-mg dose condition for one subject (subject AA) were based on three daily doses (fourth-day data missing). All other detection times were based on four daily doses.

detection times for consecutive positives at the recommended DHHS confirmation cutoff concentration of 50 ng/mL of methamphetamine (accompanied by amphetamine at LOD = 2.5 ng/mL) were 0.02 and 4.8 h (timed from start of dosing on the fourth day) for the low- and high-dose conditions, respectively. It should be noted that without accumulation, consecutive positives for oral fluid would not be expected following oral dosing due to the delay in absorption. During the high-dosing sessions, accumulated methamphetamine was sufficient to produce consecutive positives over a period of 24 h following the fourth dose for one subject (subject BB). At the lower cutoff concentration of 2.5 ng/mL, mean detection times for consecutive positive specimens following low- and high-dose sessions were 5.6 and 29.5 h, respectively. Oral fluid detection times to the last positive specimen at the recommended DHHS confirmation cutoff concentration were 4.0 and 20.6 h for the low- and high-dose conditions, respectively, and were 24.6 and 74.1 h at the lower cutoff concentration of 2.5 ng/mL.

For urine, mean detection times for consecutive positives at the DHHS confirmation cutoff concentration of 500 ng/mL of methamphetamine (accompanied by amphetamine at 200 ng/mL) were 10.8 and 32.9 h for the low- and high-dose conditions, respectively. Lowering the cutoff concentration to the recommended cutoff concentrations (250/100) increased mean detection times for consecutive positives to 21.0 and 57.0 h, respectively. Drug accumulation from prior doses appears to have been a greater factor in the production of consecutive positives in urine than was observed for oral fluid.

Mean urine detection times to the last positive at the DHHS confirmation cutoff concentration of 500/200 were 43.6 and 66.9 h for the low- and high-dose conditions, respectively. Lowering the cutoff concentration to the recommended levels (250/100) increased mean detection times to the last positive to 59.5 and 79.7 h, respectively.

Detection rates for methamphetamine in oral fluid and urine also were determined on the basis of specimens collected immediately following administration of the fourth dose in the low- and high-dose sessions (day 4). In this study, oral fluid was typically collected following the fourth dose for 8–24 h, whereas urine specimens were typically collected for a minimum of 72 h. Detection rates for oral fluid and urine are summarized in TABLE 4. For oral fluid, mean detection rates for methamphetamine at the recommended DHHS confirmation cutoff concentration (accompanied by amphetamine at LOD = 2.5 ng/mL) were 24.6% and 56.4% (timed from start of dosing on the fourth day) for the low- and high-dose conditions, respectively. As expected, higher detection rates were observed at the lower cutoff concentration of 2.5 ng/mL (75.0% and 87.5% for the low and high doses). Again, it is noted that the detection rates for oral fluid are highly influenced by accumulation (leading to higher rates) and numerous early specimen collections (leading to lower rates as a result of delay in drug absorption from oral dosing).

TABLE 4. Methamphetamine detection rates in oral fluid and urine at different confirmation cutoff concentrations following a series of four single daily doses (fourth day, mean data) of 10 and 20 mg of methamphetamine administered to five subjects. Detection times are based on time of administration of fourth dose. Oral fluid was collected on day 4 with the SAL 4

Meth dose (mg)	Specimen[a]	Time interval (h)[b]	Cutoff concentration for methamphetamine/ amphetamine (ng/mL)	%Detection rate ± SEM, h (range)
10	OF	0–24	50/2.5	24.6 ± 8.1 (0.0–44.4)
10	OF	0–24	2.5 (methamphetamine)	75.0 13.1 ± (37.5–100.0)
20	OF	0–24	50/2.5	56.4 ± 11.8 (33.3–100.0)
20	OF	0–24	2.5 (methamphetamine)	87.5 ± 12.5 (37.5–100.0)
10	Urine	0–24	500/200	48.1 ± 13.7 (25.0–100.0)
10	Urine	0–48	500/200	46.7 ± 16.8 (13.6–100.0)
10	Urine	0–72[a]	500/200	37.4 ± 21.0 (12.9–100.0)
20	Urine	0–24	250/100	75.5 ± 11.9 (41.7–100.0)
20	Urine	0–48	250/100	69.5 ± 13.9 (27.3–100.0)
20	Urine	0–72[a]	250/100	55.3 ± 15.2 (32.4–100.0)
10	Urine	0–24	2.5 (methamphetamine)	98.3 ± 1.7 (91.7–100.0)
10	Urine	0–48	2.5 (methamphetamine)	95.5 ± 4.5 (77.3–100.0)
10	Urine	0–72[a]	2.5 (methamphetamine)	92.6 ± 7.4 (70.6–100.0)
20	Urine	0–24	500/200	87.7 ± 8.1 (60.0–100.0)
20	Urine	0–48	500/200	87.8 ± 7.5 (69.2–100.0)
20	Urine	0–72	500/200	72.8 ± 5.7 (56.3–88.9)
20	Urine	0–24	250/100	100.0 ± 0.0 (NA)
20	Urine	0–48	250/100	98.3 ± 1.7 (91.3–100.0)
20	Urine	0–72	250/100	86.7 ± 4.5 (75.0–100.0)
20	Urine	0–24	2.5 (methamphetamine)	100.0 ± 0.0 (NA)
20	Urine	0–48	2.5 (methamphetamine)	100.0 ± 0.0 (NA)
20	Urine	0–72	2.5 (methamphetamine)	100.0 ± 0.0 (NA)

Meth = methamphetamine; OF = oral fluid.

[a]Detection rates for the 10-mg dose period 0–72 h are based on an $N = 4$ because of missing specimens from 48–72 h.

[b]Time interval for oral fluid collection for three subjects (S, Y, AA) was 0–8 h for the 10-mg dose and for one subject (W) at the 20-mg dose due to missing specimens or a change in type of oral fluid collection device. The remaining subjects completed the 0–24 h collection period with the same oral fluid collection device (SAL 4).

Detection rates for methamphetamine/amphetamine in urine at the DHHS confirmation cutoff concentration of 500/200 were higher during the 0–24 h period (48.1% and 87.7% for the low and high doses) than observed for oral fluid. Extending the detection periods to 0–48 and 0–72 h tended to diminish rates. At the proposed DHHS confirmation cutoff concentration of 250/100, detection rates for the 0–24 h period increased to 75.5% and 100.0% for the low and high doses and diminished only slightly for the 0–48 and 0–72 h periods. At a cutoff concentration of 2.5 ng/mL, detection rates were consistently >70% for all subjects at the low-dose and were 100% for the high-dose condition.

CONCLUSIONS

The disposition of methamphetamine and its metabolite, amphetamine, in oral fluid followed a similar time course as plasma. Proportional increases in concentration were apparent under low- and high-dosing conditions for both specimens. Higher concentrations in oral fluid were produced relative to plasma, likely as a result of the higher acidity of saliva relative to plasma. Upon repeated daily dosing, accumulation was apparent in oral fluid specimens and urine. Intrasubject variability in methamphetamine concentrations for oral fluid specimens under identical dosing and collection conditions (days 1 and 2) appeared to be relatively low, but high intersubject variability was noted. Methamphetamine detection in oral fluid and urine at conventional cutoff concentrations was assessed at the end of 4-day repeated oral dosing sessions of low- and high-dose methamphetamine. Urine testing characteristically provided relatively high detection rates for up to 3 days following drug cessation. In contrast, oral fluid testing typically provided moderate detection rates for 24 h. Detection at lower concentrations generally extended detection times for both oral fluid and urine. In general, it was observed that the use of oral fluid as a specimen for methamphetamine testing provided a shorter "window of detection." The choice between the use of oral fluid and urine in drug testing programs should include considerations of inherent differences in time course of detection, advantage of observed oral fluid collection, and program goals.

ACKNOWLEDGMENT

E. J. Cone is a consultant to OraSure Diagnostics, a company that manufactures oral fluid diagnostic products.

REFERENCES

1. SUBSTANCE ABUSE AND MENTAL HEALTH SERVICES ADMINISTRATION. Results from the 2004 National Survey on Drug Use and Health: National Findings. NSDUH Series H-28, DHHS Publication No. SMA 05-4062, 1-292. 2005. Rockville, MD, Office of Applied Studies.
2. MAXWELL, J.C. 2005. Emerging research on methamphetamine. Curr. Opin. Psychiatry **18:** 235–242.
3. BECKETT, A.H., J.A. SALMON & M. MITCHARD. 1969. The relation between blood levels and urinary excretion of amphetamine under controlled acidic and under fluctuating urinary pH values using [14C] amphetamine. J. Pharm. Pharmacol **21:** 251–258.
4. SCHEPERS, R.J., J.M. OYLER, R.E. JOSEPH JR., *et al.* 2003. Methamphetamine and amphetamine pharmacokinetics in oral fluid and plasma after controlled oral methamphetamine administration to human volunteers. Clin. Chem. **49:** 121–132.

5. KIM, I., J.M. OYLER, E.T. MOOLCHAN, *et al.* 2004. Urinary pharmacokinetics of methamphetamine and its metabolite, amphetamine following controlled oral administration to humans. Ther. Drug Monit. **26:** 664–672.
6. OYLER, J.M., E.J. CONE, R.E. JOSEPH JR., *et al.* 2002. Duration of detectable methamphetamine and amphetamine excretion in urine after controlled oral administration of methamphetamine to humans. Clin. Chem. **48:** 1703–1714.
7. DHHS. 2004. Proposed revisions to mandatory guidelines for federal workplace drug testing programs. Fed. Register **69:** 19673–19732.
8. DHHS. 1993. Mandatory guidelines for federal workplace drug testing programs. Fed. Register **58:** 6062–6072.

Salivary α-Amylase in Biobehavioral Research

Recent Developments and Applications

DOUGLAS A. GRANGER,[a] KATIE T. KIVLIGHAN,[b]
MONA EL-SHEIKH,[c] ELANA B. GORDIS,[d] AND LAURA R. STROUD[e]

[a]*Behavioral Endocrinology Laboratory, Departments of Biobehavioral Health and Human Development and Family Studies, Pennsylvania State University, University Park, Pennsylvania, USA*

[b]*Department of Population, Family, and Reproductive Health, Bloomberg School of Public Health, Johns Hopkins University, Baltimore, Maryland, USA*

[c]*Department of Human Development and Family Studies, Auburn University, Auburn, Alabama, USA*

[d]*Department of Psychology, University at Albany, SUNY, Albany, New York, USA*

[e]*Centers for Behavioral and Preventive Medicine, Brown Medical School, Providence, Rhode Island, USA*

ABSTRACT: **In the history of science, technical advances often precede periods of rapid accumulation of knowledge. Within the past three decades, discoveries that enabled the noninvasive measurement of the psychobiology of stress (in saliva) have added new dimensions to the study of health and human development. This widespread enthusiasm has led to somewhat of a renaissance in behavioral science. At the cutting edge, the focus is on testing innovative theoretical models of individual differences in behavior as a function of multilevel biosocial processes in the context of everyday life. Several new studies have generated renewed interest in salivary α-amylase (sAA) as a surrogate marker of the autonomic/sympathetic nervous system component of the psychobiology of stress. This article reviews sAA's properties and functions; presents illustrative findings relating sAA to stress and the physiology of stress, behavior, cognitive function, and health; and provides practical information regarding specimen collection and assay. The overarching intent is to accelerate the learning curve such that investigators avoid potential**

Address for correspondence: Douglas A. Granger, Behavioral Endocrinology Laboratory, Department of Biobehavioral Health, 315 Health and Human Development East, The Pennsylvania State University, University Park, PA 16802. Voice: 814-863-8402; fax: 814-863-7525.
dag11@psu.edu

Ann. N.Y. Acad. Sci. 1098: 122–144 (2007).
doi: 10.1196/annals.1384.008

pitfalls associated with integrating this unique salivary analyte into the next generation of biobehavioral research.

Keywords: salivary alpha-amylase; behavior problems; cognition; health; social relationships; psychobiology of stress; sympathetic nervous system

INTRODUCTION

Technical advances have made the noninvasive assessment of biomarkers in saliva possible and created new opportunities to study how biological and social processes interact to influence health and human behavior.[1] The ease of use and noninvasive nature of salivary measures is especially valuable because complex multilevel models of individual differences can be studied in the laboratory, or in quasi-naturalistic settings, or in response to the trials and tribulations of everyday life. Correspondingly, the last two decades have witnessed an increase in publications that report the use of salivary measures. In behavioral science, most of the empirical attention centers on the correlates and consequences of hypothalamic-pituitary-adrenal (HPA) axis activity as a result of the widespread acceptance and application of measuring cortisol in saliva.[2,3]

To advance our understanding to new limits, it has been argued that multiple measurements of stress-related biological processes should be included in these new conceptual and analytical models.[4–6] Indeed, experts have clearly emphasized that to accurately represent stress physiology, multiple measures of stress across multiple stress-related systems should be assessed.[7] Two main systems comprise the psychobiology of stress: the HPA axis and the locus ceruleus/autonomic (sympathetic) nervous system (SNS).[8] Traditional psychophysiological assessments of the SNS involve electrodes, computerized recording apparatus, and sophisticated data-reduction algorithms.[9,10] Monitoring the SNS subcomponent of the stress response in saliva would potentially afford researchers clear advantages. For instance, dissociations are frequently found among various measures of physiological reactivity, even those activated by a common system, such as the SNS,[11] because autonomic nervous system (ANS) responses are subject to the individual specificity phenomenon in which children may consistently react in one specific physiological modality.[12] Therefore, the examination of a salivary measure of SNS activity is likely to provide unique information in addition to knowledge based on other physiological measures related to the SNS (e.g., pre-ejection period). Thus, technical advances that enable us to monitor the autonomic stress response in saliva set the stage for biobehavioral science to fill several critical gaps in knowledge.

The most common serological measures of autonomic (sympathetic) activation are the catecholamines epinephrine (EPI) and norepinephrine (NE).

Unfortunately, however, the measurement of catecholamines in saliva by either high-performance liquid chromatography (HPLC) or immunoassay has proven challenging. Direct measurements of EPI or NE in saliva seem not to reflect SNS activity.[13] Consequently, researchers have been actively searching for an indirect or surrogate marker of SNS activity in saliva. NE is released at sympathetic nerve terminals in tissues and glands throughout the body (e.g., lymph nodes, salivary glands). EPI, and to a lesser extent NE, is released in response to sympathetic activation of the adrenal medulla. The salivary gland, glandular duct cells, and vascular bed of the salivary glands are rich with β-adrenoreceptors.[14] The action of norepinephrine released from sympathetic nerve terminals on adrenergic receptors influences the activity of the salivary glands by increasing the protein-to-fluid ratio in the saliva.[15] Among these proteins is salivary α-amylase, an enzyme that is produced by the salivary gland. A small, but rapidly growing literature suggests that salivary α-amylase might serve as a noninvasive and easily obtained surrogate marker of SNS activity.

The pages that follow introduce the basic properties and function of salivary α-amylase, present findings that illustrate the nature of salivary α-amylase response to stress, and characterize associations between salivary α-amylase and measures of the SNS, HPA axis, and cardiovascular psychophysiology. Sources of individual differences and intraindividual change in salivary α-amylase in relation to behavior, cognitive function, and health are reviewed. Practical information to enable investigators to collect, handle, prepare, and store samples for assay is also provided.

Salivary α-Amylase: Basic Properties and Function

Salivary biomarkers regularly employed in biobehavioral research include androgens (e.g., testosterone, dehydroepiandrosterone, androstenedione), reproductive hormones (e.g., progesterone, estradiol), glucocorticoids (e.g., cortisol), immunoglobulins (e.g., secretory IgA), controlled substances, and metabolites of nicotine use (e.g., cotinine). Salivary α-amylase is unique among these markers because it is an enzyme, officially classified as family 13 of the glycosyl hydrolases. The structure is an 8-stranded α−β barrel containing the active site, interrupted by a ∼ 70 a.a. calcium-binding domain protruding between β strand 3 and α helix 3, and a carboxyl-terminal Greek key β−barrel domain. Several unique properties distinguish salivary α-amylase from the more common saliva markers used in biobehavioral research.[16]

Production by the Salivary Gland

In contrast to many analytes present in oral fluid, salivary α-amylase is not actively transported, nor does it passively diffuse into saliva from the general circulation. Salivary α-amylase is produced locally in the oral cavity by the salivary glands. Thus, salivary α-amylase levels in the oral fluids do not represent levels in the general circulation, nor do they reflect levels in the gastrointestinal system. Under conditions of normal oral health, α-amylase is present in saliva in relatively high concentrations.

Digestive Enzyme

Glycosyl hydrolases are enzymes that hydrolyze the glycosidic bond between two or more carbohydrates, or between a carbohydrate and a noncarbohydrate. Thus, a primary biological function of salivary α-amylase is the digestion of macromolecules (e.g., carbohydrates and starch). α-Amylase is produced not only by the salivary glands but also by the pancreas within the gastrointestinal system. Levels of α-amylase in these two compartments are produced by independent sources and are not correlated. Salivary α-amylase digests a portion of ingested starch in the stomach before it enters the intestine and is exposed to pancreatic amylase. α -Amylase activity is associated with increased caloric intake from the digestion of carbohydrates (i.e., breads, potatoes, rice, and pasta) into sugars that can be absorbed through the intestinal wall.

Alpha-amylase inhibitors (AAIs), molecules that interfere with the action of the α-amylase enzyme, have been explored as dietary interventions. The use of so-called "starch blockers" has gained in popularity with the success and growth of carbohydrate-restricted diets. Theoretically, lower α-amylase activity restricts caloric intake by attenuating the breakdown of starch into sugar. Kataoka and Dimagno (1999)[17] reported that the effects of a wheat-derived AAI included slowing of the rate of digestion and gastric emptying, thereby prolonging the feeling of fullness and delaying the urge to eat.

The apparent link between salivary α-amylase and eating behavior may be of more than passing interest. Susman and colleagues (2006)[18] recently reported a positive association between salivary α-amylase activity and body mass index in adolescent males and females. It is tempting to speculate that stress-related increases in α-amylase, when combined with high carbohydrate/starch diet and sedentary lifestyle, could in part relate to the weight-related disease epidemics facing many Westernized societies.

Role in Oral Health

A secondary role of α-amylase is bacterial clearance from the mouth and prevention of bacterial attachment to oral surfaces.[19] Indeed, much of what is known about the "biobehavioral" implications of individual differences in salivary α-amylase levels is documented in the literature on oral biology and disease. Higher salivary α-amylase activity is associated with reduced risk for a variety of processes related to oral health (bacteria load, caries, and periodontal disease). Atypically low salivary α-amylase activity is associated with oral disease. Although oral health has been a relatively minor concern for mainstream biobehavioral science, the dental literature presents volumes of work that link psychosocial stress and social factors to compromised oral biology.[20]

Developmental Differences

At birth, α-amylase is not present in the oral or gastrointestinal compartment.[21] Correspondingly, newborns do not have the same capacity as do children and adults to digest complex macromolecules. This transitional physiological state is considered adaptive because generally macromolecules (such as carbohydrates) are highly immunogenic and represent a significant source of threat to the neonatal immune system. Moreover, during this period of immune immaturity, newborns are primarily protected against foreign antigens by passive immunity received from their mothers' (i.e., maternal antibodies, IgA) colostrum or breast milk. Given the absence of salivary α-amylase, maternal antibodies can be ingested without being destroyed (essentially digested) by the infant. Salivary α-amylase activity shows a sharp rise in the 0.9- to 1.9-year period, reaching maximum levels by 5 to 6 years of age.[21] The age of onset in the rise of α-amylase levels parallels the timing of the introduction of solid foods in the diet and the emergence of dentition needed to chew those solids. During middle childhood (ages 8 to 9 years), El-Sheikh *et al.* (2005)[22] report that pubertal status and age are positively associated with salivary α-amylase activity. Susman and colleagues (2006)[18] also note a relationship between α-amylase reactivity and pubertal development (Tanner Stage) in boys but not girls (ages 8 to 13 years).

Salivary α-Amylase: Relation to Physical and Psychological Stress

There is a robust pattern across studies that salivary α-amylase levels rise in response to both physical and psychological stress. Early studies reveal that levels of salivary α-amylase increase in response to stressful conditions including exercise, heat and cold stress, and written examinations.[23–25] More recent studies show large α-amylase increases in response to the Trier Social Stress Test (TSST),[26–29] to watching highly negative emotional pictures of mutilation or accidents,[30] and to participating in collegiate level individually oriented athletic competition.[31]

The typical salivary α-amylase response profile is consistent with our knowledge of the rapid activation and recovery that characterizes the response of the SNS to stress. Findings from three of our team's recent studies are illustrative.

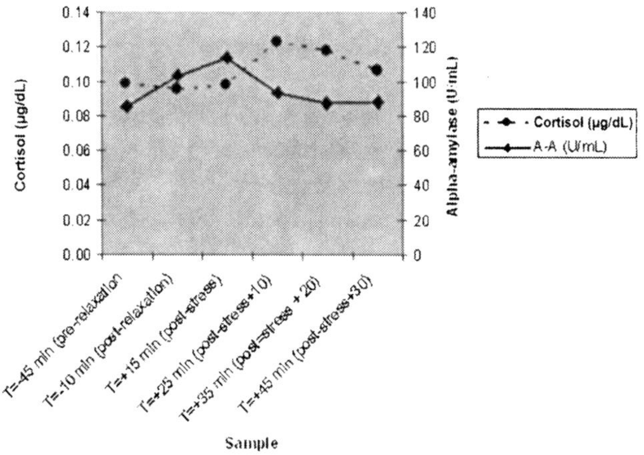

FIGURE 1. Kinetic reaction profile of salivary α-amylase (U/mL) and cortisol (µg/dL) to a modified Trier Social Stress Test in adolescents. (Modified from Gordis *et al.*[26]). A-A = α-amylase.

Gordis *et al.* (2006)[26] report that in response to the TSST, adolescents' salivary α-amylase levels increased 145% on average over pre-task levels, with 65% of the sample showing greater than 10% increases in salivary α-amylase levels from pre-task baseline to peak. Salivary α-amylase levels returned to baseline by 10 min poststressor (see FIG. 1).

Stroud *et al.* (2006)[32] examined the time course and magnitude of the salivary α-amylase response in children and adolescents to achievement and interpersonally oriented stressors. They report significant changes in salivary α-amylase across both session types with 50–60% increases from baseline. Salivary α-amylase peaked approximately 10 min following the onset of the stressors (see FIG. 2).

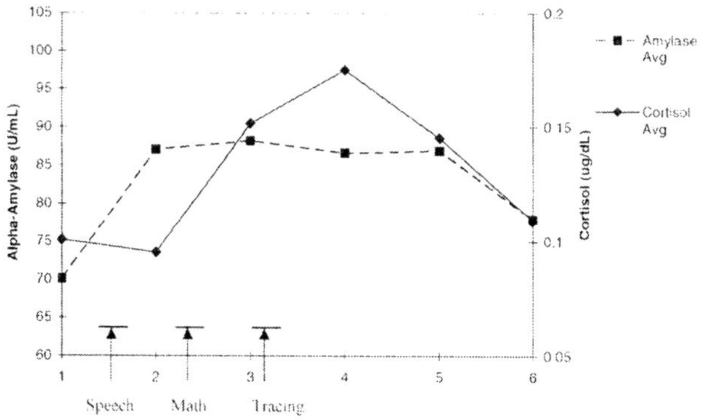

FIGURE 2. Kinetic reaction profile of salivary α-amylase (U/mL) and cortisol (µg/dL) to a modified Trier Social Stress Test in adolescents. (From Stroud *et al.*[29]).

Salivary α-Amylase and the Adrenergic Component of the Stress Response

In the late 1990s, a series of elegant studies by Chatterton and colleagues suggested a strong positive association between levels of salivary α-amylase and the SNS component of the stress response.[23–25] These studies were also among the first to show that levels of salivary α-amylase increase under both physically and psychologically stressful conditions. Salivary α-amylase concentrations were associated with baseline plasma catecholamine levels, particularly norepinephrine (NE), and were also highly correlated with NE change in response to stress.[23,33] The strong positive association between SNS activation and salivary α-amylase was corroborated by a recent placebo-controlled study showing that stress-related increases in salivary α-amylase can be inhibited by administration of the adrenergic blocker propranolol.[30] In addition, β-adrenergic agonists are capable of stimulating salivary α-amylase release without increasing salivary flow.[34] This pattern of evidence has been used to deduce that the same stimuli that result in the release of catecholamines in the blood and peripheral tissues activate sympathetic input to the salivary glands. An important methodological caveat is that use of prescription medications for angina or high blood pressure that have β-blocking properties (e.g., Inderal, Tenormin, Coreg, Lopressor) or consumables that stimulate the SNS (e.g., caffeine) should be controlled in biobehavioral studies employing salivary α-amylase measurements.

In summary, early work suggested that circulating levels of NE, associated with the stress response of the locus ceruleus/autonomic (sympathetic) nervous system, could be estimated by the concentrations of α-amylase in whole saliva specimens, and that salivary α-amylase measurements may be employed as a noninvasive measure of plasma NE concentrations in human participants. More recent studies corroborate that salivary α-amylase responds to physical and psychological stress, and relates to sympathetic tone or adrenergic activation generally. However, these studies raise hard to answer questions about the specific association between α-amylase and stress-related changes in catecholamines.[27,28]

Salivary α-Amylase and the HPA Axis Response to Stress

The profile of stress-related change in salivary α-amylase levels is distinct from the response profile measured by salivary cortisol.[35] As shown in FIGURES 1 and 2, both salivary analytes increased in response to the challenge tasks; however, salivary α-amylase reaches its peak response faster than salivary cortisol. Recovery to pre-task levels is also faster for salivary α-amylase than for cortisol. These kinetic response patterns fit our expectations of physiological differences in the timing of SNS (quicker) and HPA (slower) stress responses.[8] However, at least part of the difference may also be related to specialized issues of cortisol measurement in saliva. Cortisol must be released from the adrenal gland (above the kidneys) into the general circulation and then must passively diffuse into oral fluids. The delay between HPA axis activation and change in salivary concentration of cortisol can be as much as 15 to 20 min poststressor. On the other hand, α-amylase is released by the salivary gland directly into the mouth. Regardless of the mechanisms involved, sample collection clearly must be on a different time course to capture the response profiles of salivary cortisol and α-amylase accurately in the same study. Traditionally, studies designed to capture response and recovery of salivary cortisol would employ pre-task, 20-min post-task (peak), and 40-min post-task (recovery) designs. For salivary α-amylase measures, appropriate sample collection time points might include pre-task, 5-min post-task (peak), and 20–30-min post-task (recovery).

On the basis of our preliminary studies, salivary α-amylase may increase over pre-task levels in response to stress in a larger percentage of cases, and the magnitude of the rise in salivary α-amylase may be larger, on average, than salivary cortisol. For instance, Kivlighan and colleagues (2005)[36] and Gordis and colleagues (2006)[37] reveal that the percentage of individuals showing salivary α-amylase rise greater than 10% (a criterion applied to be sure meaningful change is only interpreted when the difference from pre- to post-task is larger than error in the assay; see below) is greater than 40%. By contrast, salivary cortisol reactivity to these socially oriented stressors was evident in a smaller percentage of cases. These differences may be due to the more sensitive threshold of reactivity in the SNS (i.e., salivary α-amylase) than in the HPA axis—theoretically, SNS reactivity would occur to challenges that are more mild than those required to activate the HPA axis.[7]

Given these differences in kinetic profiles and, potentially, sensitivity to psychosocial stress, the lack of correlation between levels of salivary cortisol and salivary α-amylase at baseline, in response to stress, or during recovery is not surprising.[26,28,31,38] This observation is key because the statistical independence of these measures suggests that individual differences in salivary α-amylase reactivity are not redundant with change in the HPA measures, and likely are indexing a different stress-response system.[23] Whereas overall the HPA and SNS systems clearly work in coordination to generate the physiologic changes associated with the stress response, the exact nature of the coordination (e.g., additive or interactive; opposing or complementary) is a subject of debate. Studies with adrenalectomized rats indicate that endogenous glucocorticoids restrain responses, such as catecholamine turnover, synthesis, and release in sympathetic nerves during immobilization stress.[39] In humans, treatment with 20 mg of prednisone reduced SNS activity and plasma NE in healthy subjects,[40] and increased catecholamine concentrations have been reported in patients with stress-related disorders characterized by hypocortisolemic stress responses. Fries *et al.* (2005)[41] suggest hypocortisolemia might result in reduced inhibitory feedback activities of cortisol on catecholamine release and synthesis.

We believe behavioral scientists should be interested in understanding how social forces and psychological states influence the *coordination* of these stress-response systems.[4,41,42] Of course, this idea is not necessarily novel,[43] but the noninvasive means of assessing both the HPA and SNS via saliva should enable the field to do so not just in the laboratory but in the context of everyday life and/or within the same individuals cross-situationally or on repeated occasions.

Salivary α-Amylase and Cardiovascular Psychophysiology

Several studies report association between salivary α-amylase and cardiovascular psychophysiology. Chatterton and colleagues (1996)[23] reported a positive relationship between salivary α-amylase and heart rate that strengthened with the intensity of physical stress. Bosch and colleagues (2003)[44] observed increases in sympathetic activity (shortened pre-ejection period, PEP) and α-amylase secretion and decreases in parasympathetic activity (decrease in heart rate variability) in response to a laboratory stressor. Nater and colleagues (2006)[28] report a positive relationship between salivary α-amylase and sympathetic tone (LF/HF) during stress. West *et al.* (2006)[45] showed that relative to resting baseline values, individuals with the largest increases in salivary α-amylase 8–10 min after a cold pressor task showed larger concurrent reductions in PEP, increases in systolic BP, increases in heart rate (HR), and increased cardiac output. The associations between salivary α-amylase and PEP are particularly noteworthy as PEP is among the few psychophysiological measures (the other is skin conductance) thought to exclusively assess sympathetic nervous system activity.[46] Klein and colleagues (2006)[47] showed that in response to caffeine administration, BP, HR, and salivary α-amylase increased, and that caffeine-induced change in salivary α-amylase was associated with increased HR. Similarly, Van Stegeren and colleagues (2005)[30] showed that β-blockade was successful in reducing HR, salivary α-amylase, and systolic BP.

Our preliminary studies have explored similar relationships in youth. In a sample of participants ages 7 to 17 years, Stroud and colleagues (2005)[29] found high magnitude positive associations between salivary α-amylase and systolic blood pressure (SBP) reactivity (defined as the difference between maximum post-stress and baseline values) to laboratory stressors, but no associations between salivary α-amylase and SBP or HR reactivity. In an elementary school–aged sample (ages 8 to 9 years), El-Sheikh *et al.* (2005)[22] studied the relationship between respiratory sinus arrhythmia (RSA) and salivary α-amylase reactivity to a stressful laboratory procedure comprising listening to an argument and performing a mirror-star tracing. Salivary α-amylase was associated with deficits in vagal suppression (lower levels of RSA during challenge tasks in comparison to baseline conditions). Whereas vagal suppression is the typical response to environmental challenges and is associated with positive in children outcomes, children with higher levels of salivary α-amylase exhibited vagal augmentation. Finally, Kivlighan *et al.* (2006)[48] examined relationships between salivary α-amylase, HR, vagal tone (VT), and PEP in response to the TSST in 33 healthy children ages 9 to 11 years. The majority of participants showed a salivary α-amylase change of at least 10% in response to the TSST, and levels had returned to baseline 10 min post task. During preparation for the TSST speech task, VT and salivary α-amylase were positively associated. Peak salivary α-amylase levels were also associated with VT levels early post task, and salivary α-amylase and VT were related during both the early and late post-speech phase of the TSST. Interestingly, there were no associations with PEP.

Summary

In general, the direction of effects is consistent with the notion that salivary α-amylase is associated with sympathetic/parasympathetic nervous system activation. Missing are studies that enable us to speculate about the meaning of these associations in relation to cardiovascular disease–related symptoms, outcomes, or treatment responsiveness.

Concomitants of Individual Differences in Salivary α-Amylase

The previous sections have documented that individual differences in salivary α-amylase are related to age and pubertal development, and may be sensitive to specific experiences that involve physical, social, and cognitive demands. Studies by our informal collaborative network also reveal that individual differences in salivary α-amylase are, independently, or in combination with salivary cortisol, associated with a wide range of behavior. We provide examples of these findings to give the reader an idea of potential patterns and themes that may be relevant to biobehavioral research.

Infant Attunement and Attachment to Caregivers

Kivlighan and colleagues (2005)[36] measured salivary α-amylase in mother–infant dyads as the infants participated in a series of tasks designed to elicit emotional distress. While the infants (6–10 months) actively participated in the task series (e.g., arm restraint), mothers were asked to watch without intervening. Three saliva samples were collected from the dyad; a baseline collected prior to administration of the challenge tasks, 20 min following the infants' peak arousal to the task, and 40 min after peak arousal. As expected, salivary α-amylase levels were higher for mothers than for infants. On average, the mothers' salivary α-amylase levels were higher 20 min post task, compared to pre task, and levels 40 min post task remained higher than pre task. By contrast, on average, salivary α-amylase levels for the infants did not change in response to the challenge tasks. When a 10% difference from pre-task is used as a criterion for change, 44.4% of the infants showed α-amylase increase, and 48.1% a decrease from pre to 20 min post task. Mothers' and infants' pre-task α-amylase levels were positively correlated.

Similarly, Shea *et al.* (2006)[49] investigated salivary α-amylase concentrations in infants in response to an acute stressor, and the impact of the mothers' depression/anxiety on this response. Mothers and their infants provided saliva samples during the course of a laboratory stressor (noise burst and infant arm restraint). Maternal and infant salivary α-amylase levels were positively correlated.

Hill and colleagues (2006)[50] measured salivary α-amylase in infants (12 months) as they participated with their mothers in the Strange Situation procedure involving brief separations and reunions. Saliva was collected from the infants before, 15 and 40 min post task. On average, the change in salivary α-amylase in response to the strange situation was not significant. However, significant differences emerged in children classified as "avoidant" versus "secure" in their α-amylase response to the strange situation. Children classified as "avoidant" had higher mean levels of salivary α-amylase than "securely" attached children.

To the best of our knowledge, these are the only studies of biobehavioral correlates of α-amylase during early development. These preliminary findings suggest that individual differences in salivary α-amylase activity are associated (or attuned) between mothers and infants, and that infants' salivary α-amylase is influenced by the nature of their social relationships with caregivers.

Preschoolers' Physical Health and Relationships with Teachers

Mize and colleagues (2005)[51] examined salivary α-amylase and cortisol as a function of stress, teacher–child relationship quality, and health in preschoolers. Children participated in a series of five developmentally appropriate challenge tasks or games intended to provoke mild frustration or disappointment (e.g., disappointment experience, impossible puzzle task, delay of gratification task). Following the tasks, the child's primary teacher helped the child make a block structure and read a book to the child. Saliva samples were collected by passive drool before and after the challenge tasks, and after the teacher interaction. Only a few children experienced cortisol increases of greater than 10% from pre- to post-challenge, and from post-challenge to follow-up (14% and 10% of children respectively). More children experienced increases in salivary α-amylase than cortisol over the same assessments (37% and 45%). Children with greater salivary α-amylase increases from pre-challenge to follow-up had more illness and less close relationships with teachers. The associations between salivary α-amylase and illness were somewhat stronger for girls than for boys.

Problem Behavior, Academic Performance, and Physical Health in Middle Childhood

El-Sheikh, Granger, and Buckhalt (paper under review),[52] examined relations between cortisol, salivary α-amylase, and problem behavior in children (ages 8 to 9 years). Saliva samples were obtained during baseline and following two laboratory stressors. Baseline levels of salivary α-amylase were positively associated with boys' aggression. Post-stress salivary α-amylase was predictive of girls' internalizing behavior problems, and higher salivary α-amylase reactivity from pre- to post-lab stressors was related to higher levels of girls' internalizing symptoms. In support of Bauer's symmetry hypothesis,[4] interaction effects between salivary α-amylase and cortisol explained unique variance in children's, especially girls', behavior problems. Symmetry in salivary α-amylase and cortisol reactivity to the lab challenge task, mostly lower levels of both salivary α-amylase and cortisol reactivity, were associated with the lowest levels of externalizing and internalizing behaviors.

In a secondary analysis using the same middle childhood sample, Buckhalt *et al.* (paper under review),[53] examined relations between salivary α-amylase, cortisol and children's cognitive and academic functioning. Children were individually administered six tests of the Woodcock–Johnson III Tests of Cognitive Abilities and two reaction-time tasks. Teachers reported on children's academic functioning, and schools provided scores from a nationally standardized, group-administered test of achievement. Higher baseline cortisol and salivary α-amylase were broadly associated with poorer cognitive/academic performance. Poststress cortisol and salivary α-amylase as well as cortisol reactivity (pre- to post-stressors) and salivary α-amylase reactivity seemed to have different relations with cognitive functioning for boys and girls. Specifically, higher levels on these variables were related to poorer performance for girls and better performance for boys. Again, consistent with Bauer's speculations,[4] a robust pattern emerged indicating worse academic/cognitive functioning among children with low levels of salivary α-amylase in conjunction with higher levels of cortisol.

Additional analyses with the same sample of healthy (no chronic or acute illnesses) 8- to 9-year-olds yielded interesting relations between salivary α-amylase levels and children's, especially girls,' physical health problems as reported by parents and teachers.[22] Higher levels of post-stress α-amylase and α-amylase reactivity from pre- to post-challenge conditions, were significantly associated with increased health problems including respiratory problems, fatigue, and frequency of illness. Further, salivary α-amylase levels were significantly and positively associated with SIgA during both post-stress and reactivity from pre- to post-stress conditions. Collectively, these findings between α-amylase and SIgA, as well as reported health problems, highlight the potential importance for examinations of α-amylase in studies focusing on various health parameters.

In summary, in studies with elementary school–aged children, individual differences in salivary α-amylase appear to relate to problem social behavior. Given the well-established association between attachment classification early in life and problem behavior later in middle childhood,[54] it is tempting to speculate about a pattern linking salivary α-amylase and social behavior/relationships across childhood. The link between salivary α-amylase and cortisol and children's academic and task-related performance is also noteworthy and is generally consistent with Frankenhauser's seminal works on SNS arousal, attention, and achievement.[55,56]

Adolescent Aggression and Antisocial Behavior

Stroud and colleagues (2006)[32] examined associations between salivary cortisol and α-amylase reactivity to the TSST. Participants were healthy children and adolescents (ages 7 to 16 years). The study involved two sessions conducted on separate days. In the first "rest" session, participants were habituated to the laboratory and physiological monitors while completing a battery of questionnaires. The second (stress) session lasted approximately 2 h, and included a baseline period, three stressors (i.e., speech, mental arithmetic, mirror tracing), and a recovery period. Greater salivary α-amylase reactivity was associated with decreases in feelings of relaxation ("relax") and increases in feeling "upset" between baseline and stress periods. Participants with low cortisol and high α-amylase responses (based on percentage change from baseline to maximum level achieved during stress) demonstrated significantly higher scores on CBCL scales consistent with positive behaviors, including the activities, social, and school subscales, compared to participants with high cortisol and low α-amylase reactivity. Similarly, participants with high cortisol and low α-amylase reactivity showed higher scores on the total and externalizing CBCL scales as well as the socfial problems, thought problems, and aggression/delinquency behavior subscales than did participants with low cortisol and high α-amylase reactivity. Participants with high cortisol and low α-amylase reactivity also scored significantly higher on the Perceived Stress Scale than subjects with low cortisol and high α-amylase reactivity.

Similarly, Gordis and colleagues (2006)[26] asked adolescents (ages 10 to 14 years) to provide 2 saliva samples before and 4 samples after a TSST. Both salivary α-amylase and cortisol increased in response to the TSST. Salivary α-amylase and cortisol stress reactivity were estimated using the area under the curve (AUC). Symmetrical cortisol and salivary α-amylase reactivity accounted for significant variance in parent-reported adolescent aggression. At lower levels of salivary α-amylase reactivity, lower cortisol reactivity corresponded to higher aggression ratings, but at high salivary α-amylase reactivity levels, cortisol reactivity was unrelated to aggression.

Susman and colleagues (2006)[18] hypothesized that the vulnerability of low salivary α-amylase reactivity to an experimental stressor would predispose adolescents (ages 9 to 13 years) to disruptive behavior. They further hypothesized that attenuation of salivary α-amylase would interact with the transitional stress of puberty, specifically, early timing of puberty, so as to predispose to disruptive behavior. Adolescents and parents reported on symptoms of disruptive behavior disorders on the Diagnostic Interview Schedule for Children (DISC-C) and DISC-P. Adolescents provided saliva samples before and after the TSST. Salivary α-amylase reactivity, based on changes from pre- to post-TSST, was negatively related to adolescent-reported symptoms of Oppositional Defiant Disorder (ODD) and Conduct Disorder (CD). That is, low change from pre- to post TSST corresponded to more ODD and CD symptoms according to adolescents' self-reports. Of note is that there were no significant findings for girls.

Studies with adolescents consistently find that salivary α-amylase increases in response to the TSST. Taken together, the findings suggest that attenuated salivary α-amylase is a potential risk for aggression, and symptoms of disruptive behavior disorders. Gender differences seem to exist in the behavioral correlates of α-amylase in studies of elementary school-aged children and adolescents. Findings with adolescents show support for Bauer and colleagues' (2002)[4] symmetry hypothesis in relation to externalizing behavior problems.

Dominance, Performance, and Mood in Adults

In adults, very few studies have explored correlates of individual differences in salivary α-amylase in relation to behavior. The exception is Kivlighan and Granger (2006),[31] who examined individual differences in salivary α-amylase response to competition in relation to previous experience, behavior, and performance. Participants were young adult members of a collegiate crew team. Salivary samples were collected before and at 20 and 40 min after an ergometer competition. Salivary α-amylase was higher across the competition for varsity than for novice athletes, and was positively associated with performance and interest in team bonding. Salivary α-amylase reactivity to competition explained individual differences in dominance, and symmetry in salivary α-amylase and cortisol reactivity to competition (low–low) was associated with high-perceived dominance.

Nater and colleagues (2006)[57] explored correlates of the diurnal pattern of salivary α-amylase activity. In two independent studies, saliva samples were collected directly after waking up, 30 and 60 min later, and each full hour until

9 pm by the subjects themselves. The compliance of the subjects was controlled by an electronic system. In order to control factors that might influence the diurnal profile of α-amylase (such as acute stress, mood, food, or body activity), at each sampling time point the subjects filled out a diary examining the activities they had carried out during the previous hour. Results from healthy male volunteers indicate that salivary α-amylase activity shows a distinct diurnal profile pattern with a trough in the morning and a steady increase of activity during the day. Multilevel modeling failed to show a within-subject association of α-amylase with acute stress, but significant associations emerged with chronic stress, mood, and alertness.

Summary

This series of preliminary studies reveals distinct associations between salivary α-amylase, behavior, and cognition, and supports the tentative conclusion that salivary α-amylase may index physiological processes that effect and are influenced by psychological, behavioral, and social processes. The association between α-amylase levels and illness susceptibility is novel. Also noteworthy is the observation that individual differences in levels of α-amylase were associated with internalizing (e.g., social withdrawal, anxiety/depression) and externalizing (e.g., aggression and symptoms of conduct disorder) problems. The consistent pattern of relationships between salivary α-amylase and cognitive function and academic performance is also worthy of note. The findings also hint that individual differences in α-amylase levels during early childhood may be regulated by social relationships (with mothers and teachers) central in young children's immediate social worlds.

Salivary α-Amylase Measurement Issues

Salivary α-amylase can be measured by a kinetic reaction. The assay system we have employed uses a chromagenic substrate, 2-chloro-*p*-nitrophenol, linked to maltotriose. The enzymatic action of α-amylase on this substrate yields 2-chloro-*p*-nitrophenol (yellow color), which can be spectrophotometrically measured at 405 nm using a standard laboratory plate reader (optical density). Higher levels of α-amylase in the sample will lead to the degradation of more maltotriose. Correspondingly, more yellow color will develop in the test well. The amount of α-amylase activity present in the sample is directly proportional to the increase (over a 2- min period) in absorbance (optical density) at 405 nm.

Assay Protocol

Saliva samples (10 μL) are diluted 1:200 in assay diluent and well mixed. Eight microliters of diluted sample or control are then pipetted into individual wells of a 96-well microtiter plate. Three hundred twenty microliters of preheated (37°C) α-amylase substrate solution (2-chloro-*p*-nitrophenol, linked to maltotriose) is added to each well and the plate is rotated at 500–600 rpm at 37°C for 3 min. Optical density (read at 405 nm) is determined exactly at the 1-min mark and again at the 3-min mark. Results are computed in U/mL of α-amylase using the formula: (Absorbance difference per minute × total assay volume [328 mL] × dilution factor [200])/ (millimolar absorptivity of 2-chloro-*p*-nitrophenol ([12.9] × sample volume [0.008 mL] × light path [0.97]).

Data Analysis

Salivary α-amylase distributions are typically leptokuric and positively skewed. In our collective experience, square root transformation (but not log transform) is sufficient to correct salivary α-amylase distributions. Even after transformation, there are wide-ranging differences between individuals in salivary α-amylase levels (range 400 to 900 U/mL). As noted above, salivary α-amylase levels are sensitive to change as a result of environmental and psychological challenge. However, within individuals, individual differences in α-amylase levels are relatively stable over time.[38]

Screening and Exclusionary Criterion

As noted above, use of any prescription or over-the-counter medication (OTC) with potential to influence the para- or sympathetic nervous system (adrenergic agonists or antagonists) should be avoided. These medications include those used to control high blood pressure and angina. Similarly, OTC use of supplements with AAI properties should be cause for exclusion. Nicotine use is associated with activation of the SNS,[58] raising the possibility that tobacco smoke exposure would be positively related to salivary α-amylase. However, studies show that the highly acidic aldehydes in tobacco smoke inactivate the α-amylase enzyme.[59] Thus, contrary to the activation effects of nicotine on the SNS, exposure to tobacco smoke is associated with lower (not higher) salivary α-amylase activity. Nicotine use (e.g., gum, patch, water) and tobacco smoke exposure should be avoided or controlled in studies involving α-amylase. The findings of Klein *et al.*, (2006)[47] suggest that caffeine doses equivalent to 1–2 8-oz cups of coffee (~200 mg) should also be of concern. Caffeine is an ingredient in many teas, sodas, waters, juices, OTC medications (e.g., NoDoz, Excedrin, Anacin) as well as most chocolate- and coffee-flavored foods. However, dosages vary widely, and for most food items, caffeine levels in single servings of foods greater than 100 mg are rare. In general, we recommend that use of caffeine should be monitored in studies measuring salivary α-amylase. Theoretically, because of the role of salivary α-amylase in the digestion of carbohydrates and starches, increased salivary α-amylase activity may be found in saliva samples collected after consumption of a high carbohydrate/starch meal.

Sample Collection, Handling, and Storage

The welcome news is that the assay of salivary α-amylase requires only 10 μL of sample (less than one-sixth of an eye-dropper drop of saliva). The most common saliva collection methods typically involve absorbing sample using cotton-based products. While this method seems appropriate for some markers (i.e., cotinine), it causes substantial interference in the assay of many others including sIgA, dehydroepiandrosterone, testosterone, estradiol, and progesterone.[60] Our preliminary data reveal that saliva samples to be assayed for α-amylase can be collected by passive drool, cotton swabs, or microsponge without compromising assay validity. Moreover, multiple freeze–thaw cycles do not have significant effects on the assay of salivary α-amylase, and samples can be stored for at least 24 h at room (RT) temperature (as if samples were mailed unrefrigerated or left on the bench in the lab inadvertently) or 4° C (as if samples were kept on ice in the mail or left in the lab refrigerated), and indefinitely at –80°C (as if samples were stored under more ideal laboratory conditions) without compromising the integrity of salivary α-amylase measurements.

Summary

Reagents are available so investigators can construct their own in-house assay and they are also commercially available in kit form, and most recently in point-of-care formats. Saliva samples to be assayed for α-amylase may be conveniently collected using the most commonly employed methods— passive drool, cotton, or hydrocellulose absorbent materials. At least in the short run (24 h), the effects of storage and transport temperatures on salivary α-amylase levels are negligible.

CONCLUDING COMMENTS

A small, but growing literature reveals age-, gender-, and stress-related differences in salivary α-amylase levels, patterns of intraindividual salivary α-amylase change in response to challenge that distinctly differs from those measured by salivary cortisol. In addition, findings suggest associations between salivary α-amylase levels and social behavior and relationships, health, negative affectivity, cognitive/academic problems, and cardiovascular activity. These findings underscore that integration of noninvasive measurements of the adrenergic component of the locus ceruleus/autonomic (sympathetic) nervous system, as indexed by salivary α-amylase, may extend our understanding of health-related biobehavioral phenomena.

Across studies a robust pattern emerges suggesting dissociation between measures of salivary cortisol and α-amylase reactivity to challenge. The overwhelming pattern extends Chatterton's conclusion that α-amylase is not responding to a stress signal related to the HPA axis. That these measures are not redundant implies that the inclusion of both salivary cortisol and α-amylase in biosocial studies has potential to improve our explanation of individual differences in stress-related vulnerability and resilience. Future studies that examine additive or interactive effects of cortisol and α-amylase in biosocial models seem well justified. Further, studies aimed to delineate features (e.g., frequency, intensity, duration) of social stressors that activate salivary cortisol only, salivary α-amylase only, or both, would seem particularly worthwhile.

A number of scattered findings here suggest links between individual differences in salivary α-amylase and health. With respect to implications for health-related research, several findings are particularly noteworthy. For example, associations emerged between α-amylase and illness susceptibility. Generally, the finding is consistent with volumes of research on the linkages between the brain, behavior, and immunity.[61] More specifically, stress-related increases in SNS activation have been associated with immune suppression. Deductive logic suggests that compromised immune function subsequently leads to increase susceptibility to negative health outcomes. The invasive nature of the measurements (venipuncture) needed to study relationships among the brain, behavior, and immunity has significantly slowed the study of these relationships in youth and special populations, and in the context of everyday social life. In addition, the earlier literature suggests a link between salivary α-amylase and eating behavior, and recent preliminary findings revealed an association with body mass index in adolescents.[18] Stress-related increases in salivary α-amylase consistently emerged across studies among adolescents in response to social challenge (TSST). Stress-related individual differences in salivary α-amylase, combined with high carbohydrate/starch diet and sedentary lifestyle, could potentially contribute to the weight-related disease epidemics facing many youth in Westernized societies. Finally, the numerous associations between individual differences and with change in salivary α-amylase and standard measures of cardiovascular psychophysiology beg questions regarding the utility of this noninvasive measure in research aimed to prevent, monitor, or remedy cardiovascular disease-related symptoms.

In conclusion, research on salivary α-amylase is burgeoning, but in-depth study is in the beginning stages. Many of the "representative findings" noted are from work that has been recently, or will soon be, presented at scientific meetings or is in the process of peer review. Thus, while we remain optimistic, we are cautious about drawing any conclusions from available empirical evidence. In particular, a very important direction for future research is to add to our knowledge regarding what individual differences in salivary α-amylase reactivity to stress are actually measuring.

ACKNOWLEDGMENTS

Components of the research were supported in part by the Behavioral Endocrinology Laboratory and the Child Youth and Families Consortium at the Pennsylvania State University, as well as the National Institute of Child Health and Development (grants PO1HD39667-01A1 and K23HD041428), National Science Foundation (grant 0126584), Alabama Agricultural Experiment Station (ALA010-008), and a Lindsey Foundation grant. Thanks are due to our colleagues who made much of the preliminary work possible: Joe Buckhalt, Kathryn Hand-werger, Leah Hibel, Laura Klein, Jared Lisonbee, Jackie Mize, Liz Susman, and Sheila West, and to Mary Curran, Becky Hamilton, Vincent Nelson, and Eve Schwartz for biotechnical support.

REFERENCES

1. Malamud, D. & L. Tabak. 1993. Saliva as a Diagnostic Fluid. Ann. N.Y. Acad. Sci. Vol. 694.
2. Kirschbaum, C. & D.H. Hellhammer. 1989. Salivary cortisol in psychobiological research: an overview. Neuropsychobiology **22:** 150–169.
3. Kirschbaum, C. & D.H. Hellhammer. 1994. Salivary cortisol in psychoneuroendocrine research: recent developments and applications. Psychoneuroendocrinology **19:** 313–333.
4. Bauer, A.M., J.A. Quas & W.T. Boyce. 2002. Associations between physiological reactivity and children's behavior: advantages of a multisystem approach. J. Dev. Behav. Pediatr. **23:** 102–113.
5. Donzella, B. *et al.* 2000. Cortisol and vagal tone responses to competitive challenge in preschoolers: associations with temperament. Dev. Psychobiol. **37:** 209–220.
6. Granger, D.A. & K.T. Kivlighan. 2003. Integrating biological, behavioral, and social levels of analysis in early child development research: progress, problems, and prospects. Child Dev. **74:** 1058–1063.
7. Lovallo, W.R. & T.L. Thomas. 2000. Stress hormones in psychophysiological research: emotional, behavioral, and cognitive implications. *In* Handbook of Psychophysiology. Second edition. J.T. Cacioppo, L.G. Tassinary & G.G. Bernston, Eds.: 342–367. Cambridge University Press. New York.
8. Chrousos, G.P. & P.W. Gold. 1992. The concepts of stress and stress system disorders. JAMA **267:** 1244–1252.

9. Cacioppo, J.T., L.G. Tassinary & G.G. Bernston. 2000. Handbook of Psychophysiology. Second edition. Cambridge University Press. New York.
10. Doussard-Roosevelt, J.A., L.A. Montgomery & S.W. Porges. 2003. Short-term stability of physiological measures in kindergarten children: respiratory sinus arrhythmia, heart period, and cortisol. Dev. Psychobiol. **43:** 230–242.
11. Quas, J.A. *et al.* 2000. Dissociations between psychobiologic reactivity and emotional expression in children. Dev. Psychobiol. **37:** 153–175.
12. Engel, B.T. 1972. Response specificity. *In* Handbook of Psychophysiology. N.S. Greenfield & R.A. Sternbach, Eds.: 571–576. Holt, Rinehart & Winston. New York.
13. Schwab, K.O., G. Heubel & H. Bartels. 1992. Free epinephrine, norepinephrine, and dopamine in saliva and plasma of healthy subjects. *In* Assessment of Hormones and Drugs in Saliva in Biobehavioral Research. C. Kirschbaum, G.F. Read & D.H. Hellhammer, Eds.: 331–336. Hogrefe & Huber. Gottingen, Germany.
14. Nedefors, T. & C. Dahlof. 1992. Effects of the beta-adrenoceptor antagonists atenolol and propranolol on human whole saliva flow rate and composition. Arch. Oral Biol. **37:** 579–584.
15. Speirs, R.L. *et al.* 1974. The influence of sympathetic activity and isoprenaline on the secretion of amylase from the human parotid gland. Arch. Oral Biol. **19:** 747–752.
16. Kirschbaum, C., G.F. Read & D.H. Hellhammer. 1992. Assessment of Hormones and Drugs in Saliva in Biobehavioral Research. Hogrefe & Huber. Gottingen, Germany.
17. Kataoka, K. & E.P. Dimagno. 1999. Effect of prolonged intraluminal alpha-amylase inhibition on eating, weight, and the small intestine of rats. Nutrition **15:** 123–129.
18. Susman, E.J., D.A. Granger, S. Dockray, et al. 2006. Alpha amylase, timing of puberty and disruptive behavior in young adolescents: a test of the attenuation hypothesis. Presented at the biennial meeting of the Society for Research on Adolescence, San Francisco, CA, March.
19. Marcotte, H. & M.C. Lavoie. 1998. Oral microbial ecology and the role of salivary immunoglobulin A. Microbiol. Molec. Biol. Rev. **62:** 71–109.
20. Quinonez, R.B. *et al.* 2001. Early childhood caries: analysis of psychosocial and biological factors in a high-risk population. Caries Res. **35:** 376–383.
21. O'Donnell, M.D. & N.J. Miller. 1980. Plasma pancreatic and salivary-type amylase and immunoreactive trypsin concentrations: variations with age and reference ranges for children. Clin. Chem. Acta **104:** 265–273.
22. El-Sheikh, M., J. Mize & D.A. Granger. 2005. Endocrine and parasympathetic responses to stress predict child adjustment, physical health, and cognitive functioning. Presented at the Biennial meeting of Society for Research in Child Development. Altanta, GA, March.
23. Chatterton, R.T. *et al.* 1996. Salivary alpha-amylase as a measure of endogenous adrenergic activity. Clin. Physiol. **16:** 433–448.
24. Chatterton, R.T. *et al.* 1997. Hormonal responses to psychological stress in men preparing for skydiving. J. Clin. Endocrinol. Metab. **82:** 2503–2509.
25. Skosnik, P.D. *et al.* 2000. Modulation of attentional inhibition by norephinephrine and cortisol after psychological stress. Int. J. Psychophysiol. **36:** 59–68.

26. GORDIS, E.B. *et al.* 2006. Asymmetry between salivary cortisol and alpha-amylase reactivity to stress: relation to aggressive behavior in adolescents. Psychoneuroendocrinology **31:** 976–987.
27. NATER, U.M. *et al.* 2005. Human salivary alpha-amylase reactivity in a psychosocial stress paradigm. Int. J. Psychophysiol. **55:** 333–342.
28. NATER, U.M. *et al.* 2006. Stress-induced changes in human salivary alpha-amylase activity—associations with adrenergic activity. Psychoneuroendocrinology **31:** 49–58.
29. STROUD, L.R. *et al.* 2006. Saliva alpha-amylase stress reactivity in children and adolescents: validity, associations with cortisol, and links behavior. Presented at the Annual Meeting of the American Psychosomatic Society. Denver, CO, March.
30. VAN STEGEREN, A. *et al.* 2006. Salivary alpha-amylase as marker for adrenergic activity during stress: effect of betablockade. Psychoneuroendocrinology **31:** 137–141.
31. KIVLIGHAN, K.T. & D.A. GRANGER. 2006. Salivary alpha-amylase response to competition: relation to gender, previous experience, and attitudes. Psychoneuroendocrinology **31:** 703–714.
32. STROUD, L.R. *et al.* 2006. Alpha amylase responses to achievement and interpersonal stressors over adolescence: developmental differences and associations with cortisol and cardiovascular responses. Presented at the Biennial Meeting of the Society for Research on Adolescence. San Francisco, CA, March.
33. ROHLEDER, N. *et al.* 2004. Psychosocial stress-induced activation of salivary alpha-amylase: an indicator of sympathetic activity? Ann. N.Y. Acad. Sci. **1032:** 258–263.
34. GALLACHER, D.V. & O.H. PETERSEN. 1983. Stimulus-secretion coupling in mammalian salivary glands. Int. Rev. Physiol. **28:** 1–52.
35. DICKERSON, S.S. & M.E. KEMENY. 2004. Acute stressors and cortisol responses: a theoretical integration and synthesis of laboratory research. Psychol. Bull. **130:** 355–391.
36. KIVLIGHAN, K.T. *et al.* 2005. Salivary alpha amylase and cortisol: levels and stress reactivity in 6-month old infants and their mothers. Presented at the Biennial Meeting of Society for Research in Child Development. Altanta, GA, March.
37. GORDIS, E.B. *et al.* 2006. Salivary alpha-amylase and cortisol responses to social stress among maltreated and comparison youth. Presented at the Biennial Meeting of the Society for Research on Adolescence. San Francisco, CA, March.
38. GRANGER, D.A. *et al.* 2006. Integrating the measurement of salivary α-amylase into studies of child health, development, and social relationships. J. Pers. Soc. Relationships. Special Issue: Physiology and Human Relationships. **23:** 267–290.
39. KVETNANSKY, R. *et al.* 1993. Endogenous glucocorticoids restrain catecholamine synthesis and release at rest and during immobilization stress in rats. Endocrinology **133:** 1411–1419.
40. GOLCZYNKA, A., J.W. LENDERS & D.S. GOLDETSEIN. 1995. Glucocorticoid-induced sympathoinhibition in humans. Clin. Pharmacol. Ther. **58:** 90–98.
41. FRIES, E. *et al.* 2005. A new view on hypocortisolism. Psychoneuroendocrinology **30:** 1010–1016.

42. HENRY, J.P. 1992. Biological basis of the stress response. Int. Physiol. Behav. Sci. **27:** 66–83.
43. FRANKENHAEUSER, M. 1982. Challenge-control interaction as reflected in sympathetic-adrenal and pituitary-adrenal activity: comparison between the sexes. Scand. J. Psychol. Suppl. **1:** 158–164.
44. BOSCH, J.A. *et al.* 2003. Innate secretory immunity in response to laboratory stressors that evoke distinct patterns of cardiac autonomic activity. Psychosom. Med. **65:** 245–258.
45. WEST, S.G. *et al.* 2006. Salivary alpha-amylase response to the cold pressor is correlated with cardiac markers of sympathetic activation. Presented at the Annual Meeting of the American Psychosomatic Society. Denver, CO, March.
46. BURNS, J.W. *et al.* 1992. Test-retest reliability of inotropic and chronotropic measures of cardiac reactivity. Int. J. Psychophysiol. 12: 165–168.
47. KLEIN, L.C. *et al.* 2006. Effects of caffeine and stress on salivary alpha-amylase in young men: a salivary biomarker of sympathetic activity. Presented at the Annual Meeting of the American Psychosomatic Society. Denver, CO, March.
48. KIVLIGHAN, K.T. *et al.* 2006. Salivary alpha amylase reactivity to the Trier Social Stress Test: relation to cortisol and autonomic responses in normally developing adolescents. Unpublished manuscript.
49. SHEA, A.K. *et al.* 2006. Maternal depression and salivary alpha amylase response to stress in their infants. Presented at the Annual Meeting of the American Psychosomatic Society. Denver, CO, March.
50. HILL, A.L. *et al.* 2006. Physiological responses to Strange Situation as a function of attachment status: vagal withdrawal, salivary alpha-amylase and salivary cortisol. Unpublished manuscript.
51. MIZE, J., J. LISONBEE & D.A. GRANGER. 2005. Stress in child care: cortisol and alpha-amylase may reflect different components of the stress response. Presented at the Biennial Meeting of Society for Research in Child Development. Altanta, GA, March.
52. EL-SHEIKH, M. *et al.* 2006. Asymmetry between salivary alpha-amylase and cortisol reactivity to stress and children's adjustment. Unpublished manuscript.
53. BUCKHALT, J.A., M. EL-SHEIKH & D.A. GRANGER. 2006. Children's cognitive functioning and academic performance: the role of cortisol and alpha-amylase. Unpublished manuscript.
54. LYONS-RUTH, K., M.A. EASTERBROOKS & C.D. CIBELLI. 1997. Infant attachment strategies, infant mental lag, and maternal depressive symptoms: predictors of internalizing and externalizing problems at age 7. Dev. Psychol. **33:** 681–692.
55. LUNDBERG, U. & M. FRANKENHAEUSER. 1980. Pituitary-adrenal and sympathetic-adrenal correlates of distress and effort. J. Psychosom. Res. **24:** 125–130.
56. RAUSTE-VON WRIGHT, M., J. VON WRIGHT & M. FRANKENHAUSER. 1981. Relationships between sex-related psychological characteristics during adolescence and catecholamine excretion during achievement stress. Psychophysiology **18:** 362–370.
57. NATER, U.M. *et al.* 2006. Determinants of diurnal course of salivary alpha-amylase activity. Presented at the Annual Meeting of the American Psychosomatic Society. Denver, CO, March.
58. GRASSI, G. *et al.* 1994. Mechanisms responsible for sympathetic activation by cigarette smoking in humans. Circulation **90:** 248–253.

59. NAGLER, R. *et al.* 2000. Effect of cigarette smoke on salivary proteins and enzyme activities. Arch. Biochem. Biophys. **379:** 229–236.
60. SHIRTCLIFF, E.A. *et al.* 2001. Use of salivary biomarkers in biobehavioral research: cotton based sample collection methods can interfere with salivary immunoassay results. Psychoneuroendocrinology **26:** 165–173.
61. ADER, R., N. COHEN & D. FELTEN. 1995. Psychoneuroimmunology. Academic Press. San Diego, CA.

The Use of Oral Fluid for Therapeutic Drug Management

Clinical and Forensic Toxicology

LORALIE J. LANGMAN

Division of Clinical Biochemistry and Immunology, Department of Laboratory Medicine and Pathology, Mayo Clinic, Rochester, Minnesota, USA

ABSTRACT: One of the underlying tenets of clinical pharmacology is that only free drugs are pharmacologically active. It is thought that only free drugs can cross biological membranes to interact with a given receptor to alter its function, and that drug responses, both efficacious and toxic, are a function of unbound concentrations. The rationale for measuring drugs in oral fluid is that the free fraction of a drug in plasma reaches equilibrium with the drug in saliva. Although reports concerning the appearance of organic solutes in saliva have been in the literature for over 70 years, it has only been in the past 30 years that there has been emphasis on the appearance of drugs. Although many assumptions for drug level monitoring in saliva are made, the primary requisite for salivary monitoring to be useful is a constant or predictable relationship between the drug concentration in saliva and the drug concentration in plasma. Measurement of oral fluid drug levels for the purpose of managing patients and making dosage adjustments may be useful for select drugs or drug classes. However, it does not appear to be useful for the majority of drugs therapeutically monitored. Some work with antipsychotic medications has indicated that although the measurement of drug concentrations themselves may not be useful for dosage adjustment, the ratio of parent drug to metabolite may reflect altered metabolic status due to either pharmacogenetic variation or other clinical conditions. Furthermore, analysis of saliva may provide a cost-effective approach for the screening of large populations.

KEYWORDS: oral fluid; free drugs; drug management; unbound drugs

Address for correspondence: Dr. Loralie Langman, Division of Clinical Biochemistry and Immunology, Department of Laboratory Medicine and Pathology, Mayo Clinic, Hilton 730, 200 First Street SW, Rochester, MN 55905. Voice: 507-284-8408; fax: 507-284-9758.
Langman.loralie@mayo.edu

Ann. N.Y. Acad. Sci. 1098: 145–166 (2007).
doi: 10.1196/annals.1384.001

INTRODUCTION

The collection of biological samples for the purpose of determining exposure to various agents is dominated by breath, blood, and urine. Each of these matrices has advantages and disadvantages with respect to ease of collection, specificity, and sensitivity to drug concentrations detected in the matrix, and interpretative value. As an alternative matrix, oral fluid has been investigated as well.

Reports concerning the appearance of organic solutes in saliva have been in the scientific literature for over 70 years.[1] One of the earliest systematic studies on the permeability of salivary glands to organic nonelectrolytes was published in 1932.[2] It demonstrated a correlation between the ability of nonelectrolytes to penetrate into saliva and their lipid solubility. It also demonstrated that compounds with low molecular weight can appear in saliva even if they are not lipid soluble. The first look at saliva testing for the presence of drugs was in 1938 with Friedemann's work investigating the excretion of ingested alcohol in saliva.[3] But it has only been in the past 30 years that there has been emphasis on the appearance of other drugs in saliva, such that more than 100 drugs have been evaluated for salivary therapeutic drug monitoring.[4]

Most studies on saliva in humans utilize whole saliva. The term "oral fluid" is preferred for the specimen collected from the mouth. Oral fluid is a complex fluid consisting not only of the secretions from the three major pairs of salivary glands (parotid, submandibular, and sublingual), but also containing secretions from the minor glands (labial, buccal, and palatal), bacteria, sloughed epithelial cells, gingival fluid, food debris, and other particulate matter.[5] The concentration of drug from each secretion and the relative contributions of the various glands to the final fluid may vary.[1]

There are several advantages to monitoring oral fluid as contrasted with monitoring plasma or serum levels. Collection of oral fluid is considered to be a noninvasive procedure and risks associated with the drawing of blood are avoided. Furthermore, the subject's fear, anxiety, and discomfort that may accompany the drawing of blood are diminished and it is easier to obtain multiple samples of saliva than of blood. There are occasions when it might not be possible to obtain multiple blood samples, but yet possible to obtain oral fluid samples. Drawing blood requires the expertise of a professional. Although some training and explanation are necessary to assure proper gathering of oral fluid samples, it does not require the level of training needed for blood sampling. Despite these advantages, monitoring drug concentration in oral fluid has been accepted clinically for only a few pharmacological agents. The principal reason for the lack of acceptance is that for most drugs there is a poor correlation between oral fluid and plasma concentrations for many substances.

Clearly, before any drug circulating in plasma can enter the oral fluid it must pass through the capillary wall, the basement membrane, and the membrane of the salivary gland epithelial cells. However, this fluid is not a simple ultrafiltrate of plasma, as has sometimes been suggested, but rather a complex fluid formed by different mechanisms including ultrafiltration through pores in the membrane, active transport against a concentration gradient, and passive diffusion.[5] Ultrafiltration allows for small polar molecules of less than about 300 Da to enter into saliva.[6,7] An active transport mechanism clearly operates for many electrolytes and for some proteins such as IgA. This mechanism has also been proven for some drugs. Because of lithium's (Li) small size (MW = 7 Da), it would be expected to appear in saliva by ultrafiltration. However, the findings of an saliva/plasma (S/P) ratio of more than 2 could indicate an active secretory mechanism or ion trapping.[8,9] Passive diffusion across the lipid bilayer is thought to be the dominant process and the rate-determining step in the passage of drugs through the lipophilic layer of the epithelial membrane. Diffusion across these membranes requires the molecule to be nonionized,[10] lipid-soluble (expressed by its octanol/water partition coefficient),[11] and predominantly unbound.[1,12] A detailed discussion of the factors affecting diffusion of drugs into oral fluid is given in other extensive reviews,[10,13–15] but the most significant factors are: salivary flow rate,[6,16,17] the fraction of protein binding in saliva and plasma, the $pK_{a,}$ or $pK_{b,}$ of the drug, and the pH of the saliva and plasma.[14,15]

Plasma contains both unbound (free) and bound drug, and it is the free drug that is generally considered to be pharmacologically active.[18] Because this free fraction excreted by the salivary gland reaches a saliva-plasma equilibrium, oral fluid drug concentrations are thought to reflect free drug plasma concentrations.[19–23] Thus, in clinical conditions in which protein binding varies, the oral fluid concentration may be more closely related to the therapeutically active drug concentration than in the plasma concentration. Also, in circumstances where the concurrent use of two or more drugs may alter drug binding, the oral fluid concentration may reflect the plasma free drug concentration.

The usefulness of oral fluid therapeutic drug monitoring (therapeutic drug management [TDM]) requires many assumptions: that the drug and metabolites are excreted in saliva; the majority of drug is free or unbound in saliva; saliva pH effects are minimal; concurrence exists for drug concentration in saliva and plasma concentration and clinical effect; the S/P drug ratio is invariate with regard to single dose and steady-state estimates; pharmacokinetic parameters are similar to those of plasma; age and physiologic or disease states do not change the fraction of drug in plasma or the concentration in saliva; concomitantly administered medications have minimal effect on saliva concentrations; the drug is stable in saliva; and that fractions and whole saliva contain similar drug concentrations.[24] However, it could be considered that the primary requisite for salivary monitoring to be useful is a constant relationship between plasma or serum concentrations and oral fluid levels over the required concentration range.[19,25] Many studies correlate saliva with plasma drug concentration such that efficacy relationships are extrapolated from those determined for plasma. The underlying pitfall for many such extrapolations is that the salivary to plasma drug concentration relationship may not comply with these assumptions.

SALIVA THERAPEUTIC DRUG MANAGEMENT APPLIED TO DRUG CLASSES

Ethanol and Drugs of Abuse

Ethanol was the first drug to be investigated in oral fluid.[3] Since then many additional studies have expanded the knowledge of this drug in saliva. Ethanol appears to reach a higher peak concentration in saliva than in peripheral blood. Because the distribution of ethanol in the body is considered to occur by passive diffusion, under equilibrium conditions, ethanol content will be dependent upon the water content of the fluid or tissue being measured. The content in saliva will therefore be higher than that found in blood or serum. On a theoretical basis, the saliva to blood (S/B) ethanol ratio should be 1.17; however, lower ratios have been found in the post absorption phase.[26,27] Although S/B ratios appear to decrease over time, corrections for differences in the water content of samples indicated that S/B ratios during the elimination phase were constant. In a study comparing ethanol concentrations in saliva versus capillary and venous blood, ethanol concentrations in saliva were found to parallel capillary blood more accurately than venous blood concentrations.[28] This phenomenon has also been described for other drugs.[29] Assuming that capillary blood was a better measure of ethanol exposure to brain cells, the authors concluded that saliva was at least as well suited as venous or capillary blood to reflect the intoxication state of an individual in the post absorption state.

Oral fluid is a beneficial tool for monitoring use of illicit substances in a variety of circumstances including pre-employment random drug testing, post accident, reasonable suspicion, return to duty, and follow-up. Unlike urine, oral fluid has the potential to show a relation between behavior/impairment and drug concentration, making it a suitable medium for monitoring drug intoxication.[88] Specimens can be monitored for cannabinoids, cocaine, opiates, amphetamines, phencyclidine, methadone, barbiturates, and benzodiazepines. The use and interpretation of the presence and concentrations of these drugs in saliva has been extensively reviewed elsewhere.[27,30–34]

Anticonvulsants

TDM of anticonvulsants received early attention and has been evaluated extensively in the treatment of seizure disorders because these drugs are used chronically; have a defined plasma therapeutic range; have a measurable and temporally coupled pharmacodynamic end point (i.e., seizures); show marked interindividual dose requirements; prediction of dose adjustments can be made from pharmacokinetic characteristics; and, most importantly, seizure control is improved when the active drug species is measured[35] and when serum data are used appropriately.[22,36,37] If oral fluid levels correspond sufficiently to levels in plasma, then a useful relationship to seizure control should be demonstrated.[38] The S/P ratio for commonly used anticonvulsant medications has been determined, with phenytoin (PTN), carbamazepine (CBZ), primidone, and ethosuximide considered potentially good candidates for therapeutic oral fluid monitoring under well-controlled standardized conditions.[39,40]

Early work emphasized a relationship between unbound drug in plasma and anticonvulsant efficacy or toxicity.[41,42] Some researchers[43] found that CBZ and PTN saliva concentrations were related to toxicity, but not to efficacy. When monotherapy was used, saliva and plasma concentrations of both PTN and CBZ showed marked interindividual variability and a respective mean S/P ratio of 0.26 and 0.13, and both drugs showed good saliva-to-plasma correlations.[44] Another study[45] found mean ratios of 0.09 for PTN, 0.33 for CBZ, and also 0.31 for phenobarbital. Both studies[44,45] found ratios similar to those reported previously.[46–49] The unbound to total serum concentration ratios were 0.08, 0.23, and 0.47, respectively.[45] The intraindividual variability was low for both the saliva/total plasma and the unbound plasma/total plasma drug concentration ratios.[45] Similar findings from previous studies[45,50] and especially those saliva versus serum correlation data reported recently[51] support the use of saliva for monitoring of these anticonvulsants.

Stable S/P ratio of PTN becomes less reliable when the drug is prescribed in conjunction with valproic acid (VPA). A drug interaction appears to be responsible for increasing PTN oral fluid concentrations (through diffusion of excess free PTN in plasma) and decreasing plasma PTN concentrations (through a decrease in protein-bound PTN).[52] Furthermore, PTN oral fluid concentrations increase when oral fluid volume is reduced, a fact that should be taken into consideration when using oral fluid for PTN therapeutic drug monitoring. Reduced oral fluid flow following atropine therapy caused an increase in PTN oral fluid concentration with no change in plasma concentration.[40] Because of a poor correlation between oral fluid and plasma concentrations and low oral fluid concentration due to high protein binding in plasma, the utility of VPA oral fluid TDM is uncertain.[53]

CBZ concentrations in oral fluid ranged from 0.1 to 5.5 mg/L, whereas free CBZ concentrations in plasma ranged from 0.7 to 4.2 mg/L.[39] S/P ratios were highly correlated (range 0.84–0.99) for CBZ and for its metabolite, CBZ epoxide. One study examined free salivary concentrations of CBZ over the therapeutic range and found a significant correlation between serum and oral fluid concentrations and a significant correlation between dose and free oral fluid concentrations.[54]

Clinical usefulness of frequent TDM with whole saliva was tested over a period of several weeks in an adult receiving CBZ to which methylphenobarbital was then added.[55] The drop in saliva CBZ and the decreasing plasma drug levels were associated with an increase in seizure frequency. Attempts to produce effective plasma levels predicted by the regression analysis of seizure frequency on saliva drug concentration gave either toxic or untolerated doses.

In one of the few reported direct approaches to use salivary TDM, seizure frequency appeared related to low anticonvulsant levels.[56] In another study,[57] a correlation of saliva-to-plasma concentration was found for both CBZ and its epoxide metabolite. Clinical effect (alleviation of mania) was not correlated with salivary drug concentration nor was there a close correspondence between effect and plasma drug concentration. Adverse effects could not be correlated with plasma levels. It would appear that drugs without a close temporal association between plasma level and effect are not candidates for saliva monitoring.

The relationships between oral fluid and plasma primidone and ethosuximide concentrations have also been examined. Primidone has an S/P ratio of approximately 1, in part due to its limited protein binding, making oral fluid a good matrix for its therapeutic drug monitoring.[58] Ethosuximide is not highly protein bound, has a good oral-fluid-to-serum correlation, and thus, oral fluid monitoring is considered potentially useful.[1,47,59,60]

Oral fluid testing may be appropriate for therapeutic drug monitoring of the new generation of antiepileptic drugs as well. Levetiracetam and topiramate concentrations were found to be highly correlated for unstimulated oral fluid and serum collections.[61,62] The mean S/P concentration ratio was approximately 0.4 and the correlation coefficient was approximately 0.87.[61] Lamotrigine (LAMO) monitoring of oral fluid also appears promising, but concentrations were found to be dependent upon the oral fluid flow rate. In one study the S/P ration was 0.46,[63] and in another 0.62.[64] Stimulated oral fluid collections produced drug concentrations that have a higher degree of correlation ($r = 0.94$) with serum than unstimulated oral fluid collections ($r = 0.85$) specimens and between total LAMO concentration in serum and the free LAMO fraction as determined by ultrafiltration ($r = 0.95$) and equilibrium dialysis ($r = 0.93$).[65] Another study found a good correlation (approximately $r = 0.8$) between unstimulated oral fluid and serum LAMO levels. However, there is wide interpatient variability in the S/P ratio, and younger patients displayed a greater degree of variability than did older patients.[64]

In patients undergoing long-term therapy,[66] concentrations of LAMO in saliva were proportional to plasma concentrations with a correlation of $r = 0.95$. The saliva concentration versus time curves for LAMO were parallel to those derived for plasma and therefore assessments of absorption, elimination and mean residence time were identical in both saliva and plasma. The S/P ratio was used to predict plasma LAMO concentrations from saliva determinations. Thus, LAMO determination in saliva could potentially provide an alternative for therapeutic drug monitoring and studying LAMO pharmacokinetics.[66]

Oxcarbazepine is rapidly reduced by cytosolic enzymes in the liver to its 10-monohydroxy metabolite (OMHC), which is primarily responsible for the pharmacological effect, and therefore is the component used as a target analyte for TDM. In one study,[67] the mean S/P OMHC concentration ratio was 0.96 ± 0.15. There was a significant positive correlation between the serum and unstimulated saliva concentrations $r = 0.941$ ($P < 0.001$). This suggests that oral fluid TDM is possible. A limitation of saliva OMHC monitoring is that individuals who have difficulty producing small quantities of saliva or who have viscous saliva should generally be avoided for this type of monitoring. It is also recommended to avoid saliva collection within 8 h after oxcarbazepine dosing to allow complete absorption and transformation of the parent drug.[67]

Benzodiazepines also can be useful therapeutic adjuncts for the treatment of epilepsy. The oral fluid concentration of diazepam reflects the unbound drug concentration in plasma, making oral fluid a useful drug-monitoring matrix.[68,69] In addition, oral fluid clobazam and nitrazepam concentrations correlated with those in plasma; however, the correlation for nitrazepam concentrations was found to be time dependent.[70] The stability of nitrazepam in oral fluid has been shown to convert into 7-aminonitrazepam at room temperature with a conversion rate that is dependent on the composition of the subject's oral fluid.[71] The disposition of benzodiazepines has been reviewed elsewhere.[27]

Bronchodilators–Theophylline

Theophylline is a methylxanthine derivative used in the treatment of asthma, bronchitis, and chronic obstructive pulmonary disease. In general, the correlation coefficients between saliva and plasma theophylline levels were high (approximately $r = 0.90$) and a S/P ratio of about 0.50 was found.[72–74] Saliva pH and saliva protein binding were not considered to be significant influences on observed variability[74] and the saliva to free plasma drug approached unity.[75] Several studies attempting to develop correlation equations to predict plasma levels from oral fluid measurement of drug concentrations have been able to predict plasma levels within 2 μg/mL.[76–78] Precision was better for the saliva to free plasma theophylline. Others found that stimulated saliva samples improved the correlation sufficiently between saliva and free fraction in serum so that saliva-based predictions of theophylline levels could be made in adults with asthma[74] or with chronic airflow obstruction.[75] In one study,[78] serum and saliva sampling concurrent with pulmonary function studies was used to estimate pharmacodynamic characteristics for theophylline in each fluid. Both serum and saliva correlated with lung function tests, but the correlation coefficient for serum was generally higher than that for saliva.[78]

Antimicrobials

Oral fluid also has been evaluated for potential use for TDM of antimicrobial medications. Concentrations of isoniazid, an antituberculosis drug, are used to assess concurrent plasma levels.[79] The excretion pattern of isoniazid was altered when ciprofloxacin was coadministered,[80] and because both isoniazid and ciprofloxacin oral fluid concentrations reflect their plasma levels, salivary monitoring of these antimicrobials could be utilized to test for drug interactions. Ofloxacin, an antibacterial used to treat Gram-positive and -negative bacteria, was found to have an S/P ratio around 0.5.[81,82] The half-lives of ofloxacin in oral fluid and plasma were similar, making ofloxacin suitable for oral fluid therapeutic drug monitoring. Gentamycin also is a good candidate for oral fluid monitoring if it is given once per day to pediatric patients; however, if the drug is administered three times daily, no correlation between oral fluid and plasma was observed.[83] Moxifloxin and clarithromycin are antibacterials used to treat upper and lower respiratory tract infections; both have been shown to have relatively constant oral fluid-serum ratios.[84]

Measurement of metronidazole concentration and its pharmacokinetics from saliva and serum were investigated.[85] The overall ratio of saliva-to-serum concentration was 0.93, and there was a significant correlation between saliva and serum concentrations. However, the correlation was better during the elimination phase ($r = 0.90$) compared to the absorption phase ($r = 0.76$). The half-life and AUC of metronidazole obtained from saliva samples was similar to that from serum. Measurement of saliva metronidazole could potentially be used to estimate pharmacokinetics.[85]

Less well studied are the antifungal and antiprotozoan drugs. A study of the antifungal drug fluconazole showed saliva concentrations were sufficient to test for compliance and even semiquantitative predictors of plasma concentrations.[86] Quinine, commonly used in the treatment of malaria, is a compound with a narrow therapeutic index, making it a good candidate for a noninvasive means of monitoring active drug concentration. It has been shown that the T_{max} and elimination half-lives ($t_{1/2}$) of quinine were similar for plasma (2.8 $\pm$ 0.2 h; 12.9 $\pm$ 2.3 h) and oral fluid (4.3 $\pm$ 0.5 h; 11.8 $\pm$ 2.9 h).[87] Even though intrasubject variability in S/P ratios was observed, the investigators suggested that mean concentrations of quinine in oral fluid were proportional to the mean levels in plasma ($r = 0.93$).

Drugs used in the treatment in HIV infections have also been studied in saliva. The concentrations of the nucleoside reverse transcriptase inhibitor, zidovudine, were determined in saliva and found to correlate ($r = 0.97$) with simultaneously collected plasma concentrations; the S/P ratio averaged 0.68.[88] The nonnucleoside reverse transcriptase inhibitors, delaverdine efavirenz and nerviapine, have also been investigated. Delavidine has been shown to have an S/P ratio of 0.06[89] and nevirapine 0.51.[90] Plasma nevirapine concentrations were strongly correlated with the salivary concentration, and the S/P concentration ratio was independent of the time after ingestion.[90] Salivary nevirapine concentrations were used to estimate the corresponding plasma concentrations and showed a bias of 4.2%, with a precision of 13.3%. Saliva could be a useful body fluid for TDM of nevirapine.[90] Indinavir, a protease inhibitor, in saliva was investigated to determine whether salivary concentrations are applicable to monitor compliance and/or predict plasma levels. Salivary indinavir concentrations showed a high correlation ($r = 0.85$, $P < 0.01$) with corresponding plasma levels. The median S/P ratio was 0.65, and was independent of the plasma concentration; however, a relation with time after ingestion was seen. It was not possible to predict plasma indinavir levels by the salivary concentrations for purposes of therapeutic drug monitoring on account of large interindividual and intraindividual variation. Nevertheless, monitoring compliance by measuring the presence of indinavir in saliva is possible.[91]

Antipyretic and Analgesic Drugs

Antipyretic and analgesic medication has also been investigated for potential monitoring in oral fluid. Acetaminophen concentrations are highly correlated in plasma and oral fluid specimens from healthy volunteers and in postoperative patients.[92,93] Paracetamol, an opioid-sparing analgesic, is used in conjunction with morphine in postoperative individuals. In one study, it was noted that some individuals did not have detectable paracetamol in oral fluid 2 h post oral dosing, perhaps because of delayed gastric emptying; yet, overall, there was an adequate correlation between oral fluid and plasma levels.[94] Oral fluid concentrations peaked earlier and were higher than those found in plasma.

TDM is not often applied to the NSAIDs (nonsteroidal anti-inflammatory drugs), possibly because the precise relationships between temporal aspects of the plasma level and effect have not been established.[95] Approaches have been developed to relate salicylate levels in saliva and plasma,[96] but these have not been applied pharmacodynamically. Ibuprofen was not detected in saliva by using a highly sensitive method,[97] and these findings are in agreement with a low distribution in breast milk for drugs of this class.[98]

Antineoplastic Agents

Antineoplastic drugs can alter the physiology of the oral cavity by reducing oral fluid production or altering the concentrations of oral fluid constituents.[99] Methotrexate concentrations in oral fluid and plasma are not highly correlated. This has been attributed to a physiological barrier to the secretion of methotrexate into oral fluid, a decrease in oral fluid production, or the rapid removal of methotrexate from oral fluid.[99] Doxorubicin, on the other hand, appears to concentrate in oral fluid, with S/P ratios of 2.8 for the free plasma fraction.[100]

Taxol, a highly protein-bound chemotherapeutic agent, has an oral fluid concentration that parallels that found in plasma.[100,101] One of the main anticancer drugs, 5-fluorouracil, is used to treat colorectal cancer. Its pharmacokinetics have been evaluated in plasma and oral fluid in cancer patients.[102] It was found that the areas under the curve (AUC) for plasma (AUCp) and saliva (AUCs) did not correlate with the toxicity experienced. In addition, the AUCs did not correlate with the AUCp, implying that oral fluid concentrations may be a poor predictor of concentrations in plasma, and there appears to be a large inter-subject variability in plasma and oral fluid concentrations of 5-fluorouracil.[103] Drug concentrations in oral fluid were explained with a biexponential half-life, with an initial rapid elimination half-life of 8 min and a terminal elimination half-life of 8 h. Cisplatin, a frequently used anticancer drug, has also been assayed in oral fluid, yet no defined correlation has been found between oral fluid and serum levels.[104,105] Serum was found to be unreliable in measuring carboplatin concentrations.[106] Topotecan (S/P ratio average of 2.3) has been shown to have similar oral fluid and plasma pharmacokinetic profiles, but a large degree of interindividual variability (80 m). It was been recommended that if monitoring topotecan using oral fluid, the first time point should include simultaneous collection of an oral fluid and plasma specimen, then only oral fluid need be collected for additional monitoring.[107] This methodology could help take into account the interindividual differences.

Antiarrhythmics

Several cardioactive drugs have been evaluated for their usefulness in TDM using oral fluid samples. In a study evaluating the efficacy of oral fluid therapeutic drug monitoring of digoxin in children, investigators found no correlation between plasma total or free digoxin levels and oral fluid concentrations, indicating that oral fluid testing is not an acceptable means of monitoring digoxin in this population.[108] In another study digoxin levels in saliva were correlated with plasma levels and with the free fraction in plasma.[109] When a constant dose of digoxin was administered orally over several days, the digoxin concentration in saliva and erythrocytes rose faster than in serum. Thus, the S/P ratio was below 1.0 after a single dose and above 1.0 in the steady state. The digoxin concentration was relatively high in unstimulated saliva and decreased with stimulation of the salivary flow rate. It is suggested that unstimulated saliva reflects the intracellular digoxin concentration and stimulated saliva reflects the free digoxin concentration of the serum. Both effects must be taken into account when interpreting the S/P ratio, and they may explain conflicting results in the literature.[110]

After repeated dosing with quinidine,[65] it was found that the S/P ratio depended on the dose number and on characteristics of the individual patient. Ventricular premature beat suppression was not related to either the saliva or plasma level of quinidine. Others found that the initial correlation between saliva and plasma quinidine levels was not sustained with repeated doses such that saliva monitoring was not useful under steady-state conditions.[111] Saliva levels of lidocaine were not found useful in patients with acute myocardial ischemia.[112]

TDM for disopyramide concentrations in oral fluid and plasma also has been investigated. Healthy volunteers were given a single dose of disopyramide followed by collection of plasma and stimulated oral fluid.[113] The concentrations of disopyramide, analyzed by enzyme-multiplied immunoassay technique, were lower in oral fluid as compared to plasma with S/P ratios between 0.09 and 0.63. The S/P ratio was found to increase as the pH of oral fluid decreased, reducing the effectiveness of oral fluid monitoring for this antiarrhythmic drug.

Psychoactive Drugs

Lithium (Li) differs from many other of the substances previously discussed as it is as efficacious as an ion. Li appears to have an inverse relationship to the rate of flow of saliva,[114] and was found to be actively secreted into the saliva.[115,116] This results in concentrations above serum levels. It was found to be due to an enzyme resembling Na^+ K^+-ATPase and Li secretion was more similar to K^+ secretion than Na^+ secretion.[115,116] There was no evidence of Li storage by the gland because salivary concentrations were responsive to changes in serum concentration, and salivary Li concentrations never decreased to serum level, supporting that it is actively transported into saliva.[114] Because secretion is by active transport, as plasma concentrations increase, salivary concentration should reach a maximum when protein carriers are saturated. Therefore, it is important to ascertain at what plasma level saturation occurs if salivary levels are to be used to indicate potential toxic plasma levels.[117,118]

Multiple studies of Li in saliva have tested its relationship to plasma levels and clinical effects. Most find a wide variation or poor correlation between saliva and plasma levels. Li serum and oral fluid concentrations for children aged 5–12 years ($n = 30$) were significantly correlated ($r = 0.78$, $P < 0.001$) at the optimal dose and overall ($r = 0.83$, $P < 0.001$), but oral fluid to serum ratios varied over a wide range (1.9–3.57). It is suggested that if oral fluid is used for TDM, occasional plasma monitoring also should be performed.[114] Other researchers have shown high inter- and intrasubject variability in adults, with greater variations in oral fluid as compared to plasma concentrations.[119–122] Even though high intersubject S/P variability has been reported, the intrasubject variability appears to be low, suggesting that salivary monitoring may be a useful predictor of plasma concentration once an S/P ratio is established for a subject.[123] Unfortunately, Li levels in saliva were too variable to use for TDM in children.[148]

Saliva TDM for psychotropic drugs has been a subject of considerable interest in attempts to determine optimum blood levels for clinical effect. Conflicting reports of a correlation between saliva and plasma levels of neuroleptic drugs (fluphenazine, haloperidol, and chlorpromazine measured as total active compounds) were found.[124,125] In one study a significant correlation was found ($r = 0.75$)[125]; however, no correlation was found in another study ($r < 0.1$).[124] The ratio between saliva and plasma levels for haloperidol was 2.2, in agreement with previous studies.[125] It was found that for chlorpromazine there was a marked concentration in saliva, 4–50 times higher when compared to plasma. Also, chlorpromazine levels in saliva and plasma were correlated during the first day of treatment ($r = 0.89$), but not after 28 days ($r = 0.25$).[126]

Although amitriptyline and its primary metabolite, nortriptyline, have been detected in oral fluid, research has shown that oral fluid levels poorly reflect concentrations in plasma for tricyclic antidepressant drugs.[127] A single dose of desipramine given to healthy volunteers showed that the S/P ratio of desipramine was time dependent and was much less than unity at 96 h.[128] A correlation existed, but the variability did not allow prediction of plasma levels from those in saliva. This is in contrast to a better correlation found at steady state and to an apparent decreased variability for ratios in the same individual.[129]

Investigation of the relationship between plasma concentrations of clozapine (CLZ) and its metabolite desmethylclozapine (DMCLZ) in saliva in patients treated with CLZ found a satisfactory correlation between plasma and saliva DMCLZ levels ($r = 0.71$), but a low correlation with CLZ ($r = 0.56$). As a consequence, whole saliva is not a satisfactory substitute for plasma when monitoring drug levels for dosage adjustment. The correlation between the CLZ/DMCLZ ratio in plasma and whole saliva ($r = 0.85$) is encouraging, but further studies are necessary to determine how useful this measure is. There may be some use in checking for compliance to treatment.[130] One study evaluating the correlation of saliva concentration and clinical outcome showed a positive correlation with clinical improvement and suggested a potential therapeutic range of 550–920 ng/mL for patients on CLZ for more than 2 years, and 400–510 ng/mL for patients with treatment for less than 2 years.[131] The usefulness of these ranges in TDM has yet to be evaluated.

Risperidone and 9-hydroxyrisperidone appear in the saliva of patients treated with risperidone. Their detection/quantification in saliva provides evidence for recent adherence with therapy.[132] Both analytes were present in the oral fluid of patients treated with risperidone, providing evidence of recent adherence with therapy. The S/P ratios of both compounds vary and the concentration in one matrix cannot be used to predict the exact concentration in the other.[132]

Immunosuppressants

Saliva may offer an alternative specimen for the therapeutic monitoring of immunosuppressants in children and patients with difficult venous access. Mean mycophenolic acid (MPA) concentrations in saliva correlated well with total ($r = 0.909$) and unbound ($r = 0.910$) MPA concentrations in plasma. Studies are required to establish the usefulness of this specimen in the clinical management of organ transplant recipients.[133] For a highly protein-bound and red blood cell–sequestered drugs such as cyclosporin (CsA), sirolimus, and Tacrolimus, saliva may also provide a practical approach for measuring the unbound concentration. The correlation coefficient value between the CsA concentration measurements in paired blood–saliva samples was 0.695 ($P = 0.006$).[134]

SPECIAL APPLICATIONS OF SALIVARY DRUG LEVELS

Although the use of oral fluid as an alternative to TDM using blood or serum may be useful for some drugs, in most cases when a patient requires a TDM evaluation, a blood sample is also required for other clinical assessments. And for many drugs, oral fluid does not appear to be useful for many therapeutically monitored drugs in serum/plasma. But because many drugs can be detected in oral fluid, it may prove to be a useful sample in assessment of compliance and use/abuse, and does not necessarily have to be coupled with other clinical monitoring.

The fact that drugs and their metabolites appear in oral fluid has resulted in investigation of this sample type as an alternative to serum analysis for determining a metabolic phenotype, the rate or degree to which a drug substrate is metabolized. The field of pharmacogenetics is trying to assess the effect of genetic mutations or polymorphisms in drug-metabolizing genes, and in many cases is predictive of phenotype. One complicating factor in predicting the metabolic effect of an altered genotype is that many factors, including a patient's existing diseases and concomitant drug administration, lead to induction or inhibition of enzyme activity and unpredictable effects. In other words, the genotype of an individual may not always predict the phenotype. The administration of probe drugs in order to characterize the pharmacokinetic behavior in an attempt to elucidate cytochrome P450 (CYP) phenotype has been well described in plasma.[135,136] The use of saliva offers a noninvasive approach to address the same question.[137,138] It has been demonstrated that hepatic clearance of substances that measure CYP activity including ethosuximide,[60] PTN,[60] antipyrine,[139] metronidazole,[85,139] and caffeine[140] can be determined from multiple saliva samples.

Antipyrine has been used as a probe of mixed CYP-mediated metabolism in multiple studies of drug metabolism.[141] Its advantages as a probe drug to assess CYP activity include complete bioavailability, blood flow–independent metabolism, and metabolism by multiple mixed function oxidases including CYP 1A2, 2B6, 2C9, 2C18, and 3A4.[141,142] These CYP isozymes are responsible in part for metabolism of as many as 70% of hepatic metabolized drugs. Several studies have evaluated the use of antipyrine in saliva to study metabolic function.[137–139,141–144]

While antipyrine has been extensively used as a probe of overall hepatic CYP metabolism, there are limitations to its use, including a lack of metabolism by other important CYP enzymes including CYP2D6.[35,145] More importantly, because antipyrine is metabolized by multiple CYP isozymes, it is not necessarily possible to directly extrapolate metabolism of antipyrine to metabolism of a therapeutic agent that is metabolized by specific CYP. For example, one cannot be assured that a twofold reduction in antipyrine metabolism is reflective of a twofold reduction in metabolism of other CYP substrates or by other CYP isozymes.

The use of drugs that are specific for a particular CYP may provide more clinically useful information. Use of saliva for assessment of a specific CYP metabolic activity, more descriptively referred to as phenotyping, has been described for CYP3A4 and 2D6. CYP3A4 phenotyping can be estimated from a single measurement of ethosuximide in saliva.[60] CYP2D6 poor (PM) and extensive metabolizers (EM) can be identified by the dextromethorphan metabolic ratio (dextromethorphan/dextrorphan).[146] In one study, urinary and salivary metabolic ratios were determined and correlated well with each other ($r = 0.704$).[146] All the PM identified by urinary metabolic ratio was also identified by the metabolic ratio in saliva. This study suggests that salivary analysis for determination of dextromethorphan metabolic phenotype is feasible.[146] Data from our laboratory (unpublished) and others[132] have suggested that risperidone may also be useful in assessing CYP2D6 phenotypes.

CONCLUSION

Although oral fluid "lacks the drama of blood, the sincerity of sweat and the emotional appeal of tears,"[147] of all the biological fluids, oral fluid is the one that offers the greatest potential to be able to substitute for blood in TDM. Oral fluid has significant utility in the detection of illicit drugs and drugs of abuse, including alcohol, but because of many factors the use of oral fluid in TDM has not widely been applied. TDM with oral fluid requires that there be a reproducible relationship between oral fluid and plasma drug concentrations and an interpretable correlation between drug concentrations and pharmacologic effects. In addition, TDM of oral fluid or plasma requires that there be a narrow margin between the drug's therapeutic and toxic effects.

Finally, there needs to be rapid and reliable methods of analysis for drug concentrations in oral fluid to justify therapeutic drug monitoring.[25] The use of oral fluid as an alternative to traditional drug monitoring using blood or serum may be applicable for some, but not all, drugs.[24] Compliance monitoring would seem to be applicable to a larger number of drugs. Because drugs and their metabolites appear in saliva, it has been suggested as an alternative matrix to measure CYP activity, with a particular focus on CYP2D6.

REFERENCES

1. SIEGEL, I.A. 1993. The role of saliva in drug monitoring. Ann. N. Y. Acad. Sci. **694:** 86–90.
2. AMBERSON, W.R. & W. HÖBER. 1932. The permeability of mammalian salivary glands for organic non-electrolytes. J. Cell. Comp. Physiol. **2:** 201–221.
3. FRIEDEMANN, T. 1938. The excretion of ingested ethyl alcohol in saliva. J. Lab. Clin. Med. **29:** 1007–1014.
4. SIEGEL, I.A. 1987. Use of saliva to monitor drug concentrations. *In* The Salivary System. L.M. Sreebny, Ed.: 157–178. CRC Press. Boca Raton, FL.
5. HOLD, K., D. DE BOER, J. ZUIDEMA & R.A. MAES. 1995. Saliva as an analytical tool in toxicology. Int. J. Drug Testing **1:** 1–36.
6. BURGEN, A.S. 1956. The secretion of non-electrolytes in the parotid saliva. J. Cell. Physiol. **48:** 113–138.
7. VINING, R.F. & R.A. MCGINLEY. 1986. Hormones in saliva. Crit. Rev. Clin. Lab. Sci. **23:** 95–146.
8. GROTH, U., W. PRELLWITZ & E. JAHNCHEN. 1974. Estimation of pharmacokinetic parameters of lithium from saliva and urine. Clin. Pharmacol. Ther. **16:** 490–498.
9. IDOWU, O.R. & B. CADDY. 1982. A review of the use of saliva in the forensic detection of drugs and other chemicals. J. Forensic Sci. Soc. **22:** 123–135.
10. KILLMANN, S.A. & J.H. THAYSEN. 1955. The permeability of the human parotid gland to a series of sulfonamide compounds, para-aminohippurate and inulin. Scand. J. Clin. Lab. Invest. **7:** 86–91.
11. HAECKEL, R. 1993. Factors influencing the saliva/plasma ratio of drugs. Ann. N. Y. Acad. Sci. **694:** 128–142.
12. MACHERAS, P. & A. ROSEN. 1984. Is monitoring of drug in saliva reliable for bioavailability testing of a protein-bound drug? A theoretical approach. Pharm. Acta. Helv. **59:** 34–36.
13. WAN, S.H., S.B. MATIN & D.L. AZARNOFF. 1978. Kinetics, salivary excretion of amphetamine isomers, and effect of urinary pH. Clin. Pharmacol. Ther. **23:** 585–590.
14. RASMUSSEN, F. 1964. Salivary excretion of sulphonamides and barbiturates by cows and goats. Acta. Pharmacol. Toxicol. (Copenh.) **21:** 11–19.
15. MATIN, S.B., S.H. WAN & J.H. KARAM. 1974. Pharmacokinetics of tolbutamide: prediction by concentration in saliva. Clin. Pharmacol. Ther. **16:** 1052–1058.
16. ARAKI, Y. 1951. Nitrogenous substances in saliva. I. Protein and non-protein nitrogens. Jpn. J. Physiol. **2:** 69–78.
17. ZUIDEMA, J. & C.A. VAN GINNEKEN. 1983. Clearance concept in salivary drug excretion. Part I: Theory. Pharm. Acta Helv. **58:** 88–93.

18. BRODIE, B.B. 1967. Physicochemical and biochemical aspects of pharmacology. JAMA **202:** 600–609.
19. RITSCHEL, W.A. & G.A. TOMPSON. 1983. Monitoring of drug concentrations in saliva: a non-invasive pharmacokinetic procedure. Meth. Find. Exp. Clin. Pharmacol. **5:** 511–525.
20. KOYSOOKO, R., E.F. ELLIS & G. LEVY. 1974. Relationship between theophylline concentration in plasma and saliva of man. Clin. Pharmacol. Ther. **15:** 454–460.
21. TROUPIN, A.S. & P. FRIEL. 1975. Anticonvulsant level in saliva, serum, and cerebrospinal fluid. Epilepsia **16:** 223–227.
22. KUTT, H. & J.K. PENRY. 1974. Usefulness of blood levels of antiepileptic drugs. Arch. Neurol. **31:** 283–288.
23. DAVIS, B.D. 1943. The binding of sulfonamide drugs by plasma proteins. A factor in determining the distribution of drugs in the body. J. Clin. Invest. **22:** 753–762.
24. WILSON, J.T. 1993. Clinical correlates of drugs in saliva. Ann. N. Y. Acad. Sci. **694:** 48–61.
25. CHOO, R.E. & M.A. HUESTIS. 2004. Oral fluid as a diagnostic tool. Clin. Chem. Lab. Med. **42:** 1273–1287.
26. JONES, A.W. 1979. Distribution of ethanol between saliva and blood in man. Clin. Exp. Pharmacol. Physiol. **6:** 53–59.
27. CONE, E.J. 1993. Saliva testing for drugs of abuse. Ann. N. Y. Acad. Sci. **694:** 91–127.
28. HAECKEL, R. & I. BUCKLITSCH. 1987. The comparability of ethanol concentrations in peripheral blood and saliva. The phenomenon of variation in saliva to blood concentration ratios. J. Clin. Chem. Clin. Biochem. **25:** 199–204.
29. POSTI, J. 1982. Saliva-plasma drug concentration ratios during absorption: theoretical considerations and pharmacokinetic implications. Pharm. Acta Helv. **57:** 83–92.
30. KADEHJIAN, L. 2005. Legal issues in oral fluid testing. Forensic Sci. Int. **150:** 151–160.
31. OYLER, J.M., E.J. CONE, R.E. JOSEPH, Jr., *et al.* 2002. Duration of detectable methamphetamine and amphetamine excretion in urine after controlled oral administration of methamphetamine to humans. Clin. Chem. **48:** 1703–1714.
32. DRUMMER, O.H. 2005. Review: pharmacokinetics of illicit drugs in oral fluid. Forensic Sci. Int. **150:** 133–142.
33. KIDWELL, D.A., J.C. HOLLAND & S. ATHANASELIS. 1998. Testing for drugs of abuse in saliva and sweat. J. Chromatogr. B Biomed. Sci. Appl. **713:** 111–135.
34. SCHEPERS, R.J., J.M. OYLER, R.E. JOSEPH, Jr., *et al.* 2003. Methamphetamine and amphetamine pharmacokinetics in oral fluid and plasma after controlled oral methamphetamine administration to human volunteers. Clin. Chem. **49:** 121–132.
35. KIVISTO, K.T. & H.K. KROEMER. 1997. Use of probe drugs as predictors of drug metabolism in humans. J. Clin. Pharmacol. **37:** 40S–48S.
36. CHOONARA, I.A. & A. RANE. 1990. Therapeutic drug monitoring of anticonvulsants state of the art. Clin. Pharmacokinet. **18:** 318–328.
37. EADIE, M.J. 1984. Anticonvulsant drugs. An update. Drugs **27:** 328–363.
38. BERAN, R.G., J.H. LEWIS, J.L. NOLTE & A.P. WESTWOOD. 1985. Use of total and free anticonvulsant serum levels in clinical practice. Clin. Exp. Neurol. **21:** 69–77.

39. LIU, H. & M.R. DELGADO. 1999. Therapeutic drug concentration monitoring using saliva samples. Focus on anticonvulsants. Clin. Pharmacokinet. **36:** 453–470.
40. DASGUPTA, A.. 2002. Clinical utility of free drug monitoring. Clin. Chem. Lab. Med. **40:** 986–993.
41. KAUFMAN, E. & I.B. LAMSTER. 2002. The diagnostic applications of saliva—a review. Crit. Rev. Oral Biol. Med. **13:** 197–212.
42. MONACO, F., S. PIREDDA, R. MUTANI, *et al.* 1982. The free fraction of valproic acid in tears, saliva, and cerebrospinal fluid. Epilepsia **23:** 23–26.
43. RYLANCE, G.W. & T.A. MORELAND. 1981. Saliva carbamazepine and phenytoin level monitoring. Arch. Dis. Child. **56:** 637–640.
44. CALLAGHAN, N. & T. GOGGIN. 1984. A comparison of plasma and saliva levels of carbamazepine and phenytoin as monotherapy. Ir. J. Med. Sci. **153:** 170–173.
45. MILES, M.V., M.B. TENNISON & R.S. GREENWOOD. 1991. Intraindividual variability of carbamazepine, phenobarbital, and phenytoin concentrations in saliva. Ther. Drug Monit. **13:** 166–171.
46. KNOTT, C. & F. REYNOLDS. 1984. The place of saliva in antiepileptic drug monitoring. Ther. Drug Monit. **6:** 35–41.
47. MCAULIFFE, J.J., A.L. SHERWIN, I.E. LEPPIK, *et al.* 1977. Salivary levels of anticonvulsants: a practical approach to drug monitoring. Neurology **27:** 409–413.
48. DANHOF, M. & D.D. BREIMER. 1978. Therapeutic drug monitoring in saliva. Clin. Pharmacokinet. **3:** 39–57.
49. WESTENBERG, H.G., R.A. DE ZEEUW, E. VAN DE KLEIJN & T.T. OEI. 1977. Relationship between carbamazepine concentrations in plasma and saliva in man as determined by liquid chromatography. Clin. Chim. Acta. **79:** 155–161.
50. WESTENBERG, H.G., E. VAN DER KLEIJN, T.T. OEI & R.A. DE ZEEUW. 1978. Kinetics of carbamazepine and carbamazepine-epoxide, determined by use of plasma and saliva. Clin. Pharmacol. Ther. **23:** 320–328.
51. HAMILTON, M.J., A.F. COHEN, A.W. YUEN, *et al.* 1993. Carbamazepine and lamotrigine in healthy volunteers: relevance to early tolerance and clinical trial dosage. Epilepsia **34:** 166–173
52. PISANI, F.D. & R.G. DI PERRI. 1981. Intravenous valproate: effects on plasma and saliva phenytoin levels. Neurology **31:** 467–470.
53. COTARIU, D. & J.L. ZAIDMAN. 1991. Developmental toxicity of valproic acid. Life Sci. **48:** 1341–1350.
54. al Za'abi M., D. DELEU & C. BATCHELOR. 2003. Salivary free concentrations of anti-epileptic drugs: an evaluation in a routine clinical setting. Acta Neurol. Belg. **103:** 19–23.
55. HERKES, G.K. & M.J. EADIE. 1989. Daily salivary anticonvulsant monitoring in patients with intractable epilepsy. Clin. Exp. Neurol. **26:** 141–149.
56. HERKES, G.K. & M.J. EADIE. 1990. Possible roles for frequent salivary antiepileptic drug monitoring in the management of epilepsy. Epilepsy Res. **6:** 146–154.
57. PETIT, P., R. LONJON, M. COCIGLIO, *et al.* 1991. Carbamazepine and its 10,11-epoxide metabolite in acute mania: clinical and pharmacokinetic correlates. Eur. J. Clin. Pharmacol. **41:** 541–546.
58. DROBITCH, R.K. & C.K. SVENSSON. 1992. Therapeutic drug monitoring in saliva. An update. Clin. Pharmacokinet. **23:** 365–379.
59. HORNING, M.G., L. BROWN, J. NOWLIN, *et al.* 1977. Use of saliva in therapeutic drug monitoring. Clin. Chem. **23:** 157–164.

60. BACHMANN, K., J. SCHWARTZ, T. SULLIVAN & L. JAUREGUI. 1986. Single sample estimate of ethosuximide clearance. Int. J. Clin. Pharmacol. Ther. Toxicol. **24:** 546–550.
61. GRIM, S.A., M. RYAN, M.V. MILES, *et al.* 2003. Correlation of levetiracetam concentrations between serum and saliva. Ther. Drug Monit. **25:** 61–66.
62. MILES, M.V., P.H. TANG, T.A. GLAUSER, *et al.* 2003. Topiramate concentration in saliva: an alternative to serum monitoring. Pediat. Neurol. **29:** 143–147.
63. RAMBECK, B. & P. WOLF. 1993. Lamotrigine clinical pharmacokinetics. Clin. Pharmacokinet. **25:** 433–443.
64. RYAN, M., S.A. GRIM, M.V. MILES, *et al.* 2003. Correlation of lamotrigine concentrations between serum and saliva. Pharmacotherapy **23:** 1550–1557.
65. UEDA, C.T., P.J. BECKMANN & B.S. DZINDZIO. 1984. Relationship between saliva and serum quinidine concentrations and suppression of ventricular premature beats. Ther. Drug Monit. **6:** 43–49.
66. TRNAVSKA, Z., H. KREJCOVA, TKACZYKOVAM, *et al.* 1991. Pharmacokinetics of lamotrigine (Lamictal) in plasma and saliva. Eur. J. Drug Metab. Pharmacokinet. Spec. No. **3:** 211–215.
67. MILES, M.V., P.H. TANG, M.A. RYAN, *et al.* 2004. Feasibility and limitations of oxcarbazepine monitoring using salivary monohydroxycarbamazepine (MHD). Ther. Drug Monit. **26:** 300–304.
68. DE GIER J.J., B.J. T HART, F.A. NELEMANS & H. BERGMAN. 1981. Psychomotor performance and real driving performance of outpatients receiving diazepam. Psychopharmacology (Berl.) **73:** 340–344.
69. HALLSTROM, C. & M.H. LADER. 1980. Diazepam and N-desmethyldiazepam concentrations in saliva, plasma and CSF. Br. J. Clin. Pharmacol. **9:** 333–339.
70. KANGAS, L., H. ALLONEN, R. LAMMINTAUSTA, *et al.* 1979. Pharmacokinetics of nitrazepam in saliva and serum after a single oral dose. Acta Pharmacol. Toxicol. (Copenh.) **45:** 20–24.
71. HART, B.J., J. WILTING & J.J. DE GIER. 1988. The stability of benzodiazepines in saliva. Meth. Find. Exp. Clin. Pharmacol. **10:** 21–26.
72. KANKIRAWATANA, P.. 1999. Salivary antiepileptic drug levels in Thai children. J. Med. Assoc. Thai. **82:** 80–88.
73. GOZAL, D., A. COLIN, H. SHACHAR & H. BEN-ARYEH. 1985. Salivary theophylline: a reliable alternative to monitoring serum levels? Isr. J. Med. Sci .**21:** 462–463.
74. KNOTT, C., M. BATEMAN & F. REYNOLDS. 1984. Do saliva concentrations predict plasma unbound theophylline concentrations? A problem re-examined. Br. J. Clin. Pharmacol. **17:** 9–14.
75. EBDEN, P., D. LEOPOLD, D. BUSS, *et al.* 1985. Relationship between saliva and free and total plasma theophylline concentrations in patients with chronic airflow obstruction. Thorax **40:** 526–529.
76. JABER, M., A.T. SCHNEIDER, S. GOLDSTEIN, *et al.* 1987. Reliability and predictive value of salivary theophylline levels. Ann. Allergy **58:** 105–108.
77. BLANCHARD, J., S. HARVEY & W.J. MORGAN. 1991. Serum/saliva correlations for theophylline in asthmatics. J. Clin. Pharmacol. **31:** 565–570.
78. IWASAKI, E. & M. BABA. 1987. Saliva and serum theophylline concentration in management of asthma in children on sustained-release therapy. J. Asthma **24:** 173–178.
79. GURUMURTHY, P., F. RAHMAN, A.S. NARAYANA & G.R. SARMA. 1990. Salivary levels of isoniazid and rifampicin in tuberculous patients. Tubercle **71:** 29–33.

80. OFOEFULE, S.I., C.E. OBODO, O.E. ORISAKWE, *et al.* 2002. Salivary and urinary excretion and plasma-saliva concentration ratios of isoniazid in the presence of co-administered ciprofloxacin. Am. J. Ther. **9:** 15–18.
81. IMMANUEL, C., A.K. HEMANTHKUMAR, P. GURUMURTHY & P. VENKATESAN. 2002. Dose related pharmacokinetics of ofloxacin in healthy volunteers. Int. J. Tuberc. Lung Dis. **6:** 1017–1022.
82. WARLICH, R., H.C. KORTING, M. SCHAFER-KORTING & E. MUTSCHLER. 1990. Multiple-dose pharmacokinetics of ofloxacin in serum, saliva, and skin blister fluid of healthy volunteers. Antimicrob. Agents Chemother. **34:** 78–81.
83. BERKOVITCH, M., M. GOLDMAN, R. SILVERMAN, *et al.* 2000. Therapeutic drug monitoring of once daily gentamicin in serum and saliva of children. Eur. J. Pediatr. **159:** 697–698.
84. BENNETT, G.A., E. DAVIES & P. THOMAS. 2003. Is oral fluid analysis as accurate as urinalysis in detecting drug use in a treatment setting? Drug Alcohol Depend. **72:** 265–269.
85. Suryawati S. MUSTOFA & B. SANTOSO. 1991. Pharmacokinetics of metronidazole in saliva. Int. J. Clin. Pharmacol. Ther. Toxicol. **29:** 474–478.
86. KOKS, C.H., K.M. CROMMENTUYN, R.M. HOETELMANS, *et al.* 2001. Can fluconazole concentrations in saliva be used for therapeutic drug monitoring? Ther. Drug Monit. **23:** 449–453.
87. BABALOLA, C.P., O.O. BOLAJI, F.A. OGUNBONA & P.A. DIXON. 1996. Relationship between plasma and saliva quinine levels in humans. Ther. Drug Monit. **18:** 30–33.
88. ROLINSKI, B., U. WINTERGERST, A. MATUSCHKE, *et al.*1991. Evaluation of saliva as a specimen for monitoring zidovudine therapy in HIV-infected patients. Aids **5:** 885–888.
89. Product Monograph, Delavidine. 2001. Agouron Pharmaceuticals, Inc. La Jolla, CA.
90. R.P. VAN HEESWIJK, A.I. VELDKAMP, J.W. MULDER, *et al.* 2001. Saliva as an alternative body fluid for therapeutic drug monitoring of the nonnucleoside reverse transcription inhibitor nevirapine. Ther. Drug Monit. **23:** 255–258.
91. HUGEN, P.W., D.M. BURGER, M. DE GRAAFF, *et al.* 2000. Saliva as a specimen for monitoring compliance but not for predicting plasma concentrations in patients with HIV treated with indinavir. Ther. Drug Monit. **22:** 437–445.
92. SMITH, M., E. WHITEHEAD, G. O'SULLIVAN & F. REYNOLDS. 1991. A comparison of serum and saliva paracetamol concentrations. Br. J. Clin. Pharmacol. **31:** 553–555.
93. HAHN, T.W., T. MOGENSEN, C. LUND, *et al.* 2000. High-dose rectal and oral acetaminophen in postoperative patients—serum and saliva concentrations. Acta Anaesthesiol. Scand. **44:** 302–306.
94. KENNEDY, J.M., N.M. TYERS & A.K. DAVEY. 2003. The influence of morphine on the absorption of paracetamol from various formulations in subjects in the supine position, as assessed by TDx measurement of salivary paracetamol concentrations. J. Pharm. Pharmacol. **55:** 1345–1350.
95. WILSON, J.T., R.D. BROWN, J.A. BOCCHINI, Jr. & G.L. KEARNS. 1982. Efficacy, disposition and pharmacodynamics of aspirin, acetaminophen and choline salicylate in young febrile children. Ther. Drug Monit. **4:** 147–180.
96. KHAN, A.Z. & L. AARONS. 1989. A note on the use of salicylate saliva concentration in clinical pharmacokinetic studies. J. Pharm. Pharmacol. **41:** 710–711.

97. ALBERT, K.S., A. RAABE, M. GARRY, *et al.* 1984. Determination of ibuprofen in capillary and venous plasma by high-performance liquid chromatography with ultraviolet detection. J. Pharm. Sci. **73:** 1487–1489.
98. SMITH, I.J., J.L. HINSON, V.A. JOHNSON, *et al.* 1989. Flurbiprofen in post-partum women: plasma and breast milk disposition. J. Clin. Pharmacol. **29:** 174–184.
99. EPSTEIN, J.B., A.H. TSANG, D. WARKENTIN & J.A. SHIP. 2002. The role of salivary function in modulating chemotherapy-induced oropharyngeal mucositis: a review of the literature. Oral Surg. Oral Med. Oral Pathol. Oral Radiol. Endod. **94:** 39–44.
100. SLAVIK, M., J. WU & C. RILEY. 1993. Salivary excretion of anticancer drugs. Ann. N. Y. Acad. Sci. **694:** 319–321.
101. SVOJANOVSKY, S.R., K.L. EGODAGE, J. WU, *et al.* 1999. High sensitivity ELISA determination of taxol in various human biological fluids. J. Pharm. Biomed. Anal. **20:** 549–555.
102. JANSMAN, F.G., J.L. COENEN, J.C. DE GRAAF, *et al.* 2002. Relationship between pharmacokinetics of 5-FU in plasma and in saliva, and toxicity of 5-fluorouracil/folinic acid. Anticancer Res. **22:** 3449–3455.
103. JOULIA, J.M., F. PINGUET, M. YCHOU, *et al.* 1999. Plasma and salivary pharmacokinetics of 5-fluorouracil (5-FU) in patients with metastatic colorectal cancer receiving 5-FU bolus plus continuous infusion with high-dose folinic acid. Eur. J. Cancer. **35:** 296–301.
104. HOLDING, J.D., W.E. LINDUP, N.B. ROBERTS, *et al.* 1999. Measurement of platinum in saliva of patients treated with cisplatin. Ann. Clin. Biochem. **36**(Pt 5):655–659.
105. TAKAHASHI, T., Y. FUJIWARA, H. SUMIYOSHI, *et al.* 1997. Salivary drug monitoring of irinotecan and its active metabolite in cancer patients. Cancer Chemother. Pharmacol. **40:** 449–452.
106. L.J. VAN WARMERDAM, O. VAN TELLINGEN, W.W. TEN BOKKEL HUININK, *et al.* 1995. Monitoring carboplatin concentrations in saliva: a replacement for plasma ultrafiltrate measurements? Ther. Drug Monit. **17:** 465–470.
107. BOUCAUD, M., F. PINGUET, S. CULINE, *et al.* 2003. Modeling plasma and saliva topotecan concentration time course using a population approach. Oncol. Res. **13:** 211–219.
108. BERKOVITCH, M., T. BISTRITZER, M. ALADJEM, *et al.* 1998. Clinical relevance of therapeutic drug monitoring of digoxin and gentamicin in the saliva of children. Ther. Drug Monit. **20:** 253–256.
109. MAHMOD, S., D.S. SMITH & J. LANDON. 1987. Radioimmunoassay of salivary digoxin by simple adaptation of a kit method for serum digoxin: saliva/serum ratio and correlation. Ther. Drug Monit. **9:** 91–96.
110. HAECKEL, R. & H.M. MUHLENFELD. 1989. Reasons for intraindividual inconstancy of the digoxin saliva to serum concentration ratio. J. Clin. Chem. Clin. Biochem. **27:** 653–658.
111. NARANG, P.K., N.H. CARLINER, M.L. FISHER & W.G. CROUTHAMEL. 1983. Quinidine saliva concentrations: absence of correlation with serum concentrations at steady state. Clin. Pharmacol. Ther. **34:** 695–702.
112. LAURIKAINEN, E. & J. KANTO. 1983. Saliva concentrations of lignocaine in patients with acute myocardial ischaemia. Br. J. Clin. Pharmacol. **16:** 199–200.
113. CORDONNIER, J., M. VAN DEN HEEDE & A. HEYNDRICKX. 1987. Saliva concentrations of disopyramide cannot substitute the drug's plasma concentrations. J. Anal. Toxicol. **11:** 179–181.

114. SPENCER, E.K., M. CAMPBELL, P. ADAMS, *et al.* 1990. Saliva and serum lithium monitoring in hospitalized children. Psychopharmacol. Bull. **26:** 239–243.
115. GROSS, S.J., T.E. WORTHY, L. NERDER, *et al.* 1985. Detection of recent cannabis use by saliva delta 9-THC radioimmunoassay. J. Anal. Toxicol. **9:** 1–5.
116. MASEDA, C., K. HAMA, Y. FUKUI, *et al.* 1986. Detection of delta 9-THC in saliva by capillary GC/ECD after marihuana smoking. Forensic Sci. Int. **32:** 259–266.
117. SPRING, K.R. & M.A. SPIRTES. 1969. Salivary excretion of lithium. II. Functional analysis. J. Dent. Res. **48:** 550–554.
118. SPRING, K.R. & M.A. SPIRTES. 1969. Salivary excretion of lithium. I. Human parotid and submaxillary secretions. J. Dent. Res. **48:** 546–549.
119. MCKEAGE M.J. & T.J. MALING. 1989. Saliva lithium: a poor predictor of plasma and erythrocyte levels. N. Z. Med. J. **102:** 559–560.
120. OBACH, R., J. BORJA, J. PRUNONOSA, *et al.*1988. Lack of correlation between lithium pharmacokinetic parameters obtained from plasma and saliva. Ther. Drug Monit. **10:** 265–268.
121. BEN-ARYEH H., A. SHALEV, R. SZARGEL, *et al.*1986. The salivary flow rate and composition of whole and parotid resting and stimulated saliva in young and old healthy subjects. Biochem. Med. Metab. Biol. **36:** 260–265.
122. WELLER, E.B., R.A. WELLER, M.A. FRISTAD, *et al.*1987. Saliva lithium monitoring in prepubertal children. J. Am. Acad. Child Adolesc. Psychiatry **26:** 173–175.
123. SANKARANARAYANAN, A., A. GOEL & V.L. PANT. 1985. Variation in the relationship between serum and saliva lithium levels. Int. J. Clin. Pharmacol. Ther. Toxicol. **23:** 365–366.
124. ZOHAR, J., B. BIRMAHER, H. SCHOENFELD & R.H. BELMAKER. 1986. Salivary and blood levels of neuroleptics during outpatient maintenance treatment. Isr. J. Psychiatry Relat. Sci. **23:** 123–128.
125. YAMAZUMI, S. & S. MIURA. 1981. Haloperidol concentrations in saliva and serum: determination by the radioimmunoassay method. Int. Pharmacopsychiatry **16:** 174–183.
126. MAY, P.R., T. VAN PUTTEN, D.J. JENDEN, *et al.* 1981. Chlorpromazine levels and the outcome of treatment in schizophrenic patients. Arch. Gen. Psychiatry **38:** 202–207.
127. BAUMANN, P., D. TINGUELY, L. KOEB, *et al.* 1982. On the relationship between free plasma and saliva amitriptyline and nortriptyline. Int. Pharmacopsychiatry **17:** 136–146.
128. PI, E.H., T.K. TRAN-JOHNSON, G.E. GRAY, *et al.* 1991. Saliva and plasma desipramine levels in Asian and Caucasian volunteers. Psychopharmacol. Bull. **27:** 281–284.
129. COOPER, T.B., N. BARK & G.M. SIMPSON. 1981. Prediction of steady state plasma and saliva levels of desmethylimipramine using a single dose, single time point procedure. Psychopharmacology (Berl.) **74:** 115–121.
130. DUMORTIER, G., A. LOCHU, A. ZERROUK, *et al.* 1998. Whole saliva and plasma levels of clozapine and desmethylclozapine. J. Clin. Pharm. Ther. **23:** 35–40.
131. YU, D. 1992. The correlation between clozapine saliva level and clinical response as well as side effect in patient with schizophrenia. Zhongua Shen Jing Jing Shen Ka Za Zhi **25:** 165–162.
132. FLARAKOS, J., W. LUO, M. AMAN, *et al.* 2004. Quantification of risperidone and 9-hydroxyrisperidone in plasma and saliva from adult and pediatric patients by liquid chromatography-mass spectrometry. J. Chromatogr. A **1026:** 175–183.

133. MENDONZA, A.E., R.Y. GOHH & F. AKHLAGHI. 2006. Analysis of mycophenolic acid in saliva using liquid chromatography tandem mass spectrometry. Ther. Drug Monit. **28:** 402–406.
134. MENDONZA, A., R. GOHH & F. AKHLAGHI. 2004. Determination of cyclosporine in saliva using liquid chromatography-tandem mass spectrometry. Ther. Drug Monit. **26:** 569–575.
135. GOODMAN & H.E. GILMAN. 2001. The Pharmacological Basis of Therapeutics, 10th ed. McGraw-Hill. New York.
136. BURTIS, C.A. *et al.* Eds. 2006. Teitz Textbook of Clinical Chemistry, 4th ed.: 2448. Elsevier Saunders. St. Louis, MO.
137. VESELL, E.S.. 1984. Noninvasive assessment *in vivo* of hepatic drug metabolism in health and disease. Ann. N. Y. Acad. Sci. **428:** 293–307.
138. LOFT, S. & POULSEN. 1990. Prediction of xenobiotic metabolism by non-invasive methods. Pharmacol. Toxicol. **67:** 101–108.
139. LOFT, S., A.J. NIELSEN, B.E. BORG & H.E. POULSEN. 1991. Metronidazole and antipyrine metabolism in the rat: clearance determination from one saliva sample. Xenobiotica **21:** 33–46.
140. BALISTRERI, W.F., H.H. AK, K.D. SETCHELL, *et al.* 1992. New methods for assessing liver function in infants and children. Ann. Clin. Lab. Sci. **22:** 162–174.
141. ENGEL, G., U. HOFMANN, H. HEIDEMANN, *et al.* 1996. Antipyrine as a probe for human oxidative drug metabolism: identification of the cytochrome P450 enzymes catalyzing 4-hydroxyantipyrine, 3-hydroxymethylantipyrine, and norantipyrine formation. Clin. Pharmacol. Ther. **59:** 613–623.
142. DANHOF, M. & D.D. BREIMER. 1979. Studies on the different metabolic pathways of antipyrine in man. I. Oral administration of 250, 500 and 1000 mg to healthy volunteers. Br. J. Clin. Pharmacol. **8:** 529–537.
143. MATZKE, G.R., R.F. FRYE, J.J. EARLY, *et al.* 2000. Evaluation of the influence of diabetes mellitus on antipyrine metabolism and CYP1A2 and CYP2D6 activity. Pharmacotherapy **20:** 182–190.
144. SVENSSON, C.K.. 1988. Is blood sampling for determination of antipyrine pharmacokinetics in healthy volunteers ethically justified? Clin. Pharmacol. Ther. **44:** 365–368.
145. POULSEN, H.E. & S. LOFT. 1988. Antipyrine as a model drug to study hepatic drug-metabolizing capacity. J. Hepatol. **6:** 374–382.
146. HOU, Z.Y., L.W. PICKLE, P.S. MEYER & R.L. WOOSLEY. 1991. Salivary analysis for determination of dextromethorphan metabolic phenotype. Clin. Pharmacol. Ther. **49:** 410–419.
147. MANDEL, I.D.. 1990. The diagnostic uses of saliva. J. Oral Pathol. Med. **19:** 119–125.
148. VITIELLO, B., D. BEHAR, R. MALONE, *et al.* 1988. J. Clin. Psychopharmacol. **8:** 355–359.

Visualization and Other Emerging Technologies as Change Makers for Oral Cancer Prevention

MIRIAM P. ROSIN,[a,d] CATHERINE F. POH,[a,c,e] MARTIAL GUILLARD,[b] P. MICHELE WILLIAMS,[c] LEWEI ZHANG,[e] AND CALUM MACAULAY[c]

[a]*Department of Cancer Control Research, British Columbia Cancer Agency (BCCA), Vancouver, BC, Canada, V5Z 1L3*

[b]*Department of Cancer Imaging, BCCA, Vancouver, BC, Canada, V5Z 1L3*

[c]*Department of Oral Oncology, BCCA, Vancouver, BC, Canada, V5Z 1L3*

[d]*School of Kinesiology, Simon Fraser University, Burnaby, BC, Canada, V5A 1S6*

[e]*Faculty of Dentistry, University of British Columbia, Vancouver, BC, Canada, V6T 1Z3*

ABSTRACT: The genomic era has fueled a rapid emergence of new information at the molecular level with a great potential for developing innovative approaches to detection, risk assessment, and management of oral cancers and premalignant disease. As yet, however, little research has been done on complementary approaches that would use different technology in conjunction with molecular approaches to create a rapid and cost-effective strategy for patient assessment and management. In our ongoing 8-year longitudinal study, a set of innovative technologies is being validated alone and in combination to best correlate with patient outcome. The plan is to use these devices in a step-by-step sequence to guide key clinicopathological decisions on patient risk and treatment. The devices include a hand-held visualization device that makes use of tissue autofluorescence to detect and delineate abnormal lesions and fields requiring follow-up, to be used in conjunction with optical contrast agents such as toluidine blue. In addition, two semi-automated high-resolution computer microscopy systems will be used to quantitate the protein expression phenotype of cell nuclei in tissue sections and exfoliated cell brushings. Previously identified risk-associated molecular changes are being used to validate these systems as well as to establish their place in a population-based triage program that will filter out high-risk cases in

Address for correspondence: Miriam P. Rosin, BC Oral Cancer Prevention Program, BC Cancer Agency/Cancer Control Research Centre, 675 West 10th Avenue, Vancouver, B.C., Canada, V5Z 1L3. Voice: 604-675-8078; fax: 604-675-8180.
miriam_rosin@shaw.ca

Ann. N.Y. Acad. Sci. 1098: 167–183 (2007). © 2007 New York Academy of Sciences.
doi: 10.1196/annals.1384.039

the community and funnel them to dysplasia clinics where higher-cost molecular tools will guide intervention. A critical development for the translation of this technology into community settings is the establishment of an effective methodology for education and training of health practitioners on the front lines.

KEYWORDS: fluorescence visualization; field alteration; oral cancer; oral premalignant lesions; prognosis; detection; loss of heterozygosity; computer imaging

OVERVIEW

Globally, there is universal recognition that a new strategy for the management of oral cancer is required. Each year nearly 300,000 new cases are identified worldwide, most often at a late stage, with significant spread reducing the probability of successful treatment. On average, half of the patients die within 5 years of diagnosis. Recurrence and formation of second cancers are frequent (10–25% of cases).[1] Even when successful, treatment of late-stage oral cancer can be devastating, associated with disfigurement, impairments in speech and eating, and an overall compromise in quality of life. Perhaps the most striking statistic, however, is the lack of significant change in prognosis for this disease over the last three decades, even in the most technologically developed countries.[2]

The irony is that in many ways the oral cavity is well suited for development of screening programs that would identify premalignant disease for early treatment. The site is readily accessible for visual inspection and oral premalignant lesions (OPLs) are known to present clinically as either white patches (leukoplakia) or, less frequently, red patches (erythroplakia). Histologically, there are early indications of the disease (i.e., dysplasia) comparable to other sites—such as the cervix—where screening has been an effective strategy for reduction of mortality. We can even potentially target such screens to known high-risk groups, such as middle-aged heavy smokers, who are more likely to have early disease.[3]

So why has there been so little progress? Several roadblocks have prevented the realization of the potential for early disease detection. A key missing link has been our limited understanding of the natural history of the disease. This is an exciting time, as molecular technology begins to identify alterations to specific genetic pathways and to unravel the process by which they have an impact on structure and function of cells in tissues as they progress to cancer. Unfortunately, the majority of this research is targeted toward cancer rather than premalignant disease. This is primarily due to the limited access of most research centers to such samples and the small size of such lesions, quite often necessitating extensive technological modification for assessment. A more pressing issue, however, is the degree of uncertainty with respect to future outcome of OPLs. Not all early disease will progress to cancer. To be effective as an early screen, molecular (and other) technologies need to target change that is strongly associated with outcome—in other words, the likelihood of progression to cancer.

This paper describes the concerted effort being made in British Columbia to lay the scientific groundwork for a province-wide oral cancer screening program that will integrate molecular features of developing OPLs with novel visualization and computer technology to identify and manage high-risk disease in community settings. The rationale for this effort is given below along with a description of the developing technology, its initial validation, and long-term plans for its implementation via knowledge transfer to community health professionals.

TARGETING DEVICE DEVELOPMENT TO KEY DECISION POINTS

The approach taken by the British Columbia Oral Cancer Prevention Program (BC OCPP) is to develop and validate technologies that are targeted toward key clinical decision points that are potential barriers to patient flow. The devices are meant not to replace conventional approaches to patient assessment, but rather to enrich them and to integrate them with current knowledge. The plan is to use these tools as overlapping sieves in a step-by-step fashion to progressively filter out patients in the community with high-risk OPLs and triage them to dysplasia clinics for management (see schematic, FIG. 1).

What are these key decision points? There are two points that represent major barriers to screening activities. The first is the visualization of the lesion. Oral premalignant changes vary considerably in clinical appearance. It can be quite challenging for the clinician to differentiate abnormalities requiring biopsy from reactive lesions associated with other causes, such as infection or trauma. The clinician may chose to watch a suspicious area for a short period of time and only to biopsy if the lesion does not resolve, or the clinician may ignore the lesion entirely. Visualization devices that facilitate the decision to biopsy (the next step in patient evaluation) could have a profound effect on outcome.

A second barrier revolves around risk prediction for OPLs. OPLs vary widely in potential for malignant transformation with only a fraction eventually progressing cancer. At present, the gold standard for prediction involves the determination of the presence and degree of dysplasia in a biopsy. Severe dysplasia or carcinoma *in situ* (CIS), grouped as high-grade premalignant lesions, are characterized by persistence, recurrence, and high likelihood of eventual progression to invasive cancer, if left untreated.[4] Even with aggressive surgical treatment, 30–40% of the lesions still recur or progress within 30 months.[5] Histology has a good predictive value for such lesions.[6] In contrast, determination of prognosis for lesions with histological changes that are less than severe is

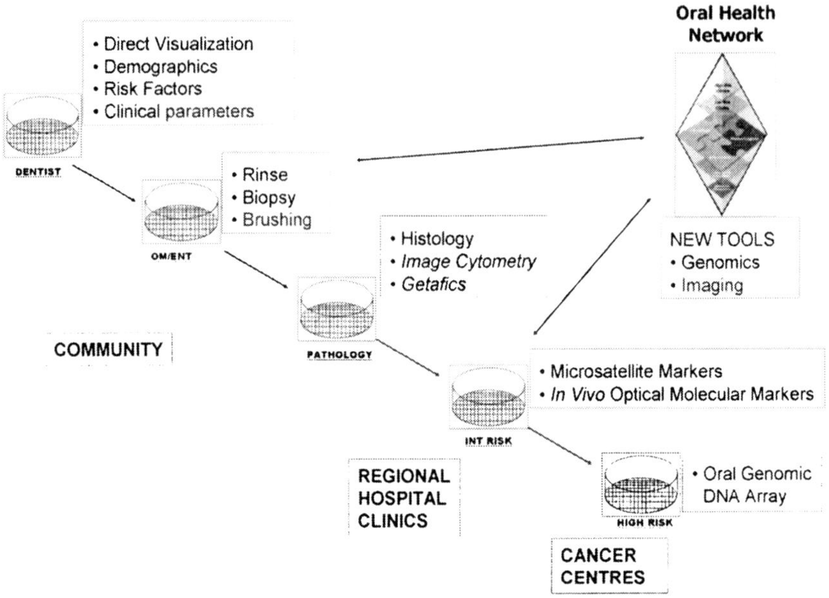

FIGURE 1. Proposed triage pathway for oral cancer screening in British Columbia. A five-step approach to patient management is proposed using devices/protocols developed and validated by the British Columbia Oral Cancer Prevention Program. Steps involve a set of "filters" to control patient flow: (1) identification of patients requiring follow-up by dentists in the community (facilitated by use of visualization devices); (2) collection of a biopsy and/or exfoliated cell samples directly by the dentist or by an oral medicine specialist/ear-nose-throat surgeon (OM/ENT) after referral; (3) assessment of biopsies and rinses at the BC Oral Biopsy Service with histology, Image Cytometry, and/or *Getafics* to triage either back to community for follow-up (little or no risk), or forward to: (4) intermediate-risk (INT) dysplasia clinics in community or (5) "high-risk" dysplasia clinics in cancer centers. Higher-cost molecular tools at these sites will guide intervention (e.g., microsatellite analysis for LOH, *in vivo* optical probes and confocal microscopes, oral genomic array comparative genomic hybridization [CGH] profiles).

more problematic. The majority of OPLs without dysplasia or with low-grade dysplasia (mild and moderate dysplasia) will not progress into cancer[6,7] and histology alone does not clearly differentiate between those that will progress and those that will not. Unfortunately, as a group, these lesions represent the bulk of leukoplakia and account for the majority of cases that later progress to cancer.[7,8] Hence, development of approaches that will differentiate cases with a high likelihood of progression from more benign lesions is of critical importance since this would allow a more aggressive, timely intervention at an earlier stage of the disease for high-risk cases while avoiding overtreatment of lesions with little risk of progression.

We have piloted two such approaches. One involves high-resolution computer microscopy, potentially a high-throughput, cost-effective filter for outcome. The second involves a final set of sieves based on molecular alterations in the tissue. The latter uses one of the more promising approaches for outcome prediction in the literature, microsatellite analysis for loss of heterozygosity (LOH). This is a robust approach, works well with DNA extracted from archival samples, and requires minute quantities of DNA (the latter is critical given small sample size of OPLs). Most importantly, several laboratories have reported, in independent studies, an association of LOH with progression.[9–12]

THE ORAL CANCER PREDICTION LONGITUDINAL (OCPL) STUDY

Central to the development and validation of the aforementioned devices is an ongoing Oral Cancer Prediction Longitudinal (OCPL) study located in British Columbia, funded by National Institute of Dental and Craniofacial Research (NIDCR) for 8 years (1999–2008). This study is following ~500 patients with high-risk oral lesions: half with cancer (at risk for recurrence) and the other half with low-grade dysplasia (at risk for progression to cancer). This is one of the richest prospective studies of oral lesions designed to date with a rigorous collection of clinical, pathological, demographic, and molecular data.

One of the aims of the OCPL study is to confirm several retrospectively observed LOH risk patterns previously associated with progression for oral dysplasia (details in the following section). Nested with the study are two other endeavors. The first involves the evaluation of three innovative, potentially complementary devices for early detection and follow-up of OPLs and oral cancer, one a fluorescence visualization device for clinical examination and the others, computer-assisted microscopy systems that are being trained to predict risk and outcome. Also nested to the study is the genome-wide profiling of specimens generated during the study using a whole genome bacterial artificial chromosome tiling set (BAC) array (~32,000 clones), which is able to determine a genetic profile of the entire genome (at the individual gene level) in a single hybridization experiment. This latter platform will be used to create a database of high-resolution genomic profiles of tumors and early OPLs, with known outcome, to ultimately identify a set of predictive markers of progression and to guide intervention strategies.

Each device is being evaluated through a series of correlations: first for association with high-risk clinical features (e.g., site and size of lesion, texture),[13] then for ability to detect cancers and high-grade OPLs (severe dysplasia/CIS), since this category represents histologically a pattern that is accepted to have a significant risk of progression. Of importance also is the association of each device with genetic risk, in this study defined as the presence of high-risk molecular clones (in our program, LOH patterns). Finally, the chief endpoint involves an assessment of the ability of each device to predict progression of lesions with minimal (i.e., low-grade) or no dysplasia to cancer. These evaluations are being done for each technology alone and in combination to determine whether the integration of these devices will facilitate the identification of critical changes earlier and with greater specificity.

Other considerations under assessment include the utility of temporal change in response of each device as a predictor of outcome. Such studies allow us to better identify conditions under which the ability to detect high-risk disease is masked by circumstances such as the presence of periodic inflammation or trauma or the temporary influence of biopsy.

A final level of evaluation in our strategy for device development involves the assessment of each technology in community clinics to obtain crucial information on the prevalence of alterations measured by these devices in such a setting and on the acceptability of the devices to patients and clinicians.

As an example of the potential value of the above strategy, a summary is given below of our experience with integration of a molecular tool (LOH analysis) with a visualization device (toluidine blue stain retention) to predict outcome for OPLs with minimal or no dysplasia.

EARLY DATA: THE TOLUIDINE BLUE/LOH PARADIGM

A few years ago, we used the large archival resources of the Oral Biopsy Service of British Columbia to begin to look for molecular patterns that would predict outcome, concentrating on those cases in which histology had the poorest predictive value—mild or moderate dysplasia and hyperplasia.[9] We selected out 116 cases, all of which had outcome data, 29 of which later progressed to cancer. All samples were assayed for LOH on seven arms (3p, 9p, 4q, 8p, 11q, 13q, and 17p) at loci previously shown to be present in OPLs and cancers. The study produced two striking observations. First, 97% of progressing lesions had LOH at 3p and/or 9p, suggesting that loss at these arms may be a progression prerequisite and that cases without this loss might have little risk of progression. This is a very significant point since it suggests that this change could be used in initial screening for cancer risk of OPLs. However, since 42% of nonprogressing lesions also showed loss on one of these arms, that loss alone is probably insufficient for malignant transformation. Indeed, cases with LOH at these arms but not others showed only a 3.8-fold increase in relative risk for cancer development compared to those that retained both arms. In contrast, additional losses at 4q, 8p, 11q, 13q, or 17p were infrequent in nonprogressing cases and individuals who showed such loss had a 33-fold increase in cancer risk. These results suggested that LOH patterns may differentiate three progression risk groups: low, with retention of 3p and 9p; intermediate, with loss at 3p and/or 9p; and high, with loss at 3p and/or 9p plus loss at 4q, 8p, 11q, 13q, or 17p. At 5 years, estimated proportions of progressing lesions for these three progression risks were: 2% (95% confidence interval [CI], 0–5%), 26% (95% CI, 5–43%), and 47% (95% CI, 26–62%) for the low-, intermediate- and high-risk group, respectively. Validation of these data is a major objective of the OCPL study.

These data support the use of LOH as a filter for outcome. However, the test still requires that a biopsy be taken and hence is dependent upon the clinician's ability to identify OPLs requiring assessment. Fortuitously, we had been following OPLs in our clinics for years—often using retention of an optical contrast agent, toluidine blue, to better differentiate lesions for biopsy. This stain has a long history of use to identify various tumors, including oral cancers. Its use with OPLs, however, was contentious, since only a portion of mild or moderate dysplasia appeared to be stained with it. Some early work from our lab and David Sidransky's at Johns Hopkins suggested that the dye might differentially stain early OPLs.[14,15]

Last year we published data from the first 100 cases of primary dysplasia (no prior head/neck cancer history) accrued to the OCPL study, with an average of 44 months of follow-up.[16] Fifteen of these cases had progressed to cancer. Of significance, there was a strong correlation of toluidine blue-positive staining with clinicopathological risk factors, molecular patterns, and with outcome. Specifically, the dye preferentially stained lesions of nonhomogeneous clinical appearance and those that were larger and which grew in size during follow-up. Staining was strongly associated with LOH patterns with increased cancer risk: multiple LOH ($P < 0.0002$), and LOH at 3p and/or 9p plus losses at any other arm ($P < 0.0001$). Histologically, it stained 16 of 17 cases of high-grade dysplasia. Although the dye was positive in only 26% (five of 19) of nondysplastic OPLs and 23% (15 of 64) of lesions with low-grade (mild/moderate) dysplasia, it was strongly correlated with lesions in these groups that later progressed to cancer. Overall, there was a >sixfold elevation in cancer risk for positive lesions, with 12 of 15 progressing lesions staining positive for the dye ($P = 0.0008$). These data support the use of the stain as a front-end filter to facilitate decisions to biopsy for OPLs and as a funnel of samples for further molecular analysis.

THE CONCEPT OF OVERLAPPING SIEVES: OTHER VISUALIZATION DEVICES

OPLs arise from a complex dynamic interaction of genetically altered and competing clones of cells for which the impact on the observable phenotype is poorly understood. Further complicating the process is the interactions of these clones with the patient's immediate and systemic environment. In recognition of this, we believe that multiple views of the tissue at different levels are critical to the recognition of the phenotype and prediction of its likely behavior.

For example, for visualization, conventional white light can perceive only a fraction of the underlying characteristics that differentiate diseased tissue from its normal counterpart. To alter our perception of the field at risk one can use optical contrast agents (toluidine blue) as previously described and/or capitalize on the explosive growth in optical technologies, particularly those based on fluorescence imaging and spectroscopy. Our early data suggest that a synergism between these approaches will likely improve our ability to detect tissue changes, such as OPLs and early cancer.

There is a wealth of literature that supports the use of tissue autofluorescence in the screening and diagnosis of precancers in the lung, uterine cervix, skin, and oral cavity.[17–23] This approach is already in clinical use in the lung[24] and the mechanism of action and interaction of tissue autofluorescence has been well described in the cervix.[25–27]

The interaction of light with tissue has generally been found to highlight changes in the structure and metabolic activity of areas of clonal activity. Specifically, the loss of autofluorescence is believed to reflect a complex mixture of alterations to intrinsic tissue fluorophore distribution, due to tissue remodeling such as the breakdown of the collagen matrix and elastin composition as well as alterations to metabolism such as the decrease in flavin adenine dinucleotide (FAD) concentration, and increase the reduction form of nicotinamide adenine dinucleotide (NADH) associated with progression of the disease. [28] Further, these structural changes in tissue morphology are associated with alterations not only in the epithelium but also in the lamina propria (e.g., thickening of the epithelium, hyperchromatin, and increased cellular/nuclear pleomorphism, or increased microvascularity). The latter changes lead to increased absorption and/or scattering of light, which in turn reduces and modifies the detectable autofluorescence signal.

We have recently developed a simple field-of-view device for the direct visualization of tissue fluorescence in the oral cavity. Our goal was to make use of the previously known changes at other sites to develop a simple portable instrument for fluorescence imaging in the oral cavity. The simplest of systems requires a light source, excitation and emission filters, and a means of detection, in this case the eye, as shown in FIGURE 2. FIGURE 3 shows a difficult-to-recognize white light area of change in one patient (A) with oral cancer at right lateral tongue and the same area under fluorescence visualization (B) and toluidine blue staining (C) with the enhancement of the field of change.

The use of fluorescence visualization results in a new, expanded definition of the altered field. In a recent study we used this device in the operating room to directly visualize subclinical field changes around oral cancers (FIG. 3 D–F).[29] We evaluated a total of 122 oral mucosa biopsies from 20 surgical specimens to address the issue of the significance of the new definition of field. Each biopsy was assessed for location outside of the clinically defined perimeter, fluorescence visualization status, histology, and LOH (10 markers on three regions: 3p14, 9p21, and 17p13).

All tumors showed loss of autofluorescence. In 19/20 tumors the loss extended in at least one direction beyond the clinically visible tumor, with the extension varying from 4–25 mm. Loss of fluorescence in these

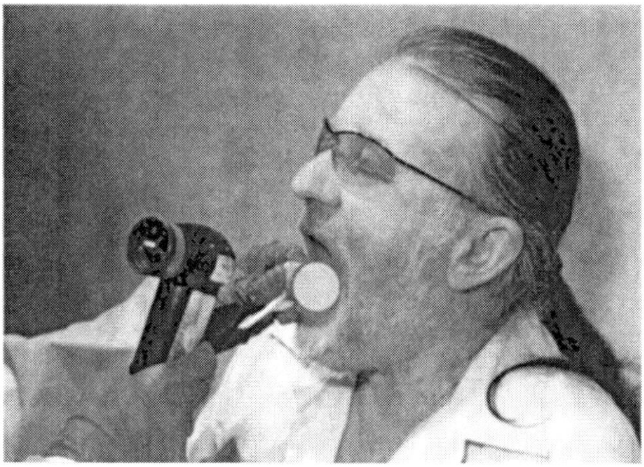

FIGURE 2. Use of a simple hand-held fluorescence visualization tool to identify field changes in the oral cavity requiring follow-up.

margins was strongly associated with histological change, present in 32 of 36 biopsies with loss (including seven with squamous cell carcinoma [SCC]/CIS, 10 severe dysplasia, 15 mild/moderate dysplasia) compared to only one of the 66 biopsies that had no loss. A significant association was observed between loss of fluorescence and molecular change, with LOH at 3p and/or 9p (previously associated with progression risk) present in 12/19 biopsies with loss of fluorescence compared with 3/13 biopsies ($P = 0.04$) in which fluorescence was retained. These data suggest that loss of fluorescence can be used to define a novel field of change around cancers that are strongly associated with histological and molecular risk and, possibly, in the future could guide decisions on surgical margins.

The device also appears to shed light on the field at risk for the earlier stages of the disease. In a preliminary analysis of the data from the ongoing OCPL study, a total of 278 cases of oral leukoplakia was examined for loss of fluorescence status at time of biopsy and its association with clinical features and histological diagnosis.[13] Lesions with loss of fluorescence tended to have high-risk clinical features (larger size, nonhomogeneous texture, location at high-risk sites, $P < 0.05$). Virtually all cancers (34/35) and severe dysplasia (34/35) were detected with the device. Also the majority of primary (no previous history of oral cancer) mild/moderate cases of dysplasia (76/134, 57%) were detected. Of interest, 15/74 (20%) of lesions with no dysplasia also had loss of fluorescence. The significance of the latter data with lesions with minimal (mild/moderate) or no dysplasia will be determined with follow-up of these lesions and with integration to LOH frequencies, both of which are ongoing. Of the few biopsy-proven normals collected so far, none has shown loss of fluorescence.

In summary, an important feature of the device is its ability to detect clinically nonapparent disease that is histologically high risk and requires further clinical

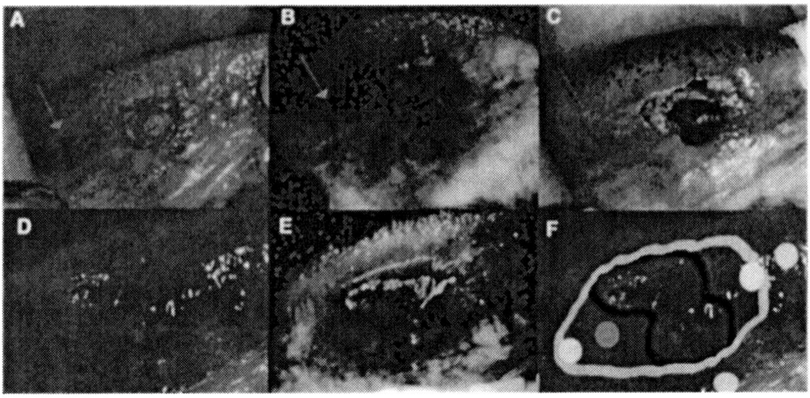

FIGURE 3. The use of novel visualization aids to define high-risk field change. (**A–C**) an ill-defined lesion on the right lateral tongue of a 65-year-old male: (**A**) white light image of the lesion; (**B**) the same lesion, viewed with a fluorescence visualization (FV) device, showing a wider well-demarcated dark area of loss of normal green fluorescence or fluorescence visualization loss (FVL, *arrow*) and a separate well-defined loss (*arrowhead*); (**C**) lesion after application of toluidine blue (TB) showing uptake of the blue stain at the nodular area only. Biopsies from the TB-stained nodular area, and from areas posterior (*arrow*) and inferior (*arrowhead*) to that region (identified by FVL) showed an invasive squamous cell carcinoma, severe dysplasia, and moderate dysplasia, respectively. (**D–F**) a second case with an ill-defined carcinoma *in situ* on the right lateral tongue of a 59-year-old male nonsmoker: (**D**) white light image of the lesion; (**E**) the same lesion, viewed with FV, showing a wider dark area of FV loss (FVL); (**F**) shows an outline of clinically apparent tumor (blue) and region with altered fluorescence (in green) as demarcated in the operating room. Biopsies from margins outside of the clinically apparent lesion that showed loss of fluorescence were severe dysplasia (orange circle) and moderate dysplasia (yellow circle). In contrast, a biopsy in margin area with normal fluorescence showed no dysplasia (green circles).

investigation. As with any screening system, there are conditions that mask detection. For the oral cavity, these could include heavy infection/inflammation and traumatic ulcers, etc. Some of these can be addressed by treatment followed by reassessment with these visualization tools. Using two visualization devices concurrently (e.g., contrast agents and autofluorescence) may also eliminate masking. We expect that over time, as we learn more about early disease and the fields delineated with the tools, these technologies will advance, further improving their ability to detect novel aspects of disease progression.

COMPUTER MICROSCOPY SYSTEMS: HIGH THROUGHPUT FILTERS FOR POPULATION-BASED STUDIES

Once one or all of the visualization technologies have identified a lesion for follow-up, current practice requires that a biopsy be taken and stained for histological assessment. This assessment has as its basis a consensus by pathologists on the types of phenotypic changes associated with risk, with a classification system based on the presence of specific features (dysplasia) and the degree of epithelial involvement (WHO, 1978).[30] Computer imaging derives a quantitative tissue phenotype (QTP) that measures these same specific features as well as the frequency and distribution of cells demonstrating these characteristics in a quantitative as opposed to qualitative fashion.[31–35] This helper tool allows the pathologist to use reproducible and repeatable measures of specific phenotypic characteristics that make up the appearance of dysplastic cells in tissues and which can individually or in combination be statistically associated with risk.

Getafics, the device in use in British Columbia, is the first almost fully automated system that assesses quantitatively stained material for association with outcome in sectioned material. The system examines ~120 nuclear features in cell nuclei in each of the hundreds of cells in a specimen. Some features assess the size, shape, and the amount of DNA within the nucleus. The majority, however, describe the distribution of DNA. For example, increased chromatin is a characteristic high-risk feature that is identified by pathologists. Conventionally this is judged as one feature, hyperchromatism. QTP measures a host of chromatin features including the distribution of DNA around the edge of the nucleus or as clustered in the center (FIG. 4), whether the nucleus is dark with light areas or light with dark areas, whether the increased chromatin is evenly distributed (euchromatin, active gene transcription) or clumped locally (heterochromatin, hypermethylated, and/or acetylated), and whether the chromatin clumps are large or small. Many of these alterations represent the phenotypic expression of specific biochemical events and could lead us to a better understanding of the association of biochemical process with phenotype. For example, the epigenetic changes associated with post-translational modification of the DNA (methylation, acetylation, and higher-order alterations by chromatin-associated proteins) could result in alterations to some of the features previously described.

Historically, studies have shown that QTP systems can characterize nuclear phenotypic changes that are strongly associated with pathological presence and degree of dysplasia at several sites, for example, lung, cervix, skin, prostate, bladder, and colon.[36–39] However, these studies have mainly explored the diagnostic potential of these devices with little data on associations with prognosis.[32] The potential of this approach is described below.

We have recently completed a pilot study of 148 specimens: 30 normals, 29 cancers, 35 high-risk pre-invasive lesions (severe dysplasia/*CIS*), and 44 low-grade lesions (lesions with no or mild/moderate dysplasia) from patients without a history of oral cancer but with known outcome. Five of the system's features were combined in a discriminant function that separated normal from cancer cell nuclei. The latter function was used to generate a nuclear phenotype score (NPS) for every biopsy that was strongly

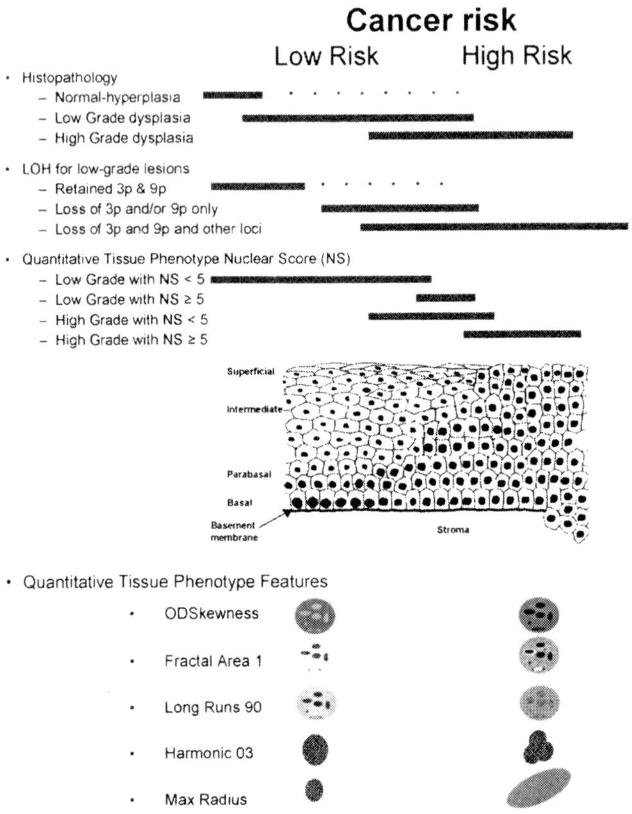

FIGURE 4. Schematic representation of risk of tissue progressing to cancer, based on outcome analysis, for a selection of biomarkers (Histopathology, LOH, and Quantitative Tissue Phenotype). *Left-hand side* indicates predominately low risk of transformation characteristics and *right-hand side* indicates characteristics strongly associated with progression to cancer. For histology, high risk is associated with the presence of severe dysplasia and/or carcinoma *in situ*. For LOH, high risk is associated with LOH at 3p and/or 9p plus loss at other loci (associated with a ~33-fold increased risk). For quantitative tissue phenotype, risk is dependent on the combination of histology (low-grade vs. high-grade dysplasia) and nuclear phenotypic scores (NPS > 5.0). Low grade represents hyperplasia and mild dysplasia and high grade represents moderate and severe dysplasia/CIS. The five features used to determine NPS are illustrated at the bottom of the figure. Low-risk *ODskewness* values associate with lightly stained areas on dark background and high-risk values denote dark areas on light background; *low-risk fractal area 1* measures indicate smooth low-contrast object and *high-risk fractal area 1* values denote irregular intensity high-contrast objects; *low-risk long runs 90* measures indicate objects with few and smooth transitions while *high-risk long runs 90* values denote objects with very rough frequent intense transitions in intensity; *low-risk harmonic 3* values are from circular objects and *high-risk values* come from objects with three prominent protrusions; *max radius* values are larger for high-risk compared to low-risk objects.

associated with histological change, the presence of high-risk molecular clones, and with outcome.[40]

There was a strong monotonic increase in NPS with increasing severity of pathology diagnosis. NPS was statistically associated with the presence of high-risk LOH patterns, as previously described: multiple LOH losses and LOH at 3p and/or 9p plus LOH at any of the arms 4q, 8p, 11q, 13q, and 17p. When OPLs were stratified into high and low NPS scores, the system was able to predict outcome. Time to progression was significantly shorter for cases with high NPS, with 57% of cases progressing to cancer in 5 years compared with only 7% for low NPS cases. Moreover, there was a ninefold increase in relative risk of progression to cancer for cases with high versus low scores (95% CI 2.0–47). When an NPS cutoff value of 5 was used, the system could correctly classify 97% of nonprogressing cases (28/29) and 73% of progressing cases (11/15).[41]

These data are extremely promising and we have begun to look at a further, less invasive, intermediate step that could have utility in a comprehensive screening program. For example, if NPS behavior can be determined in exfoliated cells, this could serve as a further bridge to biopsy decisions for clinicians. Health practitioners may be hesitant to biopsy cases identified with change either with conventional or visualization-enhanced approaches. This could result in cases with lesions at risk being missed. However, exfoliated cells can be collected from the lesion surface with a small brush quickly and noninvasively and then evaluated for nuclear changes to determine whether a biopsy is needed. Such an approach would also be high throughput and cost effective.

We are currently developing a counterpart to the *Getafics* workstation, building on our success with similar systems for lung and cervical cytologic samples.[32,34] The developing system measures in a fully automated fashion the same 120 features in all of the cells present in a cytological sample. We have completed a preliminary analysis of 196 samples, each with a brushing and a concurrent biopsy (108 from normal areas, 60 from severe dysplasia/CIS or SCC, and 28 with infection and inflammation). One hundred seventy-five of these samples (106 normal, 43 abnormal, and 26 infection/inflammation) were judged suitable for analysis (i.e., had >400 cells). All objects on the slides were analyzed by a decision tree using a subset of the aforementioned 120 features to be recognized as cells. The decision tree used was originally trained on cervical samples. This tree was used to sort the cells into normal and abnormal categories, which were then verified manually by experienced cytotechnologists. As a first crude algorithm, two categories of abnormal cells based on DNA content (both previously defined as predictive on cervical/lung data) were used to automatically classify the samples into normal and abnormal groups using the frequency of cells in these categories. Using just these two features, we were able to recognize 81% of normal and 92% of inflammation/infection as low risk and to correctly identify 86% of abnormal samples (11/13 SCC, 11/13 CIS, and 15/17 severe dysplasia). Future steps include the development of an oral cytology-specific cell classification tree and a cytological sample classification function that incorporates multiple phenotypic measures like the NPS.

It is important to note a further strength of this approach. The performance of machine learning algorithm-based diagnostic tests will improve as they are trained on larger and larger sets of samples, and several generations of tests with improving accuracy may be introduced sequentially as computerized systems refine our identification of the categories of at-risk phenotypes.

KNOWLEDGE TRANSFER

When initiating any new technology, it needs to be validated on data with balanced numbers of positives and negatives from high-risk clinics to establish the operating characteristics of the systems. Once this is done, one needs to put to trial the systems in the screening population where the disease incidence is at its natural levels, that is, one positive case per 10,000. In British Columbia, we have already set up a centralized pathology review for the community, a series of dysplasia clinics for patients to be referred to from the community with specialists trained in new technology in them, and have a linkage between these clinics and the clinicians in hospitals that will treat high-risk cases.

We are beginning the process of transfer to the community in the downtown eastside of Vancouver, one of the city's poorest neighborhoods, where drinking and smoking, two well-known risk factors for oral cancer, are widespread. We have already identified a very high risk of disease in this community using our screen: of 250 residents, two had cancers and nine had precancerous lesions. Now, we have launched a more extensive screening program in this neighborhood.

Our next step is to begin the process of seeding the use of the various sieves among community health practitioners. The plan is to integrate the 2,900 dentists of British Columbia into a screening network that will identify cases requiring follow-up and refer them forward for further assessment. We have begun to create training manuals and educational modules and to provide hands-on training for a select number of dentists who will become the peer leaders for the network and who will work within the program to create population-wide guidelines for screens.

CONCLUSIONS

The concept of using sieves with different views into early disease is supported by the fact that the disease develops over years, with continuous clonal selection that is driven by host–lesion interactions, environmental factors, and to a degree stoichastic chance. We have chosen to position our sieves at critical points at which the patients meets the health practitioner throughout their lives (dentists at hygiene appointment) and where a decision has to be made that an oral mucosal change exists that needs further examination. The sieves look at this process from a variety of different viewpoints and the interaction of these different views is expected to lead to new decision points and new models for early disease. This approach has already resulted in a paradigm shift in British Columbia, where multidisciplinary teams discuss cases around the results of the new devices. The goal is to build upon this shift to change clinical behavior to create a framework that will result in earlier and more effective treatment of oral cancer and OPLs.

ACKNOWLEDGMENTS

This work was supported by Grants R01 DE13124 and R01 DE17013 from the National Institute of Dental and Craniofacial Research.

REFERENCES

1. DAY, G.L. & W.J. BLOT. 1992. Second primary tumors in patients with oral cancer. Cancer **70:** 14–19.
2. DAY, T.A., B.K. DAVIS, M.B. GILLESPIE, *et al.* 2003. Oral cancer treatment. Curr. Treat Options Oncol. **4:** 27–41.
3. RAMADAS, K., R. SANKARANARAYANAN, B.J. JACOB, *et al.* 2003. Interim results from a cluster randomized controlled oral cancer screening trial in Kerala, India. Oral Oncol. **39:** 580–588.
4. HAYWARD, J.R. & J.A. REGEZI. 1977. Oral dysplasia and *in situ* carcinoma: clinicopathologic correlations of eight patients. J. Oral Surg. **35:** 756–762.
5. VEDTOFTE, P., P. HOLMSTRUP, E. HJORTING-HANSEN & J.J. PINDBORG. 1987. Surgical treatment of premalignant lesions of the oral mucosa. Int. J. Oral Maxillofac. Surg. **16:** 656–664.
6. WRIGHT, J.M.. 1998. A review and update of oral precancerous lesions. Tex. Dent. J. **115:** 15–19.
7. SILVERMAN, S., M. GORSKY & F. LOZADA. 1984. Oral leukoplakia and malignant transformation. A follow-up study of 257 patients. Cancer **53:** 563–568.
8. WALDRON, C.A. & W.G. SHAFER. 1975. Leukoplakia revisited. A clinicopathologic study of 3256 oral leukoplakias. Cancer **36:** 1386–1392.
9. ROSIN, M.P., X. CHENG, C. POH, *et al.* 2000. Use of allelic loss to predict malignant risk for low-grade oral epithelial dysplasia. Clin. Cancer Res. **6:** 357–362.
10. LEE, J.J., W.K. HONG, W.N. HITTELMAN, *et al.* 2000. Predicting cancer development in oral leukoplakia: ten years of translational research. Clin. Cancer Res. **6:** 1702–1710.
11. CALIFANO, J., P. van der RIET, W. WESTRA, *et al.* 1996. Genetic progression model for head and neck cancer: implications for field cancerization. Cancer Res. **56:** 2488–2492.

12. PARTRIDGE, M., G. EMILION, S. PATEROMICHELAKIS, *et al.* 1998. Allelic imbalance at chromosomal loci implicated in the pathogenesis of oral precancer, cumulative loss and its relationship with progression to cancer. Oral Oncol. **34:** 77–83.
13. NG, S.P., C.F. POH, P.M. WILLIAMS, *et al.* 2006. Identification of high-risk oral premalignant lesions (OPLs) by direct tissue fluorescence visualization (FV). *In* Annual Meeting of the American Academy of Oral Medicine, Puerto Rico.
14. GUO, Z., K. YAMAGUCHI, M. SANCHEZ-CESPEDES, *et al.* 2001. Allelic losses in OraTest-directed biopsies of patients with prior upper aerodigestive tract malignancy. Clin. Cancer Res. **7:** 1963–1968.
15. EPSTEIN, J.B., L. ZHANG, C. POH, *et al.* 2003. Increased allelic loss in toluidine blue-positive oral premalignant lesions. Oral Surg. Oral Med. Oral Pathol. Oral Radiol. Endod. **95:** 45–50.
16. LANE, P.M., T. GILHULY, P. WHITEHEAD, *et al.* 2006. Simple device for the direct visualization of oral-cavity tissue fluorescence. J. Biomed. Opt. **11:** 24006 (1–7).
17. LAM, S., C. MACAULAY, B. PALCIC. 1993. Detection and localization of early lung cancer by imaging techniques. Chest **103**(1 Suppl): 12S–14S.
18. RAMANUJAM, N., M.F. MITCHELL, A. MAHADEVAN, *et al.* 1994. *In vivo* diagnosis of cervical intraepithelial neoplasia using 337-nm-excited laser-induced fluorescence. Proc. Natl. Acad. Sci. USA **91:** 10193–10197.
19. ZENG, H., D.I. MCLEAN, C. MACAULAY & H. LUI. 2000. Autofluorescence properties of skin and applications in dermatology. *In* Proceedings of the SPIE—The International Society for Optical Engineering Biomedical Photonics and Optoelectronic Imaging. **4224:** 366–373.
20. GILLENWATER, A., R. JACOB, R. GANESHAPPA, *et al.* 1998. Noninvasive diagnosis of oral neoplasia based on fluorescence spectroscopy and native tissue autofluorescence. Arch. Otolaryngol. Head Neck Surg. **124:** 1251–1258.
21. HEINTZELMAN, D.L., U. UTZINGER, H. FUCHS, *et al.* 2000. Optimal excitation wavelengths for in vivo detection of oral neoplasia using fluorescence spectroscopy. Photochem. Photobiol. **72:** 103–113.
22. MULLER, M.G., T.A. VALDEZ, I. GEORGAKOUDI, *et al.* 2003. Spectroscopic detection and evaluation of morphologic and biochemical changes in early human oral carcinoma. Cancer **97:** 1681–1692.
23. INGRAMS, D.R., J.K. DHINGRA, K. ROY, *et al.* 1997. Autofluorescence characteristics of oral mucosa. Head Neck **19:** 27–32.
24. LAM, S., T. KENNEDY, M. UNGER, *et al.* 1998. Localization of bronchial intraepithelial neoplastic lesions by fluorescence bronchoscopy. Chest **113:** 696–702.
25. CHANG, S.K., M. FOLLEN, A. MALPICA, *et al.* 2002. Optimal excitation wavelengths for discrimination of cervical neoplasia. IEEE Trans. Biomed. Eng. **49:** 1102–1111.
26. DREZEK, R., M. GUILLAUD, T. COLLIER, *et al.* 2003. Light scattering from cervical cells throughout neoplastic progression: influence of nuclear morphology, DNA content, and chromatin texture. J. Biomed. Opt. **8:** 7–16.
27. COLLIER, T., M. FOLLEN, A. MALPICA & R. RICHARDS-KORTUM. 2005. Sources of scattering in cervical tissue: determination of the scattering coefficient by confocal microscopy. Appl. Opt. **44:** 2072–2081.
28. FOLLEN, M., S. CRAIN, C. MACAULAY, *et al.* 2005. Optical technologies for cervical neoplasia: update of an NCI program project grant. Clin. Adv. Hematol. Oncol. **3:** 41–53.

29. POH, C.F., L. ZHANG, D.W. ANDERSON, *et al.* 2006. Fluorescence visualization detection of field alterations in margins of oral cancer patients. Clin. Cancer Res. **12:** 6716–6722.
30. KRAMER, I.R., R.B. LUCAS, J.J. PINDBORG & L.H. SOBIN. 1978. Definition of leukoplakia and related lesions: an aid to studies on oral precancer. Oral Surg. Oral Med. Oral Pathol. **46:** 518–539.
31. KAMALOV, R., M. GUILLAUD, D. HASKINS, *et al.* 2005. A Java application for tissue section image analysis. Comput. Methods Programs Biomed. **77:** 99–113.
32. GUILLAUD, M., J.C. Le RICHE, C. DAWE, *et al.* 2005. Nuclear morphometry as a biomarker for bronchial intraepithelial neoplasia: correlation with genetic damage and cancer development. Cytometry A. **63:** 34–40.
33. CHIU, D., M. GUILLAUD, D. COX, *et al.* 2004. Quality assurance system using statistical process control: an implementation for image cytometry. Cell Oncol. **26:** 101–117.
34. GUILLAUD, M., D. COX, K. ADLER-STORTHZ, *et al.* 2004. Exploratory analysis of quantitative histopathology of cervical intraepithelial neoplasia: objectivity, reproducibility, malignancy-associated changes, and human papillomavirus. Cytometry A **60:** 81–89.
35. GUILLAUD, M., D. COX, A. MALPICA, *et al.* 2004. Quantitative histopathological analysis of cervical intra-epithelial neoplasia sections: methodological issues. Cell Oncol. **26:** 31–43.
36. FRANK, D.H., J.R. DAVIS, J. RANGER-MOORE, *et al.* 2005. Karyometry of infiltrating breast lesions. Anal. Quant. Cytol. Histol. **27:** 195–201.
37. RANGER-MOORE, J., D.S. ALBERTS, R. MONTIRONI, *et al.* 2005. Karyometry in the early detection and chemoprevention of intraepithelial lesions. Eur. J. Cancer. **41:** 1875–1888.
38. DA SILVA, V.D., R. MONTIRONI, D. THOMPSON, *et al.* 1999. Chromatin texture in high grade prostatic intraepithelial neoplasia and early invasive carcinoma. Anal. Quant. Cytol. Histol. **21:** 113–120.
39. SUN, X.R., J. WANG, D. GARNER & B. PALCIC. 2005. Detection of cervical cancer and high grade neoplastic lesions by a combination of liquid-based sampling preparation and DNA measurements using automated image cytometry. Cell Oncol. **27:** 33–41.
40. MACAULAY, C., M. GUILLAUD, L. ZHANG, *et al.* 2006. The dirty penny is back: image cytometry for oral cancer screening. *In* International Frontier Cancer Prevention Research; 2006, Nov. 11–14. American Association for Cancer Research: Boston.
41. GUILLAUD, M., L. ZHANG, C. POH, et al. 2005. Exploratory quantitative histopathological analysis of oral premalignant lesions (OPLs): association of morphometric index with high-risk molecular profiles and cancer development. *In* AACR Annual Meeting; April 16–20. American Association for Cancer Research: Anaheim, California.

Genomic Targets in Saliva

BERNHARD G. ZIMMERMANN,[a] NOH JIN PARK,[a]
AND DAVID T. WONG,[a,b,c,d,e]

[a]*School of Dentistry and Dental Research Institute, University of California, Los Angeles (UCLA), Los Angeles, California 90095, USA*

[b]*Jonsson Comprehensive Cancer Center, UCLA, Los Angeles, California 90095, USA*

[c]*Division of Head and Neck Surgery/Otolaryngology, David Geffin School of Medicine, UCLA, Los Angeles, California 90095, USA*

[d]*Henry Samueli School of Engineering, UCLA, Los Angeles, California 90095, USA*

ABSTRACT: Saliva, the most accessible and noninvasive biofluid of our body, harbors a wide spectrum of biological analytes informative for clinical diagnostic applications. While proteomic constituents are a logical first choice as salivary diagnostic analytes, genomic targets have emerged as highly informative and discriminatory. This awareness, coupled with the ability to harness genomic information by high-throughput technology platforms such as genome-wide microarrays, ideally positions salivary genomic targets for exploring the value of saliva for detection of specific disease states and augmenting the diagnostic and discriminatory value of the saliva proteome for clinical applications. Buccal cells and saliva have been used as sources of genomic DNA for a variety of clinical and forensic applications. For discovery of disease targets in saliva, the recent realization that there is a transcriptome in saliva presented an additional target for oral diagnostics. All healthy subjects evaluated have approximately 3,000 different mRNA molecules in their saliva. Almost 200 of these salivary mRNAs are present in all subjects. Exploration of the clinical utility of the salivary transcriptome in oral cancer subjects shows that four salivary mRNAs (OAZ, SAT, IL8, and IL1b) collectively have a discriminatory power of 91% sensitivity and specificity for oral cancer detection. Data are also now in place to validate the presence of unique diagnostic panels of salivary mRNAs in subjects with Sjögren's disease.

KEYWORDS: human saliva transcriptome analysis; mRNA biomarkers; noninvasive disease detection

Address for correspondence: David T. Wong, UCLA School of Dentistry, Dental Research Institute, 73–017 CHS, 10833 Le Conte Avenue, Los Angeles, CA 90095. Voice: 310-206-3048; fax: 310-825-7609.
dtww@ucla.edu

Ann. N.Y. Acad. Sci. 1098: 184–191 (2007).
doi: 10.1196/annals.1384.002

Saliva is a mirror of oral health and also a reservoir of analytes from systemic sources that reach the oral cavity through various pathways. The molecular composition of saliva reflects tissue fluid levels of therapeutic, hormonal, immunological, or toxicological molecules. In addition, markers for diseases (e.g., infectious and neoplastic) can be present. Consequently these fluids provide sources for assessment and monitoring of systemic health and disease states, exposure to environmental and job-related toxins, and the use of abusive or therapeutic drugs. This is the basis of our vision to develop disease diagnostics and promote human health surveillance by analysis of saliva. Proteomics, genomics, and microbial analysis will be the driving forces for disease marker development.

Our group aims to expand the toolbox of oral-fluid-based diagnostics by developing proteomic and genomic "alphabets" of healthy individuals and disease-specific signatures. Here we present the rationale and progress with mRNA-based salivary biomarkers, illustrating the presence, integrity, and potential diagnostic utility of mRNA found in saliva.

Stable, cell-free circulating DNA in plasma was first observed almost 60 years ago.[1] Increased plasma DNA levels were shown in cancer patients[2] and several groups demonstrated that plasma DNA displays tumor-specific characteristics. These include somatic mutations in tumor suppressor genes or oncogenes, microsatellite alterations, abnormal promoter methylation, mitochondrial DNA mutations, and the presence of tumor-related viral DNA.[3–9] Tumor-specific cell-free DNA in the circulation has been found in a wide range of malignancies. Genetic alterations and mRNA signatures can successfully be identified in body fluids that drain from organs affected by tumors.[10] Thus, cell-free biomarkers derived from the tumor "travel" through the body and can be detected in blood and other body fluids. As studies of tumor-derived DNA detection in plasma of cancer patients were being pursued, Lo *et al.*[11] detected "fetal" DNA in the plasma of pregnant women. In the following years, the presence of placental and tumor-specific cell-free RNA in plasma was also demonstrated.[12–15] Meanwhile many body fluids have been shown to contain cell-free nucleic acids of potential diagnostic value. Research in this area has demonstrated that these analytes are useful in noninvasive prenatal diagnosis and can detect cancer and other systemic diseases including diabetes, stroke, and myocardial infarction.[16]

The potential for saliva-based tests to detect oral cancer has been demonstrated in a number of studies utilizing the analysis of promoter hypermethylation,[17] exfoliated cells,[18] and even the microbiota.[19] Our group recently established that studying the transcript levels of mRNA in saliva presents potential diagnostic opportunities for oral cancer.[20] We are addressing the need for early cancer detection by an extensive effort to develop and validate a diagnostic test based on mRNA profiles from saliva. Four signature RNA transcripts are elevated in saliva of oral cancer patients.[21] These mRNAs were identified through microarray studies and validated by real-time quantitative polymerase chain reaction (qPCR). A distinction between patients with oral cancer and healthy control subjects demonstrated 91% specificity and 91% sensitivity, the area under the receiver operator characteristics (ROC) curve measuring 95%. These markers are being validated according to established guidelines.[22] At the same time we are developing the optimal test set-up for multicenter studies and investigating methodologies for high-throughput and on-site testing. The examination of transcriptional changes provides clear advantages in comparison to genetic changes as the subtle genetic changes (loss of heterozygosity, point mutations, and methylation changes) are translated into high copy number messages of unique sequences.

The transcriptome analysis of saliva performed by our group provides clear evidence for the clinical utility and potential of analyzing human mRNA in saliva.[23] While being easily accessible in a noninvasive manner, saliva has the advantage that the background of normal material (cells, DNA, RNA, and proteins) and inhibitory substances is much lower and less complex than in blood. This feature may prove to be an important advantage in certain instances, as our laboratory has recently shown with a comparison study of oral cancer mRNA markers.[21]

In the past decade, the potential for the use of saliva for the detection of oral cancer has been laid out by the analysis of genetic changes in the cellular compartment. However, low target concentrations, the subtleness of the genetic changes and the heterogeneity of early events in cancer development and progression have as yet prevented the translation of these findings to concepts for the early noninvasive detection of oral cancer. As such, the genomic information in saliva is mainly applied for the genetic banking for pharmacogenomic and epidemiologic studies,[24] and for a variety of identity testing situations, as, for example, in forensics, by the military for purposes of identification, and tests with possible home-based sample collection in paternity testing and genetic ancestry testing (available from http://Genetree.com).

In forensics, a molecular-based test allows the identification of body fluid stains, demonstrating the utility and validity of salivary mRNA testing: On the basis of panels of body-fluid-specific transcripts, Juusola *et al.* demonstrated the distinction of stains originating from different body fluids such as blood, urine, semen, and saliva.[25–28] It is probable that the predominant part of the RNA preserved in salivary stains is cellular.

The finding of cell-free mRNA in whole saliva has allowed us to establish a salivary core transcriptome of 185 genes present in all 10 samples analyzed, based on Affymetrix U133 array analysis.[23] An average of 3,143 probe sets on each array were assigned, representing approximately 3,000 different mRNAs per individual.

The finding of cell-free mRNA in saliva may seem surprising. Similar to cell-free RNA in plasma it is protected from degradation by association with macromolecules.[29] Since this discovery, we have pursued investigations addressing the source, integrity and stability of RNA in saliva. According to our findings from polymerase chain reaction (PCR)-based experiments, the RNA is generally fragmented. This is supported by constructing a cDNA library from salivary RNA.[30] The sequencing of the cloned fragments showed that most of the RNAs are not full length, and they stem from both nucleus and mitochondria. The sequence analysis excludes the possibility that genomic DNA contamination is a problem (or even the main contributor) in salivary RNA studies: None of the 117 sequences obtained matches the sequences of known pseudogenes or could originate from the DNA sequence of the gene. Other indications that genomic DNA does not confound our analyses of the RNA include the absence of amplification with no-RT controls[30] and the complete removal of a signal in the electrophoretic analysis of salivary RNA by RNase treatment (FIG. 1).

While salivary RNA is surprisingly stable, the accurate quantitative analysis warrants immediate sample processing or stabilization. Sample collection and processing on ice halt the "degeneration" of transcriptional patterns for several hours, and in the laboratory setting, immediate freezing of the specimen is effective. However, this is not always possible in clinical or home-based sampling. The identification of suitable stabilization reagents fitting the needs and constraints of the test is an important step toward practicability of molecular profiling. We thus compared the stability of RNA from whole saliva at room temperature with three stabilizing reagents. The preservation of RNA profiles was compared by array analysis from freshly processed sample against the RNA using samples stored at room temperature for one week either with no stabilizer, with Superase Inhibitor (20U/mL, Invitrogen [Carlsbad, CA, USA]), RNAlater (Ambion Inc., Austin, TX, USA; 1:1 mix with saliva) or the RNAprotect

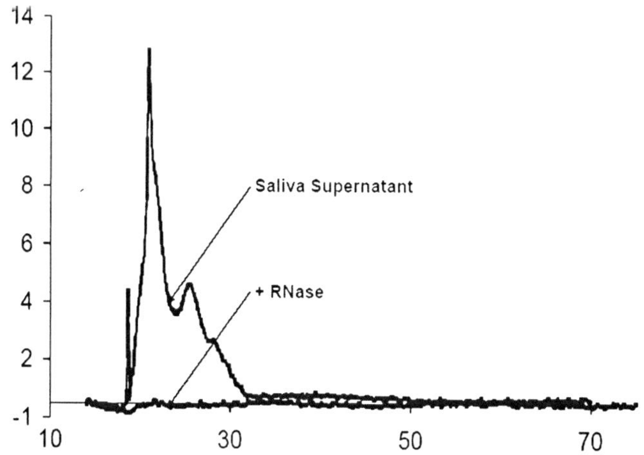

FIGURE 1. RNA extractions from saliva supernatant were analyzed with the Agilent Bioanalyzer Pico RNA Chip (Agilent Technologies, Palo Alto, CA, USA). Extracts were left untreated (Red), or digested with RNase (Green).

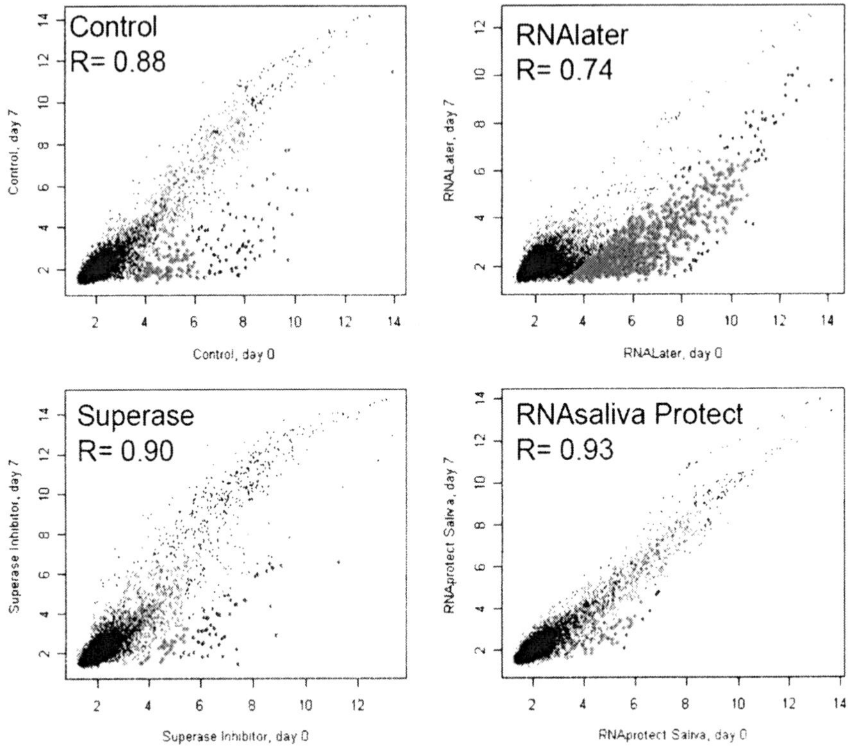

FIGURE 2. Whole saliva was left untreated or stabilized with RNAlater, Superase or RPS. RNA was extracted immediately or after incubation at room temperature for one week. The transcriptomes of fresh and stored aliquots for each stabilization reagent are compared on a scatter plot and transcripts with significantly different levels are highlighted in red. Expression profiles were established with the Human Genome U133 Plus 2.0 GeneChip Array.

Saliva reagent (RPS, Qiagen Inc., Valencia, CA, USA; 1 volume sample plus 5 volumes RPS). The samples were analyzed using Human Genome U133 Plus 2.0 GeneChip Arrays from Affymetrix (Santa Clara, CA, USA) (FIG. 2). Comparison of samples between 0 and 7 days shows a twofold decrease in signal intensity for 159 transcripts with unstabilized samples, but for only 38 transcripts with RPS-stabilized samples. The comparison clearly shows that RNAlater, which is very good for the stabilization of blood and tissue RNA, results in more salivary RNA degradation than any of the other preservatives, even worse than the control sample without any stabilizer. Superase Inhibitor is second best in this group, next to RPS, in salivary RNA stabilization. Thus it seems that RNAlater is not suited for the stabilization of salivary RNA. However, the overall expression profile is well preserved by the RPS reagent. This reagent has recently been made commercially available (Qiagen) for the

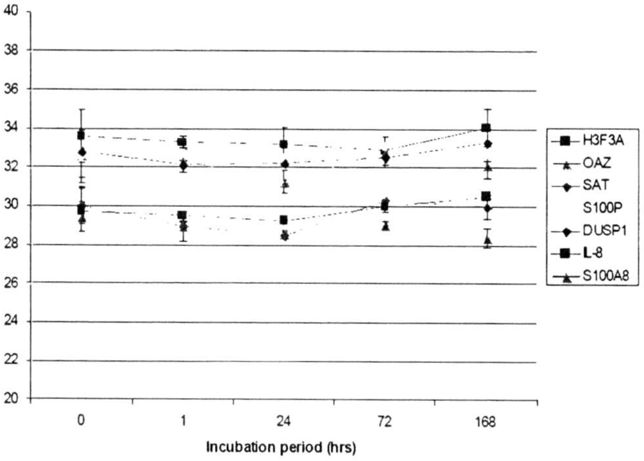

FIGURE 3. Whole saliva was stabilized with RPS and incubated for up to one week. The expression levels of seven transcripts were determined by quantitative RT-PCR, and cycle threshold values (C_T) are indicated on the y-axis.

stabilization of RNA in whole saliva. Even storage of up to 1 week at room temperature did not affect the expression profile of seven oral cancer markers compared to the fresh sample (FIG. 3). Consequently, for the analysis of mRNA from whole saliva, RPS is the preservation reagent of choice.

The clinical potential of saliva mRNA analysis has been demonstrated by providing an oral cancer-specific detection model of mRNA biomarkers.[21] The validation of the salivary oral cancer biomarkers was benchmarked against the use of serum-based prediction established with the same methodology and with samples from the same patient and control cohorts. In this comparison, the panel of four saliva-specific biomarkers possessed a slight edge over the serum-derived marker panel (ROC values of 0.95 vs. 0.88, respectively).[31] These results show that the analysis of saliva may, at least in certain instances, not only be less invasive, but also provide a better value than blood, the commonly targeted body fluid in diagnostic investigations.

Besides oral cancer, the diagnosis of other diseases of the oral cavity may benefit greatly from saliva-based RNA analysis. Sjögren's syndrome is an autoimmune disease of the salivary glands. Their secretory fluids comprise the predominant part of what we call whole saliva or oral fluid. Not only is the treatment of Sjögren's syndrome in its infancy, but also even the diagnosis and studies of the disease are inhibited by unclear, variable definitions.

Array-based analysis identified 26 potential mRNA markers that discriminate between control and individuals with Sjögren's disease. This preliminary analysis may lead to the establishment of clearer diagnostic process and definition of this autoimmune disorder affecting the salivary glands. This would be an important step toward the development of successful treatment or repression of the disease progress. Our laboratory is currently exploring the utility of the salivary transcriptome (and proteome) in other oral and systemic diseases.

In conclusion, information related to oral and systemic health and disease is embedded in a single drop of saliva. It is our intention to establish the rationale, methodology and targets to harness the potential of genomic and proteomic information for the improvement of health care and patient well-being. By moving the time of detection to an earlier stage of the disease, health costs may be decreased dramatically, while patients will be the main beneficiaries through increased health and quality of life.

ACKNOWLEDGMENTS

This work was supported by NIH Grants RO1 DE15970 & RO1DE17593 and U01 DE15018 to D.T.W.

REFERENCES

1. MANDEL, P. & P. METAIS. 1948. Les acides nucleiques du plasma sanguin chez l'homme. C. R. Acad. Sci. Paris 241–243.
2. LEON, S.A., B. SHAPIRO, D.M. SKLAROFF & M.J. YAROS. 1977. Free DNA in the serum of cancer patients and the effect of therapy. Cancer Res. **37:** 646–650.
3. STROUN, M., P. ANKER, P. MAURICE, *et al.* 1989. Neoplastic characteristics of the DNA found in the plasma of cancer patients. Oncology **46:** 318–322.
4. SORENSON, G.D., D.M. PRIBISH, F.H. VALONE, *et al.* 1994. Soluble normal and mutated DNA sequences from single-copy genes in human blood. Cancer Epidemiol. Biomarkers Prev. **3:** 67–71.
5. VASIOUKHIN, V., P. ANKER, P. MAURICE, *et al.* 1994. Point mutations of the N-ras gene in the blood plasma DNA of patients with myelodysplastic syndrome or acute myelogenous leukaemia. Br. J. Haematol. **86:** 774–779.
6. CHEN, X.Q., M. STROUN, J.L. MAGNENAT, *et al.* 1996. Microsatellite alterations in plasma DNA of small cell lung cancer patients. Nat. Med. **2:** 1033–1035.
7. NAWROZ, H., W. KOCH, P. ANKER, *et al.*1996. Microsatellite alterations in serum DNA of head and neck cancer patients. Nat. Med. **2:** 1035–1037.
8. ANKER, P., H. MULCAHY, X.Q. CHEN & M. STROUN. 1999. Detection of circulating tumour DNA in the blood (plasma/serum) of cancer patients. Cancer Metastasis Rev. **18:** 65–73.
9. CHAN, K.C. & Y.M. LO. 2002. Circulating EBV DNA as a tumor marker for nasopharyngeal carcinoma. Semin. Cancer Biol. **12:** 489–496.
10. SIDRANSKY, D. 1997. Nucleic acid-based methods for the detection of cancer. Science **278:** 1054–1058.
11. LO, Y.M., N. CORBETTA, P.F. CHAMBERLAIN, *et al.* 1997. Presence of fetal DNA in maternal plasma and serum. Lancet **350:** 485–487.
12. NG, E.K., N.B. TSUI, T.K. LAU, *et al.* 2003. mRNA of placental origin is readily detectable in maternal plasma. Proc. Natl. Acad. Sci. USA **100:** 4748–4753.

13. Chim, S.S., Y.K. Tong, R.W. Chiu, *et al.* 2005. Detection of the placental epigenetic signature of the maspin gene in maternal plasma. Proc. Natl. Acad. Sci. USA **102:** 14753–14758.
14. Kopreski, M.S., F.A. Benko, L.W. Kwak & C.D. Gocke. 1999. Detection of tumor messenger RNA in the serum of patients with malignant melanoma. Clin. Cancer Res. **5:** 1961–1965.
15. Anker, P. & M. Stroun. 2002. Progress in the knowledge of circulating nucleic acids: plasma RNA is particle-associated. Can it become a general detection marker for a cancer blood test? Clin Chem. **48:** 1210–1211.
16. Swaminathan, R., A. Butt & P. Gahan. 2006. Circulating Nucleic Acids in Plasma and Serum Ann. N. Y. Acad. Sci. **1075:** 1–353.
17. Rosas, S.L., W. Koch, M.B. da Costa Carvalho, *et al.* 2001. Promoter hypermethylation patterns of p16, O6-methylguanine-DNA-methyltransferase, and death-associated protein kinase in tumors and saliva of head and neck cancer patients. Cancer Res. **61:** 939–942.
18. Spafford, M.F., W.M. Koch, A.L. Reed, *et al.* 2001. Detection of head and neck squamous cell carcinoma among exfoliated oral mucosal cells by microsatellite analysis. Clin. Cancer Res. **7:** 607–612.
19. Mager, D.L., A.D. Haffajee, P.M. Devlin, *et al.* 2005. The salivary microbiota as a diagnostic indicator of oral cancer: a descriptive, non-randomized study of cancer-free and oral squamous cell carcinoma subjects. J. Transl. Med. **3:** 27.
20. St John M.A., Y. Li, X. Zhou, *et al.* 2004. Interleukin 6 and interleukin 8 as potential biomarkers for oral cavity and oropharyngeal squamous cell carcinoma. Arch. Otolaryngol. Head Neck Surg. **130:** 929–935.
21. Li, Y., M.A. St. John, X. Zhou, *et al.* 2004. Salivary transcriptome diagnostics for oral cancer detection. Clin. Cancer Res. **10:** 8442–8450.
22. Pepe, M.S., R. Etzioni, Z. Feng, *et al.* 2001. Phases of biomarker development for early detection of cancer. J. Natl. Cancer Inst. **93:** 1054–1061.
23. Li, Y., X. Zhou, M.A. St. John & D.T. Wong. 2004. RNA profiling of cell-free saliva using microarray technology. J. Dent. Res. **83:** 199–203.
24. Rylander-Rudqvist, T., N. Hakansson, G. Tybring & A. Wolk. 2006. Quality and quantity of saliva DNA obtained from the self-administrated oragene method—a pilot study on the cohort of Swedish men. Cancer Epidemiol. Biomarkers Prev. **15:** 1742–1745.
25. Juusola, J. & J. Ballantyne. 2003. Messenger RNA profiling: a prototype method to supplant conventional methods for body fluid identification. Forensic Sci. Int. **135:** 85–96.
26. Alvarez, M., J. Juusola & J. Ballantyne. 2004. An mRNA and DNA co-isolation method for forensic casework samples. Anal. Biochem. **335:** 289–298.
27. Juusola, J. & J. Ballantyne. 2005. Multiplex mRNA profiling for the identification of body fluids. Forensic Sci. Int. **152:** 1–12.
28. Nussbaumer, C., E. Gharehbaghi-Schnell & I. Korschineck. 2006. Messenger RNA profiling: a novel method for body fluid identification by real-time PCR. Forensic Sci. Int. **157:** 181–186.
29. Park, N.J., Y. Li, T. Yu, *et al.* 2006. Characterization of RNA in saliva. Clin. Chem. **52:** 988–994.
30. Park, N.J., X. Zhou, T. Yu, *et al.* 2007. Assessing salivary mRNA integrity by cDNA library analysis. Arch. Oral Biol. **52:** 30–35.
31. Li, Y., D. Elashoff, M. Oh, *et al.* 2006. Serum circulating human mRNA profiling and its utility for oral cancer detection. J. Clin. Oncol. **24:** 1754–1760.

Saliva and the Clinical Pathology Laboratory

MICHAEL A. PESCE AND STEVEN L. SPITALNIK

Department of Pathology, College of Physicians and Surgeons of Columbia University, New York, New York, 10032 USA

ABSTRACT: **There have been increasing numbers of applications using oral fluids, saliva in particular, as the target substrate for performing clinical diagnostic tests. These have focused primarily on point-of-care (POC) testing. These POC testing approaches range from, for example, currently available, highly specialized screening tests for the presence of antibodies recognizing HIV to the potential development of "lab-on-a-chip" platforms. Broad claims have been made that the latter will revolutionize clinical laboratory testing. From the perspective of large centralized clinical laboratories, multiple issues must be considered before implementing individual tests using saliva as the target fluid in a POC format or using saliva as a universal test fluid for measuring multiple analytes in a centralized laboratory format. The current scope of laboratory testing is large and comprehensive, involving both POC and centralized testing. Current academic laboratory programs have the ability to qualitatively identify and/or quantitatively measure several thousand analytes in various target matrices including blood, plasma, serum, urine, joint fluid, pleural fluid, peritoneal fluid, cerebrospinal fluid, and tissue. These tests fall into multiple clinical pathology disciplines, including clinical chemistry, hematology, coagulation, transfusion medicine, microbiology, cytogenetics, molecular diagnosis, and immunology. In addition, before implementing a given test, multiple issues need to be evaluated to ensure the validity of the reported result; these include considerations involving the three major phases of testing: preanalytical (e.g., patient identification and specimen collection, stability, and transport), analytical (e.g., sensitivity, specificity, accuracy, and precision), and postanalytical (e.g., reporting results, quality improvement, and turn-around-time).**

KEYWORDS: **laboratory testing; salivary diagnostics; clinical pathology**

Address for correspondence: Steven L. Spitalnik, M.D., Department of Pathology, Room PS15-408, College of Physicians and Surgeons of Columbia University, 630 West 168th Street, New York, New York 10032. Voice: 212-305-2204; fax: 212-305-3693.
ss2479@columbia.edu

Ann. N.Y. Acad. Sci. 1098: 192–199 (2007). © 2007 New York Academy of Sciences.
doi: 10.1196/annals.1384.032

INTRODUCTION

The goal of this article is to provide some insight into how clinical pathologists think about laboratory testing, in general, and about saliva testing, in particular. The field of clinical pathology (or, equivalently, laboratory medicine) is concerned with performing primarily quantitative assays on fluids or tissues obtained from patients. To this end, clinical pathology laboratories receive virtually any fluid or tissue (e.g., urine, blood, bone marrow, saliva, etc.) of virtually any amount (from a few microliters to several liters, or from several cells to the entire body) and will use any available relevant method (e.g., from gross visualization to DNA sequencing) to measure a clinically relevant analyte. Clinical pathologists (typically individuals with an M.D., Ph.D., or M.D./Ph.D.) oversee the performance of laboratory tests, manage centralized clinical laboratories and POC testing programs, and interpret the results of laboratory tests.

Why are Laboratory Tests Performed?

Laboratory tests are used to (1) diagnose the presence and type of a disease (e.g., determine that a patient is infected with HIV); (2) screen a population for the presence of disease (e.g., measuring serum levels of prostate-specific antigen (PSA) in all older men); (3) determine whether an individual is predisposed to develop a disease (e.g., identify the presence of the inherited Factor V Leiden mutation as a risk factor for developing thromboembolism); (4) provide prognostic information (e.g., an extremely high white blood cell count in peripheral blood at presentation portends a poor prognosis for patients with acute leukemia); and (5) monitor disease status or levels of therapeutic agents (e.g., measuring serum levels of human chorionic gonadotropin in patients treated for choriocarcinoma or measuring blood levels of lithium in patients with depression, respectively).

Building on a tremendous amount of research over the last 150 years, we are currently able to measure a wide variety of different types of analytes. For example, a clinical pathology program at a large academic medical center, such as that at Columbia University Medical Center–New York Presbyterian Hospital, typically offers routine testing to measure approximately 1,500 analytes. In addition, testing for an additional ~1,000 more esoteric analytes is available by sending samples to specialized referral laboratories; at Columbia, we send samples to approximately 20 different referral laboratories, on request, to obtain these results.

Virtually any type of molecule can be a clinically relevant analyte. As examples, the following classes of molecules are routinely evaluated and measured: structural proteins (e.g., troponin I), enzymes (e.g., lactate dehydrogenase), small-peptide hormones (e.g., B-natriuretic peptide), neutral lipids (e.g., cholesterol), sphingolipids (e.g., sphingomyelin), simple carbohydrates (e.g., glucose), complex carbohydrates (e.g., ABH blood group antigens), DNA (e.g., for cystic fibrosis), RNA (e.g., to quantify HIV), ions (e.g., chloride), metals (e.g., lead), "metabolites" (e.g., lactate), cell number (e.g., platelet count), cell morphology (e.g., sickle type red blood cells), crystals (e.g., urate), and "foreign objects" (e.g., bacteria). In addition, certain test methods measure the function and integrity of important physiological systems, rather than of individual analytes (e.g., the prothrombin time and partial thromboplastin time measure the function of the coagulation system; creatinine clearance measures renal function).

Evaluations of the types of molecules described above are usually grouped together into discipline-specific laboratories or test groups. Thus, there are various disciplines in laboratory medicine and it is important to determine which disciplines will benefit from the introduction of saliva testing. The commonly identified disciplines are clinical chemistry (e.g., glucose levels); hematology (e.g., complete blood counts, blood cell morphology, hemoglobin variant detection [e.g., for the diagnosis of sickle cell disease]); transfusion medicine (e.g., blood typing, crossmatching, blood transfusion, therapeutic apheresis); microbiology (containing the subfields of bacteriology, virology, mycology, parasitology (e.g., the distinction between colonization and infection, microorganism detection, susceptibility/resistance to antimicrobials); coagulation (the evaluation of bleeding disorders and thrombotic disorders [e.g., the measurement of factor levels, platelet function]); molecular diagnosis (e.g., identification of cystic fibrosis mutations); immunogenetics (e.g., HLA typing, crossmatching for organ transplantation); immunology (e.g., identification and quantification of autoantibodies); and flow cytometry (e.g., leukemia diagnosis, measurement of lymphocyte subsets).

An important concept in laboratory testing is turn-around-time; although the precise definition of this concept can be controversial, it is generally agreed that shorter turn-around-times are better than longer ones. Turn-around-time varies dramatically based upon the type of analyte being measured and the type of assay being performed, ranging from virtually immediate (e.g., pulse oximetry), to seconds (e.g., POC glucose values), minutes (e.g., troponin levels), hours (e.g., routine blood smear with differential count), days (e.g., blood culture), and weeks (e.g., antibiotic sensitivity determination for *Mycobacterium tuberculosis*).

Global Political Issues

The development of new laboratory tests and the performance and oversight of existing test methods are subject to various political concerns. For example, for better or for worse, the issue of control is very important. With particular regard to saliva testing, a point of contention relates to whether these tests are part of dentistry or medicine, and whether such a discussion is even relevant. For example, in current practice, blood tests are not the purview of hematologists and the evaluation of cerebrospinal fluid is not the purview of neurologists (or neurosurgeons). Rather, these types of testing are supervised in a centralized manner by clinical pathologists. Nonetheless, there still exist "border skirmishes," where this gentlemen's agreement breaks down. For example, in some institutions dermatopathology is the purview of medically trained dermatologists; in others it is controlled by pathology-trained dermatopathologists. Similarly, in some institutions, medically trained clinical hematologists review and interpret bone marrow smears, whereas pathology-trained hematopathologists fill this role in others.

Why Develop Oral Diagnostic Tests Based on Saliva?

In general, new diagnostic tests are developed for measuring a novel analyte, or existing test methods for a given analyte are modified, for one of the following reasons: (1) to provide improved clinical utility, (2) to increase test accuracy or clinical diagnostic accuracy, (3) to provide a less expensive diagnostic approach, (4) to decrease turn-around-time, (5) to increase convenience for the patient and/or the health care provider, and (6) to improve revenue generation (including avoiding existing patent issues [or creating new ones]).

All of the reasons provided above are also relevant to the development of saliva testing. For example, as compared to blood samples, obtaining oral fluid is easier, less invasive, and better tolerated by patients. In addition, few tests that are currently available in typical clinical pathology laboratories are relevant or useful for dentists or other oral health practitioners. Nonetheless, the use of saliva as a platform for performing diagnostic tests for systemic disease, particularly if these are POC tests performed in a dentist's office, raises the political question of whether dentists' offices are going to become general health care screening clinics. If so, it is important to consider how patients will be advised about the results of these tests and to determine the type of follow-up care that will be provided.

Multiple issues need to be considered when oral diagnostic tests for systemic disease are being developed to replace existing test methods using other types of fluid or tissue. A major issue relates to the concept of tissue specificity. Thus, at the current time, even blood or serum assays are not necessarily suitable for evaluating a tissue-based systemic disease. For example, the diagnosis of bacterial meningitis is currently made using cerebrospinal fluid. In this case, the levels of glucose, protein, and cells are quantified. In addition, the types of cells are identified by morphology, gram staining is performed to identify bacteria (if present), and a portion of the cerebrospinal fluid sample is sent to the microbiology laboratory for culture. Although there is certainly room for improvement in this approach, it is currently not even possible to make this diagnosis using, for example, a blood sample; although this would be quite useful since it is far less invasive than a lumbar puncture. Therefore, if it would ever be possible to diagnose bacterial meningitis using tests based on saliva samples, there is a high probability that a considerable amount of research and development will be required.

Laboratory Testing in Large Centralized Laboratories or in the POC Setting

The exponential growth of POC tests in the last 10 years has significantly improved health care and this is highly likely to continue in the future. For example, there is no question that measuring oxygen levels in blood by pulse oximetry is a huge improvement in speed and convenience for the patient and the health care provider as compared to performing arterial blood gas measurements. Similarly, the recent availability of saliva testing to measure antibodies to HIV has revolutionized the ability to diagnose this devastating disorder and has led to marked improvements in public health (for review, see Ref. 1).

However, many issues need to be considered when developing a test to be performed at the POC rather than in a centralized facility. For better or for worse, the performance of POC tests leads to a loss of control by medical authorities. For example, at Columbia, we expect that the operator of a POC testing instrument will perform the test as a true extension of the central laboratory; that is, they will perform the test to the same standards as if they were one of our technologists. Thus, they need to perform these tests without interruptions or distractions, particularly because the reading times after color development may be narrow for some test methods. They also need to use test kits that are not outdated. Unfortunately, these standards of performance are not always met, leading to problems in the quality of the results and also in the reporting of these results. Thus, the design of the test instruments and the robust nature of the test methods are of critical importance. These issues are particularly relevant with regards to home testing.

Issues of productivity and efficiency are also important. For example, marketing brochures and research proposals often discuss the "lab-on-a-chip." Important questions relate to the size of the "lab," the value of the tests being offered for a POC setting, and the time it takes to perform the test. Thus, if the test is being performed by a patient, then the labor is free. However, if the test is performed by a nurse, then this is skilled labor that comes at a significant cost. If no parallel processing is involved (i.e., conveyor belts and robotic instruments can be loaded with many samples in a central laboratory and then the operator can "walk away"), then serially performing individual tests on a group of patients has low throughput and efficiency, and, therefore, can be an expensive proposition. In this case, a significant value-added component needs to be justified for obtaining access to immediate results for those individual patients.

"Cycle of Laboratory Testing"

The performance of laboratory tests can be divided into three phases: preanalytical, analytical, and postanalytical. One can also think of the "cycle of laboratory testing" which encompasses these three phases. The cycle begins when someone (e.g., the physician in a hospital setting, the patient in a home testing setting) comes up with the idea that a test needs to be performed. In a classical hospital setting, the test needs to be ordered, the specimen is then collected and transported to a centralized laboratory, where it is received and accessioned. These steps, many of which are compressed together in the POC setting, constitute the preanalytical phase and need to be controlled in various ways. For example, specimen preservation, prevention of bacterial contamination, transport time, and temperature control can be important. The time of day that the specimen is collected and the posture of the patient (i.e., sitting or lying down) may also be critical. Correct specimen labeling, and the correct labeling of any aliquots derived from the initial specimen, are of particular concern.

The analytical phase involves determination of sample adequacy (e.g., is there enough sample?) and preparation of the sample, if necessary (e.g., centrifugation, pH adjustment and buffering, extraction, etc.). The specimen is then analyzed using the appropriate and relevant method. An extraordinarily wide variety of methods are used for diagnostic testing including chemical methods (e.g., spectrophotometry, atomic absorption), immunoassays (e.g., ELISA, agglutination, flow cytometry), molecular assays (e.g., PCR, sequencing), microscopy (e.g., blood smear, gram stain, fluorescence *in situ* hybridization), and culture (e.g., bacteria, viruses, fungi, and cell culture).

After specimen analysis and the evaluation of the results of quality control samples, the patient's results are verified. The analytical phase is the part of the diagnostic testing process over which the laboratory has the greatest control, which includes equipment maintenance and calibration, documented training and licensure of technologists, and quality improvement programs. This allows the laboratory to measure parameters such as analytical sensitivity, accuracy, and precision.

The postanalytical phase encompasses the reporting of verified results and the assimilation of these results by the ordering physician (in the hospital setting) or the patient (in the home testing setting). An important component of "assimilation" is determining whether the test result fits or does not fit with the patient's presentation and condition; that is, although the test result may be analytically accurate, it may not be clinically correct. For example, a markedly elevated blood glucose level does not necessarily mean that the patient has diabetes; instead the sample may have been drawn from near an intravenous line containing high levels of exogenous glucose.

The assimilation of the test result by the relevant health care provider or the patient completes the "cycle of laboratory testing." The complete turn-around-time is truly the time it takes from the idea of performing the test until the assimilation of the test results. However, in practice, turn-around-time is often defined differently.

How is the Quality of a Laboratory Test Defined?

The concept of quality is frequently used when discussing individual laboratory tests and test methods; however, this concept has different meaning in different contexts. For example, there is the analytical quality of performing the tests itself. This involves using quality control materials, ensuring that the results with quality control samples conform to expectations, and that the individuals performing the test are appropriately trained and are proficient in the given method.

One can also discuss the quality of the laboratory that performs the test. Thus, the laboratory should be enrolled in proficiency testing programs from various sources (e.g., the New York State Department of Health, the College of American Pathologists), the laboratory is subject to various state and federal regulations, and it may be inspected by various state and national agencies (e.g., the Joint Commission on Accreditation of Hospitals, the College of American Pathologists, the Food and Drug Administration, the American Association of Blood Banks, and the Foundation for the Accreditation of Cellular Therapy).

Finally, the diagnostic quality of a given test may relate to its diagnostic sensitivity and specificity (and other analogous types of calculations). Thus, an ideal test for diagnosing a particular condition would have 100% sensitivity and 100% specificity, yielding no false negative and no false positive results. Unfortunately, as yet, there are no such ideal tests. As a result, the diagnostic quality and usefulness of all currently available tests is highly dependent on the prevalence of disease in any test population. Therefore, it is important to use various approaches (e.g., history and physical examination) to increase the prior probability of disease in the patient under study.

SUMMARY

In summary, the use of oral fluid samples in diagnostic laboratory test methods has grown significantly during the recent past and POC testing is growing exponentially. Given patients' preferences for providing saliva samples, rather than blood samples, this trend is highly likely to continue. In addition, there is a high likelihood of developing high-quality tests based on saliva that will be used by various health care practitioners, including dentists, to improve oral health.

REFERENCE

1. Pesce, M.A., K.F. Chow, E. Hod & S.L. Spitalnik. 2006. Rapid HIV antibody testing: methods and clinical utilization. Am. J. Clin. Pathol. **126**(Suppl 1): 561–570.

Oral Diseases

From Detection to Diagnostics

ANTOON J. M. LIGTENBERG,[a] JOHANNES J. DE SOET,[b]
ENNO C. I. VEERMAN,[a] AND ARIE V. NIEUW AMERONGEN[a]

[a]*Department of Oral Biochemistry, Academic Centre for Dentistry, Amsterdam, the Netherlands*

[b]*Department of Oral Microbiology, Academic Centre for Dentistry, Amsterdam, the Netherlands*

ABSTRACT: In addition to saliva, other oral components such as gingival crevicular fluid, epithelial cells, bacteria, breath, and dental plaque have diagnostic potential. For oral diseases such as caries and periodontal disease, visual diagnosis is usually adequate, but objective diagnostic tests with predictive value are desired. Therefore, prediction models like the Cariogram have been developed that also include oral aspects such as saliva secretion, buffering capacity, and *Streptococcus mutans* counts for the prediction of caries. Correlation studies on salivary components and caries have not been conclusive, but correlation studies on functional aspects, such as saliva-induced bacterial aggregation and caries, look promising. Modern proteomic techniques make it possible to study simultaneously the many salivary components involved in these functions.

KEYWORDS: oral; diagnostics; caries; saliva

INTRODUCTION

Fourteen years ago the New York Academy of Sciences supported a conference organized by Drs. Daniel Malamud and Lawrence Tabak entitled "Saliva as a Diagnostic Fluid."[1] The change of title to "Oral-based diagnostics" for the present conference is, in my opinion, more accurate. First, this title circumvents semantic discussions about what saliva is, and whether it is different from oral fluid or not. Especially in the context of diagnostics it is important to realize that oral fluid is more than saliva, which by definition is the fluid secreted by the salivary glands. Oral fluid, in addition, contains contributions from crevicular fluid, epithelial cells, and bacteria. Second, this title implicitly

Address for correspondence: Antoon J. M. Ligtenberg, Department of Oral Biochemistry, Van der Boechorststraat 7, 1081 BT Amsterdam, the Netherlands. Voice: +31-(0)20-444-8674; fax: +31-(0)20-444-8685.
ajm.ligtenberg@vumc.nl

Ann. N.Y. Acad. Sci. 1098: 200–203 (2007).
doi: 10.1196/annals.1384.040

underlines that everything in the mouth in principle may be used for diagnostic purposes. Thus, in addition to saliva, serum components, epithelial cells, dental plaque, salivary bacteria, breath, and even pellicle components have diagnostic potential.

The different aspects of oral diagnostics were exemplified in this session, entitled Oral Diseases. Drs. Ira Lamster and Christoph Ramseier addressed the application of crevicular fluid for diagnosis of periodontal disease. Dr. George Preti dealt with halitosis and trimethylaminuria, a genetic disorder, for which the measurement of volatile substances in breath is important. Dr. Paul Denny and I presented data on the diagnostic value of saliva for the identification of individuals at risk of dental caries.

In many cases the visual observation of caries or periodontal disease leaves little room for doubt, but in other cases, oral diagnostics may be mobilized to: (1) find the cause, (2) monitor the activity of the process (caries and periodontal disease), (3) classify the type of disease or causative agent (periodontal disease), and (4) monitor the success of the therapy.

By definition, dental caries is the demineralization of teeth caused by acids produced by bacteria. Bacteria accumulate as an oral biofilm called dental plaque at those surfaces that cannot be cleaned properly. The consumption of fermentable carbohydrates results in the production of organic acids, primarily lactic acid, thus lowering the pH of the oral biofilm. This low pH favors the outgrowth of *Streptococcus mutans* and lactobacilli, bacteria that are still metabolically active below pH 5.5, the critical pH for demineralization. Thus, it is the prolonged exposure to low pH that results in an ecological shift in the tooth biofilm, making it more acidogenic.[2]

About 30 years ago, the need for a reliable caries prediction model was less urgent than it is today. In those days, sooner or later almost everybody developed dental caries, making periodic visits to the dentist necessary and economically effective. At that time, caries diagnostics was limited to caries detection. Nowadays most visits to the dentist are only for periodic check-up, not treatment. This has raised the question whether, for economical reasons, the frequency of these periodic check-ups can be reduced. However, reduction of periodic checks enhances the risk that developing caries are missed or discovered too late. This has stimulated the need for reliable caries diagnostics with a better predictive value. Caries is a complex, multifactorial disease, involving host, environmental, and behavioral risk factors, making caries prediction very complex. An example of such a caries prediction model is Cariogram[3] (FIG.1; http://www.db.od.mah.se/car/cariogram/cariograminfo.html). Cariogram is a computer model that illustrates a caries risk profile based on weighted interpretation of available information on different etiological factors. This model contains several characteristics that can be demonstrated by oral diagnostics, and also some parameters that have nothing to do with oral fluid.

In this model, saliva secretion, buffering capacity, and counts of mutans streptococci and lactobacilli have proven to be sensitive parameters for

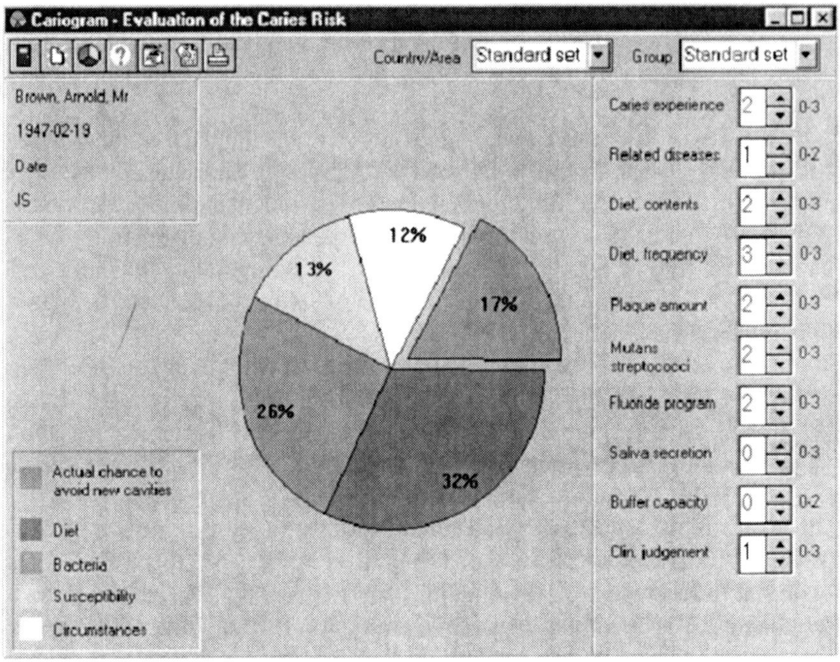

FIGURE 1. Cariogram, example of a computerized caries prediction model. It weighs different etiological factors resulting in a percentage chance to avoid new caries.

determining caries risk. Kits for *S. mutans* and *Lactobacillus* counts are widely used in dental practice and can be conducted without laboratory facilities. They can identify patients at risk for caries, motivate patients, and help in the monitoring of food habits. Individuals with high *Lactobacillus* counts are frequent consumers of sugars.

For salivary secretion, a picture has emerged that as long as the salivary secretion rate is within the normal range, there is no direct correlation with dental caries. For the determination of the salivary buffering capacity, commercial kits are available.

Numerous studies have been published aimed at finding a correlation between caries and the protein composition of saliva, obviously without too much success, which has tempted Colin Dawes to make the statement that "a moratorium should probably be placed on further studies of whole saliva and caries risk."[4] There are several reasons for this poor correlation.

First, saliva output and composition are only two links in a whole chain of events that cause dental caries as is illustrated above in the caries prediction models.

Second, in the context of dental caries, the properties of the plaque fluid, in particular the degree of undersaturation with respect to calcium hydroxyapatite, are relevant, rather than those of whole saliva. Since plaque accumulates primarily at surfaces that are poorly accessible to saliva, the mineral composition of plaque fluid poorly reflects the salivary mineral composition. In addition, saliva composition shows large variations. The circadian rhythm and variation in salivary film velocity in the mouth create the situation that whole saliva composition is only a weak reflection of the local situation.

Third, salivary proteins show overlap in function and, in addition, many proteins have more than one function, making it difficult to correlate dental caries to specific salivary components.[5] In addition, salivary components *in vivo* are present as heterotypic complexes of multiple salivary proteins. It has been shown that the functional properties of such complexes are different from those of the individual proteins. Also, the composition of salivary proteins on the tooth surface is different from that in the soluble phase of saliva, both quantitatively and qualitatively. Therefore, the ratio between different saliva components might well be more important than their concentration. Studies that have focused on functional aspects of whole saliva, rather than studying the quantities of individual proteins, have yielded promising results.[6–8] For example, high bacterial aggregation by saliva has been associated with low incidence of caries. Paul Denny discusses this in his presentation.

Conclusively, we can say that the measurement of secretion rate, buffer capacity, and bacteria in oral fluid already make it an important diagnostic tool in caries prediction models. Modern proteomic techniques give the opportunity to look at many salivary components simultaneously. This may greatly stimulate not only future developments in the field of caries diagnostics, but also enhance our understanding of the role played by saliva in protection against caries.

REFERENCES

1. MALAMUD, D. 1992. Saliva as a diagnostic fluid. Br. Med. J. **305:** 207–208.
2. BRADSHAW, D.J. & P.D. MARSH. 1998. Analysis of pH-driven disruption of oral microbial communities *in vitro*. Caries Res. **32:** 456–462.
3. BRATTHALL, D. & G. HANSEL PETERSSON. 2005. Cariogram—a multifactorial risk assessment model for a multifactorial disease. Community Dent. Oral Epidemiol. **33:** 256–264.
4. DAWES, C. 1998. Non-bacterial salivary predictors of caries risk. *In* Kariesdynamik and Kariesrisiko. L. Stosser, Ed.: 152–156. Quitessenz Verlags-Gmbh. Berlin.
5. RUDNEY, J.D. 1995. Does variability in salivary protein concentrations influence oral microbial ecology and oral health? Crit. Rev. Oral Biol. Med. **6:** 343–367.
6. STENUDD, C., A. NORDLUND, M. RYBERG, *et al.* 2001. The association of bacterial adhesion with dental caries. J. Dent. Res. **80:** 2005–2010.
7. EMILSON, C.G., J.E. CIARDI, J. OLSSON & W.H. BOWEN. 1989. The influence of saliva on infection of the human mouth by mutans streptococci. Arch. Oral Biol. **34:** 335–340.
8. ROSAN, B., B. APPELBAUM, E. GOLUB, *et al.* 1982. Enhanced saliva-mediated bacterial aggregation and decreased bacterial adhesion in caries-resistant versus caries-susceptible individuals. Infect. Immun. **38:** 1056–1059.

A Novel Caries Risk Test

PAUL C. DENNY,[a,b] PATRICIA A. DENNY,[a,b] JONA TAKASHIMA,[a]
JOYCE GALLIGAN,[a] AND MAHVASH NAVAZESH[a,b]

[a]*School of Dentistry, University of Southern California, Los Angeles, California, USA*

[b]*Proactive Oral Solutions, Inc., Los Alamitos, California, USA*

ABSTRACT: A diagnostic test is particularly beneficial if it reveals the level of susceptibility prior to onset of a disease process. In the case of childhood caries, such a diagnostic test affords the opportunity for preventive measures to be implemented before caries begins. Salivary glycoproteins contain a wealth of individually specific oligosaccharide motifs. Depending on microbial compatibilities and individual genotypes, the glycoproteins that form the pellicle coating of teeth may provide attachment sites that foster colonization leading to cariogenesis. Alternatively, certain oligosaccharides, when present in nonpellicle glycoproteins, can interact with planktonic bacteria and lower their ability to interact with the tooth surface. We have found that in young adults the ratio of the two classes of oligosaccharides present in resting saliva exhibits a strong correlation with caries history (DFT: number of decayed and filled teeth). Oligosaccharide moieties associated with the test are quantitated in dried spots of whole saliva on nitrocellulose using commercially available biotinylated lectins with a variety of reporters. A combination of multiple linear regression and neural net analyses were used to develop the algorithms that describe the relationship between oligosaccharide patterns and DFT. During test development several different groups of adults and children have been studied. The correlation algorithms routinely exceed an R^2 (coefficient of determination) of 0.96. When the test is applied to the saliva of children, it yields a projection of their future caries history. Modifying the test result metric to reflect the groups of teeth with caries in young adults, the test identifies those teeth at risk for future caries in children. This test outcome can then be accompanied with suggested specific preventive measures for each tooth group–based risk level.

KEYWORDS: saliva; dental caries risk test; prediction; children; lectins; oligosaccharides; glycoproteins; mucin; caries history; DFT; tooth groups

Address for correspondence: Dr. Paul Denny, School of Dentistry, Rm. 4114A, University of Southern California, 925 W. 34th St., Los Angeles, CA 90089-0641. Voice: 213-740-1406; fax: 213-740-1402. pdenny@usc.edu

**Ann. N.Y. Acad. Sci. 1098: 204–215 (2007). © 2007 New York Academy of Sciences.
doi: 10.1196/annals.1384.009**

INTRODUCTION

Caries risk tests by intent are predictive and may be instrumental in modifying the future course of the disease in an individual by prescribing preventive measures that are beyond usual dental practice. There is need for such a test because caries is not uniformly distributed among the population. Approximately 75% of all childhood caries in the United States occurs in about 25% of children.[1] Thus, the ability to target those at risk is a benefit both from the point of restricting caries formation in those who are susceptible and preventing unwanted overtreatment in those who do not need it.

There have been many studies designed to identify predictors of future caries development, and those predictors should be incorporated into a risk test. In a review spanning more than 20 years of studies, it was concluded that "past caries history" was the best predictor of risk of future caries in every age group analyzed.[2] The next best indicator is titers of oral pathogens in saliva, which was significant for some age groups. Past caries history, whether described as decayed, missing, and filled teeth (DMFT) or decayed, missing teeth, and filled surfaces (DMFS), does not provide an explanation for the heightened risk, and thus may not suggest the best avenue for prevention. Most of all, this metric is not a satisfactory predictor because the level of disease that provides the predictor is no longer preventable. On the other hand, the predictive quality of past caries history indicates that personally related factors, which were associated with the formation of the historical caries, are most likely to be continuing and to be the genesis of the predicted future caries. All in all, this validates the concept that prediction is possible and that it can be based on factors that vary between individuals.

Abusive oral health practices and nutritional habits can lead to elevated caries levels, and indeed, are addressed in popular caries risk assessment questionnaires.[3,4] Similarly, certain health-related conditions, whether naturally incurred or medically induced, can elevate the level of caries acquisition.[5,6] In addition, it has long been suspected that caries susceptibility was also connected to inherent host factors. This appears to be true, but is evident primarily in studies that have focus on the genetic ties of identical twins.[7,8] The failure to show a clear genetic relationship in familial studies, which even include significant numbers of nonidentical twins, suggests that the genetic factors may exhibit a complexity that is beyond current analytical protocols.

With regard to genetics and saliva, it is well known that there are saliva-specific proteins that reflect specific gene expression by one or more of the salivary glands.[9,10] In a more global sense, it has long been known that the oligosaccharides of glycoproteins, which represent common blood types, are present in saliva.[11] Although based on more limited information, it also appears that concentrations of saliva components, such as MUC7 (MG2) and MUC5B (MG1) mucins, and salivary flow rates, show individualized characteristics.[12] The potential for these saliva components, and properties, to exhibit personalized expression is consistent with the above criteria of host genetic or epigenetic control and of variation between individuals.

Of all the salivary analytes, the mucins appear to be prime candidates involved in cariogenesis at the host level. Mucins are specific to epithelial secretions, carry many of the host's oligosaccharide-based blood types, and show host-specific concentrations in resting saliva.[11-13] In addition, MUC7 mucin is known to possess a variety of antimicrobial activities, and to interact with specific oral bacteria in a manner that is consistent with the infectious disease process in other tissues.[14,15] Finally, concentrations of MUC7 mucin in the saliva of elderly subjects were shown to have a significantly negative correlation with the salivary titers of *S. mutans*, a known cariogenic pathogen.[16]

The caries risk test presented in this article has its origins in a relationship that exists between MUC7 mucin, and to a lesser extent MUC5B mucin, and past caries history in young adults. This relationship with these mucins was inadequate for an accurate predictive test in a high percentage of individuals. Thus, there was an evolution of the test to the oligosaccharides present on the salivary mucins, and then ultimately to the oligosaccharides of the entire salivary proteome. It will become evident that this test does not use caries history in its prediction algorithm, but rather, uses oligosaccharide patterns and saliva flow rates to predict caries history. This report presents the current status of this predictive test and discusses some of the steps taken to address the saliva-related issues.

METHODS AND MATERIALS

Subjects

The young adults accrued for this study were volunteer students from the Dental and Dental Hygiene programs at the University of Southern California, School of Dentistry and represented the reference group for development of the caries risk assessment algorithms. The subjects ranged in age from 24 to 34 years. This age range represented a stable statistical endpoint for first-time caries acquisition as judged by a lack of significant correlation of their ages with caries history. The subjects included a variety of races and ethnicities. Although detailed medical histories were taken, the only exclusions from the algorithm development reference group occurred if dental sealants had been applied during childhood, or if the subject had taken a medication with known saliva-altering properties within the last 24 h. The latter subjects were requested to return another day. It is likely that the majority of these subjects were raised in households that practiced mainstream oral health standards and caries-neutral nutritional habits throughout their formative years. Thus, the endpoint of the test, which has been calibrated with these subjects, will reflect caries and risk levels that are largely related to host factors.

Children 7 to 10 years old were selected because they represent an end-point for the caries history of their deciduous teeth (DFS-decayed and filled surfaces) and also because an early record of caries in their permanent teeth is already evident. These children were equally divided between Asians and Hispanics.

Saliva Collection

Resting whole saliva was collected by drooling after a fast of at least 1 h.[17] All dental examinations were visual and followed the saliva collection. For many of the subjects, annual dental examinations spanning 4–5 years were made, and saliva samples were collected and archived at –80°C.

Mucin Quantitation

Equivalent quantities of resting saliva were separated by SDS-PAGE and stained with Stains-All.[18] The mucin bands were quantitated by image analysis. Stains-All binding capacity of MUC7 mucin was abolished by pretreatment of the saliva sample with neuraminidase, suggesting that the amount of sialic acid associated with the mucin was the target analyte.[16]

Oligosaccharide Quantitation

The oligosaccharides were quantitated by dot blot technology. Dried spots containing whole saliva were probed with commercially available lectins. The amount of lectin binding was proportional to the amount of a particular oligosaccharide present in the spot of saliva. The conditions of the assay were limited to an approximate 10-fold range of quantitations. Where the range of a particular oligosaccharide motif exceeded the linear range of the assay, the quantitation range was extended by saliva dilution. For a few lectins, the quantitative range was extended in the opposite direction by assaying more than the usual equivalent of saliva contained in the spot. The bound lectin was then visualized and quantitated relative to the blood glycoprotein, glycophorin, using a streptavidin–alkaline phosphatase-based reporter system. The variation between subjects for different oligosaccharides within the reference group ranged from 3-fold to more than 5,200-fold. The combinations of lectins that are used in the test are proprietary.

Mathematical Procedures

Linear regression has been reported to be one of the best approaches for evaluating correlation of caries risk factors.[2] All results presented here were achieved by multiple linear regression analyses. However, development of the most recent algorithms involved a combination of multiple linear regression and neural net mathematics. SigmaStat™ (SPSS, Inc., Chicago, IL, USA) was used for multiple linear regression, significance, and coefficient of determination (R^2), and the AI Trilogy™ (Ward Systems Group, Inc., Frederick, MD, USA) was used for the neural net mathematics.

RESULTS

MUC7 Mucin and Past Caries History

The first indication that MUC7 mucin was one of the important variables in saliva came from a comparison of mucin concentrations in saliva of Caucasian elderly versus young adult subjects.[18] The young adults displayed an approximate 500-fold range of MUC7 mucin concentrations in their resting saliva. Later, in a series of biweekly saliva collections, the range of personal variation of MUC7 mucin, MUC5B mucin, and resting flow rates, all were more limited than the range of total variation in the test population.[12] This prompted further examination of the original data set,[18] looking for correlation between MUC7 mucin concentrations and the past caries history of the young adults.[19] The significant correlation (FIG. 1) provides the rudiments of a predictive test that has three nonoverlapping zones of significance. As the test applicability was explored beyond Caucasian subjects, we found that the correlation between MUC7 mucin and past caries history did not hold up well in Asian subjects.

Assessment of Specific Oligosaccharide Motifs

The lectin MAL II was used to assess one-dimensional saliva glycoprotein profiles from selected Caucasian and Asian subjects for the protein associations and relative amounts of *sialic acidα-2,3galactose*. This oligosaccharide motif is one of the most prevalent motifs on MUC7 mucin and was shown to be involved in mucin-induced oral bacteria aggregation.[20] Although the total amounts of *sialic acidα-2,3galactose* appeared to be similar for those with apparently similar caries histories in both groups, some of the Asian subjects had the sugar group on glycoproteins other than MUC7 mucin. Test correlations greatly improved for all races by supplementing mucin determination with quantitations for specific oligosaccharides with lectins. Eventually, the test came to rely solely on lectin-based specific oligosaccharide quantitations of

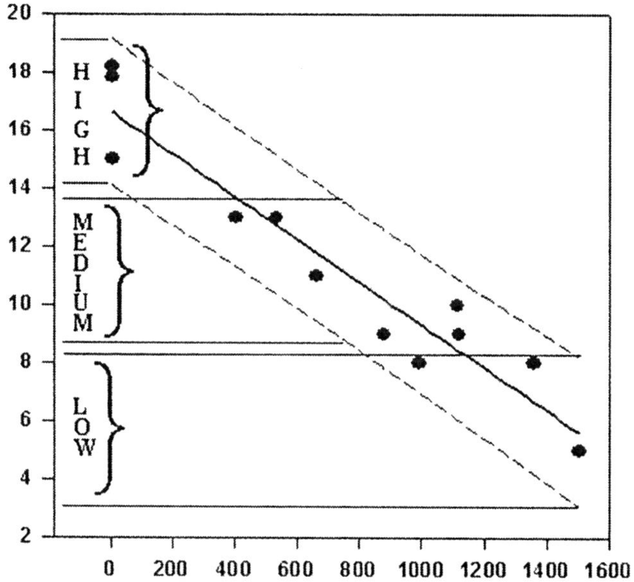

FIGURE 1. MUC7 mucin and caries history. The x axis represents MUC7 mucin concentration (units/mL) and the y axis represents DFT (number of decayed and filled permanent teeth). The regression (*solid line*) and 95% confidence interval (*broken lines*) were derived from the correlations of DFT and MUC7 mucin concentrations in unstimulated saliva collected from 24- to 33-year-old adults of Caucasian ethnicity ($n = 12$). Each pair of horizontal lines indicates the range of DFT within a 95% confidence interval of prediction.

whole resting saliva, using dot blot technology. FIGURE 2 illustrates the relative differences in the amounts of total *sialic acidα-2,3galactose* in saliva samples from more than 60 young adults. There is an approximate 50-fold range in the amount of this sugar motif within the group.

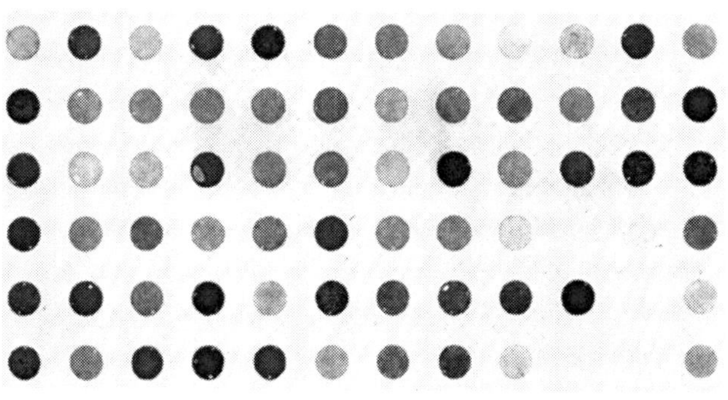

FIGURE 2. Dot blot system for measuring individual amounts of a specific sugar motif. Spots of saliva from a panel of young adults, were assayed for SAα-2,3Gal, using the lectin, MAL II.

Relationship between Oligosaccharide Motifs in Saliva and Past Caries History

After several years of working to improve and simplify the lectin-based test, it appears that the most direct metric linking sugar motifs on the salivary glycoproteins with past caries history is DFT (FIG. 3). This excludes missing teeth (M) because in the young adults of our reference group, this was invariably due to trauma rather than rampant caries. DFS was not used because discretionary decisions in treatment planning can lead to a variety of treatments that range from aggressive to minimalist for the same lesion. Although we use groups of teeth as our test metric, it too probably does not best represent the fundamental relationship between glycoprotein oligosaccharides in saliva and past caries history. Groups of teeth, as the test outcome metric, do largely avoid the subjective issues raised above. However, this metric represents a mathematical simplification of the relationship and relies heavily on the pattern of the developmental appearance of permanent teeth. In spite of the reservations, all the test outcome metrics are correlated with each other, and also show good correlations with the input oligosaccharide data regardless of which metric was used for development of the test algorithm.

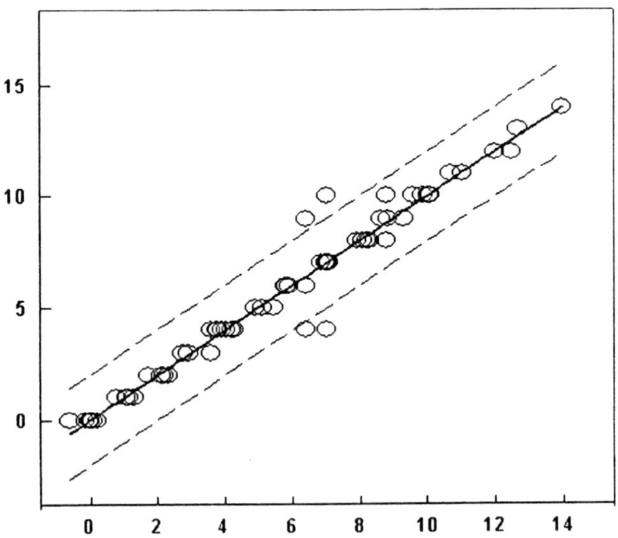

FIGURE 3. Correlation of caries history (DFT) with a select set of quantitated oligosaccharides in resting saliva ($n = 66$) by multiple linear regression. The regression line (*solid*) and confidence interval of 98% (*broken lines*) represent a significant ($P \leq 0.001$) and predictive ($R^2 = 0.961$) relationship.[21]

Method for Creating the Tooth Group Risk Algorithm and Its Application to Saliva of Children

The tooth group outcome metric was chosen because the result of the test points directly to a tooth-specific preventive treatment plan. Development of the caries risk test, which yields the groups of teeth at risk in children, began with an evaluation of the diagnostic information for the reference group of young adults. The caries history information of the reference subjects, which had been catalogued on an individual tooth basis, was grouped into a variety of developmental- and positional-based configurations. Each configuration was tested against the oligosaccharide motif patterns present in the saliva of the reference subjects. The grouping that provided the best fit with lectin affinity patterns was a general back-to-front correlation with increasing caries history or risk level. Risk levels related to tooth groups in this orientation were chosen because of the possibility of targeted preventive treatment plans. Other positional parameters, such as right, left, upper, and lower quadrants, showed no statistical correlation with caries history in the adult reference group.

The finalized metric for groups of teeth is shown in FIGURE 4. The screening exercise that was pursued identified a sixth group, which included caries in molars, premolars, and incisors. However, it was folded back into the fifth risk level because there are no specific preventive measures that could be added beyond those recommended for risk level 5. The screening process also identified a distinguishable risk level 2. This is the low-slow risk level because it only involves one or two teeth, and the earliest the first cavity appeared at this risk level within the reference group was at 15 years of age. This risk level was added because the approach to preventive treatment is different than for the higher levels. Once these groupings of teeth were determined, the algorithm was further optimized and streamlined to maximize accuracy, using the fewest number of lectins. Part of this optimization included testing saliva flow rates. Somewhat surprisingly, normal flow rates contributed substantially to simplification of the algorithm developed to describe the relationship portrayed in FIGURE 4.

The algorithm that was developed in the reference adult group is shown in FIGURE 5 as applied to children. The first step was to test the children's saliva samples with the same lectins used to develop the algorithm, and then to fold in their saliva flow rates. The result of this exercise was to group the children into the same tooth group–based risk levels that emerged from the adult reference group. However, in this case, the risk levels make a projection of the groups of teeth that are likely to acquire caries by 24 years of age, if preventive measures are not applied before the caries first appear. In fact, nearly half of the 7- to 10-year-old children in this group already had at least one cavity in their first set of four permanent molars. The caries risk test correctly assigned those children with caries to risk levels 3 to 6.

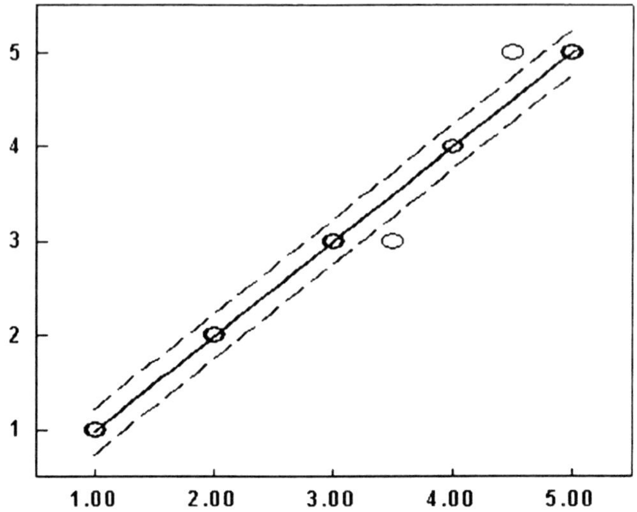

FIGURE 4. Prediction of risk levels in children: Test development. Resting saliva was collected from a group of young adults ($n = 66$) and assessed for a variety of sugar motifs with a select group of lectins. These results were correlated with the caries history of each adult, expressed as groups of teeth with caries that then equated to a risk level. The derived relationship was highly significant ($P \leq 0.001$) and individually predictive ($R^2 = 0.995$).[21] Risk levels: 1, no caries; 2, caries limited to one or two teeth; 3, caries only in molars; 4, caries in molars and adjacent premolars (p2); 5, caries in molars and both sets of premolars.

DISCUSSION

Development of a predictive test that uses saliva as the diagnostic medium presents a variety of issues that must be addressed. Some of the concerns are largely specific to saliva and others more generally related to the predictive aspect of the test. Criteria for selection of the reference subjects and for establishing specific risk levels were discussed above. Chief among the saliva-specific questions is whether to focus on resting or stimulated saliva samples for test development and application. Focus of development of the caries risk test has been on resting saliva. We view resting saliva as a representative of the natural oral fluid environment that exists during the major portion of a daily cycle. By contrast, stimulated saliva is more transient and variable in composition, depending on the nature of the stimulus.[22] Though still in early development, we anticipate that the final version of the caries risk test will require approximately 0.1 mL of saliva. Collection of this amount of resting saliva in 5 minutes or less has been accomplished for all but a rare few who either have chronic xerostomia or have taken a medication with this side effect.

Another equally important property of saliva that enters into test performance is the normal variation of analytes. An accurate caries risk test must

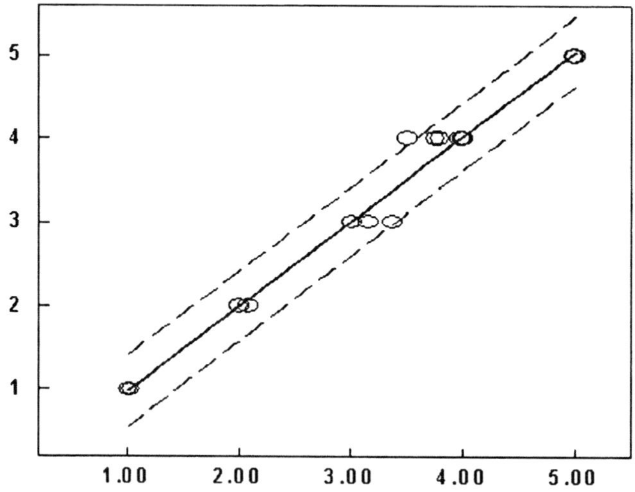

FIGURE 5. Prediction of risk levels in children: Test application. Resting saliva was collected from 7- to 10-year-old children ($n = 27$) and assessed with the battery of lectins used to develop the predictive algorithm in young adults. The results from each child's saliva were processed with the predictive algorithm, yielding a risk level. The predictive risk levels refer to the groups of teeth that are destined to have caries by 24 years of age.[21] Risk levels are described in FIGURE 4.

accommodate ranges of variation from an individual and still yield the same risk level. This is presently a work in progress; however, we believe that there are mathematical solutions to the problem. One example may be viewed relative to the correlation of MUC7 mucin and caries history (FIG. 1). Extension of horizontal lines from the 95% confidence interval lines to the vertical axis, as drawn, yields ranges of DFT that are statistically related to a single MUC7 mucin concentration point on the slope of the regression line. On the other hand, if vertical lines are dropped from the two confidence interval intersects with one of the horizontal lines, a range of mucin concentrations is inscribed.

The range of mucin concentrations that fall within this 95% confidence interval could reflect variation introduced from several areas. However, this range of mucin concentrations matches the maximum individual variation of MUC7 mucin concentration observed in biweekly collections of saliva from a group of young adult subjects.[12] We do not believe that this is a coincidence but rather a statistical solution that encompasses all the variations associated with MUC7 mucin concentrations, of which individual variation is the greatest contributor. We have recently extended the concept of individually limited variation to the analytes used in the caries risk test for periods of 3 or 4 years in a group of 30 young adults and children.[23]

In the example of variation in the MUC7 mucin concentration, a test that uses this information might be improved by modifying the concentration scale to a metric that is keyed to the maximum variation, such as the full or one-half width of the 95% confidence interval. As was discussed earlier relative to development of risk levels, this type of data management can easily be evaluated for best fit against the selected output metric.

ACKNOWLEDGMENT

This project is supported by NIDCR STTR phase I and phase II grants, R42 DE014650, "Saliva Test for Caries Risk" to Proactive Oral Solutions, Inc. (P.C. Denny, PI). Clinical studies were performed under subcontract to USC School of Dentistry (J. M. Galligan, PI).

CARE test™ is the registered trademark of Proactive Oral Solution, Inc. The commercial test is in mid-phase development and will soon be market-ready.

REFERENCES

1. U.S. Dept. of Health & Human Services. 2001. National Institutes of Health Consensus Development Conference Statement. Diagnosis and Management of Dental Caries Throughout Life. NIH Consensus Statement Online 2001 March 26–28. **18:** 1–24.
2. Powell, L.V. 1998. Caries prediction: a review of the literature. Community Dent. Oral Epidemiol. **26:** 361–371.
3. Featherstone, J.D.B. *et al.* 2003. Caries management by risk assessment: consensus statement, April 2002. J. Cal. Dent. Assoc. **31:** 257–269.
4. American Academy of Pediatric Dentistry, Oral Health Policies. 2004. Pediatr. Dent. **26:** 16–61.
5. Anderson, M.H. *et al.* 1993. Modern management of dental caries: the cutting edge is not the dental bur. J. Am. Dental Assoc. **124:** 37–44.
6. Featherstone, J.D.B. 2000. The science and practice of caries prevention. J. Am. Dental Assoc. **131:** 887–899.
7. Bretz, W.A. *et al.* 2005. Longitudinal analysis of heritability for dental caries traits. J. Dent. Res. **84:** 1047–1051.
8. Hassell, T.M. & E.L. Harris. 1995. Genetic influences in caries and periodontal diseases. Crit. Rev. Oral Biol. Med. **6:** 319–342.
9. Cohn, R.E. *et al.* 1991. Immunocytochemistry and immunogenicity of low molecular weight human salivary mucin. Arch. Oral. Biol. **36:** 347–356.
10. Troxler, R.F. *et al.* 1995. Molecular cloning of a novel high molecular weight mucin (MG1) from human sublingual gland. Biochem. Biophys. Res. Commun. **217:** 1112–1115.
11. Prakobphol, A. *et al.* 1993. The high-molecular-weight human mucin is the primary salivary carrier of ABH, Le^a, and Le^b blood group antigens. Crit. Rev. Oral Biol. Med. **4:** 325–333.
12. Denny, P.A. *et al.* 2002. Individual variation of mucin concentration in human saliva. J. Dent. Res. **81:** A–501.
13. Nieuw Amerongen, A.V. *et al.* 1995. Salivary mucins: protective functions in relation to their diversity. Glycobiology **5:** 733–740.

14. PRAKOBPHOL, A. *et al.* 1999. Separate oligosaccharide determinants mediate interactions of the low-molecular-weight salivary mucin with neutrophils and bacteria. Biochemistry **38:** 6817–6825.
15. SHARON, N. 1996. Carbohydrate-lectin interactions in infectious disease. Adv. Exp. Med. Biol. **408:** 1–8.
16. BAUGHAN, L.W. *et al.* 2000. Salivary mucin as related to oral *Streptococcus mutans* in the elderly. Oral Micro. Immunol. **15:** 10–14.
17. NAVAZESH, M. & C.M. Christensen. 1982. A comparison of whole mouth resting and stimulated salivary measurement procedures. J. Dent. Res. **61:** 1158–1162.
18. DENNY, P.C. *et al.* 1991. Age-related changes in mucins from human whole saliva. J. Dent. Res. **70:** 1320–1327.
19. DENNY, P.C. *et al.* 2003. Correlation of saliva mucin concentrations with caries history in Caucasian young adults. J. Dent. Res. **82**(Special Iss. B): 250.
20. MURRAY, P.A. *et al.* 1986. Preparation of a sialic acid-binding protein from *Streptococcus mitis* KS32AR. Infect. Immun. **53:** 359–365.
21. DENNY, P.C. *et al.* 2006. A novel saliva test for caries risk assessment. C. D. A. Journal **34:** 287–294.
22. AGUIRRE, A. *et al.* 1993. Sialochemistry: a diagnostic tool? Crit. Rev. Oral Biol. Med. **4:** 343–350.
23. DENNY, P.C. 2006. Unpublished results.

Analysis of Gingival Crevicular Fluid as Applied to the Diagnosis of Oral and Systemic Diseases

IRA B. LAMSTER AND JOSEPH K. AHLO

Columbia University, College of Dental Medicine, New York, New York, USA

ABSTRACT: Gingival crevicular fluid (GCF), a serum transudate or inflammatory exudate, can be collected from the gingival crevice surrounding the teeth. As such, the fluid reflects the constituents of serum, the cellular response in the periodontium, and contributions from the gingival crevice. The study of GCF has focused on defining the pathophysiology of periodontal disease, and identification of a potential diagnostic test for active periodontitis. The majority of markers that have been identified as potential candidates for such a test are measures of inflammation (i.e., prostaglandin E2 (PGE2), neutrophil elastase, and the lysosomal enzyme β-glucuronidase). Further, analysis of inflammatory markers in GCF may assist in defining how certain systemic disorders (e.g., diabetes mellitus) can modify periodontal disease, and how periodontal disease/periodontal inflammation can influence certain systemic disorders (i.e., cardiovascular/cerebrovascular diseases). Methodological concerns related to the collection and analysis of GCF are important factors that need to be considered when studying GCF. Practical concerns argue against the widespread clinical application of GCF as an adjunct to periodontal diagnosis. Rather, analysis of GCF-derived mediators in saliva may serve as a means of rapid screening for periodontal disease.

KEYWORDS: gingival crevicular fluid; diagnosis; periodontal disease

INTRODUCTION

Gingival crevicular fluid (GCF) can be collected from the gingival sulcus surrounding the teeth, and exists as either a serum transudate or more commonly as an inflammatory exudate. The constituents of the fluid are derived from a variety of sources. GCF contains substances from the host as well as from microorganisms in the subgingival and supragingival plaque. Constituents from the host include molecules from blood, and contributions from cells and

Address for correspondence: Ira B. Lamster, D.D.S., M.M.Sc., Columbia University College of Dental Medicine, 630 West 168th Street, Box 20, New York, New York 10032. Voice: 212-305-4511; fax: 212-305-7134.

IBL1@columbia.edu

Ann. N.Y. Acad. Sci. 1098: 216–229 (2007). © 2007 New York Academy of Sciences.
doi: 10.1196/annals.1384.027

tissues of the periodontium. The latter includes the vasculature, epithelium, nonmineralized and mineralized connective tissues, as well as the inflammatory and immune cells that have infiltrated into the periodontal tissues. Among the important host-derived constituents in GCF are markers of inflammation, including enzymes, cytokines, and interleukins. Further, products of tissue breakdown can also be detected in crevicular fluid.

GINGIVAL CREVICULAR FLUID

The potential diagnostic value of GCF and the dynamic nature of the fluid was recognized a half century ago. In the late 1950s Brill and Krasse[1] demonstrated that filter paper placed in the gingival sulcus of teeth in experimental animals could detect a dye that was injected into the systemic circulation. From a physiological perspective, Egelberg *et al.*[2] observed that following gingival stimulation there was an increase in vascular permeability of the blood vessels, leading to an accumulation of fluid in the sulcus.

Studies in the 1970s began to identify enzymes and other indicators of the host response in crevicular fluid,[3,4] but interest in GCF was modest until longitudinal clinical trials identified the episodic nature of periodontitis,[5] and the importance of the host response to the progression of periodontal disease came into focus.[6] A natural outgrowth of these studies was the search for a diagnostic test for the active phases of periodontitis that could be used in a clinical environment. Before this could be accomplished, certain methodological issues needed to be addressed.

Traditionally, methods of collection of GCF include the use of micropipettes, appliances to isolate and collect fluid from the gingival margin, and small filter paper strips. The most clinically applicable method uses precut methylcellulose filter paper strips that are placed in the sulcus (FIG. 1). The fluid absorbed on the strips is then eluted and analyzed. This approach offers the potential of a noninvasive means of assessing the host response in periodontal disease. The filter strip method of fluid collection can, however, be time consuming and technique sensitive. Attention must be given to strip placement so that the sample is relatively free of plaque (supragingival plaque is removed), and there is no contamination by saliva and blood. It has been shown that both plaque and saliva on the test strip can influence the volume of fluid that is collected.[7] Also, the strip must remain in the sulcus long enough to obtain an adequate sample of fluid. Concern has also been raised about the approach to reporting data that examined the levels of host mediators in GCF. The problem is related to the available volume of GCF. Unlike the analysis of serum, where the sample of fluid is a small part of the total fluid volume, sampling of GCF often collects the entire volume of fluid at the sampled site, and this volume varies from tooth site to tooth site. As a result, Lamster *et al.*[8,9] developed an approach to GCF sampling that standardizes the time of collection and reports the data as total amount (or activity) in the timed sample.

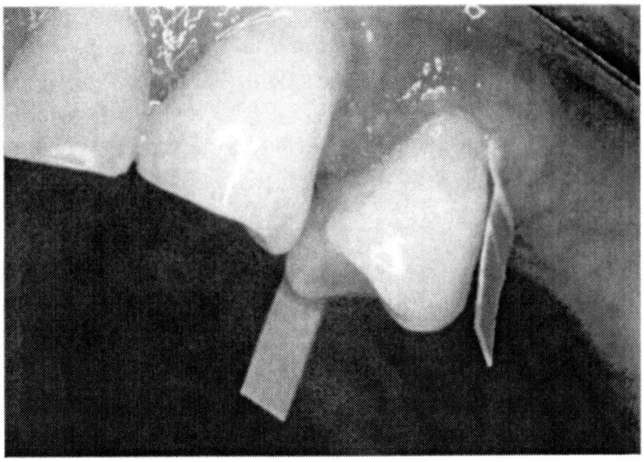

FIGURE 1. Gingival crevicular fluid is most often collected with methylcellulose filter paper strips inserted into the crevice. Standardization is obtained with timed sampling (30 seconds).

GCF ANALYSIS IN PERIODONTAL DISEASE

The volume of GCF present at a given site may be directly related to tissue inflammation (including vascular permeability and the nature of the inflammatory infiltrate) as well as permeability and ulceration of the crevicular epithelium. Sites characterized as being moderately or severely inflamed demonstrate a greater volume of GCF than less inflamed sites.[10] Nevertheless, no studies have demonstrated that an increased volume of fluid in the crevice is related to the risk for periodontal tissue destruction.

With the realization that clinical measures alone are limited as disease predictors, attention was placed on developing a reliable and clinically applicable diagnostic test to evaluate the risk of active periodontal disease. While GCF analysis has included studies of gingivitis and periodontitis, much of the attention has focused on the more severe of the two conditions. Similar to other chronic diseases, periodontitis is characterized by periods of active tissue destruction and quiescence. Consequently, a practical, predictive diagnostic test would be an important clinical advance.[11] According to Armitage,[11] more than 65 GCF constituents have been evaluated as potential diagnostic markers of periodontal disease progression. These markers can be divided into three groups: host-derived enzymes and their inhibitors, inflammatory mediators and host-response modifiers, and byproducts of tissue breakdown.

Further, Loos and Tjoa[12] reviewed potential diagnostic markers of periodontitis present in GCF. Their review identified eight potentially valuable markers, including alkaline phosphatase, β-glucuronidase (βG), cathepsin B, collagenase-2 (matrix metalloproteinase, MMP-8), gelatinase (MMP-9), dipeptidyl peptidase (DPP) II and III, and elastase. The predominance of enzymes associated with tissue breakdown clearly emphasizes the importance of the inflammatory response to the pathogenesis of periodontitis. A number of these markers, as well as a few others, have received particular attention in terms of clinical application and attempts to develop commercial test kits.

Alkaline phosphatase is a membrane-bound glycoprotein that is involved in maintenance of alveolar bone and renewal of the periodontal ligament. Produced by many cells, alkaline phosphatase in GCF is believed to originate primarily from polymorphonuclear leukocytes (PMNs). Similar levels of alkaline phosphatase in GCF have been found in gingival health and experimental gingivitis,[13] but a longitudinal study demonstrated that elevated alkaline phosphatase levels preceded clinical attachment loss and that the total amount of alkaline phosphatase in GCF was significantly higher in active sites.[14] MMP-8 is a collagenolytic enzyme that is found in many cells, but PMNs are considered the primary source of this enzyme in GCF. In a longitudinal study of patients with gingivitis and stable and progressive periodontitis, MMP-8 activity was significantly elevated in patients with active periodontitis.[15] Additional data from patients receiving treatment supports the diagnostic application of MMP-8 in GCF.[16] A recent study by Beklen *et al.*[17] examined levels of MMP-3 in periodontal patients. Specifically, activated gingival tissue fibroblast-derived MMP-3 was found to activate neutrophil-derived pro-MMP-8 and pro-MMP-9 in GCF, suggesting the importance of MMP cascades in the pathogenesis of periodontitis.

Cathepsin B is a cysteine protease involved in proteolysis. Kunimatsu *et al.*[18] observed that levels of cathepsin B were increased in periodontitis when compared to gingivitis, despite similar GCF flow. The source of cathepsin B in GCF is mainly macrophages, and analysis of cathepsin B in GCF appears to differentiate chronic gingivitis from periodontitis.[19] Furthermore, GCF levels of cathepsin B correlate significantly with clinical parameters before and after periodontal treatment, suggesting a use for this enzyme in assessment of treatment outcomes.[20] Eley and Cox[21] evaluated DPP II and IV activity in GCF in relationship to changes in the probing attachment level in patients with periodontitis. DPP II is a lysosomal enzyme that is found in macrophages and fibroblasts, while DPP IV is found in macrophages, fibroblasts, and CD4 and CD8 lymphocytes. High sensitivity, specificity, and negative predictive values (>99%), as well as moderate positive predictive values (55–70%) were seen with the use of these enzymes to identify sites at risk for disease activity.

In addition, elastase and βG were also included by Loos and Tjoa[12] and aspartate aminotransferase (AST) and prostaglandin E2 (PGE2) must also be considered because of attention focused on these markers.

Neutrophil elastase (or elastase) is a potent proteolytic enzyme found in lysosomal granules. Elastase levels in GCF increase with induction of experimental gingivitis, and decrease when plaque removal is reinsitituted.[22] In a longitudinal study, Eley and Cox[23] demonstrated that increased elastase in GCF was predictive of periodontal attachment loss. Long-term observation of adult patients with periodontitis undergoing supportive periodontal therapy showed a positive correlation of elastase in GCF with clinical attachment loss.[24]

βG, a lysosomal enzyme that degrades proteoglycans and ground substance, and serves as a marker for primary grade release from PMNs, has been studied as a possible diagnostic marker in GCF. Lamster *et al.*[25] reported βG to have a high sensitivity and specificity when related to occurrence of clinical attachment loss. In a multicenter, 6-month study of the relationship of βG to progression of periodontal disease, Lamster *et al.*[26] related elevated βG in GCF to various thresholds of increased probing depth or probing attachment loss. The relative risk of disease progression when elevated levels of the enzyme were present ranged from 6 to 14 times. This analysis reported data for the patient but not the individual sites. Nevertheless, this study emphasizes the association of a prominent influx of PMN cells into the crevice (as a measure of the intensity of the acute inflammatory response) to destructive periodontitis.

AST is an enzyme found at increased concentration in red blood cells, liver, heart, and muscle. Formerly referred to as serum glutamic oxaloacetic transaminase, AST is a cytoplasmic enzyme, and its presence in a biologic fluid is indicative of tissue damage and cellular necrosis. Chambers *et al.*[27] found that there was an increase in GCF levels of AST during the induction of experimental periodontitis using a beagle dog model. Furthermore, Persson *et al.*[28] analyzed the relationship between GCF levels of AST and active periodontal tissue destruction in patients who were being treated for chronic periodontitis. Median AST levels were significantly higher for the sites experiencing clinical attachment loss in comparison to sites not experiencing attachment loss. Similar results were reported by Chambers *et al.*[29]

PGE2 is a byproduct of arachidonic acid metabolism and is an important proinflammatory mediator. It is released from various cell membranes by action of the enzyme cyclooxygenase. Some of its effects include inflammatory cell chemotaxis, induction of collagenase release, vasodilation, and osteoclast activation, all of which can directly or indirectly contribute to the loss of the supporting tissues of the teeth. The role of PGE2 in inflammation is also seen in its effect on pain receptors, resulting in an enhanced response to normal pain stimuli. PGE2 was originally identified in GCF in the mid 1970s[4] and subsequently studied in relation to periodontal diseases. Offenbacher *et al.*[30] showed that there were differences in the GCF concentration of PGE2 in patients with gingivitis compared with periodontitis. Subsequently, it was found that there was a correlation between increased PGE2 concentration and clinical attachment loss in patients who were diagnosed with moderate-to-severe periodontitis.[31]

In addition, evidence exists suggesting that proinflammatory cytokines in particular IL-1β, may play an integral role in the etiology of periodontal disease. Liu *et al.*[32] found that with an increase in gingival index and probing, there was a corresponding increase in IL-1β in both the gingival tissue and

GCF. A longitudinal study by Engebretson *et al.*[33] evaluated IL-1β levels in GCF from patients with varying degrees of periodontal disease following scaling and root planing. An important observation was that, compared to shallow sites in patients with mild/moderate periodontitis, shallower sites in patients with severe periodontitis demonstrated increased levels of IL-1β. This suggests that GCF IL-1β expression is genetically influenced and not solely a result of local clinical parameters.

Breakdown products, such as glycosaminoglycans (GAG) and pyridinoline cross-linked carboxy terminal telopeptide of type I collagen (ICTP), have been analyzed using experimental animal studies. Shibutani *et al.*[34] were able to demonstrate an association between an increase in GCF content of GAG and periodontal disease destruction, while Giannobile *et al.*[35] demonstrated a significant increase in ICTP in GCF in an animal model of ligature-induced periodontal destruction.

In addition, fluid can be collected from the gingival cuff that surrounds dental implants (peri-implant fluid, PIF).[36] PIF has been analyzed for the presence of inflammatory mediators primarily identified in GCF, including elastase, α-2 macroglobulin, and alkaline phosphatase,[37] as well as other markers of inflammation (calprotectin) and measures of osteolysis (cross-linked N-terminal telopeptides).[38] Nevertheless, there have not been any longitudinal trials examining the relationship of specific markers in PIF to the loss of osseous support about the implant or loss of the implant.

Although these GCF markers are promising as diagnostic tests, limitations to the application of a GCF-based diagnostic test clearly exist. First, by the nature of filter paper-based GCF collection, there is a need to collect multiple samples from each patient. Second, selection of the teeth and sites at risk for disease progression is often difficult. Third, laboratory diagnostic tests to manage periodontal disease are not routinely employed for dental disease, and an in-office system to analyze multiple patient samples is unlikely to be easily adopted by clinicians.

GCF RELATED TO SYSTEMIC CONDITIONS

GCF analysis can be used to study how systemic diseases and conditions may influence the progression of periodontal disease, and may eventually be used to assess how periodontal disease influences the progression of certain systemic diseases. In addition, GCF can be analyzed to determine whether specific markers of systemic disease can be identified in the oral cavity. In this last category, most of the studies have focused on identification of markers of infectious diseases.

Alterations Associated with Systemic Disorders

Within the past 15 years, the importance of diabetes mellitus as a risk factor for periodontitis has been defined. Studies have been conducted that examine mediators in GCF from patients with diabetes mellitus. Levels of PGE2 and IL-1β in GCF from patients with insulin-dependent diabetes mellitus (IDDM; now type 1 diabetes) have been examined.[39] These mediators were higher in the IDDM group (regardless of the severity of periodontal disease) versus the systemically healthy group. This may be as a result of a heightened systemic inflammatory response associated with diabetes. Kurtis *et al.*[40] measured IL-6 levels in GCF from patients with non-insulin-dependent diabetes mellitus (NIDDM; now type 2 diabetes), adult periodontitis, and healthy controls. Higher IL-6 levels were observed in the NIDDM and periodontitis patients versus the healthy group. Further, neutrophil elastase in GCF has been shown to be related to clinical measures of periodontitis in patients with type 1 and type 2 diabetes mellitus, but HbA1c levels (as an indicator of metabolic control) did not correlate with clinical measures of periodontal status or GCF elastase.[41] Vascular endothelial growth factor (VEGF) in GCF was increased in periodontally inflamed tissues as well as healthy sites of patients with diabetes mellitus, implying that the VEGF effect may be involved with either the pathology or healing phase of periodontal disease.[42] A study by Engebretson *et al.*[43] found that there was a correlation between poor glycemic control and increased levels of IL-1β levels in GCF, emphasizing a link between poorly controlled diabetes and periodontal disease severity.

The established role of MMP in the progression of periodontal disease led to the study of these enzymes in patients with diabetes. Safkan-Seppala *et al.*[44] examined GCF levels of total collagenase in patients with type 1 diabetes and suggested that periodontal tissue destruction in these patients may be mediated by collagenase activity. GCF enzyme activity in patients with poorly controlled diabetes was slightly, but not significantly, elevated as compared to patients without diabetes but with chronic periodontitis.

Smoking tobacco is considered to be the most important environmental risk factor for periodontal disease. Because periodontal pathogens in smokers have been found to be similar to what is seen in nonsmokers, the role of the host response has been examined. Gingival microcirculation, as assessed by both gingival blood flow and volume of GCF, was reduced in smokers, and recovered following smoking cessation.[45] Cytokine profiles of patients with experimental gingivitis have been studied in relationship to smoking.[46] Throughout the study, smokers demonstrated higher levels of IL-8, but low levels of IL-4, than were found in nonsmokers. Cytokine profiles in GCF from patients with aggressive periodontitis suggested that smoking has an effect on the cytokine network.[47] In particular, IL-6 was elevated in patients with periodontitis who smoked. Prior to therapy, transforming growth factor beta-1 (TGF-β1) levels in GCF from smokers were greater than in nonsmokers. After initial periodontal treatment, smokers exhibited less of a reduction in GCF volume, suggesting an altered vascular response in these patients.[48] In contrast, the IL-1α concentration in GCF has been found to be lower in smokers versus nonsmokers.[49] Further,

ICTP levels have been analyzed in PIF from patients who smoke. These levels were found to be elevated in smokers, implying that smoking could negatively affect osseous support around implants.[50]

The relationship of cerebrovascular disease/cardiovascular disease and periodontal disease is an area of special interest in periodontology. Though only limited data are currently available, levels of inflammatory mediators in GCF in relation to risk for cerebrovascular disease have been analyzed. Back *et al.* evaluated leukotriene levels in GCF in patients with atherosclerosis.[51] It was found that there was an increase in leukotriene B4 levels in periodontitis patients, and higher cysteinyl–leukotriene levels in GCF in atherosclerotic patients with and without periodontitis, suggesting that cysteinyl–leukotrienes in GCF could prove to be an important inflammatory marker for an increase in risk for atherosclerosis associated with periodontal disease. Other studies have examined the constituents of GCF in patients with other systemic disorders, but many of these are case reports or isolated cross-sectional trials.[52–55]

In HIV-infected patients, Grbic *et al.*[56] observed an increase in GCF IgG antibody and IL-1β in deep periodontal sites together with a decrease in βG activity in HIV-positive patients. The absence of a correlation between PMN activity in the gingival crevice and clinical measures of existing periodontal disease suggested a local manifestation of immune dysregulation.[57] High levels of IL-1β, IL-6, and TNF-α in GCF were found to be associated with periodontal disease in HIV-1-infected patients versus uninfected patients, suggesting the enhanced cytokine levels may be responsible for the advanced periodontal lesions observed in HIV+ patients.[58] Also, Alpagot *et al.*[59] reported that higher GCF levels of IFN-γ in patients with HIV infection were found at progressing sites versus nonprogressing sites, suggesting the importance of this cytokine to the progression of periodontitis in seropositive patients.

Evidence of Systemic Disease in GCF

Although most of the findings relating to GCF and HIV infection have involved alterations of mediators in GCF, Suzuki *et al.*[60] were able to observe the presence of HIV in leukocytes present in GCF. These data suggested that these cells could be the intraoral source of the virus.

Hepatitis B (HBV) and C (HCV) markers in GCF have also been studied. Ben-Aryeh *et al.*[61] were able to detect HBV surface antigen (HBsAg) in GCF and whole saliva in approximately 90% of samples from seropositive patients, implying that GCF is likely the source of HBsAg in saliva. Anti-HCV antibodies were detected in the GCF of HCV-positive patients,[62] again reflecting the contribution of serum to GCF. Recently, analysis of GCF and saliva for the concentration of HCV RNA revealed a higher detection rate and greater levels of RNA in GCF versus saliva.[63,64] These data support the concept that GCF may be a significant source of hepatitis virus in saliva.

MEDIATORS IN GCF DETECTED IN SALIVA

Whole or mixed saliva is a combination of oral fluids that originates from secretions of the minor and major salivary glands, bronchial and nasal secretions, serum and blood derivatives from oral wounds, bacteria and bacterial byproducts, viruses, fungi, desquamated epithelial cells, food, cellular components, and GCF.[65,66] GCF-derived constituents in saliva have been analyzed as an approach to development of a diagnostic test for periodontal disease.

As noted, analysis of GCF collected from individual sites/teeth offers the potential of diagnostic information for individual sites and teeth. Nevertheless, collection can be technically challenging and time consuming. Further, while the periodontal disease can be localized to sites or teeth, patients tend to demonstrate patterns of affected teeth, and can be characterized as such.[67,68] Consequently, since GCF flows into the oral cavity, analysis of mediators in whole saliva may provide a basis for the development of a simple patient-based diagnostic test for periodontal disease. From a practical standpoint, analysis of saliva offers some distinct advantages when used for diagnostic purposes. Whole saliva can easily be collected in a noninvasive manner, without the need for special equipment. Further, salivary analysis may provide a feasible, cost-effective approach for large-scale screening of patients. An example will be provided of an inflammatory marker identified in GCF, which has been studied in saliva.

GCF levels of βG have been shown in a 6-month trial to identify patients at risk for active periodontal disease as defined by an increase in attachment loss or increase in probing depth at a number of sites.[25,26] Subsequently, an analysis of salivary levels of βG was evaluated in relationship to periodontal parameters.[69] Unstimulated whole saliva samples were available from 380 patients who also had a comprehensive periodontal examination. Analysis revealed strong positive correlations between salivary levels of βG and mean probing depth, mean gingival index, and the number of sites with a probing depth 5 mm and higher. When βG was elevated (defined as ≥ 100 units), logistic regression revealed that the odds ratio that periodontitis (defined as ≥ 4 sites with ≥ 5 mm of probing depth) was present was 3.77. In comparison, the odds ratio for current and former smokers was 3.15 and 2.29, respectively. These data provide evidence that increased salivary levels of βG, representative of a GCF marker that has been linked to an increased risk of periodontitis, could be used as a diagnostic marker/screening test for periodontitis.

CONCLUSIONS

Study of GCF has contributed to the understanding of the host response in periodontal disease. In the past, individual assays were used to analyze individual GCF constituents. Today, automated methods can analyze a panel of inflammatory markers in a sample of GCF. Mass spectrometry is now being used for the characterization of various peptides and proteins in GCF. Reverse-phase high-performance liquid chromatography coupled with electrospray ionization mass spectrometry has been used to analyze samples for the presence of multiple peptides in GCF.[70] In addition, GCF cytokine profiling using a cytokine antibody array has also been introduced. Advantages of this array include a higher sensitivity and reduced variability relative to enzyme-linked immunosorbent assay (ELISA). Sakai *et al.*[71] were able to detect 36 inflammatory cytokines in GCF from patients with periodontitis.

While analysis of GCF has contributed to our understanding of the role of the inflammatory response in periodontitis, traditional analysis using filter strips to collect multiple samples from the mouth is impractical in the clinical setting. Saliva collection is less technique-sensitive and more amenable to chairside utilization. A patient-based assessment of disease can be obtained as saliva contains contributions from crevices about all of the teeth. These studies may assume added significance when considering the reported relationship of periodontal disease and certain systemic disorders.

REFERENCES

1. Brill, N. & B. Krasse. 1958. The passage of tissue fluid into the clinically healthy gingival pocket. Acta Odontol. Scand. **16:** 233–245.
2. Egelberg, J. 1966. Permeability of the dento-gingival blood vessels. IV. Effect of histamine on vessels in clinically healthy and chronically inflamed gingivae. J. Periodontal Res. **1:** 297–302.
3. Attstrom, R. & J. Egelberg. 1970. Emigration of blood neutrophils and monocytes into the gingival crevices. J. Periodontal Res. **5:** 48–55.
4. Goodson, J. M., F. E. Dewhirst & A. Brunetti. 1974. Prostaglandin E2 levels and human periodontal disease. Prostaglandins **6:** 81–85.
5. Socransky, S. S. *et al.* 1984. New concepts of destructive periodontal disease. J. Clin. Periodontol. **11:** 21–32.
6. Genco, R. J. & J. Slots. 1984. Host responses in periodontal diseases. J. Dent. Res. **63:** 441–451.
7. Griffiths, G. S., J. M. Wilton & M. A. Curtis. 1992. Contamination of human gingival crevicular fluid by plaque and saliva. Arch. Oral Biol. **37:** 559–564.
8. Lamster, I. B. *et al.* 1988. A comparison of 4 methods of data presentation for lysosomal enzyme activity in gingival crevicular fluid. J. Clin. Periodontol. **15:** 347–352.
9. Lamster, I. B. 1997. Evaluation of components of gingival crevicular fluid as diagnostic tests. Ann. Periodontol. **2:** 123–137.
10. Ozkavaf, A. *et al.* 2000. Relationship between the quantity of gingival crevicular fluid and clinical periodontal status. J. Oral Sci. **42:** 231–238.
11. Armitage, G. C. 2004. Analysis of gingival crevice fluid and risk of progression of periodontitis. Periodontology 2000 **34:** 109–119.
12. Loos, B. G. & S. Tjoa. 2005. Host-derived diagnostic markers for periodontitis: do they exist in gingival crevice fluid? Periodontology 2000 **39:** 53–72.

13. NAKASHIMA, K., N. ROEHRICH & G. CIMASONI. 1994. Osteocalcin, prostaglandin E2 and alkaline phosphatase in gingival crevicular fluid: their relations to periodontal status. J. Clin. Periodontol. **21:** 327–333.
14. NAKASHIMA, K. *et al.* 1996. A longitudinal study of various crevicular fluid components as markers of periodontal disease activity. J. Clin. Periodontol. **23:** 832–838.
15. LEE, W. *et al.* 1995. Evidence of a direct relationship between neutrophil collagenase activity and periodontal tissue destruction in vivo: role of active enzyme in human periodontitis. J. Periodontal Res. **30:** 23–33.
16. KINANE, D. F. *et al.* 2003. Changes in gingival crevicular fluid matrix metalloproteinase-8 levels during periodontal treatment and maintenance. J. Periodontal Res. **38:** 400–404.
17. BEKLEN, A. *et al.* 2006. Gingival tissue and crevicular fluid co-operation in adult periodontitis. J. Dent. Res. **85:** 59–63.
18. KUNIMATSU, K. *et al.* 1990. Granulocyte medullasin levels in gingival crevicular fluid from chronic adult periodontitis patients and experimental gingivitis subjects. J. Periodontal Res. **25:** 352–357.
19. KENNETT, C. N., S. W. COX & B. M. ELEY. 1997. Investigations into the cellular contribution to host tissue proteases and inhibitors in gingival crevicular fluid. J. Clin. Periodontol. **24:** 424–431.
20. CHEN, H. Y., S. W. COX & B. M. ELEY. 1998. Cathepsin B, alpha2-macroglobulin and cystatin levels in gingival crevicular fluid from chronic periodontitis patients. J. Clin. Periodontol. **25:** 34–41.
21. ELEY, B. M. & S. W. COX. 1995. Correlation between gingival crevicular fluid dipeptidyl peptidase II and IV activity and periodontal attachment loss. A 2-year longitudinal study in chronic periodontitis patients. Oral Dis. **1:** 201–213.
22. GIANNOPOULOU, C. *et al.* 1992. Neutrophil elastase and its inhibitors in human gingival crevicular fluid during experimental gingivitis. J. Dent. Res. **71:** 359–363.
23. ELEY, B. M. & S. W. COX. 1996. A 2-year longitudinal study of elastase in human gingival crevicular fluid and periodontal attachment loss. J. Clin. Periodontol. **23:** 681–692.
24. BADER, H. I. & R. L. BOYD. 1999. Long-term monitoring of adult periodontitis patients in supportive periodontal therapy: correlation of gingival crevicular fluid proteases with probing attachment loss. J. Clin. Periodontol. **26:** 99–105.
25. LAMSTER, I. B. *et al.* 1988. Enzyme activity in crevicular fluid for detection and prediction of clinical attachment loss in patients with chronic adult periodontitis. Six month results. J. Periodontol. **59:** 516–523.
26. LAMSTER, I. B. *et al.* 1995. The relationship of beta-glucuronidase activity in crevicular fluid to probing attachment loss in patients with adult periodontitis. Findings from a multicenter study. J. Clin. Periodontol. **22:** 36–44.
27. CHAMBERS, D. A. *et al.* 1984. Aspartate aminotransferase increases in crevicular fluid during experimental periodontitis in beagle dogs. J. Periodontol. **55:** 526–530.
28. PERSSON, G. R., T. A. DEROUEN & R. C. PAGE. 1990. Relationship between gingival crevicular fluid levels of aspartate aminotransferase and active tissue destruction in treated chronic periodontitis patients. J. Periodontal Res. **25:** 81–87.
29. CHAMBERS, D. A. *et al.* 1991. A longitudinal study of aspartate aminotransferase in human gingival crevicular fluid. J. Periodontal Res. **26:** 65–74.
30. OFFENBACHER, S., D. H. FARR & J. M. GOODSON. 1981. Measurement of prostaglandin E in crevicular fluid. J. Clin. Periodontol. **8:** 359–367.

31. Offenbacher, S., B. M. Odle & T. E. Van Dyke. 1986. The use of crevicular fluid prostaglandin E2 levels as a predictor of periodontal attachment loss. J. Periodontal Res. **21:** 101–112.
32. Liu, C. M. *et al.* 1996. Relationships between clinical parameters, interleukin 1B and histopathologic findings of gingival tissue in periodontitis patients. Cytokine **8:** 161–167.
33. Engebretson, S. P. *et al.* 2002. GCF IL-1beta profiles in periodontal disease. J. Clin. Periodontol. **29:** 48–53.
34. Shibutani, T. *et al.* 1993. ELISA detection of glycosaminoglycan (GAG)-linked proteoglycans in gingival crevicular fluid. J. Periodontal Res. **28:** 17–20.
35. Giannobile, W. V. *et al.* 1995. Crevicular fluid osteocalcin and pyridinoline cross-linked carboxyterminal telopeptide of type I collagen (ICTP) as markers of rapid bone turnover in periodontitis. A pilot study in beagle dogs. J. Clin. Periodontol. **22:** 903–910.
36. Apse, P. *et al.* 1989. Microbiota and crevicular fluid collagenase activity in the osseointegrated dental implant sulcus: a comparison of sites in edentulous and partially edentulous patients. J. Periodontal Res. **24:** 96–105.
37. Plagnat, D. *et al.* 2002. Elastase, alpha2-macroglobulin and alkaline phosphatase in crevicular fluid from implants with and without periimplantitis. Clin. Oral Implants Res. **13:** 227–233.
38. Friedmann, A. *et al.* 2006. Calprotectin and cross-linked N-terminal telopeptides in peri-implant and gingival crevicular fluid. Clin. Oral Implants Res. **17:** 527–532.
39. Salvi, G. E. *et al.* 1997. Inflammatory mediator response as a potential risk marker for periodontal diseases in insulin-dependent diabetes mellitus patients. J. Periodontol. **68:** 127–135.
40. Kurtis, B. *et al.* 1999. IL-6 levels in gingival crevicular fluid (GCF) from patients with non-insulin dependent diabetes mellitus (NIDDM), adult periodontitis and healthy subjects. J. Oral Sci. **41:** 163–167.
41. Alpagot, T. *et al.* 2001. Crevicular fluid elastase levels in relation to periodontitis and metabolic control of diabetes. J. Periodontal Res. **36:** 169–174.
42. Guneri, P. *et al.* 2004. Vascular endothelial growth factor in gingival tissues and crevicular fluids of diabetic and healthy periodontal patients. J. Periodontol. **75:** 91–97.
43. Engebretson, S. P. *et al.* 2004. Gingival crevicular fluid levels of interleukin-1beta and glycemic control in patients with chronic periodontitis and type 2 diabetes. J. Periodontol. **75:** 1203–1208.
44. Safkan-Seppala, B. *et al.* 2006. Collagenases in gingival crevicular fluid in type 1 diabetes mellitus. J. Periodontol. **77:** 189–194.
45. Morozumi, T. *et al.* 2004. Smoking cessation increases gingival blood flow and gingival crevicular fluid. J. Clin. Periodontol. **31:** 267–272.
46. Giannopoulou, C., I. Cappuyns & A. Mombelli. 2003. Effect of smoking on gingival crevicular fluid cytokine profile during experimental gingivitis. J. Clin. Periodontol. **30:** 996–1002.
47. Kamma, J. J. *et al.* 2004. Cytokine profile in gingival crevicular fluid of aggressive periodontitis: influence of smoking and stress. J. Clin. Periodontol. **31:** 894–902.
48. Stein, S. H., B. E. Green & M. Scarbecz. 2004. Augmented transforming growth factor-beta1 in gingival crevicular fluid of smokers with chronic periodontitis. J. Periodontol. **75:** 1619–1626.

49. PETROPOULOS, G., I. J. MCKAY & F. J. HUGHES. 2004. The association between neutrophil numbers and interleukin-1alpha concentrations in gingival crevicular fluid of smokers and non-smokers with periodontal disease. J. Clin. Periodontol. **31:** 390–395.
50. OATES, T. W., D. CARAWAY & J. JONES. 2004. Relation between smoking and biomarkers of bone resorption associated with dental endosseous implants. Implant Dent. **13:** 352–357.
51. BACK, M. *et al.* 2006. Increased leukotriene concentrations in gingival crevicular fluid from subjects with periodontal disease and atherosclerosis. Atherosclerosis In press. doi: 10,1016/j. atherosclerosis. 2006.07.003.
52. LAMSTER, I. B., R. L. OSHRAIN & D. S. HARPER. 1987. Infantile agranulocytosis with survival into adolescence: periodontal manifestations and laboratory findings. A case report. J. Periodontol. **58:** 34–39.
53. BARR-AGHOLME, M. *et al.* 1997. Prostaglandin E2 level in gingival crevicular fluid from patients with Down syndrome. Acta. Odontol. Scand. **55:** 101–105.
54. PEREZ, L. A. *et al.* 2002. Treatment of periodontal disease in a patient with Ehlers-Danlos syndrome. A case report and literature review. J. Periodontol. **73:** 564–570.
55. TSILINGARIDIS, G., T. YUCEL-LINDBERG & T. MODEER. 2003. Enhanced levels of prostaglandin E2, leukotriene B4, and matrix metalloproteinase-9 in gingival crevicular fluid from patients with Down syndrome. Acta Odontol. Scand. **61:** 154–158.
56. GRBIC, J. T., I. B. LAMSTER & D. MITCHELL-LEWIS. 1997. Inflammatory and immune mediators in crevicular fluid from HIV-infected injecting drug users. J. Periodontol. **68:** 249–255.
57. LAMSTER, I. B. *et al.* 1997. Epidemiology and diagnosis of HIV-associated periodontal diseases. Oral Dis. **3**(Suppl 1): S141–S148.
58. BAQUI, A. A. *et al.* 2000. Enhanced interleukin 1 beta, interleukin 6 and tumor necrosis factor alpha in gingival crevicular fluid from periodontal pockets of patients infected with human immunodeficiency virus 1. Oral Microbiol. Immunol. **15:** 67–73.
59. ALPAGOT, T., K. FONT & A. LEE. 2003. Longitudinal evaluation of GCF IFN-gamma levels and periodontal status in HIV+ patients. J. Clin. Periodontol. **30:** 944–948.
60. SUZUKI, T. *et al.* 1997. Characterization of HIV-related periodontitis in AIDS patients: HIV-infected macrophage exudate in gingival crevicular fluid as a hallmark of distinctive etiology. Clin. Exp. Immunol. **108:** 254–259.
61. BEN-ARYEH, H., I. UR & E. BEN-PORATH. 1985. The relationship between antigenaemia and excretion of hepatitis B surface antigen in human whole saliva and in gingival crevicular fluid. Arch. Oral Biol. **30:** 97–99.
62. MONTEBUGNOLI, L. & G. DOLCI. 2000. Anti-HCV antibodies are detectable in the gingival crevicular fluid of HCV positive subjects. Minerva Stomatol. **49:** 1–8.
63. MATICIC, M. *et al.* 2001. Detection of hepatitis C virus RNA from gingival crevicular fluid and its relation to virus presence in saliva. J. Periodontol. **72:** 11–16.
64. SUZUKI, T. *et al.* 2005. Quantitative detection of hepatitis C virus (HCV) RNA in saliva and gingival crevicular fluid of HCV-infected patients. J. Clin. Microbiol. **43:** 4413–4417.
65. MANDEL, I. D. & S. WOTMAN. 1976. The salivary secretions in health and disease. Oral Sci. Rev. 25–47.

66. SREEBNY, L. M. 1989. Salivary flow in health and disease. Compend. Suppl. S461–S469.
67. HIRSCHFELD, L. & B. WASSERMAN. 1978. A long-term survey of tooth loss in 600 treated periodontal patients. J. Periodontol. **49:** 225–237.
68. MCFALL, W. T., JR. 1982. Tooth loss in 100 treated patients with periodontal disease. A long-term study. J. Periodontol. **53:** 539–549.
69. LAMSTER, I. B. *et al.* 2003. Beta-glucuronidase activity in saliva: relationship to clinical periodontal parameters. J. Periodontol. **74:** 353–359.
70. PISANO, E. *et al.* 2005. Peptides of human gingival crevicular fluid determined by HPLC-ESI-MS. Eur. J. Oral Sci. **113:** 462–468.
71. SAKAI, A. *et al.* 2006. Profiling the cytokines in gingival crevicular fluid using a cytokine antibody array. J. Periodontol. **77:** 856–864.

Oral Fluid–Based Biomarkers of Alveolar Bone Loss in Periodontitis

JANET S. KINNEY,[a] CHRISTOPH A. RAMSEIER,[a]
AND WILLIAM V. GIANNOBILE[a,b]

[a]*Department of Periodontics and Oral Medicine and Michigan Center for Oral Health Research, University of Michigan, Ann Arbor, Michigan, USA*

[b]*Department of Biomedical Engineering, College of Engineering, University of Michigan, Ann Arbor, Michigan, USA*

ABSTRACT: Periodontal disease is a bacteria-induced chronic inflammatory disease affecting the soft and hard supporting structures encompassing the teeth. When left untreated, the ultimate outcome is alveolar bone loss and exfoliation of the involved teeth. Traditional periodontal diagnostic methods include assessment of clinical parameters and radiographs. Though efficient, these conventional techniques are inherently limited in that only a historical perspective, not current appraisal, of disease status can be determined. Advances in the use of oral fluids as possible biological samples for objective measures of current disease state, treatment monitoring, and prognostic indicators have boosted saliva and other oral-based fluids to the forefront of technology. Oral fluids contain locally and systemically derived mediators of periodontal disease, including microbial, host-response, and bone-specific resorptive markers. Although most biomarkers in oral fluids represent inflammatory mediators, several specific collagen degradation and bone turnover-related molecules have emerged as possible measures of periodontal disease activity. Pyridinoline cross-linked carboxyterminal telopeptide (ICTP), for example, has been highly correlated with clinical features of the disease and decreases in response to intervention therapies, and has been shown to possess predictive properties for possible future disease activity. One foreseeable benefit of an oral fluid–based periodontal diagnostic would be identification of highly susceptible individuals prior to overt disease. Timely detection and diagnosis of disease may significantly affect the clinical management of periodontal patients by offering earlier, less invasive, and more cost-effective treatment therapies.

KEYWORDS: periodontal disease; oral fluids; saliva; disease progression; diagnosis; bone resorption

Address for correspondence: William V. Giannobile, D.D.S., D.Med.Sc., Michigan Center for Oral Health Research, University of Michigan Clinical Center, 24 Frank Lloyd Wright Dr., Lobby M, Box 422, Ann Arbor, MI 48106, Voice: 734-998-1468; fax: 734-998-7228.
wgiannob@umich.edu

**Ann. N.Y. Acad. Sci. 1098: 230–251 (2007). © 2007 New York Academy of Sciences.
doi: 10.1196/annals.1384.028**

PERIODONTAL DISEASES: BACKGROUND

Chronic infectious diseases of the oral cavity include dental caries and periodontal disease, the former causing destruction of the teeth, while the latter is, a group of inflammatory conditions, affects the supporting structures of the dentition.[1] The unequivocal role of the microbial challenge in the etiology of periodontal disease has been well studied. However, it is the paradoxical impact of the susceptible host's inflammatory response to the microbial challenge that ultimately leads to the destruction of the periodontal structures and subsequent tooth loss.[2–4]

Periodontal diseases are further divided into reversible and nonreversible categories. Gingivitis is a reversible inflammatory reaction of the marginal gingiva to dental plaque biofilms. Gingivitis is characterized by an initial increase in blood flow, enhanced vascular permeability, and influx of cells (polymorphonuclear leukocytes [PMNs] and monocyte-macrophages) from the peripheral blood into the periodontal connective tissue. Overt soft tissue alterations during the state of gingivitis include redness, edema, bleeding, and tenderness. The feature distinguishing gingivitis from the destructive form of periodontal disease is the intact anatomical location of the junctional epithelium on the root surface.

Periodontitis, the destructive category of periodontal disease, is a nonreversible inflammatory state of the supporting structures. After its initiation, the disease progresses with the loss of collagen fibers and attachment to the cemental surface, apical migration of the pocket epithelium, formation of deepened periodontal pockets, and the resorption of alveolar bone. If left untreated, the disease continues to progressive bone destruction, leading to tooth mobility and subsequent tooth loss.[5]

Chronic periodontitis is the most prevalent form of destructive periodontal disease and typically progresses at a slow, steady pace with bouts of extensive disease destruction separated by quiescent periods of bone loss.[6–8] Albandar *et al.* examined the prevalence and severity of chronic periodontitis in the United States adult population (30 years and older) and found that from 1988 to 1994 approximately 35% presented with chronic periodontitis. After adjusting for measurement error due to the partial readings, it was determined that approximately half of U.S. adults had chronic periodontitis. Further breakdown of these findings indicate that 31% of the U.S. population exhibit mild forms of the disease, 13% display moderate severity, and 4% suffer advanced disease.[9,10]

More than 600 different bacteria are capable of colonizing the human mouth with any individual typically harboring 150 to 200 varying species. Of these many bacteria, it is estimated that approximately 10% play a causal role in the initiation of periodontal disease.[2,11–13] Three organisms in particular, *Tanerella forsythensis, Porphyromonas gingivalis,* and *Treponema denticola* have been directly associated with chronic periodontitis.[13–15] *Actinobacillus actinomycetemcomitans,* another virulent gram-negative bacterium, has been observed in early-onset forms of periodontal disease and aggressive periodontitis.[12] Inherent virulence factors of these pathogenic species enable the bacteria to colonize on the tooth and in the gingival sulcus, defend itself from the host's antibacterial defense mechanisms, and cause tissue damage by producing potent substances that subsequently trigger the host's innate inflammatory response.[16] Although pathogen-based diagnostic tests are imperative for the initiation of periodontal disease, their utility has not been successful for prediction of periodontal disease.[17] At best, pathogen-based tests serve as adjuncts to traditional diagnostic methods by assessing vulnerability of patient and site, classifying disease category, and assisting in treatment modality.

NEED FOR A PERIODONTAL DIAGNOSTIC INDICATOR

A periodontal diagnostic tool provides pertinent information for differential diagnosis, localization of disease, and severity of infection. These diagnostics, in turn, serve as a basis for planning treatment and provide a means for assessing the effectiveness of periodontal therapy.[18] Current clinical diagnostic parameters that were introduced more than 50 years ago continue to function as the basic model for periodontal diagnosis in clinical practice today. They include probing pocket depths, bleeding on probing, clinical attachment levels, plaque index, and radiographs that quantify alveolar bone levels.[19,20] Albeit easy to use, cost-effective, and relatively noninvasive, clinical attachment loss evaluation by the periodontal probe measures damage from past episodes of destruction and requires a 2- to 3-mm threshold change before a site can be deemed as having experienced significant breakdown. Recent revisions in the design of automated periodontal probes have improved the accuracy and long-term tracking of disease progression. Furthermore, the use of subtraction radiography also offers a method to detect minute changes in the height of alveolar bone. However, both of the above-mentioned techniques are most often seen in the research setting and seldom in clinical practice. In addition to these limitations, conventional disease diagnosis techniques lack the capacity to identify highly susceptible patients who are at risk for future breakdown.[21–23] Researchers are confronted then with the need for an innovative diagnostic test that focuses on the early recognition of the microbial challenge to the host. Optimal innovative approaches would correctly determine the presence of current disease activity, predict sites vulnerable for future breakdown, and assess the response to periodontal interventions. A new paradigm for periodontal diagnosis would ultimately affect improved clinical management of periodontal patients.

ORAL FLUID BIOMARKERS OF PERIODONTAL DISEASE

In response to requests from the Office of the Surgeon General and the National Institute of Dental and Craniofacial Research (NIDCR) for concentrated research in salivary diagnostics, significant advancements have been achieved within the past 10 years using saliva, gingival crevicular fluid (GCF), and mucosal transudate as biological samples for the detection of oral and systemic illnesses.[24] Easily collected and containing local- and systemic-derived biomarkers of periodontal disease, oral fluids may offer the basis for a patient-specific diagnostic test for periodontal disease.[25–28] A biomarker is an objective measure that has been evaluated and confirmed either as an indicator of physiologic health, a pathogenic process, or a pharmacologic response to a therapeutic intervention.[29] Oral fluid biomarkers that have been studied for periodontal diagnosis include proteins of host origin (e.g., enzymes and immunoglobulins), phenotypic markers, host cells (e.g., PMNs), hormones, bacteria and bacterial products, ions, and volatile compounds.[20,25,30–32] Because of the complex, multifaceted nature of periodontal disease, it is highly unlikely that a single biomarker will prove to be a stand-alone measure for periodontal disease diagnosis. More probable may be the development of an oral fluid-based diagnostic using a combination of host- and site-specific markers that accurately assess periodontal disease status.[31,33–36]

During the initiation of an inflammatory response in the periodontal connective tissue (FIG. 1), numerous cytokines, such as prostaglandin E2 (PGE2), interleukin (IL)-1β, IL-6, or tumor necrosis factor (TNF)-α are released from cells of the junctional epithelia, connective tissue fibroblasts, macrophages, and PMNs. Subsequently, T and B cells emerge at the infection sites and secrete immunoglobulins as an antigen-specific response.[37] Additionally, a number of enzymes, such as matrix metalloproteinase (MMP)-8, MMP-9, or MMP-13 are produced by PMNs and osteoclasts, leading to the degradation of connective tissue collagen and alveolar bone.[38] As a consequence of bone resorption, pyridinoline cross-linked carboxyterminal telopeptide (ICTP) and osteocalcin are released into the periodontal tissues. During the inflammatory process intercellular products are created and migrate toward the gingival sulcus or periodontal pocket (FIG.1). These mediators of disease activity have been identified and sampled from various biological fluids, such as saliva and GCF.[39]

GCF is an inflammatory exudate originating from the gingival plexus of blood vessels in the gingival corium, subjacent to the epithelium lining of the dentogingival space. As GCF traverses through inflamed periodontal tissues en route to the sulcus, biological molecular markers are gathered from the surrounding site.[40] GCF sampling methods have been shown to accurately capture inflammatory and connective tissue breakdown mediators.[31,33] As recently reviewed by Loos and Tjoa, more than 90 different components in GCF

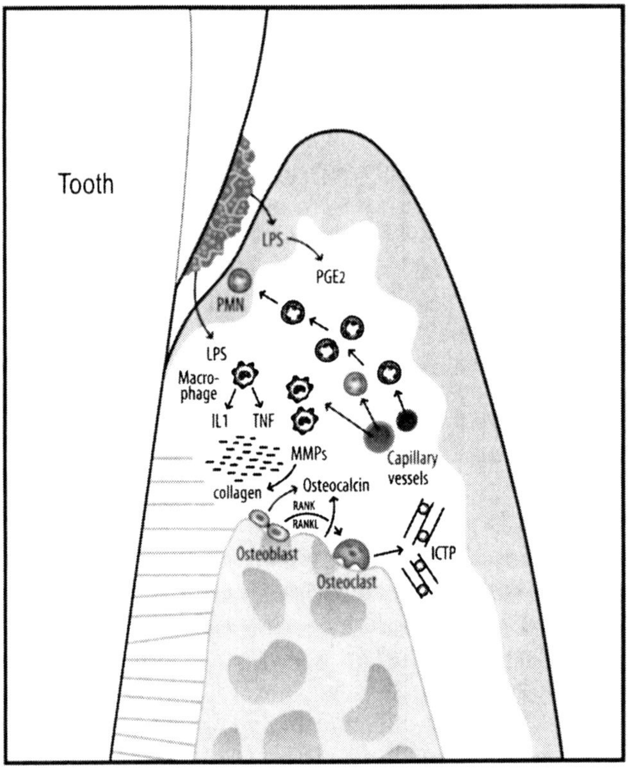

FIGURE 1. Schematic overview of the pathogenic processes in periodontal disease. Initial events are triggered by LPS from gram-negative plaque biofilms on the tooth root surfaces. As a first line of defense, PMNs are recruited to the site. Monocytes and activated macrophages respond to endotoxin by releasing cytokines TNF and IL-1, which direct further destructive processes. MMP, powerful collagen-destroying enzymes, are produced by fibroblasts and PMNs. TNF, IL-1, and receptor activator of NF-κB ligand (RANKL) are elevated in active sites and mediate osteoclastogenesis and bone breakdown. Bone-specific markers, such as pyridinoline cross-linked carboxyterminal telopeptide of type I collagen (ICTP), are released into the surrounding area and transported by way of GCF into the sulcus or pocket and serve as potential biomarkers for periodontal disease detection.

have been evaluated to date for periodontal diagnosis.[41] Of the numerous constituents in GCF, however, the vast majority constitute soft tissue inflammatory events, while only a few are regarded as specific biomarkers of alveolar bone destruction (TABLE 1).

In the early 1970s several researchers demonstrated a correlation between collagenolytic activity of GCF and severity of periodontal disease.[42–45] Golub *et al.* and co-workers discovered that collagenase activity was more highly correlated with pocket depth than with inflammation and therefore "may reflect the degradative activity of the gingival tissues lining the pocket and

TABLE 1. Bone-related biomarkers from oral fluids associated with periodontal diseases

Name	Association	Reference
ALP	T	32,34,62,64–66
Cathepsin B	S, T	68–72
Collagenase-2 (MMP-8)	S, T	79,81,83–89
Gelatinase (MMP-9)	T	90
Collagenase-3 (MMP-13)	T	79,87
Calprotectin	S	98
Osteocalcin	S	34,53,79,102–104
Pyridinoline cross-links (ICTP)	S, T	79,102,109–116
Osteonectin	S	119
Osteopontin	S	128,129

NOTE: Association with severity of periodontal disease (S) and treatment planning and outcome (T).

could, therefore, be of diagnostic value."[42] Villela and Birkedal-Hansen also observed a connection between collagenolytic activity and active disease.[46] As a result from these early observations, thought was given to the possible use of collagenases as plausible biochemical markers for disease progression.

Prostaglandins are arachidonic acid metabolites composed of 10 classes, of which D, E, F, G, H, and I are of main importance. Of this group, PGE2 is one of the most extensively studied mediators of periodontal disease activity.[47–52] During the host's innate defense response to bacterial lipopolysaccharide (LPS), monocytes, PMNs, macrophages, and other cells release IL-1, TNF, and PGE2. PGE2 acts as a potent vasodilator and increases capillary permeability, which elicits clinical signs of redness and edema. PGE2 also stimulates fibroblasts and osteoclasts to increase production of MMPs. Ultimately, MMPs affect the remodeling and degradation of the periodontium.[33] Offenbacher and co-workers demonstrated that patients with periodontitis had higher levels of GCF-PGE2 than patients with gingivitis.[53,54] This group subsequently performed a retrospective analysis of GCF-PGE2 by examining the longitudinal relationship of PGE2 concentrations in GCF to attachment loss in adult patients with periodontitis. Results showed that elevated PGE2 was detectable in GCF 6 months before the identification of periodontal disease activity and significantly decreased 1 month after scaling and root planing was provided.[55] Although PGE2 has shown much promise as a biomarker of periodontitis, PGE2-based diagnostics have not entered the clinical arena.

Another oral fluid that has recently gained significant recognition is saliva. Technologies are emerging that use minute amounts of saliva as reliable diagnostic fluids for identification, monitoring, and prediction of various diseases. Saliva-based diagnostic tests are currently being used in a broad range of applications, such as autoimmune disorders, cardiovascular disease, infectious diseases, and in monitoring drugs of abuse.[24] Much work is currently under way in the field of salivary diagnostics for periodontal disease. Pederson *et al.*

measured quantities of host-response indicators—cathepsin G, elastase, elastase inhibitors, and C-reactive protein (CRPs)—to determine whether their levels were directly related to the individual's periodontal status. Forty-five participants were categorized according to periodontal status (healthy, gingivitis, mild-to-moderate periodontitis, or moderate-to-severe periodontitis) and whole-saliva samples were collected. With the exception of α_1-antitrypsin, an increase in salivary levels for all of the targeted host-response markers correlated with increasing severity of disease.[56] Other early saliva-based investigations detected significantly increased levels of collagenase 2 (MMP-8) in periodontally diseased patients.[57] Subsequent explorations of immunoglobulins found in whole saliva directed against periodontal pathogens have indicated some correlations with the status of periodontal disease.[58–61] Although saliva-based clinical testing shows much potential, more-extensive research is required to identify the best candidate markers that can be simultaneously evaluated to obtain a profile for diagnosis, monitoring, and prediction of oral diseases.

BIOMARKERS OF BONE RESORPTION OR TURNOVER

Several biomarkers have been studied as applicable to the diagnostics of periodontal bone loss (TABLE 1). These components, which are evaluated next, are the potential candidates for oral fluid–based diagnostics of periodontal disease.

Alkaline Phosphatase (ALP)

Enzymes found in whole saliva originate from three main sources: (1) the actual salivary secretions *per se*; (2) the GCF, stemming from PMNs and tissue degradation; and (3) disposed bacterial cells from dental biofilms and mucosal surfaces. ALP is a catalyzing enzyme that accelerates the removal of phosphate groups in the 5 and 3 positions from a variety of molecules, including nucleotides, proteins, and alkaloids. Although present in all tissues, ALP is particularly concentrated in the bone, liver, bile duct, kidney, and placenta. Of interest in oral health, of course, is the association between ALP and periodontal disease.

Early investigations of ALP and periodontal disease in an experimental gingivitis model showed a significant correlation between ALP and pocket depth and between ALP and inflammation.[62] Nakamura and Slots studied a total of 76 enzyme activities in mixed whole saliva and noted higher enzyme activity in individuals with periodontal disease than nondiseased individuals.[32] Gibert *et al.* analyzed serum levels of ALP from patients with chronic periodontal disease and compared the findings with those of control patients.[63] Results showed a relationship between attachment loss in the periodontal group and a drop in ALP activity in serum. Contrary to these results, Totan *et al.* investigated the influence of periodontal disease on ALP, aminotransferase (AST), aminopeptidase, and glucuronidase.[64] Salivary samples from patients with confirmed periodontal disease were analyzed and revealed that periodontal destruction by measurement of probing depth, gingival bleeding, and suppuration were related to higher ALP levels in saliva. Supporting these results are the findings by Todorovic *et al.* that increased activity of salivary ALP is seen in patients with periodontal disease in relation to a nondisease control group.[65] This group further showed a positive correlation between the salivary enzyme activity and gingival index values. As a predictive indicator for future periodontal breakdown, ALP has not been supported by research findings and therefore may best serve as a marker in periodontal treatment planning and monitoring.[34,66]

Cathepsin B

As an enzyme belonging to the class of cysteine proteinases, cathepsin B functions in proteolysis. In GCF, macrophages are the main producers of cathepsin B.[67] GCF concentrations of cathepsin B were found to be elevated in patients with periodontal disease, but lower in patients with gingivitis.[68] Ichimaru *et al.* concluded that cathepsin B may be correlated with the severity of periodontitis.[69] Further investigators have shown positive correlations between cathepin B levels and the severity of periodontal disease, while noting a reduction of cathepsin B levels after periodontal intervention therapies were provided.[70,71] Eley and Cox studied cathepsin B and evaluated its use as a predictor of attachment loss.[72] Forty-nine patients were monitored after initial periodontal therapy for 2 years. A total of 121 sites were found with attachment loss (90 with rapid loss and 31 with gradual loss). Cathepsin B levels were higher in the sites with rapid loss than in the paired control sites. Moreover, in the sites with gradual attachment loss, cathepsin B levels were elevated when compared with the paired control sites. With a cut-off value of 7.5 μU/30 sec GCF sample for the total cathepsin B activity and 30 μU/uL for enzyme concentration, remarkable results of 100% sensitivity and 99.8% specificity for both cathepsin B parameters were reported. Cathepsin B may have a potential use in distinguishing periodontitis from gingivitis and in planning treatment and monitoring treatment outcomes.[41]

Collagenase-2 (MMP-8)

MMPs are host proteinases responsible for both tissue degradation and remodeling.[73–77] During progressive periodontal breakdown, gingival and periodontal ligament collagens are cleaved by host cell–derived interstitial collagenases. One vital interstitial collagenase capable of degrading the triple helical structures of native types I, II, and III collagens found in alveolar bone matrix is collagenase-2. Collagenase-2, also referred to as MMP-8, is released during the maturation of PMNs in the bone marrow. Once produced, it becomes glycosylated and is prestored in the sub-cellular-specific granules, where it is subsequently released in large quantities as the PMNs are recruited to a site of inflammation. Chubinskaya *et al.* demonstrated the ability of non-neutrophil-lineage mesenchymal cells, such as human gingival and periodontal ligament fibroblasts and chondrocytes, to also be able to produce MMP-8.[78]

MMP-8 is the most prevalent MMP found in diseased periodontal tissue and GCF.[79–81] Nomura *et al.* found no difference in MMP-8 levels from patients with periodontal disease when compared to patients with gingivitis.[82] From this early investigation, it was believed that MMP-8 may serve as a proinflammatory marker, but not as a discriminating marker for chronic periodontitis and gingivitis. However, Mancini and co-workers found an 18-fold increase of MMP-8 in patients experiencing active periodontal tissue breakdown as compared with patients under stable conditions. Conclusions from this investigation indicated the potential use of MMP-8 as a screening test for detection of active disease progression.[83] Elevated MMP-8 levels in active disease progression were observed by Lee *et al.* in a longitudinal study using patients with gingivitis, nonprogressive, and progressive periodontitis. The total collagenase activity was observed to be 50% higher in the disease progression group.[81]

Golub *et al.* introduced a 20-mg low-dose doxycycline (LDD) capsule, which preserved its proteinase-inhibitory ability to suppress connective tissue breakdown, but without antibiotic/antimicrobial capabilities. The group went on to conduct several studies demonstrating that LDD can function as an MMP by way of suppressing the collagenase activity in GCF and gingival tissues of patients with adult periodontitis.[84,85] To test the hypothesis that LDD could lower GCF levels of bone-type collagen fragments, clinical parameters (gingival inflammation, pocket depth, and radiographic evidence of bone loss) that predicted excessive MMP activity in periodontal pockets of 18 adult patients were evaluated. All patients received supragingival scaling 1 month before the baseline appointment. At the baseline visit and at the subsequent 1- and 2-month visits, GCF samples were collected. Conventional clinical measures (gingival index, plaque index, probing depth, and attachment level) were taken at each time point in the study. Western blots analyses determined that neutrophil-type collagenase (MMP-8) was increased in disease and substantially reduced by approximately 60% during the 2-month protocol of LDD.[79] MMP-8 may have some future value as a diagnostic marker for periodontal disease, an indicator for disease progression, and as a signal to determine the efficacy of treatment.[41]

MMP-8 has also been detected in elevated amounts in peri-implant sulcular fluid (PISF) from peri-implantitis lesions. Teronen *et al.* identified higher collagenase-2 levels in failing dental implants compared to nonmobile implants.[86] Ma *et al.* went on to explore for the presence of MMP-8 and collagenase-3, MMP-13, in peri-implant sulcus fluid. Forty-nine randomly selected dental implant sites in 13 patients were studied. Implants were categorized into three groups according to the amount of bone loss in the vertical dimension: <1 mm, from 1 to 3 mm, or >3 mm. Results from this investigation showed that both MMP-8 and MMP-13 levels were significantly higher in the >3 mm bone loss group when compared to the groups that had less bone loss.[87] Additional studies were conducted by Kivela-Rajamaki *et al.*, looking at MMP-8 levels in combination with laminin-5 and, during a separate study, with MMP-7. Conclusions drawn from both investigations indicate that elevated levels of MMP-8 can be seen in diseased PISF as compared to healthy PISF.[88,89] Collectively, these findings offer hope for the use of MMP-8 as a marker for active phase of peri-implant disease. Longitudinal studies are required to evaluate MMP-8 either alone or in conjunction with other molecular biomarkers to predict the risk of future disease occurrence and to monitor treatment interventions.

Gelatinase (MMP-9)

Gelatinase (MMP-9), another member of the collagenase family, is produced by neutrophils and degrades collagen extracellular ground substance. In a longitudinal study conducted by Teng *et al.*, patients were asked to rinse and expectorate, providing subject-based instead of individual site-based GCF rinse samples.[90] When analyzed, a twofold increase in mean MMP-9 levels was reported in patients with recurrent attachment loss. Once given systemic metronidazole, mouthrinse samples from patients with initial elevated MMP-9 concentrations markedly dropped. Given these results, future use of MMP-9 in oral diagnostics may best serve as a guide in periodontal treatment monitoring.

Collagenase-3 (MMP-13)

Collagenase-3, referred to as MMP-13, is another collagenolytic MMP with an exceptionally wide substrate specificity.[91,92] MMP-13 is expressed during bone formation and gingival wound healing and at heightened quantities during pathological tissue destructive states, such as arthritis, chronic ulcers, atherosclerosis, and several types of malignant tumors.[93,94] Uitto *et al.* examined MMP-13 in chronically inflamed oral mucosa and found that during the course of prolonged inflammation undifferentiated epithelial cells produce significant concentrations of MMP-13.[95] Tervaharitiala *et al.* examined diseased gingival sulcular epithelium of patients with adult periodontitis (AD) and patients with localized juvenile periodontitis (LJP) for evidence of MMP and found sulcular epithelium expressing detectible levels of MMP-13, MMP-8, and MMP-2 *in vivo*.[93] Golub *et al.* were the first to discover MMP-13 in

GCF of periodontal patients, albeit in only a small proportion (3–4%) of the total amount of GCF collagenase. Golub further investigated the effects of LDD as a MMP inhibitor and found that LDD reduced GCF concentrations of MMP-13 faster and more efficiently than MMP-8 levels.[79] MMP-13 has also been implicated in peri-implantitis. Ma *et al.* concluded that elevated levels of both MMP-13 and MMP-8 correlated with irreversible peri-implant vertical bone loss around loosening dental implants.[87] In the future, MMP-13 may be useful for diagnosing and monitoring the course of periodontal disease as well as tracking the efficacy of therapy.

Calprotectin

Calprotectin is a 36-kDa protein composed of a dimeric complex of 8- and 14-kDa subunits. Neutrophils are the primary source of calprotectin although other cells, such as activated monocytes and macrophages and specific epithelial cells, are also capable of manufacturing the protein. Calprotectin acts as a calcium- and zinc-binding protein with both antimicrobial and antifungal activities. Furthermore, calprotectin plays a role in immune regulation through its ability to inhibit immunoglobulin production and, of particular interest, its role as a proinflammatory protein for neutrophil recruitment and activation.

Current research is using calprotectin as a marker for medical conditions such as ulcerative colitis and Crohn's disease.[96,97] In periodontology, Kido *et al.* identified calprotectin in GCF and found that GCF concentration levels in patients with periodontal disease were higher than those in GCF from healthy subjects.[98] The expression of calprotectin from inflammatory cells appears to offer protection of the epithelial cells against binding and invasion by *P. gingivalis*. In periodontal disease, calprotectin appears to improve resistance to *P. gingivalis* by boosting the barrier protection and innate immune functions of the gingival epithelium.[99]

Osteocalcin

Elevated serum osteocalcin levels have been found during periods of rapid bone turnover, such as osteoporosis, multiple myeloma, and fracture repair.[100,101] Therefore, studies have investigated the relationship between GCF osteocalcin levels and periodontal disease.[34,53,79,102–104] Kunimatsu *et al.* reported a positive correlation between GCF osteocalcin N-terminal peptide levels and clinical parameters in a cross-sectional study of patients with periodontitis and gingivitis.[103] The authors also reported that osteocalcin could not be detected in patients with gingivitis. In contrast, Nakashima *et al.* reported significant GCF osteocalcin levels from both periodontitis and gingivitis patients.[53] Osteocalcin levels were also significantly correlated with pocket depth and gingival index scores, as well as GCF levels of ALP and PGE2. In a longitudinal study of untreated periodontitis patients with ≥1.5 mm attachment loss during the monitoring period, GCF osteocalcin levels alone were unable to discriminate between active and inactive sites.[34] However, when a combination of the biochemical markers osteocalcin, collagenase, PGE2, α-2 macroglobulin, elastase, and ALP was evaluated, increased diagnostic sensitivity and specificity values of 80% and 91%, respectively, were reported.[34]

A longitudinal study using an experimental periodontitis model in beagle dogs reported a strong correlation between GCF osteocalcin levels and active bone turnover as assessed by bone-seeking radiopharmaceutical uptake (BSRU).[102] However, osteocalcin was shown to possess only modest predictive value for future bone loss measured by computer-assisted digitizing radiography. Moreover, treatment of chronic periodontitis patients with sub-anti-microbial doxycycline failed to reduce GCF osteocalcin levels,[79] and a cross-sectional study of periodontitis patients reported no differences in GCF osteocalcin levels between deep and shallow sites in the same patients.[105] Moreover, osteocalcin levels in the GCF during orthodontic tooth movement were highly variable between subjects and lacked a consistent pattern related to the stages of tooth movement.[106] In summary, the results of these studies show a role for intact osteocalcin as a bone-specific marker of bone turnover.

Pyridinoline Cross-Linked Carboxyterminal Telopeptide of Type I Collagen (ICTP)

Given the specificity and sensitivity for bone resorption, pyridinoline cross-links represent a potentially valuable diagnostic aid for periodontal disease, since biomarkers specific for bone degradation may be useful in differentiating between the presence of gingival inflammation and active periodontal or peri-implant bone destruction.[107] Several investigations have explored the ability of pyridinoline cross-links to detect bone resorption in periodontitis and peri-implantitis as well as in response to periodontal therapy.[79,102,105,108–116]

Palys *et al.* related ICTP levels to the subgingival microflora of various disease states on GCF.[110] Subjects were divided into groups representing health, gingivitis, and chronic periodontitis, and GCF and plaque samples were collected from each subject. The samples were analyzed for ICTP levels and the presence of 40 subgingival species by using checkerboard DNA–DNA hybridization techniques. ICTP levels differed significantly between health, gingivitis, and periodontitis subjects, and related modestly to several clinical disease parameters. ICTP levels were also strongly correlated with whole subject levels of several periodontal pathogens including *T. forsythensis, P. gingivalis, P. intermedia,* and *T. denticola*. In a subsequent study, Oringer *et al.* examined the relationship between ICTP levels and subgingival species around implants and teeth in 20 partially edentulous and two fully edentulous patients. No significant differences were found among ICTP levels and subgingival plaque composition between implants and teeth. Strong correlations between elevated ICTP levels at implant sites and colonization with organisms associated with failing implants, such as *P. intermedia, F. nucleatum ss vincentii,* and *S. gordonii* were found.[109]

Golub *et al.* found that treatment of chronic periodontitis patients with non-surgical periodontal therapy and LDD resulted in a 70% reduction in GCF ICTP levels after 1 month, concomitant with a 30% reduction in collagenase levels.[79] An investigation of periodontitis patients treated with scaling and root planing also demonstrated significant correlations between GCF ICTP levels and clinical periodontal disease parameters, including attachment loss, pocket depth, and bleeding on probing.[107] In addition, elevated GCF ICTP levels at baseline, especially at shallow sites, were found to be predictive for future attachment loss as early as 1 month after sampling. Furthermore, treatment of a group of periodontitis subjects by SRP and locally delivered minocycline led to rapid reductions in GCF ICTP levels.[79]

In summary, studies assessing the role of GCF ICTP levels as a diagnostic marker of periodontal disease activity have produced promising results to date. ICTP has been shown to be a promising predictor of both future alveolar bone and attachment loss. Furthermore, ICTP was strongly correlated with clinical parameters and putative periodontal pathogens, and demonstrated significant reductions after periodontal therapy. Controlled human longitudinal trials are needed to fully establish the role of ICTP as a predictor of periodontal tissue destruction, disease activity, and response to therapy in periodontal patients.

Therefore, the measurement of connective tissue–derived molecules, such as ICTP or osteocalcin, may lead to a more accurate assessment of tissue breakdown.[117]

Osteonectin

Also referred to as secreted protein acidic and rich in cysteine and basement membrane protein (BM-40), osteonectin is a single-chain polypeptide that binds strongly to hydroxyapatite and other extracellular matrix proteins including collagens. Because of its affinity for collagen and hydroxylapatite, osteonectin has been implicated in the early phases of tissue mineralization.[118] In a cross-sectional study by Bowers *et al.*, GCF samples were analyzed from patients with gingivitis, at moderate or severe periodontal disease states. Using a dot blot assay, both osteonectin and N-propeptide alpha I type I collagen were significantly increased in patients with periodontal disease. Furthermore, the protein concentrations found in GCF were elevated as probe depth measures increased in the sites evaluated.[119] At the final analysis of this study, osteonectin appeared to be the more sensitive marker for detection of periodontal disease status, when compared with N-propeptide alpha I type I collagen.

Osteopontin (OPN)

OPN is a single-chain polypeptide having a molecular weight of approximately 32,600.[120] It is found in the kidney, blood, mammary gland, salivary glands, and bone. In bone matrix, OPN is highly concentrated at sites where osteoclasts are attached to the underlying mineral surface, that is, the clear zone attachment areas of the plasma membrane.[121,122] However, since OPN is produced by both osteoblasts and osteoclasts, it holds a dual function in bone maturation and mineralization as well as bone resorption.[123–127] Kido *et al.* investigated the presence of OPN in GCF and the correlation between these levels and probing depth measures of periodontally healthy and diseased patients. Results from this study revealed that OPN could be detected in GCF, and increased OPN levels coincided with increased probing depth measures.[128] Sharma *et al.* recently published findings from an investigation of GCF OPN. A total of 45 subjects were divided into three groups (healthy, gingivitis, and chronic periodontitis) based on clinical examination, modified gingival index, Ramfjord periodontal disease index scores, and radiographic evidence of bone loss. The chronic periodontitis group subsequently received nonsurgical therapy and GCF samples were collected again 6 to 8 weeks after treatment. Results indicated that GCF OPN concentrations increased proportionally with the progression of disease and when nonsurgical periodontal treatment was provided, GCF OPN levels were significantly reduced. Although additional long-term prospective studies are needed, at this point OPN appears to hold promise as a possible biomarker of periodontal disease progression.[129]

FUTURE DIRECTIONS FOR PERIODONTAL ORAL FLUID–BASED DIAGNOSTICS

Researchers involved in delivery of periodontal therapy are currently investigating the possible use of oral fluids in the diagnosis of oral diseases and drug development. The movement of the pharmaceutical industry toward pharmacogenomics will require the tailored design of specific diagnostic profiles of patients for individualized dental therapy. Professionals in seemingly unrelated arenas, such as the insurance industry, Environment Protection Agency, and Homeland Security, are interested in the possibility of oral fluid use as well for rapid screening of oral and systemic health status.

As it relates to periodontology, the great need for periodontal diagnostics has become increasingly evident. From physical measurements by periodontal probing to sophisticated genetic susceptibility analysis and molecular arrays for the detection of biomarkers on the different stages of the disease, substantial improvements have been made in the understanding of the mediators implicated on the initiation, pathogenesis, and progression of periodontitis. Through the biomarker discovery process, new therapeutics have been designed linking therapeutic and diagnostic approaches together, especially in the area of host modulatory drugs for periodontal disease treatment. Moreover, new diagnostic technologies, such as microarray and microfluidics, are now currently available for risk assessment and comprehensive screening of biomarkers. The future is bright for the use of rapid, easy-to-use diagnostics that will provide an enhanced patient assessment that can guide and transform customized therapies for dental patients, leading to more individualized, targeted treatments for oral health.

ACKNOWLEDGMENTS

This work was supported by NIH/NIDCR grant U01-DE14961 and NCRR grant M01-RR000042. The authors acknowledge Dr. Thiago Morelli and Noah Smith for their assistance in the preparations of the manuscript and Mr. Chris Jung for design of the figure.

REFERENCES

1. ARMITAGE, G.C. 1999. Development of a classification system for periodontal diseases and conditions. Ann. Periodontol. **4:** 1–6.
2. SOCRANSKY, S.S. & A.D. HAFFAJEE. 1992. The bacterial etiology of destructive periodontal disease: current concepts. J. Periodontol. **63**(4 Suppl): 322–331.
3. GENCO, R.J. 1992. Host responses in periodontal diseases: current concepts. J. Periodontol. **63**(4 Suppl): 338–355.
4. PAGE, R.C. & K.S. KORNMAN. 1997. The pathogenesis of human periodontitis: an introduction. Periodontol. 2000 14: 9–11.
5. OFFENBACHER, S. 1996. Periodontal diseases: pathogenesis. Ann. Periodontol. **1:** 821–878.
6. ALBANDAR, J.M. 1990. A 6-year study on the pattern of periodontal disease progression. J. Clin. Periodontol. **17**(7 Pt 1): 467–471.
7. ALBANDAR, J.M. & D.K. ABBAS. 1986. Radiographic quantification of alveolar bone level changes. Comparison of 3 currently used methods. J. Clin. Periodontol. **13**: 810–813.
8. SOCRANSKY, S.S., A.D. HAFFAJEE, J.M. GOODSON & J. LINDHE. 1984. New concepts of destructive periodontal disease. J. Clin. Periodontol. **11:** 21–32.
9. ALBANDAR, J.M. 2002. Periodontal diseases in North America. Periodontol 2000. **29:** 31–69.
10. ALBANDAR, J.M., J.A. BRUNELLE & A. KINGMAN. 1999. Destructive periodontal disease in adults 30 years of age and older in the United States, 1988–1994. J. Periodontol. **70:** 13–29.

11. Socransky, S.S. & A.D. Haffajee. 2002. Dental biofilms: difficult therapeutic targets. Periodontol. 2000. **28:** 12–55.
12. Zambon, J.J. 1996. Periodontal diseases: microbial factors. Ann. Periodontol. **1:** 879–925.
13. Socransky, S.S., A.D. Haffajee, M.A. Cugini, *et al.* 1998. Microbial complexes in subgingival plaque. J. Clin. Periodontol. **25:** 134–144.
14. Loesche, W.J., D.E. Lopatin, J. Giordano, *et al.* 1992. Comparison of the benzoyl-DL-arginine-naphthylamide (BANA) test, DNA probes, and immunological reagents for ability to detect anaerobic periodontal infections due to *Porphyromonas gingivalis*, *Treponema denticola*, and *Bacteroides forsythus*. J. Clin. Microbiol. **30:** 427–433.
15. Loesche, W.J., C.E. Kazor & G.W. Taylor. 1997. The optimization of the BANA test as a screening instrument for gingivitis among subjects seeking dental treatment. J. Clin. Periodontol. **24:** 718–726.
16. Listgarten, M.A. 1976. Structure of surface coatings on teeth. A review. J. Periodontol. **47:** 139–147.
17. Listgarten, M.A. & P.M. Loomer. 2003. Microbial identification in the management of periodontal diseases. A systematic review. Ann. Periodontol. **8:** 182–192.
18. Armitage, G.C. 1996. Periodontal diseases: diagnosis. Ann. Periodontol. **1:** 37–215.
19. Armitage, G.C. 2004. The complete periodontal examination. Periodontol. 2000 **34:** 22–33.
20. Lamster, I.B. & J.T. Grbic. 1995. Diagnosis of periodontal disease based on analysis of the host response. Periodontol. 2000. **7:** 83–99.
21. Haffajee, A.D., S.S. Socransky & J.M. Goodson. 1983. Clinical parameters as predictors of destructive periodontal disease activity. J. Clin. Periodontol. **10:** 257–265.
22. Goodson, J.M. 1992. Diagnosis of periodontitis by physical measurement: interpretation from episodic disease hypothesis. J. Periodontol. **63**(4 Suppl): 373–382.
23. Haffajee, A.D., S.S. Socransky, J. Lindhe, *et al.* Clinical risk indicators for periodontal attachment loss. J. Clin. Periodontol. **18:** 117–125.
24. Streckfus, C.F. & L.R. Bigler. 2002. Saliva as a diagnostic fluid. Oral Dis. **8:** 69–76.
25. Mandel, I.D. 1990. The diagnostic uses of saliva. J. Oral Pathol. Med. **19:** 119–125.
26. Mandel, I.D. 1993. Salivary diagnosis: promises, promises. Ann. N. Y. Acad. Sci. **694:** 1–10.
27. Mandel, I.D. 1993. A contemporary view of salivary research. Crit. Rev. Oral Biol. Med. **4:** 599–604.
28. Mandel, I.D. 1993. Salivary diagnosis: more than a lick and a promise. J. Am. Dent. Assoc. **124:** 85–87.
29. Biomarkers Definitions Working Group. 2001. Biomarkers and surrogate endpoints: preferred definitions and conceptual framework. Clin. Pharmacol. Ther. **69:** 89–95.
30. Ferguson, D.B. 1987. Current diagnostic uses of saliva. J. Dent. Res. **66:** 420–424.
31. Kaufman, E. & I.B. Lamster. 2000. Analysis of saliva for periodontal diagnosis–a review. J. Clin. Periodontol. **27:** 453–465.

32. NAKAMURA, M. & J. SLOTS. 1983. Salivary enzymes. Origin and relationship to periodontal disease. J. Periodontal Res. **18:** 559–569.
33. OZMERIC, N. 2004. Advances in periodontal disease markers. Clin. Chim. Acta. **343:** 1–16.
34. NAKASHIMA, K., C. GIANNOPOULOU, E. ANDERSEN, *et al.* 1996. A longitudinal study of various crevicular fluid components as markers of periodontal disease activity. J. Clin. Periodontol. **23:** 832–838.
35. SINGH, A.K., A.E. HERR, A.V. HATCH, *et al.* 2006. Integrated microfluidic platform for oral diagnostics (IMPOD) [abstract]. New York Academy of Sciences: Oral-Based Diagnostics—A New York Academy of Sciences Meeting.
36. RAMSEIER, C.A., J.S. KINNEY, A.E. HERR, *et al.* 2006. Salivary diagnostics for inflammatory periodontal diseases [abstract]. New York Academy of Sciences: Oral-Based Diagnostics—A New York Academy of Sciences Meeting.
37. KORNMAN, K.S., R.C. PAGE & M.S. TONETTI. 1997. The host response to the microbial challenge in periodontitis: assembling the players. Periodontol. 2000. **14:** 33–53.
38. MANTYLA, P., M. STENMAN, D.F. KINANE, *et al.* 2003. Gingival crevicular fluid collagenase-2 (MMP-8) test stick for chair-side monitoring of periodontitis. J. Periodontal Res. **38:** 436–439.
39. TABA, M., JR, J. KINNEY, A.S. KIM & W.V. GIANNOBILE. 2005. Diagnostic biomarkers for oral and periodontal diseases. Dent. Clin. North Am. **49:** 551–571, vi.
40. CIMASONI, G. 1983. Crevicular fluid updated. Monogr. Oral Sci. **12:** III–VII, 1–152.
41. LOOS, B.G. & S. TJOA. 2005. Host-derived diagnostic markers for periodontitis: do they exist in gingival crevice fluid? Periodontol 2000. **39:** 53–72.
42. GOLUB, L.M., K. SIEGEL, N.S. RAMAMURTHY & I.D. MANDEL. 1976. Some characteristics of collagenase activity in gingival crevicular fluid and its relationship to gingival diseases in humans. J. Dent. Res. **55:** 1049–1057.
43. GOLUB, L.M. & I. KLEINBERG. 1976. Gingival crevicular fluid: a new diagnostic aid in managing the periodontal patient. Oral Sci. Rev. **8:** 49–61.
44. ROBERTSON, P.B., H.E. GRUPE, JR, R.E. TAYLOR, *et al.* 1973. The effect of collagenase-inhibitor complexes on collagenolytic activity of normal and inflamed gingival tissue. J. Oral Pathol. **2:** 28–32.
45. KOWASHI, Y., F. JACCARD & G. CIMASONI. 1979. Increase of free collagenase and neutral protease activities in the gingival crevice during experimental gingivitis in man. Arch. Oral Biol. **24:** 645–650.
46. VILLELA, B., R.B. COGEN, A.A. BARTOLUCCI & H. BIRKEDAL-HANSEN. 1987. Collagenolytic activity in crevicular fluid from patients with chronic adult periodontitis, localized juvenile periodontitis and gingivitis, and from healthy control subjects. J. Periodontal. Res. **22:** 381–389.
47. AIRILA-MANSSON, S., B. SODER, K. KARI & J.H. MEURMAN. 2006. Influence of combinations of bacteria on the levels of prostaglandin E2, interleukin-1beta, and granulocyte elastase in gingival crevicular fluid and on the severity of periodontal disease. J. Periodontol. **77:** 1025–1031.
48. IKARASHI, F., K. YAMAZAKI, K. HARA & H. NOHARA. 1990. Production of prostaglandin E2 by polymorphonuclear neutrophils isolated from gingival crevicular fluid and peripheral blood of dogs in periodontal health and disease. Nippon Shishubyo Gakkai Kaishi **32:** 121–128.

49. WEYNA, E. 1985. Comparative evaluation of histological studies of the gingiva and levels of prostaglandin-like substances in cases of deep inflammatory periodontal disease. Czas. Stomatol. **38:** 53–56.
50. WEYNA, E. 1983. Levels of prostaglandin-like substances in the gingiva of subjects with healthy periodontium or in patients with inflammatory periodontal disease as determined by biological methods. Czas. Stomatol. **36:** 289–293.
51. LONING, T., H.K. ALBERS, B.P. LISBOA, *et al.* 1980. Prostaglandin E and the local immune response in chronic periodontal disease. Immunohistochemical and radioimmunological observations. J. Periodontal. Res. **15:** 525–535.
52. GOODSON, J.M., F.E. DEWHIRST & A. BRUNETTI. 1974. Prostaglandin E2 levels and human periodontal disease. Prostaglandins **6:** 81–85.
53. NAKASHIMA, K., N. ROEHRICH, G. CIMASONI. 1994. Osteocalcin, prostaglandin E2 and alkaline phosphatase in gingival crevicular fluid: their relations to periodontal status. J. Clin. Periodontol. **21:** 327–333.
54. OFFENBACHER, S., B.M. ODLE, R.C. GRAY & T.E. VAN DYKE. 1984. Crevicular fluid prostaglandin E levels as a measure of the periodontal disease status of adult and juvenile periodontitis patients. J. Periodontal. Res. **19:** 1–13.
55. OFFENBACHER, S., B.M. ODLE & T.E. VAN DYKE. 1986. The use of crevicular fluid prostaglandin E2 levels as a predictor of periodontal attachment loss. J. Periodontal. Res. **21:** 101–112.
56. PEDERSON, E.D., S.R. STANKE, S.J. WHITENER, *et al.* 1995. Salivary levels of alpha 2-macroglobulin, alpha 1-antitrypsin, C-reactive protein, cathepsin G and elastase in humans with or without destructive periodontal disease. Arch. Oral Biol. **40:** 1151–1155.
57. IIJIMA, K., K. ANDO, M. KISHI, *et al.* 1983. Collagenase activity in human saliva. J. Dent. Res. **62:** 709–712.
58. BASU, M.K., E.C. FOX & J.F. BECKER. 1976. Salivary IgG and IgA before and after periodontal therapy. A preliminary report. J. Periodontal. Res. **11:** 226–229.
59. SANDHOLM, L. & E. GRONBLAD. 1984. Salivary immunoglobulins in patients with juvenile periodontitis and their healthy siblings. J. Periodontol. **55:** 9–12.
60. SANDHOLM, L., K. TOLO & I. OLSEN. 1987. Salivary IgG, a parameter of periodontal disease activity? High responders to *Actinobacillus actinomycetemcomitans* Y4 in juvenile and adult periodontitis. J. Clin. Periodontol. **14:** 289–294.
61. SCHENCK, K., D. POPPELSDORF, C. DENIS & T. TOLLEFSEN. 1993. Levels of salivary IgA antibodies reactive with bacteria from dental plaque are associated with susceptibility to experimental gingivitis. J. Clin. Periodontol. **20:** 411–417.
62. ISHIKAWA, I. & G. CIMASONI. 1970. Alkaline phosphatase in human gingival fluid and its relation to periodontitis. Arch. Oral Biol. **15:** 1401–1404.
63. GIBERT, P., P. TRAMINI, V. SIESO & M.T. PIVA. 2003. Alkaline phosphatase isozyme activity in serum from patients with chronic periodontitis. J. Periodontal. Res. **38:** 362–365.
64. TOTAN, A., M. GREABU, C. TOTAN & T. SPINU. 2006. Salivary aspartate aminotransferase, alanine aminotransferase and alkaline phosphatase: possible markers in periodontal diseases? Clin. Chem. Lab. Med. **44:** 612–615.
65. TODOROVIC, T., I. DOZIC, M. VICENTE-BARRERO, *et al.* 2006. Salivary enzymes and periodontal disease. Med. Oral Patol. Oral Cir. Bucal. **11:** E115–E119.
66. MCCAULEY, L.K. & R.M. NOHUTCU. 2002. Mediators of periodontal osseous destruction and remodeling: principles and implications for diagnosis and therapy. J. Periodontol. **73:** 1377–1391.

67. KENNETT, C.N., S.W. COX & B.M. ELEY. 1997. Investigations into the cellular contribution to host tissue proteases and inhibitors in gingival crevicular fluid. J. Clin. Periodontol. **24:** 424–431.
68. KUNIMATSU, K., K. YAMAMOTO, E. ICHIMARU, *et al.* 1990. Cathepsins B, H and L activities in gingival crevicular fluid from chronic adult periodontitis patients and experimental gingivitis subjects. J. Periodontal. Res. **25:** 69–73.
69. ICHIMARU, E., M. TANOUE, M. TANI, *et al.* 1996. Cathepsin B in gingival crevicular fluid of adult periodontitis patients: identification by immunological and enzymological methods. Inflamm. Res. **45:** 277–282.
70. CHEN, H.Y., S.W. COX & B.M. ELEY. 1998. Cathepsin B, alpha2-macroglobulin and cystatin levels in gingival crevicular fluid from chronic periodontitis patients. J. Clin. Periodontol. **25:** 34–41.
71. COX, S.W. & B.M. ELEY. 1992. Cathepsin B/L-, elastase-, tryptase-, trypsin- and dipeptidyl peptidase IV-like activities in gingival crevicular fluid. A comparison of levels before and after basic periodontal treatment of chronic periodontitis patients. J. Clin. Periodontol. **19:** 333–339.
72. ELEY, B.M. & S.W. COX. 1996. The relationship between gingival crevicular fluid cathepsin B activity and periodontal attachment loss in chronic periodontitis patients: a 2-year longitudinal study. J. Periodontal. Res. **31:** 381–392.
73. WOESSNER, J.F., JR. 1991. Matrix metalloproteinases and their inhibitors in connective tissue remodeling. FASEB J. **5:** 2145–2154.
74. BIRKEDAL-HANSEN, H. 1993. Role of matrix metalloproteinases in human periodontal diseases. J. Periodontol. **64**(5 Suppl): 474–484.
75. SALO, T., M. MAKELA, M. KYLMANIEMI, *et al.* 1994. Expression of matrix metalloproteinase-2 and -9 during early human wound healing. Lab. Invest. 70: 176–182.
76. LLANO, E., A.M. PENDAS, V. KNAUPER, *et al.* 1997. Identification and structural and functional characterization of human enamelysin (MMP-20). Biochemistry **36:** 15101–15108.
77. PIRILA, E., N. RAMAMURTHY, P. MAISI, *et al.* 2001. Wound healing in ovariectomized rats: effects of chemically modified tetracycline (CMT-8) and estrogen on matrix metalloproteinases -8, -13 and type I collagen expression. Curr. Med. Chem. **8:** 281–294.
78. CHUBINSKAYA, S., K. HUCH, K. MIKECZ, *et al.* 1996. Chondrocyte matrix metalloproteinase-8: up-regulation of neutrophil collagenase by interleukin-1 beta in human cartilage from knee and ankle joints. Lab. Invest. **74:** 232–240.
79. GOLUB, L.M., H.M. LEE, R.A. GREENWALD, *et al.* 1997. A matrix metalloproteinase inhibitor reduces bone-type collagen degradation fragments and specific collagenases in gingival crevicular fluid during adult periodontitis. Inflamm. Res. **46:** 310–319.
80. LEE, W., S. AITKEN, G. KULKARNI, *et al.* 1991. Collagenase activity in recurrent periodontitis: relationship to disease progression and doxycycline therapy. J. Periodontal. Res. **26:** 479–485.
81. LEE, W., S. AITKEN, J. SODEK & C.A. MCCULLOCH. 1995. Evidence of a direct relationship between neutrophil collagenase activity and periodontal tissue destruction *in vivo*: role of active enzyme in human periodontitis. J. Periodontal. Res. **30:** 23–33.
82. NOMURA, T., A. ISHII, Y. OISHI, *et al.* 1998. Tissue inhibitors of metalloproteinases level and collagenase activity in gingival crevicular fluid: the relevance to periodontal diseases. Oral Dis. **4:** 231–240.

83. MANCINI, S., R. ROMANELLI, C.A. LASCHINGER, *et al.* 1999. Assessment of a novel screening test for neutrophil collagenase activity in the diagnosis of periodontal diseases. J. Periodontol. **70:** 1292–1302.
84. GOLUB, L.M., S. CIANCIO, N.S. RAMAMAMURTHY, *et al.* 1990. Low-dose doxycycline therapy: effect on gingival and crevicular fluid collagenase activity in humans. J. Periodontal. Res. **25:** 321–330.
85. GOLUB, L.M., M. WOLFF, S. ROBERTS, *et al.* 1994. Treating periodontal diseases by blocking tissue-destructive enzymes. J. Am. Dent. Assoc. **125:** 163–169; discussion 169–171.
86. TERONEN, O., Y.T. KONTTINEN, C. LINDQVIST, *et al.* 1997. Human neutrophil collagenase MMP-8 in peri-implant sulcus fluid and its inhibition by clodronate. J. Dent. Res. **76:** 1529–1537.
87. MA, J., U. KITTI, O. TERONEN, *et al.* 2000. Collagenases in different categories of peri-implant vertical bone loss. J. Dent. Res. **79:** 1870–1873.
88. KIVELA-RAJAMAKI, M.J., O.P. TERONEN, P. MAISI, *et al.* 2003. Laminin-5 gamma2-chain and collagenase-2 (MMP-8) in human peri-implant sulcular fluid. Clin. Oral Implants. Res. **14:** 158–165.
89. KIVELA-RAJAMAKI, M., P. MAISI, R. SRINIVAS, *et al.* 2003. Levels and molecular forms of MMP-7 (matrilysin-1) and MMP-8 (collagenase-2) in diseased human peri-implant sulcular fluid. J. Periodontal. Res. **38:** 583–590.
90. TENG, Y.T., J. SODEK & C.A. MCCULLOCH. 1992. Gingival crevicular fluid gelatinase and its relationship to periodontal disease in human subjects. J. Periodontal. Res. **27:** 544–552.
91. JOHANSSON, N. & V.M. KAHARI. 2000. Matrix metalloproteinases in squamous cell carcinoma. Histol. Histopathol. **15:** 225–237.
92. KONTTINEN, Y.T., T. SALO, R. HANEMAAIJER, *et al.* 1999. Collagenase-3 (MMP-13) and its activators in rheumatoid arthritis: localization in the pannus-hard tissue junction and inhibition by alendronate. Matrix Biol. **18:** 401–412.
93. TERVAHARTIALA, T., E. PIRILA, A. CEPONIS, *et al.* 2000. The *in vivo* expression of the collagenolytic matrix metalloproteinases (MMP-2, -8, -13, and -14) and matrilysin (MMP-7) in adult and localized juvenile periodontitis. J. Dent. Res. **79:** 1969–1977.
94. KIILI, M., S.W. COX, H.Y. CHEN, *et al.* 2002. Collagenase-2 (MMP-8) and collagenase-3 (MMP-13) in adult periodontitis: molecular forms and levels in gingival crevicular fluid and immunolocalisation in gingival tissue. J. Clin. Periodontol. **29:** 224–232.
95. UITTO, V.J., K. AIROLA, M. VAALAMO, *et al.* 1998. Collagenase-3 (matrix metalloproteinase-13) expression is induced in oral mucosal epithelium during chronic inflammation. Am. J. Pathol. **152:** 1489–1499.
96. FAGERHOL, M.K. 2000. Calprotectin, a faecal marker of organic gastrointestinal abnormality. Lancet. **356:** 1783–1784.
97. POULLIS, A., R. FOSTER, M.A. MENDALL & M.K. FAGERHOL. 2003. Emerging role of calprotectin in gastroenterology. J. Gastroenterol. Hepatol. **18:** 756–762.
98. KIDO, J., T. NAKAMURA, R. KIDO, *et al.* 1999. Calprotectin in gingival crevicular fluid correlates with clinical and biochemical markers of periodontal disease. J. Clin. Periodontol. **26:** 653–657.
99. NISAPAKULTORN, K., K.F. ROSS & M.C. HERZBERG. 2001. Calprotectin expression *in vitro* by oral epithelial cells confers resistance to infection by *Porphyromonas gingivalis*. Infect. Immun. **69:** 4242–4247.

100. Bataille, R., P. Delmas & J. Sany. 1987. Serum bone gla-protein in multiple myeloma. Cancer **59:** 329–334.
101. Slovik, D.M., C.M. Gundberg, R.M. Neer & J.B. Lian. 1984. Clinical evaluation of bone turnover by serum osteocalcin measurements in a hospital setting. J. Clin. Endocrinol. Metab. **59:** 228–230.
102. Giannobile, W.V., S.E. Lynch, R.G. Denmark, *et al.* 1995. Crevicular fluid osteocalcin and pyridinoline cross-linked carboxyterminal telopeptide of type I collagen (ICTP) as markers of rapid bone turnover in periodontitis. A pilot study in beagle dogs. J. Clin. Periodontol. **22:** 903–910.
103. Kunimatsu, K., S. Mataki, H. Tanaka, *et al.* 1993. A cross-sectional study on osteocalcin levels in gingival crevicular fluid from periodontal patients. J. Periodontol. **64:** 865–869.
104. Lee, A.J., T.F. Walsh, S.J. Hodges, A. Rawlinson. 1999. Gingival crevicular fluid osteocalcin in adult periodontitis. J. Clin. Periodontol. **26:** 252–256.
105. Williams, R.C., D.W. Paquette, S. Offenbacher, *et al.* 2001. Treatment of periodontitis by local administration of minocycline microspheres: a controlled trial. J. Periodontol. **72:** 1535–1544.
106. Griffiths, G.S., A.M. Moulson, A. Petrie & I.T. James. 1998. Evaluation of osteocalcin and pyridinium crosslinks of bone collagen as markers of bone turnover in gingival crevicular fluid during different stages of orthodontic treatment. J. Clin. Periodontol. **25:** 492–498.
107. Giannobile, W.V., K.F. Al-Shammari & D.P. Sarment. 2003. Matrix molecules and growth factors as indicators of periodontal disease activity. Periodontol. 2000 **31:** 125–134.
108. Armitage, G.C., M.K. Jeffcoat, D.E. Chadwick, *et al.* 1994. Longitudinal evaluation of elastase as a marker for the progression of periodontitis. J. Periodontol. **65:** 120–128.
109. Oringer, R.J., M.D. Palys, A. Iranmanesh, *et al.* 1998. C-telopeptide pyridinoline cross-links (ICTP) and periodontal pathogens associated with endosseous oral implants. Clin. Oral Implants. Res. **9:** 365–373.
110. Palys, M.D., A.D. Haffajee, S.S. Socransky & W.V. Giannobile. 1998. Relationship between C-telopeptide pyridinoline cross-links (ICTP) and putative periodontal pathogens in periodontitis. J. Clin. Periodontol. **25**(11 Pt 1): 865–871.
111. Shibutani, T., Y. Murahashi, E. Tsukada, *et al.* 1997. Experimentally induced periodontitis in beagle dogs causes rapid increases in osteoclastic resorption of alveolar bone. J. Periodontol. **68**: 385–391.
112. Talonpoika, J.T. & M.M. Hamalainen. 1994. Type I collagen carboxyterminal telopeptide in human gingival crevicular fluid in different clinical conditions and after periodontal treatment. J. Clin. Periodontol. **21:** 320–326.
113. Gapski, R., J.L. Barr, D.P. Sarment, *et al.* 2004. Effect of systemic matrix metalloproteinase inhibition on periodontal wound repair: a proof of concept trial. J. Periodontol. **75:** 441–452.
114. Perez, L.A., K.F. Al-Shammari, W.V. Giannobile & H.L. Wang. 2002. Treatment of periodontal disease in a patient with Ehlers-Danlos syndrome. A case report and literature review. J. Periodontol. **73:** 564–570.
115. Oringer, R.J., K.F. Al-Shammari, W.A. Aldredge, *et al.* 2002. Effect of locally delivered minocycline microspheres on markers of bone resorption. J. Periodontol. **73:** 835–842.

116. AL-SHAMMARI, K.F., W.V. GIANNOBILE, W.A. ALDREDGE, *et al.* 2001. Effect of non-surgical periodontal therapy on C-telopeptide pyridinoline cross-links (ICTP) and interleukin-1 levels. J. Periodontol. **72:** 1045–1051.
117. GIANNOBILE, W.V. 1999. C-telopeptide pyridinoline cross-links. Sensitive indicators of periodontal tissue destruction. Ann. N. Y. Acad. Sci. **878:** 404–412.
118. TERMINE, J.D., H.K. KLEINMAN, S.W. WHITSON, *et al.* 1981. Osteonectin, a bone-specific protein linking mineral to collagen. Cell **26**(1 Pt 1): 99–105.
119. BOWERS, M.R., L.W. FISHER, J.D. TERMINE & M.J. SOMERMAN. 1989. Connective tissue-associated proteins in crevicular fluid: potential markers for periodontal diseases. J. Periodontol. **60:** 448–451.
120. OLDBERG, A., A. FRANZEN & D. HEINEGARD. 1986. Cloning and sequence analysis of rat bone sialoprotein (osteopontin) cDNA reveals an Arg-Gly-Asp cell-binding sequence. Proc Natl. Acad. Sci. USA **83:** 8819–8823.
121. RODAN, G.A. 1995. Osteopontin overview. Ann. N. Y. Acad. Sci. **760:** 1–5.
122. REINHOLT, F.P., K. HULTENBY, A. OLDBERG & D. HEINEGARD. 1990. Osteopontin—a possible anchor of osteoclasts to bone. Proc. Natl. Acad. Sci. USA **87:** 4473–4475.
123. KADONO, H., J. KIDO, M. KATAOKA, *et al.* 1999. Inhibition of osteoblastic cell differentiation by lipopolysaccharide extract from Porphyromonas gingivalis. Infect. Immun. **67:** 2841–2846.
124. HEINEGARD, D., G. ANDERSSON, F.P. REINHOLT. 1995. Roles of osteopontin in bone remodeling. Ann. N. Y. Acad. Sci. **760:** 213–222.
125. OWEN, T.A., M. ARONOW, V. SHALHOUB, *et al.* 1990. Progressive development of the rat osteoblast phenotype *in vitro*: reciprocal relationships in expression of genes associated with osteoblast proliferation and differentiation during formation of the bone extracellular matrix. J. Cell. Physiol. **143:** 420–430.
126. IKEDA, T., S. NOMURA, A. YAMAGUCHI, *et al.* 1992. In situ hybridization of bone matrix proteins in undecalcified adult rat bone sections. J. Histochem. Cytochem. **40:** 1079–1088.
127. HORTON, M.A., M.A. NESBIT & M.H. HELFRICH. 1995. Interaction of osteopontin with osteoclast integrins. Ann. N. Y. Acad. Sci. **760:** 190–200.
128. KIDO, J., T. NAKAMURA, Y. ASAHARA, *et al.* 2001. Osteopontin in gingival crevicular fluid. J. Periodontal. Res. **36:** 328–333.
129. SHARMA, C.G. & A.R. PRADEEP. 2006. Gingival crevicular fluid osteopontin levels in periodontal health and disease. J. Periodontol. **77:** 1674–1680.

Human Breath Odors and Their Use in Diagnosis

CHRIS L. WHITTLE,[a] STEVEN FAKHARZADEH,[b] JASON EADES,[a] AND GEORGE PRETI[a,b]

[a]*Monell Chemical Senses Center, 3500 Market Street, Philadelphia, Pennsylvania 19104, USA*

[b]*Department of Dermatology, School of Medicine, University of Pennsylvania, Philadelphia, Pennsylvania, USA*

ABSTRACT: **Humans emit a complex array of volatile and nonvolatile molecules that are influenced by an individual's genetics, health, diet, and stress. Olfaction is the most ancient of our distal senses and may be used to evaluate food and environmental toxins as well as recognize kin and potential predators. Many body odors evolved to be olfactory messengers, which convey information between individuals. Consequently, those practicing the healing arts have used olfaction to aid in their diagnosis of disease since the dawn of medical practice. Studies using modern instrumental analyses have focused upon analysis of breath volatiles for biomarkers of internal diseases. In these studies, a subject's oral health status appears to seldom be considered. However, saliva and properly collected alveolar air samples must pass over or come in contact with the posterior dorsal surface of the tongue, a site of bacterial plaque development and source of halitosis-related volatiles. Because of our basic research into the nature of human body odors, our lab has received referrals of people with idiopathic malodor production, from either the oral cavity or body. We developed a protocol to help differentiate individuals with chronic halitosis from those with the genetic, odor-producing metabolic disorder trimethylaminuria (TMAU). In our referred population, TMAU is the largest cause of undiagnosed body odor. Many TMAU-positive individuals present with oral symptoms of dysguesia and halitosis as well as body odor. We present data regarding the presentation of our referred subjects as well as the analytical results from a small number of these subjects regarding their oral levels of halitosis-related malodorants and trimethylamine.**

KEYWORDS: **human breath; biomarkers; breath condensate; disease; pathology; trimethylamine; trimethylaminuria; choline; flavin-containing monooxygenase 3 (FMO3); SPME; GC/MS**

Address for correspondence: George Preti, Ph.D., Monell Chemical Senses Center, 3500 Market Street, Philadelphia, PA 19104. Voice: 215-898-4713; fax: 215-898-2084.
preti@monell.org

Ann. N.Y. Acad. Sci. 1098: 252–266 (2007).
doi: 10.1196/annals.1384.011

INTRODUCTION

Humans emit a complex array of nonvolatile and volatile molecules. The metabolic processes of an individual, and hence the compounds emitted into the environment, may be influenced by genetics, diet, stress, and the immune status of the individual. Human olfaction is the most ancient of our distal senses and does provide information from distant sources in real time. Olfactory information may be used to detect and evaluate food sources and environmental toxins as well as to recognize kin and potential predators. In addition, many body odors evolved to be olfactory cues that convey information between individuals.[1] Large numbers of volatile compounds may be emitted from several areas of the body that are prone to odor production; these include the scalp, axillae, feet, groin, and oral cavity.[2] Consequently, it is not surprising that physicians have used their olfactory and gustatory senses to aid in the differential diagnosis of diseases since the beginning of medical practice.[3–5]

Hippocrates is reported to have used exhaled breath and smelled the breath of patients as part of his assessment.[6] Breath testing as a scientific tool was pioneered in the 18th century by Lavoisier and Laplace.[7] These pioneers established the presence of CO_2 in exhaled breath air. In the second half of the 20th century, the development of more sophisticated analytical techniques, such as gas chromatography, has allowed the separation and identification of volatile compounds from complex biological matrices, such as exhaled breath. In the 1960s, the separating power of gas chromatography was combined with mass spectrometry to create combined gas chromatography/mass spectrometry (GC/MS); see Watson for a review.[8] This development allowed the separation of complex mixtures as well as structural identification of separated, volatile components.

Numerous investigators have captured and concentrated breath and salivary volatiles. Studies by these researchers have focused upon establishing the normal breath constituents and searching for biomarkers from the oral cavity to aid in the diagnosis of illness or severity of disease.[9–13]

Analysis of exhaled breath provides a unique opportunity to examine the organic constituents of blood because alveolar breath reflects the concentration of metabolites that have passively diffused across the pulmonary alveolar membrane. Breath primarily consists of nitrogen, oxygen, carbon dioxide, inert gases, water vapor, and a trace amount of volatile organic compounds (VOCs) (e.g., acetone, isoprene, and pentane) that are present in nano- to picomolar concentrations.[14] Because the organic constituents of exhaled breath are representative of the blood-borne concentrations of metabolites, breath analysis provides a noninvasive means to examine blood-borne constituents relative to using blood and/or urine samples. Some of the advantages of using exhaled breath include the fact that lung air volatiles reflect the arterial concentrations of biological substances; also the VOCs are removed from complex fluid matrices, such as blood and urine; consequently, the complete sample of all compounds are present in the collected sample with no work-up required prior to analysis of the sample. However, breath analysis also has limitations. One issue that hinders routine breath tests in clinical practice is a lack of standardized collection and analysis methods and the need for complex and unfamiliar (for clinicians) instrumentation. In addition, simple and automated chemical tests that are routinely used for blood and urine analyses are relatively inexpensive when compared to the most commonly used instrumentation for breath analyses, GC/MS. The latter requires an expensive initial investment (> \$70,000) and maintenance (>\$10,000/year) as well as skilled operators who must be able to interpret data from large complex data sets.[9,15]

A large number of studies using modern instrumental analyses have focused upon analysis of breath volatiles in search of biomarkers for disease states. The vast majority of these studies have examined respiratory system diseases such as asthma. Nitric oxide is the most extensively exhaled marker investigated, and has been linked to a host of respiratory ailments including chronic obstructive pulmonary disease,[16,17] rhinitis,[18,19] rhinorrhea,[20] chronic cough (primary),[21,22] asthma,[23,24] cystic fibrosis,[25] bronchiectasis,[26,27] and lung cancer.[28,29] Numerous exhaled markers of disease have been studied, such as several leukotrienes and nitrogenous compounds.[30–33]

Many recent studies focused upon branched or methylated hydrocarbons or VOCs with broad diagnostic potential. These have been postulated to increase in breath effluvia as a function of oxidative stress or damage (see Miekisch *et al.* for a review).[11] Although hydrocarbons are commonly found in the air of urban areas, where many of the above-cited studies have taken place, investigators appear certain that these products are products of human biological activity and not exogenous, environmental contaminants. These include the straight-chained hydrocarbons, ethane and pentane, which are markers of oxidative damage and have also been linked to diseases, such as asthma,[34,35] lung and breast cancer,[33,36] interstitial lung disease,[37] chronic obstructive pulmonary disease,[35] and heart rejection.[38] Numerous other volatile compounds have been found in exhaled breath and have been linked to different diseases. Although not exhaustive, TABLE 1 illustrates the range of VOCs that may serve as biomarkers for disease states.

In these studies, however, a subject's oral health status appears to seldom be considered. Halitosis is one of the most frequent complaints expressed by dental patients. Approximately 90% of oral malodor is thought to originate from the oral cavity, with the remaining 10% originating from distal points in the digestive and respiratory systems.[39–41] Persistent halitosis may be indicative of underlying medical conditions, such as diabetes, leukemia, gastrointestinal ulcers, lung cancer, trimethylaminuria (TMAU), and several other idiopathic conditions.[39,42] Even correctly collected alveolar air and saliva samples must pass over or come in contact with the posterior dorsal surface of the tongue. This is a site of bacterial plaque development that is a principal

TABLE 1. Oral/breath volatiles identified in patients with systemic disease

Pathologic condition	Compound(s)	Reference
Diabetes mellitus	Acetone, other ketones	Booth and Ostenson, 1966[63] Walsh, 2004[64]
	Breath methylated alkane contour (BMAC)	Phillips *et al.*, 2004[38]
Sleep apnea	Interleukin IL-6, 8-isoprostane	Carpagnano *et al.*, 2002[65]
H. pylori infection	Nitrate, cyanide	Lechner *et al.*, 2005[66]
	Carbon dioxide	Pathak *et al.*, 1994[67]
Sickle cell disease	Carbon monoxide	Sylvester *et al.*, 2005[68]
Methionine adenosyl-transferase deficiency	Dimethylsulfide	Chamberlin *et al.*, 1996[46]
Asthma	Leukotrienes	Montuschi and Barnes, 2002;[30] Hanazawa *et al.*, 2000;[69] Cap *et al.*, 2004[70]
Breast cancer	2-propanol, 2,3-dihydro-1-phenyl-4 (1H)-quinazolinone, 1-phenyl-ethanone, heptanal	Phillips *et al.*, 2006[33]
Lung carcinoma	Acetone, methylethylketone, *n*-propanol	Gordon *et al.*, 1985[71]
	Aniline, *o*-toluidine	Preti *et al.*, 1988[72]
	Alkanes, mono-methylated breath alkanes, alkenes	Phillips *et al.*, 2003[28]
Chronic obstructive pulmonary disease	Hydrogen peroxide	Dekhuijzen *et al.*, 1997[73]
	Nitrosothiols	Corradi *et al.*, 2001[31]
	Nitrosothiols nitric oxide	Liu and Thomas, 2005[74]
Cystic fibrosis	8-isoprostane	Montuschi *et al.*, 2000[75]
	Leukotriene B(4), interleukin-8	Bodini *et al.*, 2005[76]
Liver disease	Hydrogen disulfide, limonene	Friedman *et al.*, 1994[47]
Noncholestatic Primary biliary cirrhosis	Hydrogen disulfide	Friedman *et al.*, 1994[47]
Decompensated cirrhosis of the liver (*foetor hepaticus*)	C_2–C_5 aliphatic acids, methylmercaptan	Chen *et al.*, 1970a[48]; Kaji *et al.*, 1978[77]
	Ethanethiol, dimethylsulfide	Tangerman *et al.*, 1994[78]
Uremia/kidney failure	Dimethylamine, trimethylamine	Simenhoff *et al.*, 1977[79]
Trimethylaminuria	Trimethylaminine	Leopold *et al.*, 1990[56] Preti *et al.*, 1995[12]

source of halitosis-related volatiles.[43] The bad breath mixture has been studied principally for its volatile sulfur compounds (VSCs): hydrogen sulfide, methylmercaptan, dimethyl sulfide, carbonyl sulfide. These compounds provide much of the impact odor of halitosis; however, the breath of individuals with halitosis does contain a variety of volatile organic odorants, not just VSCs.[44,45] Consequently, VSCs as well as many other volatiles may be indicative of oral-related health issues rather than non-oral disorders and disease. Studies addressing markers from exhaled breath may want to address the subject's oral health status.

Several authors examining exhaled breath of patients with liver diseases for VSCs have not addressed their subjects' oral health.[46,47] These authors found elevated levels of VSCs in these patients versus controls. However, the historical link of liver diseases with foul breath odor (*foetor hepaticus*) appears to exclude an oral cause for breath odor in these patients. In addition, Chen *et al.* used a methionine challenge to elicit VSC production in their subjects.[48]

Because of our basic research into the nature and origin of human body odors, our lab has become the focal point for a large number of referrals of people with idiopathic malodor production from either the body or oral cavity. Regardless of presenting symptoms and to help differentiate individuals with bad breath from other possible disorders, all individuals are examined for odor production from the oral cavity and upper body by the same protocol first described by Preti *et al.*[12] A central part of this work-up is the choline challenge test for TMAU developed by Tjoa and Fennessey.[49]

Trimethylaminuria was first described by Humbert,[50] and is a metabolic disorder characterized by the inability of individuals to oxidize and convert dietary derived trimethylamine (TMA) to trimethylamine *N*-oxide (TMAO) in the liver. This disorder results from an inherited autosomal recessive trait in the gene, which codes for the flavin-containing monooxygenase enzyme 3 (FMO3). The genetic changes range from gene mutations associated with the most severe cases to the more common single nucleotide polymorphic changes in the FMO3 gene that may be associated with the less severe cases.[51,52]

Malodorous TMA is formed in the gut by bacterial metabolism of dietary constituents, mainly choline. In normal individuals, TMA is converted/oxidized to TMAO at $>95\%$ efficiency by FMO3. Individuals that have FMO3 metabolic capacity $<90\%$ conversion of TMA to TMAO are considered positive for TMAU.[53,54] TMAO is nonodorous, more polar and water-soluble than TMA, and readily excreted in the urine.[55] Individuals suffering from TMAU have a reduced capacity to oxidize TMA to TMAO.

TMA is a gas at body temperature and has a foul, rotten fish odor. At low concentrations it may be perceived as unpleasant or garbage-like. The inability to efficiently oxidize TMA results in the sporadic production of a body odor that is perceived as foul, unpleasant, and in its most extreme cases fish-like. This odor is caused by excess, unmetabolized TMA present in the circulatory system that is excreted in urine, sweat, breath, and saliva. Because there are many foods that are rich in choline (i.e., eggs, certain legumes, and organ meats), TMAU-affected individuals, family members, friends, and physicians are unlikely to associate the odor with food intake.

Symptoms may include foul body odor, halitosis, and/or dysguesia that can produce social embarrassment and may only be temporarily relieved by normal hygienic procedures.[56] The main difficulties experienced by TMAU-affected individuals are psychosocial ones that are caused by sporadic, undiagnosed odor production.[57]

To enable an easier diagnosis for TMAU and to examine whether or not elevated salivary levels of TMA might accompany oral symptoms in our referred subjects, we began collecting saliva from all subjects reporting to our lab with malodor production problems to examine this fluid for TMA.

METHODS

All procedures were approved by the Office of Regulatory Affairs at the University of Pennsylvania and informed consent was obtained from each subject before study participation.

We adapted the method of Mills and Walker[58] that employs the techniques of solid-phase microextraction (SPME) and gas chromatography-mass spectrometry (GC/MS) to quantify salivary TMA levels.[57] This adaptation included the use of perdeuterated trimethylamine (D_9-TMA) as an internal standard to aid in quantitation. We report herein on the data obtained from the saliva of several subjects using SPME-GC/MS. The linear range of the method was investigated by obtaining a calibration curve in the concentration ranges of interest. Trimethylamine concentrations (as trimethylamine hydrochloride) were made up from a stock solution of 2 mg/mL in acidified water (pH$\sim$1). The following concentrations were used: 2 mL each of acidified, distilled water containing TMA concentrations of 0.006 mg/mL, 0.003 mg/mL, 0.0012 mg/mL, 0.0006 mg/mL, 0.0003 mg/mL, and 0.00012 mg/mL. TMA-HCl were combined with 2 mL of 0.002 mg/mL perdeuterated (D_9)-TMA-DCl and 0.5 mL of 12 M NaOH. Each solution was equilibrated for 15 min at 50°C prior to exposing the SPME fiber to collect volatiles for 15 min. Each solution was analyzed a minimum of three times for both the calibration curve and patient samples.

Gas Chromatography/Mass Spectrometry

After collecting salivary volatiles containing the TMA, the SPME fiber was inserted into the hot injector of the GC/MS (held at 230°C) and exposed for 1 min to desorb volatiles collected on the fiber. A Thermoquest/Finnigan Voyager GC/MS with Xcalibur software (ThermoElectron Corp., San Jose, CA, USA) was used for all analyses. A polar, Stabilwax column, 30 M × 0.32 mm with 1.0-μ coating, (Restek Corp., Bellefonte, PA, USA) was used for separation and analysis of the volatiles extracted from the samples. The separation of TMA from other components was done isothermally by holding the column at 50°C for 5 min. TMA elutes within the first 5 min; consequently, we rapidly increased the column temperature at 40°C/min to the final temperature of 220°C to bake-off undesired volatiles. The column was recycled back to the starting temperature of 50°C for the next analysis. The injection port was set at 230°C. Helium carrier gas was used at a constant column flow rate of 2.5 mL/min throughout the analysis.

Data acquisition and operating parameters for the mass spectrometer were set as follows: scan rate 2/sec; scan range *m/z* 41 to *m/z* 440; ion source temperature 200°C; ionizing energy 70eV. Identification of structures/compounds was performed using both the NIST '02 library, as well as a manual interpretation of mass spectra compared with those reported in the literature.

Gas Chromatography with Flame Photometric Detection to Measure Oral Volatile Sulfur Compounds

We used gas chromatography with flame photometric detection (GC/FPD) to detect and quantify the amount of volatile sulfur compounds in the oral cavity of all subjects. The instrument used for these analyses was a Finnigan 9001 fitted with a flame photometric detector. We employed the method of Tonzetich,[59] albeit with modifications previously reported by Kostelc *et al.* and Preti *et al.*[60,61] Analysis of the subject's mouth air was performed in duplicate. Each analysis employed 10 mL of the subject's mouth air pulled into the chromatograph via an atmospheric sampling loop using a gas tight syringe.

FMO3 Gene Sequence Analysis

Genomic DNA was extracted from patient peripheral blood samples using the QIAamp DNA Mini Kit (Qiagen, Valencia, CA, USA). FMO3 coding exons 2-9 and flanking intron sequences were amplified by PCR using primers and conditions reported by Dolphin *et al.*[62] PCR amplification products were purified from 1.5% agarose gels prior to sequencing. All samples were submitted to the DNA Sequencing Facility at the University of Pennsylvania School of Medicine for analysis. Sequencing reactions were performed in an ABI GeneAmp® 9700 thermal cycler, resolved with an ABI 3730 DNA sequencer, and analyzed using ABI Sequencing Analysis software v 5.1 (ABI, Applied Biosystems, Incorporated, Foster City, CA, USA).

RESULTS

We have seen and tested more than 300 individuals in our laboratory using the protocol outlined in TABLE 2. One hundred two of these have been diagnosed with some form of TMAU using the choline challenge test.[49] The presenting

TABLE 2. Diagnostic protocol for referred individuals seen at the Monell Chemical Senses Center

1. First morning voided urine (base line for choline challenge test).
2. Individual presents in fasted condition (no cologne, cosmetics, fragrances, teeth and tongue plaque brushing); questionnaire and interview.
3. Organoleptic evaluation of individual's body odor and breath: 3 judges, scale of 1–10.
4. Collect axillary odors by placing a 4× 4-inch cotton pad in individual's axillae.
5. Analysis of individual's mouth air by GC/FPD for volatile sulfur compounds (VSCs): H_2S, CH_3SH, $(CH_3)_2S$.
6. Swab tongue for bacterial plaque and use for collection and analysis of volatiles.
7. Determination of resting whole mouth saliva flow rate using the methods described in Christensen and Navazesh, 1982.[80]
8. Determine and collect two, whole mouth-stimulated saliva samples using flavorless gum base. One of these is collected into a vial with 0.2 mL of 6N HCl for analysis of precholine challenge TMA. Other volatiles are examined by SPME-GC/FPD and SPME-GC/MS.
9. Lung air collection (10 L) for analysis of volatiles by GC/FPD.
10. Analyze tongue plaque volatiles by SPME-GC/MS and SPME-GC/FPD.
11. Administer 5 g of choline in 12–16 oz of juice for choline challenge[a].
12. Remove axillary pads and perform organoleptic evaluation.
13. For the next 24 h individual must:
    (A) Collect urine in three 8-h aliquots
    (B) Collect two stimulated whole mouth saliva samples during each 8-h time period using vials with 0.2 mL of 6N HCl.
    Next day:
    Individual returns saliva and urine samples.
    Blood sample taken for genotyping of FMO3 gene.

[a]All urine samples collected during choline challenge testing were analyzed for trimethylamine and triemthylamine oxide in the laboratory of Dr. Paul Fennessey and Ms. Susan Tjoa,[49] at the University of Colorado Health Sciences Center.

symptoms for the TMAU-positive individuals are illustrated in FIGURE 1. A majority of individuals had oral symptoms, complaining of either bad breath or bad taste, many in conjunction with body odors.

In addition, as illustrated in FIGURE 2, the majority of our TMAU-positive individuals were females. However, regardless of gender, in our experience, TMAU was the largest cause for undiagnosed body odor. The large number of TMAU-positive individuals who listed oral complaints led us to hypothesize that salivary concentrations of TMA were responsible for these symptoms.

During most individuals' visits, saliva was collected in conjunction with urine as described in TABLE 2. This has resulted in an archive of more than 270 frozen (at –10°C) saliva samples. However, only a small number of the saliva samples collected have been examined thus far for TMA levels: six individuals each from TMAU-positive and -negative categories. We report these preliminary results here.

TABLE 3 lists the mean salivary concentrations of TMA measured in each 8-h interval for six TMAU-positive and six TMAU-negative subjects. Clearly, the saliva of the positive subjects contains far more TMA than the saliva from

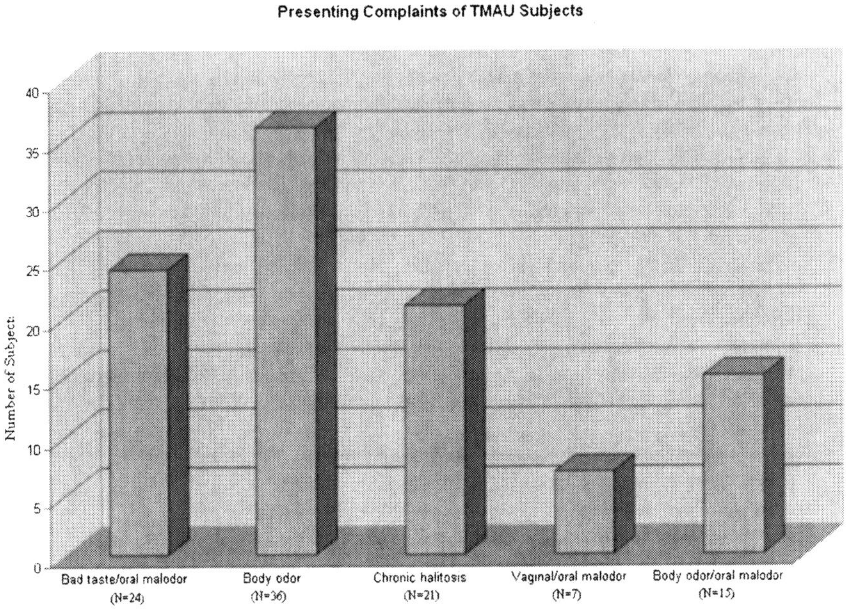

FIGURE 1. Presenting symptoms and the numbers of subjects reporting each symptom, in the subjects' own words. "Body odor" is the most common presenting symptom found in the literature pertaining to these patients. The symptoms were not always confirmed during subject evaluation using the protocol in TABLE 2.

TMAU-negative individuals, but there is a great deal of variation in the data, particularly from the positive subjects. Two of the positive subjects account for much of this variation because they demonstrated more than a 100-fold increase in salivary TMA levels from their precholine challenge base line to their highest levels in the 2nd or 3rd 8-h interval: Positive male 3 went from a base line of 25 ng/mL to 4,812 ng/mL in the 3rd, 8-hour segment; and positive male 6 went from a base line of 46.7 ng/mL to 4, 511 ng/mL in the 2nd, 8-h segment. Each of these subjects had a very low conversion of TMA to TMAO, indicative of one or more genetic mutations present in the gene for FMO3: male 3 had 25% TMAO conversion; male 6 had only 11% TMAO conversion.

We also measured the volatile sulfur compounds in the oral cavity associated with each of these subjects. Results are summarized in TABLE 4. In the subjects chosen for these analyses, we found that, on average, TMAU-negative subjects had greater concentrations of each VSC measured; these concentrations were also converted to parts per billion (ppb) levels and are presented in TABLE 4. The difference in VSC levels between the TMAU-negative and positive individuals is a result of the subjects chosen for these analyses: in our clinical experience, both TMAU-positive and -negative individuals may have halitosis. However, the two conditions, TMAU and chronic halitosis caused by bacterial tongue plaque, are independent of each other.

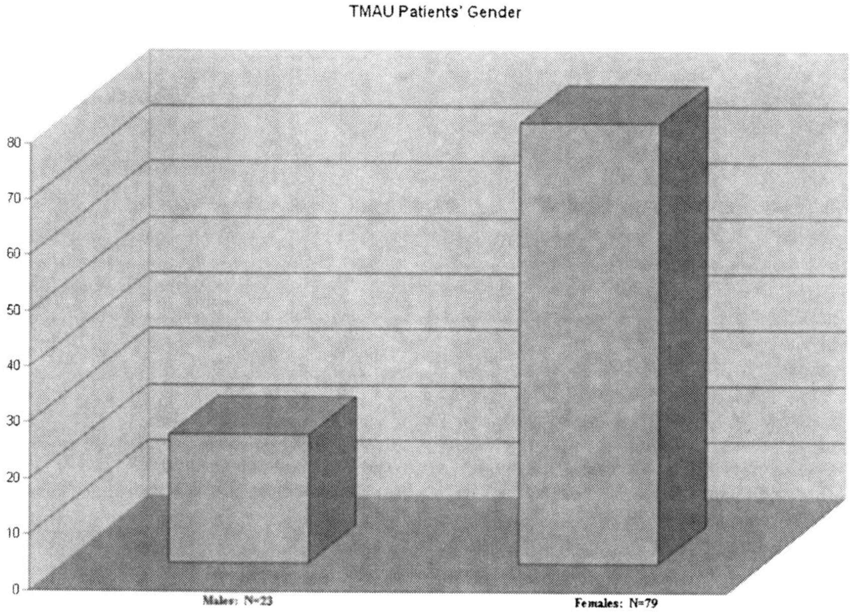

FIGURE 2. The gender distribution of TMAU-positive subjects seen in our laboratory is shown in this figure. We do not know whether or not the larger number of females affected is due to hormonal differences in the regulation of the FMO3 gene. Females report symptoms before and after menopause.

DISCUSSION

Our findings regarding the presenting symptoms of TMAU-affected individuals are in contrast to results found in most of the medical literature. Articles discussing TMAU suggest to the reader that all sufferers have a fishy body odor presentation. In our population, all of whom have been seen in person, the fish odor presentation was present in only about 10% of individuals who are TMAU-positive. Further to this point, these individuals emitted a strong fish odor recognizable at social distances only after choline challenge. Consequently, the assumption that the individual with TMAU will always smell "like fish" is incorrect and is often the reason that many TMAU-affected individuals

TABLE 3. Mean salivary trimethylamine concentrations: precholine and for each 8-h period (ng/mL)

	Precholine	1st 8 hour	2nd 8 hour	3rd 8 hour
Negative ($n = 6$)	37.41	48.37	61.07	68.40
SE	15.27	19.74	24.93	27.92
Positive ($n = 6$)	31.2	595.61	799.20	1736.57
SE	12.74	243.10	742.39	823.88

TABLE 4. Volatile sulfur compounds (VSCs) in the mouth air of referred subjects: ng/10 mL of mouth air

	COS	H_2S	CH_3SH	$(CH_3)_2S$
TMAU-negative subjects: $N = 5^a$				
Mean	1.08	2.17	1.97	0.21
SE	0.67	1.38	1.04	0.10
Mean parts per billion (ppb) of total VSC in the mouth air of TMAU-negative subjects = 453				
	COS	H_2S	CH_3SH	$(CH_3)_2S$
TMAU-positive subjects: $N = 5^a$				
Mean	0.416	0.83	0.0	0.0
SE	0.415	0.57	0.0	0.0
Mean ppb of total VSC in the mouth air of TMAU-positive subjects = 104				

[a] Instrument malfunction caused one subject in each group not to be sampled for VSC levels.

ABBREVIATION: COS = carbonyl sulfide; H_2S = hydrogen sulfide; CH_3SH = methylmercaptan; $(CH_3)_2S$ = dimethylsulfide.

are sent from one clinical specialist to another: quite often they are sent to a psychiatrist since their reported symptoms are thought to be subjective.

The choline challenge test for TMAU provides a recognized means for diagnosing this disorder. The diagnosis currently relies upon a 24-h urine collection, divided into three 8-h aliquots.[49] In our initial attempt to extend this diagnosis to saliva, we collected whole mouth-stimulated saliva to determine salivary TMA levels. Our hypothesis that the oral symptoms of many TMAU-affected individuals appears to be supported by the preliminary results presented here, although the numbers of subjects analyzed is still small. The data in TABLE 3 show much larger variation in the salivary TMA concentrations of TMAU-positive versus TMAU-negative individuals. TMAU is known to be caused by a "spectrum" of genetic changes to the gene that codes for FMO3;[51] consequently, this variation may be due, in part, to differences in the genotype of each of the TMAU-positive individuals. This is supported by our clinical observations regarding the odor of different individuals as well as genotyping data. Two of the six TMAU-positive individuals whose saliva was analyzed presented with overt fish odor from their upper body and oral cavity after ($\sim$22 h) choline challenge. As noted above, each of these male subjects had a low conversion of TMA to TMAO (<25%) and documented mutations in their FMO3 gene (data not shown).[51]

ACKNOWLEDGMENTS

This research was supported, in part, by the NIH Institutional Training Grant 2T32DC00014 as well as unrestricted funds from the Monell Chemical Senses Center.

REFERENCES

1. Wysocki, C.J. & G. Preti. 2004. Facts, fallacies, fears, and frustrations with human pheromones. Anat. Rec. **281A:** 1201–1211.
2. Wysocki, C.J. & G. Preti. 2000. Human body odors and their perception. Jpn. J. Taste Smell Res. **7:** 19–42.
3. Liddell, K. 1976. Smell as a diagnostic marker. J. Postgrad. Med. **52:** 136–138.
4. Phillips, M. 1992. Breath tests in medicine. Sci. Am. **267:** 74–79.
5. Penn, D. & W.K. Potts. 1998. Chemical signals and parasite-mediated sexual selection. Trends Ecol. Evol. **13:** 391–396.
6. Geist H.C.I. 1957. Halitosis in ancient literature. Dent. Abst. **2:** 417–418.
7. Caspary, W.F. 1978. Breath tests. Clin. Gastroenterol. **7:** 351–74.
8. Watson, J.T. 1998. A historical perspective and commentary on pioneering developments in gas chromatography/mass spectrometry at MIT. J. Mass Spect. **33:** 103–108.
9. Cao, W. & Y. Duan. 2006. Breath analysis: potential for clinical diagnosis and exposure assessment. Clin. Chem. **52:** 800–811.
10. Rahman, I. & F. Kelly. 2003. Biomarkers in breath condensate: a promising new non-invasive technique in free radical research. Free Rad. Res. **37:** 1253–1266.
11. Miekisch, W., J.K. Schubert & G.F.E. Noeldge-Schomburg. 2004. Diagnostic potential of breath analysis—focus on volatile organic compounds. Clin. Chim. Acta **347:** 25–39.
12. Preti, G., H.J. Lawley, C.A. Hormann, *et al.* 1995. Non-oral and oral aspects of oral malodor. *In* Bad Breath Research Perspectives. M. Rosenberg, Ed.: 149–174. Ramot Publishing. Tel Aviv, Israel.
13. Kharitonov, S.A. & P.J. Barnes. 2002. Biomarkers of some pulmonary diseases in exhaled breath. Biomarkers **7:** 1–32.
14. Mukhopadhyay, R. 2004. Don't waste your breath. Anal. Chem. **76:** 273A–276A.
15. Wolfram, M., J.K. Schubert & G.F.E. Noeldge-Schomburg. 2004. Diagnostic potential of breath analysis—focus on volatile organic compounds. Clin. Chim. Acta **347:** 25–39.
16. Kharitonov, S.A. & P.J. Barnes. 2001. Exhaled markers of pulmonary disease. Am. J. Resp. Crit. Care Med. **163:** 1693–1722.
17. Marczin, N., S.A. Kharitonov & S.M.H. Yacoub. 2003. Disease Markers in Exhaled Breath. Dekker. New York.
18. Baraldi, E., N.M. Azzolin, S. Carra, *et al.* 1998. Effect of topical steroids on nasal nitric oxide production in children with perennial allergic rhinitis: a pilot study. Resp. Med. **92:** 558–561.
19. Henriksen, A.H., M. Sue-Chu, T. Lingaas Holmen, *et al.* 1999. Exhaled and nasal NO levels in allergic rhinitis: relation to sensitization, pollen season, and bronchial hyperresponsiveness. Eur. Resp. J. **13:** 1301–1306.
20. Franklin, P.J., S.W. Turner, G.L. Hall, *et al.* 2005. Exhaled nitric oxide is reduced in infants with rhinorrhea. Pediat. Pulmonol. **39:** 117–119.
21. Dupont, L.J., F. Rochette, M.G. Demedts & G.M. Verleden. 1998. Exhaled nitric oxide correlates with airway hyperresponsiveness in steroid-naive patients with mild asthma. Am. J. Resp. Crit. Care Med. **157:** 894–898.
22. Chatkin, J.M.A.K., P.E. Silkoff, P. McClean, *et al.* 1999. Exhaled nitric oxide as a noninvasive assessment of chronic cough. Am. J. Resp. Crit. Care Med. **159:** 1810–1813.

23. Dupont, L.J.F., M.G. Demedts & G.M. Verleden. 1999. Prospective evaluation of the accuracy of exhaled nitric oxide for the diagnosis of asthma. Am. J. Resp. Crit. Care Med. **159:** 861.
24. Dupont, L.J., M.G. Demedts & G.M. Verleden. 2003. Prospective evaluation of the validity of exhaled nitric oxide for the diagnosis of asthma. Chest **123:** 751–756.
25. Thomas, S.R., S.A. Kharitonov, S.F. Scott, *et al.* 2000. Nasal and exhaled nitric oxide is reduced in adult patients with cystic fibrosis and does not correlate with cystic fibrosis genotype. Chest **117:** 1085–1089.
26. Tracey, W.R., C. Xue, V. Klinghofer, *et al.* 1994. Immunocytochemical detection of inducible NO synthase in human lung. Am. J. Physiol. **266:** L722–L727.
27. Ho, L.P., J.A. Innes & A.P. Greening. 1998. Exhaled nitric oxide is not elevated in the inflammatory airways diseases of cystic fibrosis and bronchiectasis. Eur. Resp. J. **12:** 1290–1294.
28. Phillips, M., R.N. Cataneo, B.A. Ditkoff, *et al.* 2003. Volatile markers of breast cancer in the breath. Breast J. **9:** 184–191.
29. Phillips, M., K. Gleeson, J.M. Hughes, *et al.* 1999. Volatile organic compounds in breath as markers of lung cancer: a cross-sectional study. Lancet **353:** 1930–1933.
30. Montuschi, P. & P.J. Barnes. 2002. Exhaled leukotrienes and prostaglandins in asthma. J. Allergy Clin. Immunol. **109:** 615–620.
31. Corradi, M., P. Montuschi, L.E. Donnelly, *et al.* 2001. Increased nitrosothiols in exhaled breath condensate in inflammatory airway diseases. Am. J. Resp. Crit. Care Med. **163:** 854–858.
32. Balint, B., S.A. Kharitonov, T. Hanazawa, *et al.* 2001. Increased nitrotyrosine in exhaled breath condensate in cystic fibrosis. Eur. Resp. J. **17:** 1201–1207.
33. Phillips, M., R.N. Cataneo, B.A. Ditkoff, *et al.* 2006. Prediction of breast cancer using volatile biomarkers in the breath. Breast Cancer Res. Treat. **99:** 19–21.
34. Olopade, C.O., M. Zakkar, W.I. Swedler & I. Rubinstein. 1997. Exhaled pentane levels in acute asthma. Chest **111:** 862–865.
35. Paredi, P., S.A. Kharitonov, D. Leak, *et al.* 2000. Exhaled ethane, a marker of lipid peroxidation is elevated in chronic obstructive pulmonary disease. Am. J. Resp. Crit. Care Med. **162:** 69–73.
36. Phillips, M., R.N. Cataneo, A.R.C. Cummin, *et al.* 2003. Detection of lung cancer with volatile markers in the breath. Chest **123:** 2115–2123.
37. Kanoh, S., H. Kobayashi & K. Motoyoshi. 2005. Exhaled ethane: an *in vivo* biomarker of lipid peroxidation in interstitial lung diseases. Chest **128:** 2387–2392.
38. Phillips, M., J.P. Boehmer, R.N. Cataneo, *et al.* 2004. Heart allograft rejection: detection with breath alkanes in low levels (the HARDBALL study). J. Heart Lung Transplant. **23:** 701–708.
39. Tonzetich, J. 1995. Preface. *In* Bad Breath Research Perspectives. M. Rosenberg, Ed.: xi–xviii. Ramot Publishing. Tel Aviv, Israel.
40. Scully, C., M. el-Maaytah, S.R. Porter & J. Greenman. 1997. Breath odor: etiopathogenesis, assessment and management. Eur. J. Oral Sci. **105:** 287–293.
41. Feller, L. & E. Blignaut. 2005. Halitosis: a review. SADJ. **60:** 17–19.
42. Messadi, D.V. 1997. Oral and non-oral sources of halitosis. J. Calif. Dent. Assoc. **25:** 127–131.

43. LOESCHE, W. & C. KAZOR. 2002. Microbiology and treatment of halitosis. Periodontol. 2000 **28:** 256–279.
44. KOSTELC, J.G., G. PRETI, P.R. ZELSON, *et al.* 1980. Salivary volatiles as indicators of periodontitis. J. Periodontal Res. **15:** 185–192.
45. PAYNE, R.K., J.N. LABOWS & X. LIU. 1999. Released oral malodors measured by solid phase microextraction/gas chromatography mass spectrometry (HS-SPME-GC-MS). ACS Symp. Series **76:** 373–386.
46. CHAMBERLIN, M.E., U. TSUNEYUKI, S.H. MUDD, *et al.* 1966. Demyelination of the brain is associated with methionine adenosylation I/II deficiency. J. Clin. Invest. **98:** 1021–1027.
47. FRIEDMAN, M.I., G. PRETI, R.O. DEEMS, *et al.* 1994. Limonene in expired lung air of patients with liver disease. Digest. Dis. Sci. **39:** 1672–1676.
48. CHEN, S., L. ZIEVE & V. MAHADEVAN. 1970a. Mercaptans and dimethyl sulfide in the breath of patients with cirrhosis of the liver. J. Lab. Clin. Med. **75:** 628–635.
49. TJOA, S. & P.V. FENNESSEY. 1991. The identification of trimethylamine excess in man: quantitative analysis and biochemical origins. Anal. Biochem. **197:** 77–82.
50. HUMBERT, J.R., K. B. HAMMOND, W.E. HATHAWAY, J.G. MARCOUX & D. O'BRIEN. 1970. Trimethylaminuria: the fish-odour syndrome. Lancet **2:** 770–771.
51. CASHMAN, J.R., K. CAMP, S.S. FAKHARZADEH, *et al.* 2003. Biochemical and clinical aspects of the human flavin-containing monooxygenase form 3 (FMO3) related to trimethylaminuria. Curr. Drug Metab. **4:** 151–170.
52. LATTARD, V., J. ZHANG, Q. TRAN, *et al.* 2003. Two new polymorphisms of the FMO3 gene in Caucasian and African-American populations: comparative genetic and functional studies. Drug Metab. Disp. **31:** 854–860.
53. ZHANG, A.Q., S. MITCHELL & R. SMITH. 1995. Fish odour syndrome: verification of carrier detect test. J. Inherit. Metab. Dis. **18:** 669–674.
54. ZSCHOCKE, J., D. KOHLMUELLER, E. QUAK, *et al.* 1999. Mild trimethylaminuria caused by common variants in FMO3 gene. Lancet **354:** 834–835.
55. CASHMAN, J.R. & J. ZHANG. 2002. Interindividual differences of human flavin-containing monooxygenase 3: genetic polymorphisms and functional variation. Drug Metab. Disp. **30:** 1043–1052.
56. LEOPOLD, D.A., G. PRETI, M.M. MOZELL, *et al.* 1990. Fish-odor syndrome presenting as dysosmia. Arch. Otolaryngol. Head Neck Surg. **116:** 354–355.
57. WALKER, V. 1993. The fish odour syndrome: the problems are psychosocial. Br. Med. J. **307:** 539.
58. MILLS, G.A., V. WALKER & H. MUGHAL. 1999. Quantitative determination of trimethylamine in urine by solid-phase microextraction and gas chromatography-mass spectrometry. J. Chromat. B. **723:** 281–285.
59. TONZETICH, J. 1971. Direct gas chromatographic analysis of sulphur compounds in mouth air in man. Arch. Oral Biol. **16:** 587–597.
60. KOSTELC, J.G., G. PRETI, P.R. ZELSON, *et al.* 1984. Oral odors in early experimental gingivitis. J. Periodontal Res. **19:** 303–312.
61. PRETI, G., L. CLARK, B.J. COWART, *et al.* 1992. Non-oral etiologies of oral malodor. J. Periodontol. **63:** 790–796.
62. DOLPHIN, C.T., J.H. RILEY, R.L. SMITH, *et al.* 1997. Structural organization of the human flavin-containing monooxygenase 3 gene (FMO3), the favored candidate for fish-odor syndrome, determined directly from genomic DNA. Genomics **46:** 260–267.
63. BOOTH, G. & S. OSTENSON. 1966. Acetone in alveolar air and the control diabetes. Lancet **2:** 1102–1105.

64. WALSH, J. 2004. A breath of fresh air in diabetes detection. Spectroscopy **19:** 50.
65. CARPAGNANO, G.E., S.A. KHARITONOV, O. RESTA, *et al.* 2002. Increased 8-isoprostane and interleukin-6 in breath condensate of obstructive sleep apnea patients. Chest **122:** 1162–1167.
66. LECHNER, M., A. KARLSEDER, D. NIEDERSEER, *et al.* 2005. *H. pylori* infection increases levels of exhaled nitrate. Helicobacter **10:** 385–390.
67. PATHAK, C.M., D.K. BHASIN, D. PANIGRAHI & R.C. GOEL. 1994. Evaluation of ^{14}C-urinary excretion and its comparison with $^{14}CO2$ in breath after ^{14}C-urea administration in *Helicobacter pylori* infection. Am. J. Gastroenterol. **89:** 734–738.
68. SYLVESTER, K.P. 2005. Exhaled carbon monoxide levels in children with sickle cell disease. Eur. J. Pediat. **164:** 162–165.
69. HANAZAWA, T., S.A. KHARITONOV, W. OLDFIELD, *et al.* 2000. Nitrotyrosine and cysteinyl leukotrienes in breath condensates are increased after withdrawal of steroid treatment in patients with asthma. Am. J. Resp. Crit. Care Med. **161:** A919.
70. CAP, P., J. CHLADEK, F. PEHAL, *et al.* 2004. Gas chromatography/mass spectrometry analysis of exhaled leukotrienes in asthmatic patients. Thorax **59:** 465–470.
71. GORDON, S.M., J.P. SZIDON, B.K. KROTOSZYNSKI, *et al.* 1985. Volatile organic compounds in exhaled air from patients with lung cancer. Clin. Chem. **31:** 1278–1282.
72. PRETI, G., J.N. LABOWS, J.G. KOSTELC, *et al.* 1988. Analysis of lung air from patients with bronchogenic carcinoma and controls using gas chromatography-mass spectrometry. J. Chromat. B: Biomed. Appl. **432:** 1–11.
73. DEKHUIJZEN, P.N., K.K. ABEN, I. DEKKER, *et al.* 1996. Increased exhalation of hydrogen peroxide in patients with stable and unstable chronic obstructive pulmonary disease. Am. J. Resp. Crit. Care Med. **154:** 813–816.
74. LIU, J. & P.J. THOMAS. 2005. Exhaled breath condensate as a method of sampling airway nitric oxide and other markers of inflammation. Med. Sci. Monitor **11:** MT53–MT62.
75. MONTUSCHI, P., S.A. KHARITONOV, G. CIABATTONI, *et al.* 2000. Exhaled 8-isoprostane as a new non-invasive biomarker of oxidative stress in cystic fibrosis. Thorax **55:** 205–209.
76. BODINI, A., C. D'ORAZIO, D. PERONI, *et al.* 2005. Biomarkers of neutrophilic inflammation in exhaled air of cystic fibrosis children with bacterial airway infections. Pediat. Pulmonol. **40:** 494–499.
77. KAJI, H., M. HISAMURA, N. SATO & M. MURAO. 1978. Gas chromatographic determination of volatile sulfur compounds in the expired alveolar air in the hepatopathic subjects. J. Chromat. B: Biomed. Appl.
78. TANGERMAN, A., M.T. MEUWESE-ARENDS & J.B. JANSEN. 1994. Cause and composition of *foetor hepaticus*. Lancet **343:** 730.
79. SIMENHOFF, M.L., J.F. BURKE, J.J. SAUKKONEN, *et al.* 1977. Biochemical profile of uremic breath. N. Engl. J. Med. **247:** 132–135.
80. CHRISTENSEN, C. & M.A. NAVAZESH. 1982. A comparison of whole mouth resting and stimulated salivary measurement procedures. J. Dent. Res. **61:** 1158–1162.

Molecular and Protein Markers of Disease

MICHAEL GLICK

Department of Oral Medicine, Arizona School of Dentistry and Oral Health, and College of Osteopathic Medicine–Mesa, Mesa, Arizona, USA

ABSTRACT: Oral fluid diagnostics are becoming a part of routine care. The key to their use is to correctly interpret their findings so that patients' well-being can be optimized.

KEYWORDS: oral fluid markers; saliva

The ability to detect specific molecular and protein markers in oral fluids unlocks a plethora of opportunities to expand and improve existing paradigms of health-care delivery. This new-found capability also brings to light numerous questions and concerns that should be addressed to properly use this rapidly advancing technology. Establishing appropriate parameters for the practical utilization of oral fluid diagnostics will guide and encourage further inquiries, as well as prevent inappropriate and unsuitable uses.

The ease and expedience of using noninvasive means to determine health or disease for large numbers of individuals will greatly influence the pervasiveness of both acute and chronic conditions. Oral fluid diagnostics can be integrated within existing health-care delivery systems but can also be incorporated within nontraditional medical care settings. As an example, recent data indicate that fewer than 40% of adult Americans visit a physician on an annual basis, yet more than 60% will visit a dental office.[1] The inclusion of oral health-care professionals in screening and monitoring of diseases could, for example, provide an additional venue to diminish the impact of many chronic illnesses, including cardiovascular diseases.[2] Although nonphysicians can use oral fluids to screen and monitor different diseases, they can also be used by physicians for point-of-care diagnosis, disease progression, and decisions on treatment end points.

Before oral fluid diagnostics go mainstream, several important questions need to be answered.

Address for correspondence: Michael Glick, D.M.D., Professor of Oral Medicine, Arizona School of Dentistry and Oral Health, College of Osteopathic Medicine, 5850 E. Still Circle, Mesa, Arizona 85206, USA. Voice: 480-219-6103; fax: 480-219-6102.
mglick@atsu.edu

Ann. N.Y. Acad. Sci. 1098: 267–268 (2007). © 2007 New York Academy of Sciences.
doi: 10.1196/annals.1384.044

Are specific molecules and proteins found in the oral cavity directly linked to a specific disease? Is there a causative relationship? What is the clinical relevance? Are salivary markers predictive? What is the temporal relation between the appearance of the oral biomarkers and disease progression?

There are additional topics that need to be addressed such as factors that may alter the interpretation of oral fluid markers. These may include differences in salivary flow, the influence of oral microbes, and medications. Habits such as smoking, and drug and alcohol use may change the composition of oral samples, as would possibly different medical treatments, such as hemodialysis, local pathogenic conditions, or psychological states such as depression and anxiety.

The key to the incorporation of oral fluid diagnostics into routine medical care will be the interpretation of the oral composition in relation to disease and well being, and the consequences of its use.

REFERENCES

1. National Center for Health Statistics. 2006. Health, United States, 2006. NCHS. Hyattsville, MD.
2. Glick, M. & B.L. Greenberg. 2005. Primary cardiovascular risk screening by dentists. J. Am. Dent. Assoc. **136:** 1541–1546.

Subclinical Cardiovascular Disease Markers Applicable to Studies of Oral Health

Multiethnic Study of Atherosclerosis

DAVID R. JACOBS, JR.[a,b] AND RICHARD S. CROW[a]

[a]*Division of Epidemiology, School of Public Health, University of Minnesota, Minneapolis, Minnesota, USA*

[b]*Department of Nutrition, University of Oslo, Oslo, Norway*

ABSTRACT: Recent findings associate periodontal disease with established coronary heart disease (CHD) and with disorders of the carotid artery. Besides measures of the carotid artery, a number of other noninvasive subclinical markers of cardiovascular disease exist and are summarized here. Included are computed tomography (CT) of the coronary arteries, ultrasound of the carotid arteries, echocardiography, magnetic resonance imaging (MRI), ankle–brachial index, microalbuminuria, and other biochemical measures of kidney dysfunction, flow-mediated dilation in the brachial artery, and pulse wave form analysis. Use of these measures may simplify and add depth to studies of oral health and cardiovascular disease. However, it is noted that the measures are not highly correlated with each other (based on 6,814 persons in the Multiethnic Study of Atherosclerosis, Pearson correlations among the above subclinical measures, range from about 0.1–0.4), do not include propensity for the important atherosclerotic phase of plaque rupture, and do not fully substitute for studies of clinical cardiovascular disease endpoints.

KEYWORDS: periodontitis; coronary calcium; carotid artery wall thickness; magnetic resonance imaging; urinary albumin; ankle–brachial index

INTRODUCTION

Consequences of arterial disease, which over lifetimes kill about one-third to one half of people, are devastating. Understanding, ameliorating, and preventing such disease is a worldwide priority. Despite its ubiquity, study of arterial disease is complicated because incident clinical disease is rare in young and

Address for correspondence: David R. Jacobs, Jr., Ph.D., Division of Epidemiology, School of Public Health, University of Minnesota, 1300 South 2nd Street, Suite 300, Minneapolis, MN. Voice: 55454-612-624-4196; fax: 612-624-0315.
jacobs@epi.umn.edu

Ann. N.Y. Acad. Sci. 1098: 269–287 (2007).
doi: 10.1196/annals.1384.029

middle age, even though atherosclerosis is seen in young adulthood and has its roots in behaviors established in childhood.[1]

Consequently, clinical events accumulate slowly and large, long, and expensive studies are needed to investigate incidence. Sample sizes for clinical trials or prospective observational studies of clinical cardiovascular disease outcomes are typically in the thousands with follow-up for 5 to 20 years. It is apparent that atherosclerosis and related arterial disease is complex and develops in various arterial beds over many years. Knowledge of subclinical stages can inform treatment and potentially provide intermediate markers as endpoints in smaller, simplified studies. The aim of this article is to describe several subclinical disease markers that have recently become available.

Both clinical and subclinical markers are highly relevant to oral health researchers, particularly those focusing on periodontitis. There is an important inflammatory component in the pathogenesis of atherosclerosis, which could well be activated by chronic infectious diseases, including periodontitis. This topic was well reviewed by Demmer and Desvarieux.[2] Although some studies use a prospective design, many studies relating periodontitis and heart disease have used a simple case–control design. Such studies reported that periodontitis is more common in the participants who have heart disease. By way of example, we provide details from three recent studies.

Geismar *et al.*[3] studied 110 Danes with coronary heart disease (CHD) verified from a Department of Cardiovascular Medicine and 140 controls without CHD from the Copenhagen City Heart Study. Radiographically determined alveolar bone loss was greater among the CHD patients than controls, though only for participants aged <60 years; bone loss was partly explained by diabetes and smoking. Briggs *et al.*[4] studied middle-aged males in Northern Ireland; the 92 patients were aged >40 years with angiographically proven CHD, while the 79 CHD-free controls were age and residence matched. Periodontal disease (whether represented by more teeth removed, more dental plaque, more bleeding on probing, or greater probing depths) was more common in the CHD cases, even when adjusted for potential confounders. In the German study, Coronary Event and Periodontal Disease (CORODONT), Spahr *et al.*[5] studied 263 patients with angiographically confirmed, stable CHD and 526 population-based, age- and sex-matched controls without a history of CHD. Subgingival biofilm samples were analyzed for the periodontal pathogens *Actinobacillus actinomycetemcomitans*, *Tannerella forsythensis*, *Porphyromonas gingivalis*, *Prevotella intermedia*, and *Treponema denticola*. They found a statistically significant association between the periodontal pathogen burden and the presence of CHD and considered *A. actinomycetemcomitans* to be of special importance. Such findings in a case–control design are provocative but not etiologically definitive. Specifically, there is no indication of temporality of the association, namely whether the excess periodontal disease occurred before or after the CHD.

The use of subclinical cardiovascular disease markers has added a dimension to this research area. We identified several studies in which subclinical carotid artery disease, either intimal media thickening (IMT) or presence of atherosclerotic plaque, has been found to be associated with periodontal disease. Most of these studies have a cross-sectional design, which, like the case–control study, does not admit inference about temporality of association. A positive feature, compared to the case–control design, is that the cross-sectional design is easily amenable to follow-up, converting to a prospective design, in which temporality becomes clearer. Beck *et al.*[6] studied 6,017 persons aged 52 to 75 years in the Atherosclerosis Risk in Communities (ARIC) Study 1996 to 1998 examination. Greater attachment loss was associated with increased prevalence of carotid arterial IMT$\geq$1 mm. The Oral Infections and Vascular Disease Epidemiology Study (INVEST), conducted in northern Manhattan in 711 participants (mean age 66 years) found that greater tooth loss was associated with higher prevalence of carotid artery plaque.[7] They interpreted tooth loss as a marker of periodontal disease. A related article in a subset of INVEST participants found that greater radiologically measured alveolar bone loss was associated with carotid artery plaque.[8] The Study of Health in Pomerania (SHIP) assessed clinical attachment loss in 1,710 German participants aged 45–75 years with no history of CHD. In men only, there was a positive association of both tooth loss and attachment loss with carotid artery plaque, with parallel findings for carotid artery IMT.[9] The INVEST study assessed 11 microbes in dental plaque from 657 subjects.[10] Increased mean carotid artery IMT was associated with higher levels of four microbes assumed to be causative of periodontitis,[11] but with lower levels of *Actinomyces naeslundii* and *Veillonella parvula*, both associated with periodontal health.[11] The ARIC study then reported on the association of elevated carotid artery IMT with serum IgG antibodies to 17 periodontal microbes in 4,585 participants.[12] Most microbes were positively associated with elevated IMT, including *Veillonella parvula*, which the INVEST study had found to be inversely associated with IMT.

In the last study, Söder *et al.*[13] used a different design: they examined carotid artery IMT in 2001 in patients who had been examined in 1985: 82 individuals, randomly chosen from a larger study, in whom the presence of periodontal disease was documented in 1985 and confirmed between 2001 and 2003, and 31 individuals periodontally healthy in 1985 and 2001–2003. They found that mean IMT was higher in the patients with periodontal disease. Whether the increased IMT was present in 1985 was not known, but it can be stated that thickened IMT was either already present then or persisted in periodontal patients 16 years after periodontal examination.Thus assessment of subclinical cardiovascular disease has contributed to an understanding of oral health in relation to atherosclerosis.

Natural History of Atherosclerosis in the Coronary Arteries

Atherosclerotic disease progression begins in childhood (FIG. 1). It is slow and asymptomatic for many years. Clinical evaluation at this stage may find

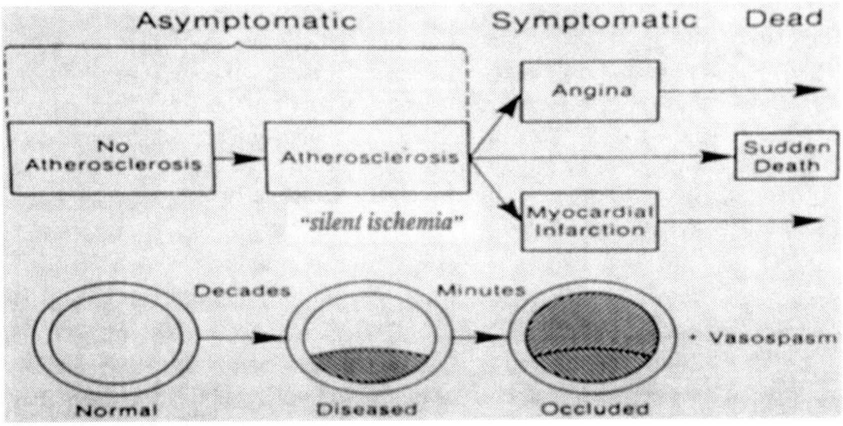

FIGURE 1. Natural history of atherosclerosis in the coronary arteries.

evidence of elevated risk factors as well as indices of "silent ischemia" (subclinical disease). Subclinical markers of atherosclerosis have been developed to understand the pathophysiologic course that leads to heart attack, ischemic stroke, and related atherosclerotic diseases and to reduce sample size and thus increase feasibility of research.

As shown in FIGURE 2, muscular arteries have three layers: intima, media, and adventitia. The structure of the vascular wall, as described by Schoen and Cotran,[14] includes blood supply provided by diffusion and from arterioles that arise from outside the vessel and penetrate the external elastic membrane. In the intima, the endothelium is supported by the basal lamina, a noncellular adhesive sheet consisting of endothelial cell–secreted glycoproteins, which acts as a selective filter and scaffolding along which cells can migrate to repair

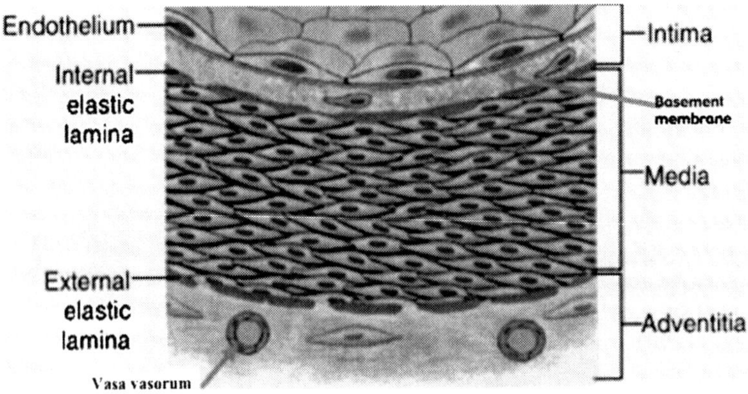

FIGURE 2. Normal vascular wall–muscular artery. (From Schoen and Cotran.[14] Reproduced by permission.)

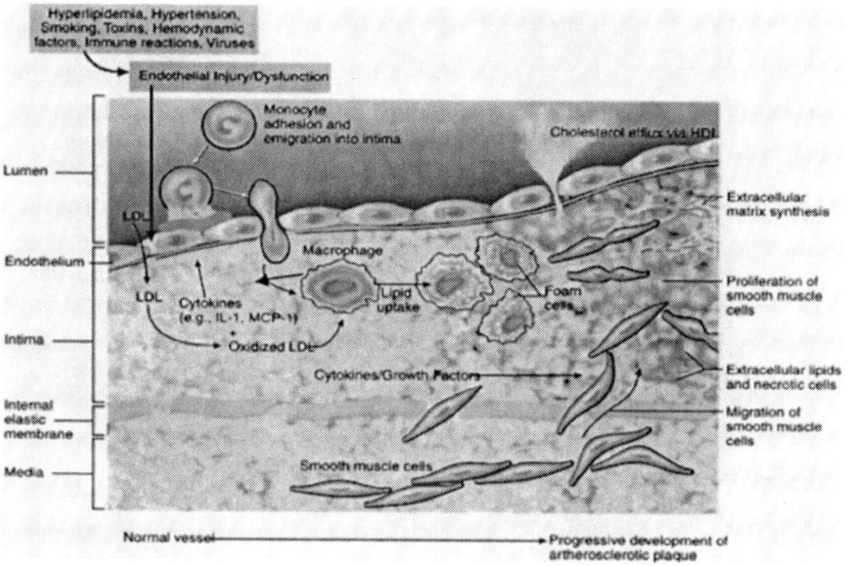

FIGURE 3. Summary of events in pathogenesis of atherosclerosis. (From Schoen and Cotran.[14] Reproduced by permission.)

an injury. The relative amounts of basic structural constituents vary along the arterial system because of adaptations to different mechanical and metabolic needs.

Atherosclerosis is a disease in which there is deposition of lipid-laden plaque in the intima of medium and large elastic arteries. It is the primary cause of myocardial infarction, ischemic stroke, and peripheral vascular disease. Arteriolosclerosis is the result of thickening of the basement membrane, proliferative changes in the vessel wall, hyaline and hyperplastic deposits, and breakdown of the media, especially in small arteries (arterioles). It is the primary cause of intraparenchymal hemorrhagic stroke. Also of special importance are diabetes-related changes in which generalized glycosylation-related changes modify cell charge, osmotic capability, and basement membrane thickness, especially in the small arteries.

Cholesterol, as a predominant feature of atherosclerotic plaque, plays a central role in atherogenesis. A comprehension of normal cholesterol function is therefore essential to understand its role in atherosclerosis. Cholesterol is an essential constituent of all mammalian cell membranes and as such is central to life. Cholesterol is water-insoluble and carried in the blood in protein-covered packages, called lipoprotein particles. Particles of different densities have different compositions and functions. A simplification is that low-density lipoprotein (LDL) carries cholesterol for nourishment to the cells, entering and leaving the intima many times per second. In contrast, high-density lipoprotein (HDL) carries cholesterol from the cells (reverse transport).

FIGURE 3 depicts plaque deposition and accumulation, the central hypothesis of the pathogenesis of atherosclerosis. Hyperlipidemia and other risk factors cause injury to the endothelium, leading to adhesion of platelets/monocytes.[14] These cells then insinuate themselves into the intima, introducing oxidized or other modified LDL particles. These are seen as invaders and are taken up by macrophages, thereby getting trapped in the subendothelial space. There is migration of smooth muscle cells into the intima and increased cell proliferation. Trapped macrophages form foam cells through uncontrolled particle uptake by scavenger receptors and intracellular accumulation. These particles have proatherogenic activities, including promotion of oxidative damage, inflammation, release of growth factors, chemotactic activity, cytotoxicity, and endothelial dysfunction. There is insudation of extracellular lipid from the vessel lumen and degenerating foam cells. At first, fatty streaks form, then atherosclerotic plaque. Advanced plaque is lipid-laden and often has a fibrous cap. Calcium deposition occurs throughout the process.

In this first phase of atherosclerosis, there are relatively fixed atherosclerotic plaques,[15] as depicted in FIGURE 4; this sets the stage for the second phase, consisting of a surface defect, hematoma or hemorrhage, and plaque rupture, leading to thrombus or embolus,[16] as depicted in FIGURES 5 and 6. As described by Wissler,[16] fissuring of atherosclerotic plaques (FIG. 6) leads to highly thrombotic plaque material interacting with blood, leading to the development of a thrombus. Resultant clinical symptoms depend on the extent of luminal obstruction.

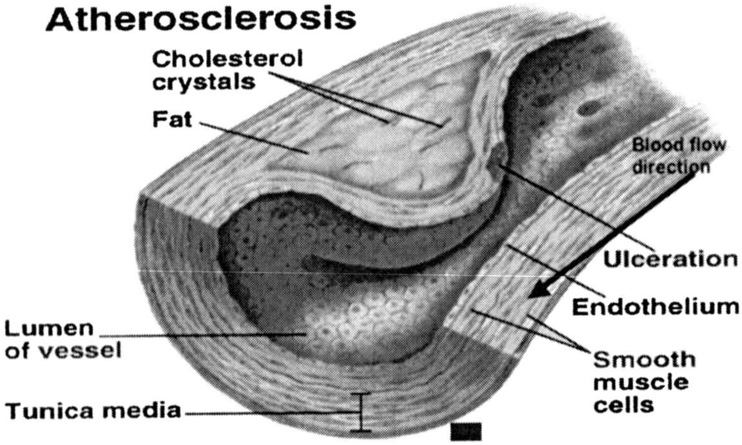

FIGURE 4. Relatively fixed atherosclerotic coronary obstruction. (From Fox.[15] Reproduced by permission.)

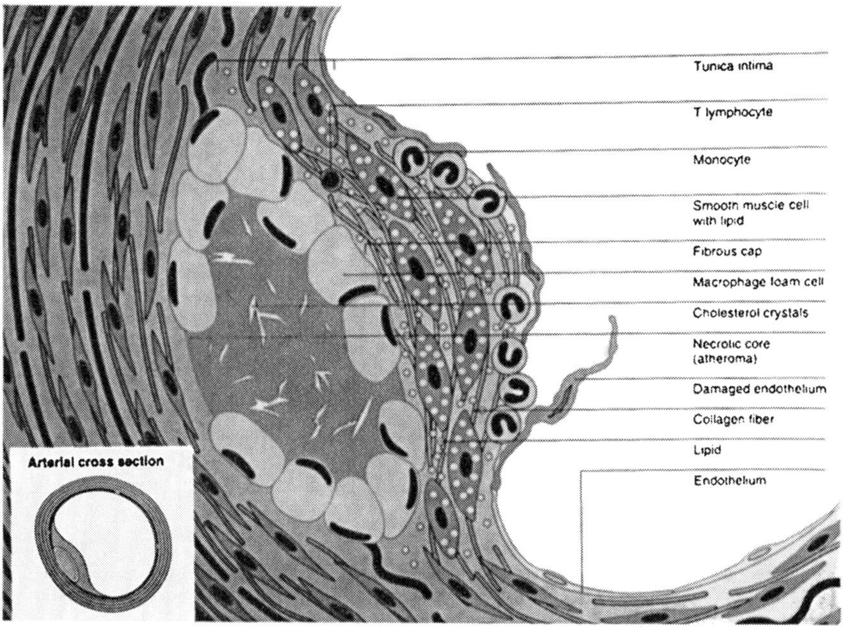

FIGURE 5. Evolving atherosclerotic plaque. (From Wissler.[16] Reproduced by permission.)

Markers of Subclinical Atherosclerotic Disease

Early studies used risk factor endpoints as substitutes for clinical disease events. Typical markers were blood pressure and blood lipids, with results interpreted as related to heart disease risk. Later, invasive coronary angiography endpoints were used, for example, in studies of regression of atherosclerosis under statin treatment.[17] Such studies were influential in deciding to carry out major statin trials with clinical endpoints. A current observational study[18] is the Multiethnic Study of Atherosclerosis (MESA), focusing on the natural history of newer noninvasive measures of subclinical atherosclerosis, including that in the coronary arteries, the carotid artery, peripheral arteries, and the kidney. The primary MESA goal is to study whether subclinical measures predict subsequent clinical events and whether such prediction adds to that of risk factors alone. Pending such follow-up, the subclinical markers assessed in MESA are not currently regarded as definitive with respect to clinical disease and may be simply another form of risk factor.

Computed tomography (CT) of the coronary arteries assesses calcium deposits, indicative of coronary artery atherosclerosis (FIG. 7).[19] As noted above, calcium in plaque occurs throughout the large arteries. CT measurement is done quickly to overcome artifacts due to heart motion. The electron beam CT completes its image in 0.1 sec, late in diastole; electron beam machines have

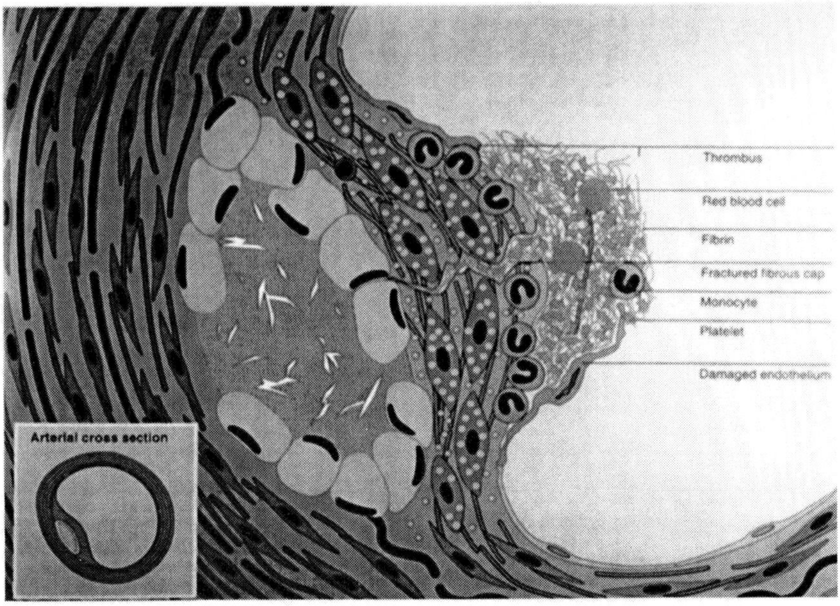

FIGURE 6. Fissuring of atherosclerotic plaque. (From Wissler.[16] Reproduced by permission.)

now mostly been replaced with multidetector CT machines, in which the CT takes 0.5 sec, starting early in diastole. The method assesses calcium deposits, a subset of coronary artery atherosclerosis. The calcium deposits may be present in tiny quantities or may be extensive. The extent of coronary artery calcification is expressed in the Agatston score,[20] which sums over all observed plaques the product of the area × an indicator of peak density. Tiny quantities may be indicative of an early stage of atherosclerosis, while large quantities may indicate extensive coronary atherosclerosis, and could be capped-off plaque. The method misses fatty streaks and noncalcified plaque. This method is expensive and not transportable, but reproducible over follow-up. Appropriate CT machines currently exist in many medical centers.

Ultrasound of the carotid artery is used to assess both carotid artery plaque and thickening of the intima and media of the vessel wall. Current research is attempting to measure atherosclerotic plaque composition (lipid-laden soft plaque vs. hard calcium-rich plaque). On the basis of flow characteristics, plaque is seen to be much more common in the internal carotid artery and around the bifurcation than in the common carotid artery. It is not known whether thickening of the intima and media always has the same pathologic basis as does deposition of atherosclerotic plaque. Therefore, if IMT is of primary interest, it may be best to restrict analysis to the common carotid artery. FIGURE 8 provides a schematic of carotid artery ultrasound testing.[21] FIGURE 9 presents an image of a normal artery[21] and FIGURE 10 presents an

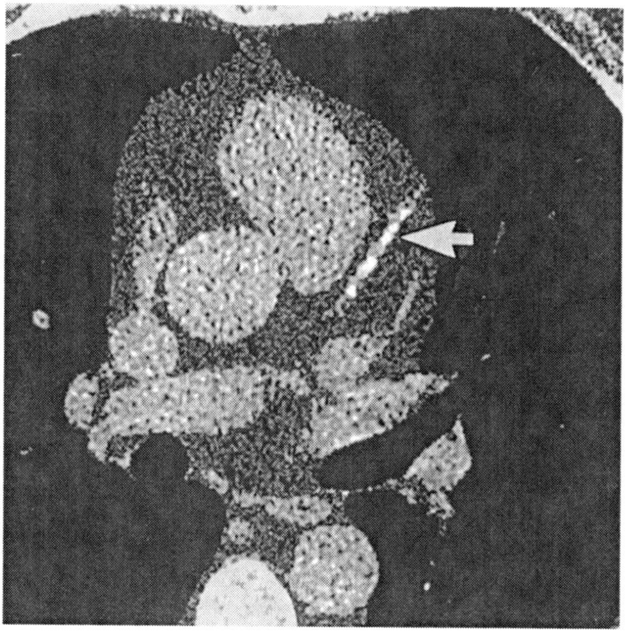

FIGURE 7. Computed tomographic image of the heart showing (*arrow*) calcification in the left anterior descending coronary artery. (From Becker *et al.*[19] Reproduced by permission.)

image of plaque near the arterial bifurcation and thickened intima media in the arterial wall.[21] Repeatability of this examination can be problematic, as it is difficult to define edges and to return to the exact same anatomic location with the same probe angle on repeat testing. The method is not too expensive and is transportable.

Echocardiography and magnetic resonance imaging (MRI) assesses many parameters, including left ventricular size, wall motion abnormalities, arterial distensibility, and cardiac output. These methods capture the heart in motion. MRI produces clear pictures,[22] as shown in FIGURE 11, depicting the right coronary artery,[23] and FIGURE 12, a 3D image depicting plaque in the carotid artery. Among interesting features that can be measured are left ventricular size, which may increase pathologically, especially in hypertension. Wall motion abnormalities, occurring in congestive heart failure and after myocardial infarction, are detectable. Cardiac output, aortic distensibility, plaque composition and conformation (with potential to predict likelihood of rupture) are all measurable. Important parameters that could relate to potential for plaque rupture would be atherosclerotic plaque with a thin cap, eccentric luminal distribution, and a large lipid core. The study of vulnerable versus stable plaques can be done with MRI using a contrast agent. Echocardiography is widely available and has a modest cost. MRI is available in many medical centers, but

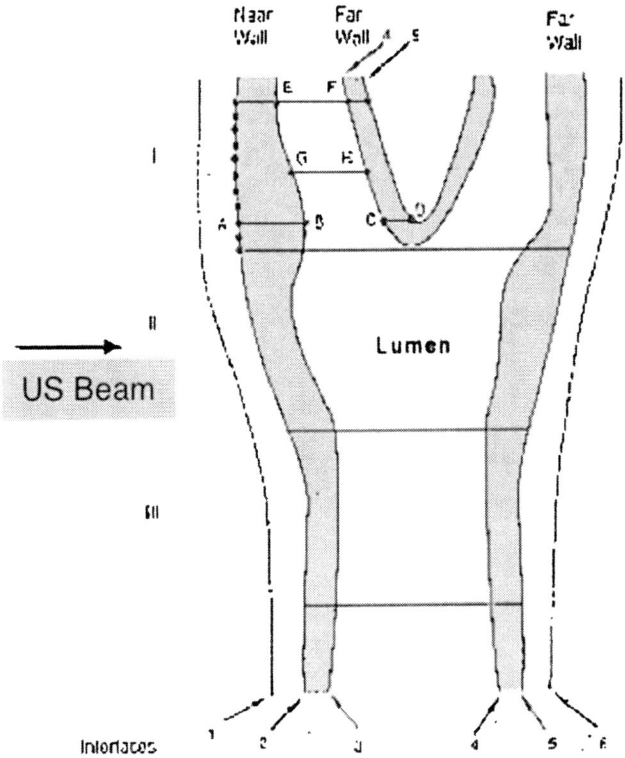

FIGURE 8. Schematic overview of carotid artery B-mode ultrasound measurements. (Accessed from http://www.cscc.unc.edu/ARIC[21])

remains very expensive. The technology is evolving rapidly; therefore keeping methods constant over years of follow-up could be a problem. Some subjects are excluded from MRI because they have claustrophobia.

Ankle–brachial index compares blood pressure in the arms versus legs as a measure of potentially reduced peripheral blood flow, most often due to atherosclerosis.[24] The ratio of ankle to arm blood pressure is well below 0.9 in clinical peripheral artery disease; values close to 0.9 suggest subclinical disease. The meaning of different values in the normal range, for example, above 1.0, is not clear. This method is inexpensive, easy to do, and transportable.

Urinary albumin excretion rate is used to assess changes in the kidney, a primarily vascular organ. In health, albumin excretion is low. Microalbuminuria assesses adverse changes in endothelial cells in the kidney, predictive of subsequent cardiovascular and renal disease. At least one-third of microalbuminuria is secondary to hypertension and diabetes.[25,26] It is distinct from other forms of glomerular dysfunction, for example due to kidney infection or *arteriosclerotic changes in the glomeruli*. Albumin excretion rate is preferably measured in a timed urine sample, but considerable success has been achieved using a single untimed urine, dividing by urinary creatinine, which tends to have a constant

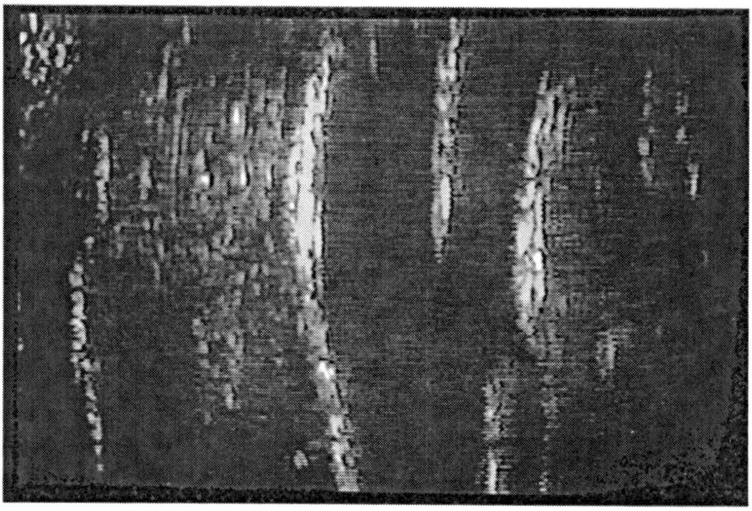

FIGURE 9. Carotid ultrasound normal carotid bifurcation. (Accessed from the ARIC web site.[21])

flow rate throughout the day. Similarly, inadequate glomerular filtration will prevent passage of large molecules, such as creatinine or cystatin-C. Small serum creatinine elevations may also indicate arteriosclerotic changes in the kidney. Because most people with chronic kidney disease die of cardiovascular disease rather than reaching a stage of kidney failure and dialysis, all these measures of kidney inadequacy are subclinical cardiovascular disease markers.

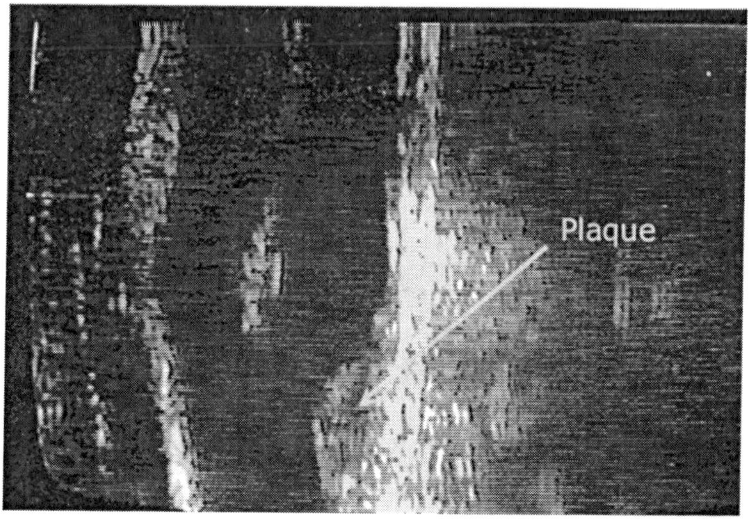

FIGURE 10. Carotid ultrasound diseased carotid bifurcation. (Accessed from the ARIC web site.[21])

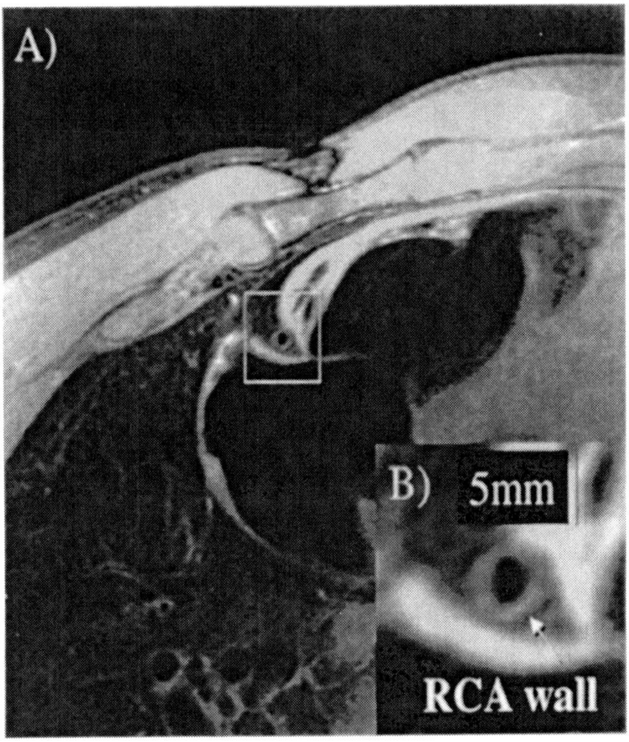

FIGURE 11. MRI image of right coronary artery (box in A, enlarged in B). (From Botnar *et al.*[22] Reproduced by permission.)

Flow-mediated dilation in the brachial artery assesses the ability of the artery to recover after occlusion, presumably a function of the health of the endothelium. Related measures are *arterial distensibility* (MRI or ultrasound of aorta, carotid, or brachial arteries) and *pulse wave form analysis* (performed at the

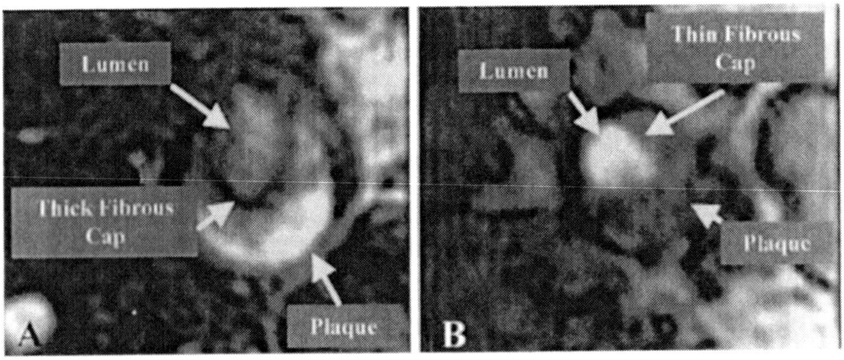

FIGURE 12. MRI image of carotid artery plaque. (From Hatsukami *et al.*[23] Reproduced by permission.)

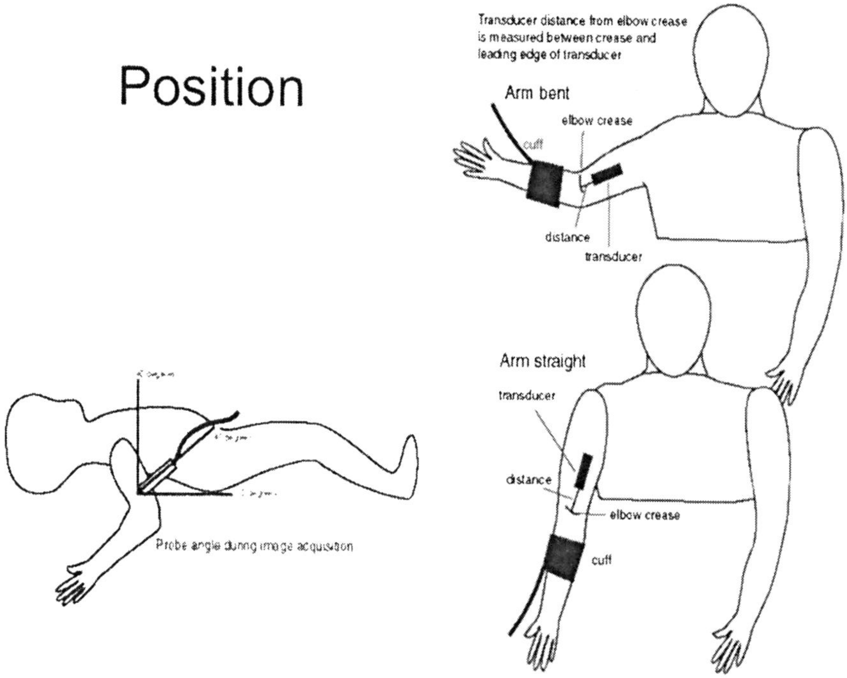

FIGURE 13. Flow-mediated brachial artery reactivity. (Courtesy of Joseph Polak, M.D.)

brachial artery and resulting in arterial compliance measures that are interpreted as pertaining to the pool of large arteries and to the pool of small arteries).[27] Flow-mediated dilation in the brachial artery, depicted in FIGURES 13–15 assesses the ability of the artery to recover after occlusion, presumably a function of the health of the endothelium, and perhaps NO release.[28] It measures arterial diameter before and after 5-min occlusion by a cuff placed above the elbow. It is inexpensive, but not transportable. Furthermore, there have been some problems with this measure because small changes in arterial dimension are difficult to measure reproducibly. Pulse wave form analysis appears to be easier to do and to be more repeatable.

ASSOCIATIONS AMONG SUBCLINICAL MEASURES

MESA[18] is a study of 6,814 white (38.5%), black (28%), Hispanic (22%), and Chinese (12%) people living in 6 U.S. cities. Participant mean age is 62 ± 10 years (range 45–84 years), and 53% of the participants are men. Many of the subclinical measures were implemented in MESA, as described in detail in MESA manuals (more information is available at http://www.mesa-nhlbi.org/). The subclinical measures discussed above were not found to be

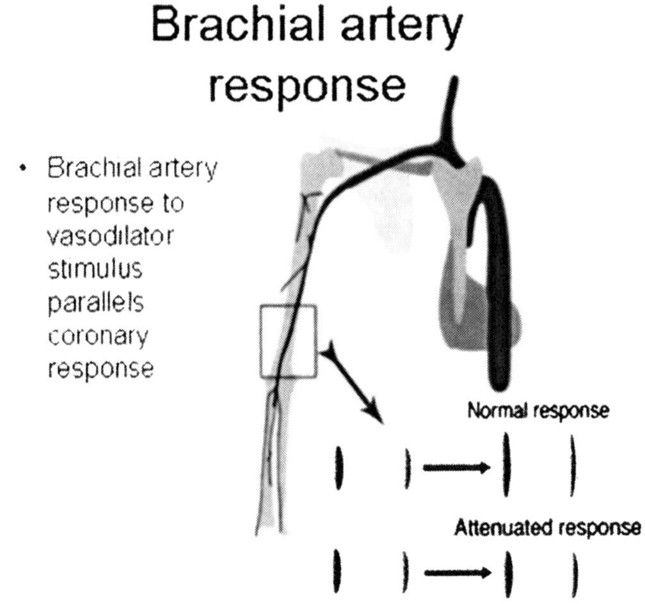

FIGURE 14. Measurement of brachial artery diameter. (Courtesy of Joseph Polak, M.D.)

highly correlated: for example, only three of the seven measures in TABLE 1 have correlations that exceed 0.3, and the highest correlation is 0.45 between IMT measured in the two sites of the carotid artery. Correlations were similar or less

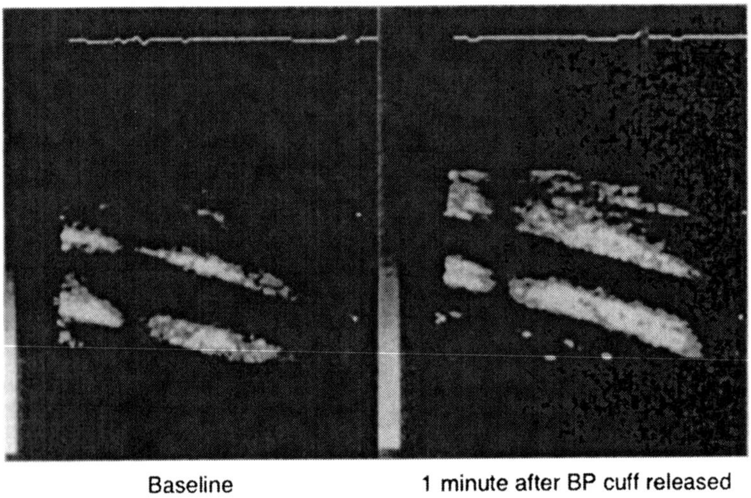

FIGURE 15. Image obtained during brachial artery reactivity. (From Corretti *et al.*[28] Reproduced by permission.)

TABLE 1. Unadjusted Pearson correlation coefficients and available sample sizes in pairs of subclinical measures: Multi-Ethnic Study of Atherosclerosis, 2000–2002

	Computed tomography of the coronary arteries:	Carotid artery ultrasound		Peripheral arteries:	Urinary albumin:	MRI	
	Coronary artery calcification	Common carotid artery	Internal carotid artery	Ankle–Brachial Index	Albumin excretion rate	Left ventricular mass index	Aortic distensibility
Coronary artery calcification	1.00 6,814	0.33 6,726	0.40 6,629	−0.12 6,735	0.21 6,775	0.12 5,004	−0.17 3,719
Common carotid intima media thickness		1.00 6,726	0.45 6,618	−0.17 6,649	0.19 6,693	0.23 4,957	−0.24 3,682
Internal carotid intima media thickness			1.00 6,629	−0.18 6,553	0.15 6,596	0.16 4,891	−0.16 3,630
Ankle-brachial Index				1.00 6,735	−0.09 6,699	0.03 4,953	0.10 3,684
Albumin excretion rate					1.00 6,775	0.29 4,981	−0.17 3,704
Left ventricular mass index						1.00 5,004	−0.12 3,703
Aortic distensibility							1.00 3,719

Notes on variable representation: Coronary artery calcification: ln (1+Agatston score, phantom adjusted). Common and internal carotid arteries: ln (1+intima media thickness). Ankle–brachial index: continuous.

when the measures were represented as dichotomies (coronary artery calcification above its 90th percentile (Agatston score 395), common carotid artery IMT above its 90th percentile (1.11 mm), internal carotid artery IMT above its 90th percentile (1.94 mm), plaque visualizable in the carotid artery (which was mostly found in the internal carotid artery and had a correlation of 0.7 with internal carotid artery IMT, ankle–brachial index suggestive of peripheral artery disease ($\leq$0.9), or urinary albumin classified as microalbumuria or macroalbuminuria.

GENERAL ASSESSMENT OF SUBCLINICAL MEASURES

Most of the subclinical measures discussed above relate to some phase of plaque deposition and accumulation, which is a long-term process. All (except flow-mediated dilation and arterial distensibility) are related to adverse clinical outcomes and probably add information beyond standard risk factors.[29–40] However, atherosclerosis develops at different rates in different parts of the vasculature, apparently somewhat independently of each other. Plaque rupture resulting in thrombus or embolus is highly relevant to clinical events, but occurs in seconds and is hard to capture. Not much progress has been made in assessing the propensity for a plaque to rupture. All of these subclinical measures require much smaller sample sizes and shorter follow-up than do clinical endpoints. While it is undoubtedly beneficial to have less atherosclerosis or to see regression of plaque or related characteristics, it is possible to be asymptomatic with fairly extensive atherosclerosis. The finding of a relationship between periodontitis and carotid IMT and atherosclerotic plaque clearly establishes an oral health/cardiovascular association that deserves additional study. Additional studies of oral health in relation to other subclinical measures of cardiovascular disease will help fill in the picture of possible pathologic connections. However, studies of subclinical endpoints do not fully substitute for studies of clinical endpoints.

ACKNOWLEDGMENT

This paper is an expanded version of a talk first presented at a National Institute of Dental and Craniofacial Research conference, "Methods for Enhancing the Efficiency of Dental/Oral Health Clinical Trials: Current Status, Future Possibilities," held May 6–7, 2004.

REFERENCES

1. Ross, R. 1999. Atherosclerosis: an inflammatory disease. N. Engl. J. Med. **340:** 115–126.

2. DEMMER, R.T. & M. DESVARIEUX. 2006. Periodontal infections and cardiovascular disease: the heart of the matter. JADA **137**(10 Suppl): 14S–20S.
3. GEISMAR, K., K. SOLTZE, B. SIGURD, *et al.* 2006. Periodontal disease and coronary heart disease. J. Periodontol. **77:** 1547–1554.
4. BRIGGS, J.E., P.P. MCKEOWN, V. L. CRAWFORD, *et al.* 2006. Angiographically confirmed coronary heart disease and periodontal disease in middle-aged males. J. Periodontol. **77:** 95–102.
5. SPAHR, A., E. KLEIN, N. KHUSEYINOVA, *et al.* 2006. Periodontal infections and coronary heart disease: role of periodontal bacteria and importance of total pathogen burden in the Coronary Event and Periodontal Disease (CORODONT) study. Arch. Intern. Med. **166:** 554–559.
6. BECK, J.D., J.R. ELTER, G. HEISS, *et al.* 2001. Relationship of periodontal disease to carotid artery intima-media wall thickness: the atherosclerosis risk in communities (ARIC) study. Arterioscler. Thromb. Vasc. Biol. **21:** 1816–1822.
7. DESVARIEUX, M., R.T. DEMMER, T. RUNDEK, *et al.* 2003. Relationship between periodontal disease, tooth loss, and carotid artery plaque: the Oral Infections and Vascular Disease Epidemiology Study (INVEST). Stroke **34:** 2120–2125.
8. ENGEBRETSON, S.P., I.B. LAMSTER, M. S. ELKIND, *et al.* 2005. Radiographic measures of chronic periodontitis and carotid artery plaque. Stroke **36:** 561–566.
9. DESVARIEUX, M., C. SCHWAHN, H. VOLZKE, *et al.* 2004. Gender differences in the relationship between periodontal disease, tooth loss, and atherosclerosis. Stroke **35:** 2029–2035.
10. DESVARIEUX, M., R.T. DEMMER, T. RUNDEK, *et al.* 2005. Periodontal microbiota and carotid intima-media thickness: the Oral Infections and Vascular Disease Epidemiology Study (INVEST). Circulation **111:** 576–582.
11. SOCRANSKY, S.S., A.D. HAFFAJEE, M. A. CUGINI, *et al.* 1998. Microbial complexes in subgingival plaque. J. Clin. Periodontol. **25:** 134–144.
12. BECK, J.D., P. EKE, D. LIN, *et al.* 2005. Associations between IgG antibody to oral organisms and carotid intima-medial thickness in community-dwelling adults. Atherosclerosis **183:** 342–348.
13. SÖDER, P.O., B. SÖDER, J. NOWAK & T. JOGESTRAND. 2005. Early carotid atherosclerosis in subjects with periodontal diseases. Stroke **36:** 1195–1200.
14. SCHOEN, F.J. & R. S. COTRAN. 2003. The blood vessels. *In* Robbins Basic Pathology, seventh edition. V. Kumar, R. S. Cotran, S. L. Robbins, Eds.: 326. Figure 10.1. Saunders, Philadelphia, PA.
15. FOX, S.I. 2002. Heart and circulation. *In* Human Physiology, seventh edition: 397. Figure 13.30. McGraw-Hill Higher Education, New York.
16. WISSLER, R.W. 1992. Important points in the pathogenesis of atherosclerosis. *In* Atherosclerosis: 54, 55. Figures 7a, 7b. Upjohn. Kalamazoo, MI.
17. BROWN, G., J.J. ALBERS, L. D. FISHER, *et al.* 1990. Regression of coronary artery disease as a result of intensive lipid-lowering therapy in men with high levels of apolipoprotein B. N. Engl. J. Med. **323:** 1289–1298.
18. BILD, D.E., D.A. BLUEMKE, G. L. BURKE, *et al.* 2002. Multi-ethnic study of atherosclerosis: objectives and design. Am. J. Epidemiol. **156:** 871–881.
19. BECKER, C.R., T.F. JAKOBS, S. AYDEMIR, *et al.* 2000. Helical and single-slice conventional CT versus electron beam CT for the quantification of coronary artery calcification. Am. J. Roentgenol. **174:** 543–547.
20. AGATSTON, A.S., W.R. JANOWITZ, F. J. HILDNER, *et al.* 1990. Quantification of coronary artery calcium using ultrafast computed tomography. J. Am. Coll. Cardiol. **15:** 827–832.

21. Atherosclerosis Risk In Communities: Manual of Operations. http://www.cscc.unc.edu/ARIC, accessed October 9, 2006
22. BOTNAR, R.M., M. STUBER, K. V. KISSINGER, *et al.* 2000. Noninvasive coronary vessel wall and plaque imaging with magnetic resonance imaging. Circulation **102:** 2582–2587.
23. HATSUKAMI, T.S., R. ROSS, N. L. POLISSAR & C. YUAN. 2000. Visualization of fibrous cap thickness and rupture in human atherosclerotic carotid plaque in vivo with high-resolution magnetic resonance imaging. Circulation **102:** 959–964.
24. HIRSCH, A.T., M.H. CRIQUI, D. TREAT-JACOBSON, *et al.* 2001. Peripheral arterial disease detection, awareness, and treatment in primary care. JAMA **286:** 1317–1324.
25. GOETZ, F.C., D.R. JACOBS, B. CHAVERS, *et al.* 1997. Risk factors for kidney damage in the adult population of Wadena, Minnesota. A prospective study. Am. J. Epidemiol. **145:** 91–102.
26. JACOBS, D.R., M.A. MURTAUGH, M. STEFFES, *et al.* 2002. Gender- and race-specific determination of albumin excretion rate using albumin-to-creatinine ratio in single, untimed urine specimens: the Coronary Artery Risk Development in Young Adults Study. Am. J. Epidemiol. **155:** 1114–1119.
27. COHN, J.N. 2006. Arterial stiffness, vascular disease, and risk of cardiovascular events. Circulation **113:** 601–603.
28. CORRETTI, M.C., G.D. PLOTNICK & R. A. VOGEL. 1995. Correlation of cold pressor and flow-mediated brachial artery diameter responses with the presence of Coronary Artery Disease. Am. J. Cardiol. **75:** 783–787.
29. ARAD, Y., L.A. SPADARO, K. GOODMAN, *et al.* 1996. Predictive value of electron beam computed tomography of the coronary arteries: 19-month follow-up of 1173 asymptomatic subjects. Circulation **93:** 1951–1953.
30. KEELAN, P.C., L.F. BIELAK, K. ASHAI, *et al.* 2001. Long-term prognostic value of coronary calcification detected by electron-beam computed tomography in patients undergoing coronary angiography. Circulation **104:** 412–417.
31. WONG, N.D., J.C. HSU, R. C. DETRANO, *et al.* 2000. Coronary artery calcium evaluation by electron beam computed tomography and its relation to new cardiovascular events. Am. J. Cardiol. **86:** 495–498.
32. CHAMBLESS, L.E., G. HEISS, A. R. FOLSOM, *et al.* 1997. Association of coronary heart disease incidence with carotid arterial wall thickness and major risk factors: the Atherosclerosis Risk in Communities (ARIC) Study, 1987-1993. Am. J. Epidemiol. **146:** 483–449.
33. WATTANAKIT, K., A.R. FOLSOM, L. E. CHAMBLESS & F. J. NIETO. 2005. Risk factors for cardiovascular event recurrence in the Atherosclerosis Risk in Communities (ARIC) study. Am. Heart J. **149:** 606–612.
34. ZHENG, Z.J., A.R. SHARRETT, L. E. CHAMBLESS, *et al.* 1997. Associations of ankle brachial index with clinical coronary heart disease, stroke and preclinical carotid and popliteal atherosclerosis: the Atherosclerosis Risk in Communities (ARIC) Study. Atherosclerosis **131:** 115–125.
35. GERSTEIN, H.C., J.F. MANN, Q. YI, *et al.* 2001. Albuminuria and risk of cardiovascular events, death, and heart failure in diabetic and nondiabetic individuals. JAMA **286:** 421–426.
36. HILLEGE, H.L., V. FIDLER, G. F. DIERCKS, *et al.* 2002. Urinary albumin excretion predicts cardiovascular and noncardiovascular mortality in general population. Circulation **106:** 1777–1782.

37. ARNLOV, J., J.C. EVANS, J. B. MEIGS, *et al.* 2005. Low-grade albuminuria and incidence of cardiovascular disease events in nonhypertensive and nondiabetic individuals: the Framingham Heart Study. Circulation **112:** 969–975.
38. DE ZEEUW, D., G. REMUZZI, H. H. PARVING, *et al.* 2004. Albuminuria, a therapeutic target for cardiovascular protection in type 2 diabetic patients with nephropathy. Circulation **110:** 921–927.
39. JAGER, A., P.J. KOSTENSE, H. G. RUHE, *et al.* 1999. Microalbuminuria and peripheral arterial disease are independent predictors of cardiovascular and all-cause mortality, especially among hypertensive subjects: five-year follow-up of the Hoorn Study. Arterioscler. Thromb. Vasc. Biol. **19:** 617–624.
40. GREY, E., C. BRATTELI, S. P. GLASSER, *et al.* 2003. Reduced small artery but not large artery elasticity is an independent risk marker for cardiovascular events. Am. J. Hypertens. **16:** 265–269.

Do Salivary Antibodies Reliably Reflect Both Mucosal and Systemic Immunity?

PER BRANDTZAEG

Laboratory for Immunohistochemistry and Immunopathology (LIIPAT), Department and Institute of Pathology, University of Oslo, Rikshospitalet-Radiumhospitalet Medical Center, N-0027 Oslo 1, Norway

ABSTRACT: Two major antibody classes operate in saliva: secretory IgA (SIgA) and IgG. The former is synthesized as dimeric IgA by plasma cells (PCs) in salivary glands and is exported by the polymeric Ig receptor (pIgR). Most IgG in saliva is derived from serum (mainly via gingival crevices), although some is locally produced. Gut-associated lymphoid tissue (GALT) and nasopharynx-associated lymphoid tissue (NALT) do not contribute equally to mucosal PCs throughout the body. Thus, enteric immunostimulation is an inadequate mode of stimulating salivary IgA antibodies, which are poorly associated with the intestinal SIgA response, for instance after enteric cholera vaccination. Nevertheless, the IgA response in submandibular/sublingual glands is better related to B cell induction in GALT than the parotid response. Such disparity is suggested by the elevated levels of IgA in submandibular secretions of AIDS patients, paralleling their highly upregulated intestinal IgA system. Moreover, in patients with active celiac disease, IgA antibodies to disease-precipitating gliadin are reliably represented in whole saliva but not in parotid secretion. Parotid SIgA may be more consistently linked to immune induction in palatine tonsils and adenoids (human NALT), as supported by the homing molecule profile of NALT-derived B cell blasts. Also several other variables influence the levels of antibodies in oral secretions. These include difficulties with reproducibility and standardization of immunoassays, the impact of flow rate, acute or chronic stress, protein loss during sample handling, and uncontrolled admixture of serum-derived IgG and monomeric IgA. Despite such problems, saliva remains an interesting biological fluid with great scientific and clinical potentials.

KEYWORDS: IgA; IgG; mucosa-associated lymphoid tissue; MALT; gut-associated lymphoid tissue; GALT; nasopharynx-associated lymphoid tissue; NALT; salivary glands; crevicular fluid; polymeric Ig receptor; pIgR; secretory component; SC; vaccination; celiac disease

Address for correspondence: Prof. Per Brandtzaeg, LIIPAT, Rikshospitalet, N-0027 Oslo 1, Norway. Voice: 47-23072743; fax: 47-23071511.
per.brandtzaeg@medisin.uio.no

**Ann. N.Y. Acad. Sci. 1098: 288–311 (2007). © 2007 New York Academy of Sciences.
doi: 10.1196/annals.1384.012**

INTRODUCTION

A molecular basis for the presence of antibodies in external body fluids became available when it was shown in 1960 that salivary secretions contain immunoglobulin (Ig) molecules.[1] Conclusive information to this end was not obtained, however, until specific identification of different Ig classes was possible, and several laboratories reported that IgA predominates in many external secretions.[2] The discovery of unique properties of secretory IgA (SIgA) in 1965 further intensified investigation of local immunity.[3] SIgA was shown to be polymeric (mainly dimers) and associated with an 80-kDa epithelial glycoprotein initially called "transport piece" and later named "secretory component" (SC). Also importantly, the same year it was found that the Ig class distribution of intestinal plasma cells (PCs) is strikingly different from that observed in lymph nodes and bone marrow[4]; in normal mucosal tissues, IgA^+ plasmablasts and PCs are approximately 20 times as numerous as IgG^+ PCs. In 1973 our laboratory provided the first direct evidence that human mucosal IgA^+ PCs produce mainly dimers and some larger polymers (collectively called pIgA) rather than monomers,[5] and in 1974 this characteristic was found to be associated with the expression of the joining (J) chain by the same cells.[6]

SECRETORY IMMUNITY

In the late 1960s our laboratory demonstrated that not only pIgA, but also pentameric IgM is selectively exported to the secretions, apparently because of a common epithelial transport mechanism.[7,8] Secretory IgM (SIgM) was a few years later shown to be associated with the SC and to follow the same intracellular route through secretory epithelia as SIgA.[9,10] At the same time a shared receptor-mediated mechanism involving endocytosis and transcytosis, was proposed for the generation of SIgA and SIgM.[5,6,11,12] Our transport model was based on the suggested crucial cooperation between J chain–expressing mucosal IgA^+ and IgM^+ PCs and SC-expressing serous-type of secretory epithelial cells (FIG. 1).

Transmembrane SC is a carbohydrate-rich glycoprotein of ~100 kDa constitutively expressed on the basolateral surface membrane of the epithelial cells, where it exhibits strong noncovalent affinity for J chain–containing pIgA and pentameric IgM.[13] It belongs to the Ig supergene family and is now usually referred to as the polymeric Ig receptor (pIgR). Its human gene has been cloned and characterized,[14] and several DNA elements in its promoter and first intron responsible for a remarkably high constitutive as well as cytokine-enhanced expression have been identified.[15]

At the apical epithelial surface, SIgA and SIgM are exocytosed after cleavage of the receptor (FIG. 1); only the C-terminal smaller pIgR segment remains

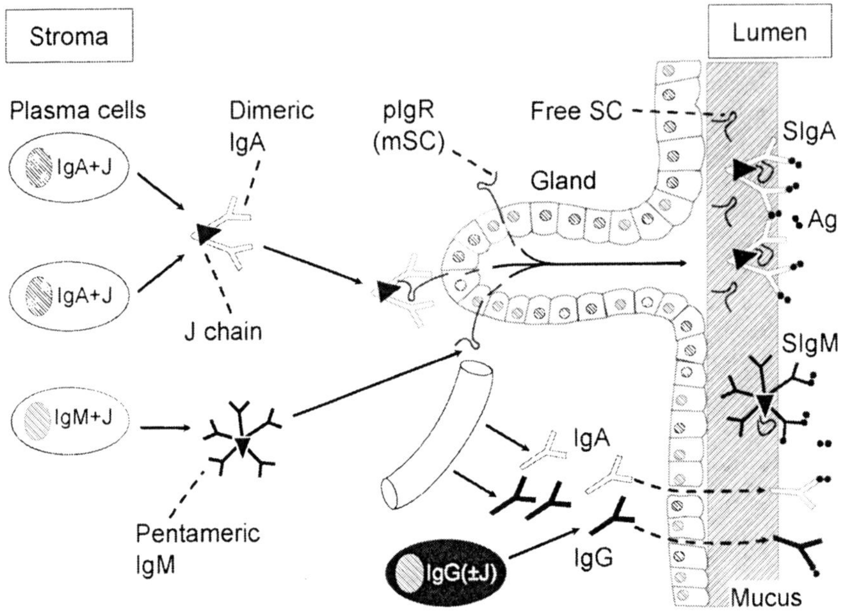

FIGURE 1. Simplified depiction of receptor-mediated export of dimeric IgA and pentameric IgM. Membrane secretory component (mSC) functions as polymeric Ig receptor (pIgR) and is expressed basolaterally on serous-type glandular epithelial cells. The polymeric Ig molecules are produced with incorporated J chain (IgA + J and IgM + J) by plasma cells residing in the glandular stroma. The resulting secretory Ig molecules (SIgA and SIgM) act in a first line of defense by performing immune exclusion of antigens (Ag) in the mucus layer on the epithelial surface. Although J chain is often produced by mucosal IgG^+ plasma cells (70–90%), it does not combine with IgG (nor with IgD or IgE; not shown) but is degraded intracellularly as denoted by (±J) in the figure. Locally produced and serum-derived IgG and monomeric IgA can be transmitted paracellularly to the lumen in amounts depending on the integrity of the epithelial barrier (*broken arrows*). Free SC (depicted in mucus) is generated when unoccupied pIgR (*top symbol*) is cleaved at the apical face of the epithelial cell in the same manner as bound SC incorporated into SIgA and SIgM. (Based on the epithelial transport model proposed in Refs. 11 and 12.)

for intracellular degradation, whereas the extracellular part (~80 kDa) is incorporated into the SIg molecules as bound SC, thereby endowing particularly SIgA with resistance against proteolytic degradation.[16,17] Excess of unoccupied pIgR is released in the same manner by proteolytic cleavage to form so-called free SC according to the recommended nomenclature.[18] This 80-kDa glycoprotein can be found in most secretions including saliva.[8] By equilibrium with bound SC, free SC exerts a stabilizing effect on the quaternary structure of SIgM in which SC is not covalently linked in contrast to SIgA.[9] Both free SC and bound SC show, in addition, several innate functions, such as inhibition of epithelial adhesion of certain Gram-negative bacteria and neutralization of certain bacterial toxins.[16,17]

The binding sites of pIgA and pentameric IgM initially contacting the first extracellular domain of pIgR have largely been defined.[19] In addition, it has been shown that the 15-kDa J chain is crucial for the stabilization of the initial noncovalent complexing between the Ig polymers and pIgR or free SC.[13,19] Our original suggestion that the J chain and membrane SC are involved in a "lock and key" mechanism in the selective epithelial export of pIgA and pentameric IgM is now firmly established.[20–22]

IMMUNOLOGICAL ACTIVITY OF SALIVARY GLANDS

Origin of Salivary Immunoglobulins

Salivary secretions are not particularly rich in IgA,[8] which in fact represents a minor fraction of total protein compared with the dominating enzyme amylase.[23,24] Nevertheless, the parotid IgA-to-IgG ratio is about 500 times increased compared with that in serum (TABLE 1), reflecting the selective export of pIgA (FIG. 1). The same transport mechanism explains that also the IgM-to-IgG ratio is substantially increased in normal parotid fluid compared with that in serum; but because of the diffusion advantage of the relatively small IgG molecules, active export of IgM to whole saliva is virtually masked.[8,24] The monomeric fraction of salivary IgA is generally small, that is, about 10% in parotid fluid and 13–17% in whole saliva, depending on the clinical state of the gingiva.[24] It has been estimated that up to 77% of monomeric IgA in saliva is serum derived.[25]

These observations and the significant association of IgG in whole saliva with the product of the serum IgG concentration and the extent of gingival–periodontal inflammation ($r = 0.85$) show that monomeric IgA and IgG mainly enter the oral cavity via crevicular fluid.[8,24] Also of note, a substantial proportion (50–60%) of parotid IgA exists in >25S complexes in the secretion;

TABLE 1. Variations in mean results of salivary IgA determinations performed by the same laboratory (LIIPAT, 1970–1991)

Samples (no. of adult subjects)	Conc. (μg/mL) IgA	Conc. (μg/mL) IgG	IgA secretion rate (μg/min)
Stim. parotid secretion ($n = 9$)[a]	40	0.36	27
Stim. parotid secretion ($n = 27$)[a]	36	ND[c]	34
Stim. parotid secretion ($n = 19$)[b]	27	ND	14
"Unstim." parotid secretion ($n = 5$)[a]	120	ND	10
"Unstim." whole saliva:			
Healthy individuals ($n = 8$)[a]	194	14.4	ND
Periodontitis patients ($n = 13$)[a]	371	69.7	ND

[a]Single radial immunodiffusion; [b]ELISA; [c]ND = not determined.
Adapted from Ref. 24.

this fraction is even higher for SIgA dissolved from the sedimented mucus clot obtained by centrifugation,[26] most likely reflecting the mucophilic properties of bound SC.[17]

Variability of Salivary IgA Levels

Because of the slow development of the salivary IgA system,[24] age is an important variable in studies of Igs in oral fluids. Also, various stressors reportedly[27] influence the IgA levels in different manners (TABLE 2). In addition, quantification of salivary IgA is afflicted with many methodological problems. It is difficult to standardize the quantitations due to problems with sample collection, processing, and storage, as reflected in different normal values obtained for total IgA and IgM concentrations, even in studies performed by the same laboratory (TABLE 1). Thus, the results vary strikingly between studies based on single radial immunodiffusion or enzyme-linked immunosorbent assay (ELISA).

It is particularly important to be aware of the striking impact of the secretory flow rate on the salivary IgA level (TABLE 1), which may partly explain differences among studies. "Unstimulated" parotid secretion thus contains at least three times more IgA than the stimulated counterpart,[23,28] and a similar proportional difference has been reported for whole saliva.[29] Some investigators have tried to avoid this problem by reporting salivary IgA related to total protein or albumin, but this will also be misleading because the secretory response of individual parotid proteins is quite different, with large individual variations,[23,24,30] and the salivary level of albumin will depend on leakage from serum in a manner similar to IgG and monomeric IgA (FIG. 1).

Some studies recommend reporting the output of salivary proteins in secretion rates (μg/min),[8] whereas others have suggested that the actual concentration (μg/mL) may be a better alternative for IgA.[31] Both parameters show considerable variations over time in the same individual, and between the left and the right parotid gland.[8,24] Therefore, it is probably best to record both the absolute concentration and the secretion rate, and to sample from the same

TABLE 2. Effect of different stressors on salivary IgA levels

Definition of stressors	Salivary IgA
Chronic academic stress (e.g., during exam period)	Reduced
Acute academic stress (e.g., just before or after exam)	Increased
Acute "naturalistic stress" (e.g., work shift)	Increased
Laboratory stressors:	
"Acute coping" of challenges (sympathetic activation followed by parasympathetic rebound)	Reduced (?) Increased
"Passive coping," feeling of disgust	Reduced

Adapted from Ref. 27.

side at the same time of the day on every occasion in longitudinal studies. Relatively large study populations are clearly required to obtain reliable results. Stimulated secretion may be preferable as a test sample[31,32]; it is also more easily collected and less adversely affected by storage than unstimulated fluid.[23]

Although the secretion rate of parotid IgA in an individual appears to be more stable over time than the actual IgA concentration,[8,24] SIgA is more subjected to short-term variation than other salivary proteins.[33,34] This may reflect differences in the glandular structures involved in the secretion of various protein components[35] and also the fact that SIgA is mainly a product of adaptive immunity. Thus, studies in inbred mice have suggested that fluctuations in glandular IgA^+ PCs may contribute to variations in salivary Ig levels.[36] Diurnal and seasonal variations should also be considered, as should relation to meals, cigarette smoking, and pregnancy,[31,37,38] in addition to various stressors as mentioned above (TABLE 2).

Standardized Ig quantitations are even more difficult in whole saliva than in parotid fluid. First, the contributions to whole saliva from the minor, submandibular, sublingual and parotid glands vary greatly according to the rate of flow,[39] and contamination with nasal secretions (and tears) may be difficult to avoid, particularly in uneasy children.[40] Notably, the mean IgA concentration in secretion from labial glands has been reported to be three times higher than that in parotid fluid.[41] On this basis it can be estimated that the minor glands contribute 30–35% of the total salivary IgA.[41] Second, the flow rate of whole saliva cannot be so accurately measured as that of the parotid secretion. Third, whole saliva samples usually require centrifugation before quantification, and the sediment represents a variable Ig loss, probably because SIgA binds not only to mucus,[17,26] but also to oral bacteria.[42] Centrifugation may be avoided by collecting the fluid by careful suction from the floor of the mouth, and even more controlled sampling might be obtained by the use of absorbing discs.[43]

Despite various disadvantages, whole saliva is commonly used as a "representative" external secretion because it is easily obtained. To increase fluid volume, chewing on paraffin wax or Parafilm is often recommended. Although this is convenient, it entails certain pitfalls, however; the wax adsorbs organic material[44] and the chewing may increase leakage of serum proteins into the oral cavity, for instance from inflamed gingivae. Therefore, oral health should be considered when whole saliva is used for immunological studies.

Local Ig Production in Salivary Glands

IgA^+ PCs are normally found scattered among the acini of major salivary glands and often in clusters adjacent to ducts.[24] Interestingly, the submandibular glands contain on average approximately two times more IgA^+ PCs per tissue unit than the parotid,[45,46] in accordance with a larger output of SIgA.[31] It is tempting to speculate that antigens gain easier access to submandibular glands, thereby inducing a more active local immune system. The daily output of IgA/kg wet weight of glandular tissue is similar for salivary and lactating mammary glands, so the superiority of the latter as an SIgA source depends on the organ size and ductal storage system.[46]

Because the minor salivary glands are numerous and have close proximity to the oral mucosal surface, they are probably quite important in the defense of the oropharynx. This is supported by the observation of numerous IgA^+ PCs adjacent to their ducts[47,48] and an abundant output of SIgA from these glands.[41] In fact, the density of IgA^+ PCs in the labial glands has been reported to be three times that in the parotid.[49]

The subclass IgA2 is more stable than IgA1 because of its resistance to certain bacterial proteases.[50] Therefore, it is interesting that a relatively large proportion (35–38%) of the IgA^+ PCs in salivary glands produce IgA2.[48,51] In this respect the salivary glands are intermediate between the upper airways and the distal gut, a disparity that clearly reflects regional immunoregulatory differences.[52] In agreement with the similar affinity of IgA1 and IgA2 for free SC,[53] both subclasses appear to be equally well exported by pIgR into the secretion.[26] This transport not only takes place primarily through the intercalated ducts but also through serous-type acini.[35,47]

Although SIgM is a result of pIgR-mediated export (FIG. 1), it is not secondarily stabilized with SC by covalent bonding,[9] and its resistance to proteolytic degradation is inferior compared to SIgA. Also, when comparing the proportions of parotid PC classes and the IgA:IgM ratio in the secretion, the glandular export of pIgA is favored over that of pentameric IgM by a factor of approximately five (or 12-fold on a molar basis).[54] This is not explained by different handling of the two polymers by pIgR but is due to diffusion restriction for the relatively large IgM through stromal matrix and basement membranes.[54]

HOMING OF MUCOSAL B CELLS TO SALIVARY GLANDS

Multiple Sites of B Cell Activation

The initial stimulation of mucosal pIgA-expressing B cells takes place mainly in mucosa-associated lymphoid tissue (MALT), particularly in Peyer's patches of the distal ileum as well as in other parts of gut-associated lymphoid tissue (GALT), such as the numerous isolated lymphoid follicles and the appendix.[52] From these organized structures, activated B cells reach peripheral blood and migrate to secretory effector sites, where their extravasation depends on complementary adhesion molecules and chemokine–chemokine receptor pairs.[52]

It is not well delineated which part of MALT is most important for induction of immune responses subsequently expressed as salivary IgA antibody production, but there is convincing evidence both in animals and humans that activated B cells can migrate from GALT to salivary glands.[55–58] Nevertheless, in subjects immunized orally with a cholera toxin (CT) B subunit–whole-cell *Vibrio cholerae* vaccine, the specific IgA antibody detection sensitivity in saliva was not better than in serum and only about 50% of that in intestinal lavage (FIG. 2). After infection with *V. cholerae* or enterotoxigenic *Escherichia coli*, the detection sensitivity for antibodies against the respective toxins increased, but not as much as that seen in serum.[59] These results suggest that intestinal immune induction is not so well reflected in the salivary IgA system. Also notably, this study probably overestimated the enteric–oral B cell homing axis because CT and enteropathogens breach the gut mucosa to become unduly disseminated to the systemic immune system, as reflected by the high levels of specific serum IgG antibodies (FIG. 2).

Recent studies point to considerable compartmentalization of B cell homing.[52] Thus, nasopharynx-associated lymphoid tissue (NALT), such as the adenoids and palatine tonsils in humans, may be relatively more important than GALT as inductive sites for B cells destined to the salivary glands.[60,61] Although it remains uncertain to what extent these lymphoepithelial structures of Waldeyer's ring are functionally comparable to NALT of rodents,[18] they are indeed strategically located to orchestrate regional immune functions against both airborne and alimentary antigens. Moreover, they have obvious advantages as immune-inductive organs because of antigen-retaining crypts and absence of antigen-degrading digestive enzymes.

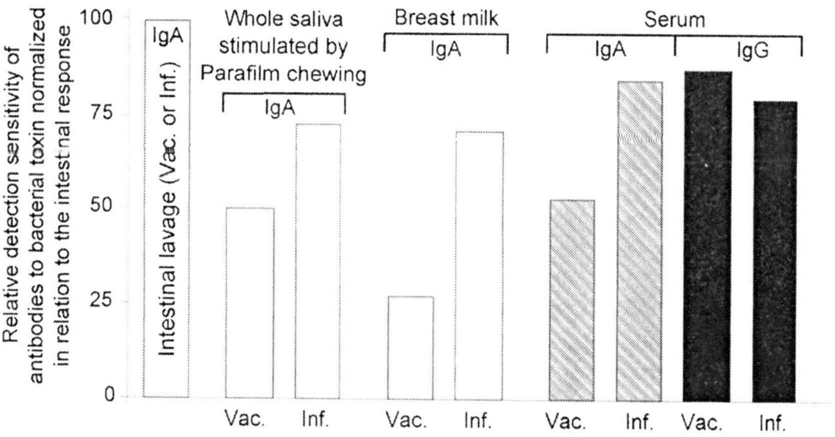

FIGURE 2. Antibody responses against CT or *Escherichia coli* heat-labile toxin in adult Bangladeshi volunteers after peroral *Vibrio cholerae* vaccination (Vac.), and after infection (Inf.) with *V. cholerae* or enterotoxigenic *E. coli*. Vaccination was performed twice by oral administration of a CT B⁻ subunit–whole-cell vaccine approved for human use. The antibody detection sensitivity in stimulated whole saliva, breast milk, and serum for antibodies to the relevant bacterial toxin was normalized in relation to the detection of specific IgA antibody in intestinal lavage (set to 100% for each individual). (Based on data from Ref. 59.)

Evidence for Homing of B Cells from NALT to Salivary Glands

Regionalization of the mucosal immune system is supported by B cell homing studies in rats as reviewed elsewhere.[60] Also, preferential appearance of specific IgA antibodies in saliva has been noted in rabbits after tonsillar antigen exposure.[62,63] In keeping with the latter observation, direct immunization of human palatine tonsils, and particularly nasal vaccination, gave rise to local B cell responses in palatine tonsils and adenoids as well as circulating specific B cells that apparently did not enter the intestinal mucosa.[64] Moreover, in babies dying of the sudden infant death syndrome (SIDS), the tonsillar germinal centers were shown to be overstimulated as revealed by an increased number of IgG^+ and IgA^+ PCs,[65] probably reflecting airway infection; such activated B cells were apparently distributed in excessive numbers to regional secretory effector sites including the parotid glands,[66] thereby giving rise to increased levels of salivary IgA and IgM in SIDS.[67]

Altogether, there is accumulating evidence to support the notion that human NALT supplies secretory effector sites of the upper aerodigestive tract with activated pIgA precursor cells.[24,68] The reason for the suggested homing dichotomy between this region and the small intestine appears to be differences in the employed homing molecules.[69] The integrin $\alpha 4\beta 7$ is important for B cell extravasation into the intestinal lamina propria by interaction with the mucosal addressin cell adhesion molecule (MAdCAM)-1 expressed on the mucosal microvascular endothelium, but this integrin does not appear to be important for homing to the airways and salivary glands.[52] Also the involved chemokine receptor–chemokine interactions (CCR9–CCL25 vs. CCR10–CCL28) show a striking dichotomy between the two body regions.[52,69]

FUNCTION OF SALIVARY ANTIBODIES

Various Defense Mechanisms of SIgA

The remarkable stability of SIgA makes it well suited to function in protease-containing secretions, such as whole saliva.[70] Nevertheless, several oral bacteria produce enzymes that can selectively cleave SIgA1 in its extended hinge region, especially certain strains of *Streptococcus sanguis* and *S. mitior*, but also *Bacteroides* and *Capnocytophaga* species that are involved in periodontal disease.[50] On average, at least 60% of salivary SIgA consists of the IgA1 isotype,[26] and parotid antibodies to *S. mutans* occur predominantly in this subclass,[71] whereas activity to lipoteichoic acid from *S. pyogenes* and to lipopolysaccarides from *Bacteroides gingivalis, B. fragilis,* and *E. coli* is carried mainly by the SIgA2 isotype.[71]

Although Fabα fragments released by the IgA1 proteases may retain antigen-binding capacity,[72] this immune reaction may be adverse rather than protective.

Such fragments may shield microorganisms from the defense function of SIgA antibodies and may even enhance epithelial colonization,[73] whereas intact SIgA can specifically inhibit cellular attachment and penetration of influenza virus in contrast to monomeric IgA- or IgG-neutralizing antibodies.[74]

The chief defense function of SIgA appears simply to be binding of soluble or particulate antigens (FIG. 1). *In vivo* coating of oral bacteria with SIgA can be directly demonstrated by immunofluorescence staining.[42] This is no proof of antibody reactivity because many strains of group A or B streptococci possess Fcα receptors,[75] but SIgA coating of bacteria via Fc interactions may nevertheless be of similar functional importance.

Many identified mechanisms may contribute to simple mucosal immune exclusion of soluble or particulate antigens.[16,24] In addition to efficient antigen binding, complexing and neutralization, SIgA antibodies show better agglutinating properties than monomeric IgA[16] and may render bacteria mucophilic in saliva.[76] The SIgA function appears to be considerably enhanced by a high level of cross-reactivity observed for salivary antibodies.[77]

Salivary Immunoglobulins in Relation to Disease

Numerous studies have attempted to relate salivary IgA to a variety of oral as well as systemic diseases. Reviews of such reports have been published, but far from all available data can be considered conclusive.[24,38]

Dental Caries

How SIgA antibodies might protect against dental disease has for long been a matter of dispute.[78] The balance of evidence indicates that there is an inverse relationship between caries suspectibility and the output of salivary IgA in children and young adults.[79] An inverse relationship between salivary IgA antibodies to *S. mutans* and its early oral colonization, or the colonized individual's caries experience, has also been reported; the mechanistic interest in this respect is focused on bacterial adhesins, glucosyltransferases, and glucan-binding proteins.[80]

Parotid secretion of young adults regularly contains inhibiting IgA antibodies to glucosyltransferase, and this enzyme may play a major role in dental plaque formation. Oral vaccination with killed *S. mutans* of host origin induced salivary IgA antibodies that inhibited glucosyltransferase and reduced the numbers of viable *S. mutans* organisms in whole saliva and dental plaque.[81] Also interestingly, an experimental study in young Americans suggested that a low parotid IgA antibody level to *S. mutans* serotype *c* was associated with enhanced colonization on molar tooth surfaces, whereas rapid clearance of serotype *d* was associated with relatively higher levels of corresponding parotid IgA antibodies.[82] Characterization of the salivary IgA response to this bacterium continues to be of interest[80] because it is of high relevance in strategies aiming at a future active caries vaccine.[83]

Periodontal Disease

Several studies have shown a positive relationship between the concentration of salivary IgA (TABLE 1) and periodontal disease.[8,79] Accumulation of dental plaque may stimulate IgA production by increasing the amounts of swallowed bacteria that activate B cells in NALT and GALT. Notably, elevated levels of parotid IgA antibodies to *Actinobacillus actinomycetemcomitans* were seen in subjects whose subgingival plaque harbored this microorganism.[84] Secretory immunity may hence be involved in host resistance to periodontal disease; but SIgA antibodies probably have little or no effect on the growth of an established dental plaque.

Mucous Membrane Diseases

Low capacity for IgA antibody production as reflected in serum apparently predisposes to recurrence and complications of herpes simplex infection, particularly development of intraoral lesions.[24] Parotid IgA antibody production to *Candida albicans* may similarly be low in some patients with mucocutaneous candidiasis.[24] Conversely, chronic alcoholics and/or heavy smokers have increased IgA levels in whole saliva, and the same is true for patients with oropharyngeal carcinoma.[85] As molecular characterization was not performed, the results might be explained by leakage of monomeric serum IgA into the oral cavity. This possibility was suggested by a study including measurements of whole saliva IgG as well as parotid and submandibular IgA.[86]

Chronic Sialadenitis and Sjögren's Syndrome

IgG and albumin are relatively much more elevated than IgA in parotid fluid during the early phase of parotitis.[87] This suggests admixture with serum proteins from an inflammatory exudate. In addition, some IgG appears to be of local origin, which harmonizes with the presence of an increased number of IgG^+ PCs in chronic sialadenitis.[24] Alterations of the parotid Ig pattern in patients with Sjögren's syndrome are less pronounced and mimic those found late in the recovery phase of recurrent parotitis, but the fluid flow rate is more reduced.[87]

Most PCs in the benign lymphoepithelial or myoepithelial parotid lesions of Sjögren patients occur in areas of remaining secretory epithelial elements, including proliferating ducts, and are often dominated by the IgG class.[24] In contrast to the myoepithelial islands, the secretory epithelium shows increased expression of pIgR/SC and uptake of IgA, indicating enhanced pIgR-mediated pIgA export despite extensive pathology.[88] Some of the IgA$^+$ and IgM$^+$ PCs may be involved in local production of rheumatoid factors,[89] which have been detected in whole saliva of patients with Sjögren's syndrome with or without associated rheumatoid arthritis.[90]

Also, labial glands of Sjögren patients often show increased production of IgG and to a lesser extent IgM, apparently as a function of disease severity.[91] Immunohistochemistry of labial biopsy specimens may thus aid the diagnostic evaluation. However, it has been shown that labial glands from patients with rheumatoid arthritis or systemic lupus erythematosus may similarly contain increased numbers of IgG$^+$ and IgM$^+$ PCs, even in the absence of overt Sjögren's syndrome.[92]

Non-Hodgkin's Malignant B Cell Lymphoma

Patients with Sjögren's syndrome have a 40- to 50-fold increased risk of developing malignant lymphoma, but early recognition of such monoclonal B cell proliferations are often difficult by conventional histopathological examination. Proliferative areas with immunoblasts and plasmacytoid cells have been reported in up to 90% of myoepithelial lesions.[93] However, molecular studies of Ig heavy chain gene rearrangements showed a lower frequency of B cell monoclonality in parotid (68%), and particularly in labial (~15%), salivary glands.[94,95] It would have been of diagnostic interest if the monoclonal IgM product could be identified in secretions from the affected glands. In Waldenström's macroglobulinemia (which may be associated with Sjögren's syndrome), IgM is increased in parotid fluid[87] and its monoclonal nature can be identified.[96] Synthesis of monoclonal IgM has been demonstrated in salivary glands of such patients.[89]

Celiac Disease

Patients with untreated celiac disease, or gluten-sensitive enteropathy, have elevated levels of IgA antibodies against wheat gluten, or more specifically against the ethanol-soluble gliadin fraction of gluten, both in intestinal secretions and serum; these antibodies disappear during a successful treatment with a gluten-free diet.[97] As a reflection of gluten-induced GALT activation, untreated patients also have a raised number of circulating IgA-producing gliadin-specific plasmablasts, which can be detected in an enzyme-linked immunospot (ELISPOT) assay.[98] The number of such activated B cells is reduced to a normal level after successful diet treatment, but increases again in most celiac patients after 1–2 weeks of challenge with ordinary gluten-containing food.[98]

These results show that the gluten-specific IgA response is disseminated by B cell trafficking throughout the body. It could therefore be anticipated that IgA antibodies to gliadin would occur in saliva, and in the 1990s this possibility was explored in search of a noninvasive approach to screen for celiac disease (FIG. 3). In two studies employing samples of unstimulated whole saliva, the discrimination of confirmed celiac patients was quite satisfactory,[99] or even better than by measurement of IgA antibodies in serum.[100] However, comparable studies employing stimulated parotid secretion showed only little,[101] or virtually no[102] elevation of IgA antibody levels to gliadin in contrast to the results obtained with serum (FIG. 3).

Such observations suggest that GALT-induced B cells home to the parotid glands with only poor efficiency. Comparable homing to the submandibular, sublingual, and minor salivary glands appears to be more consistent, as deemed from the levels of IgA antibodies to gliadin in whole saliva (FIG. 3). This disparity points to compartmentalization of the local B cell system even among various components of the salivary gland system, which would not be surprising in view of the fact that the mucosal homing mechanism also differs between the small and large bowel.[52] In view of the above information, it should be determined whether whole saliva could provide a convenient and noninvasive screening medium for IgA autoantibody activity to tissue transglutaminase, which at present represents the most reliable nonbiopsic indicator of celiac disease when performed with blood serum samples.[97]

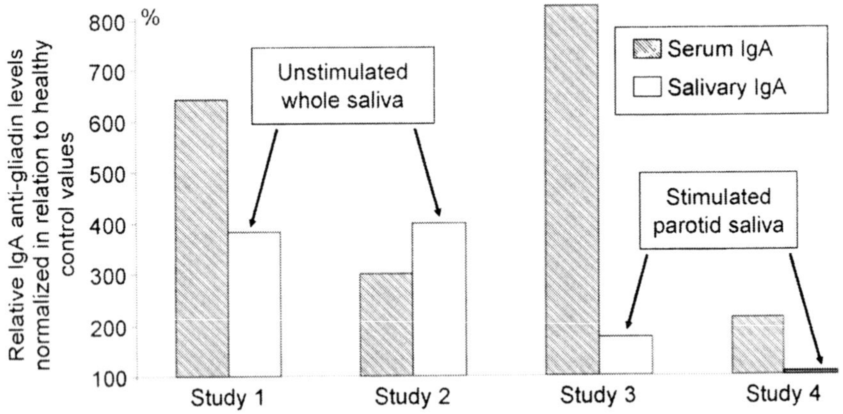

FIGURE 3. IgA antibody levels to gliadin in serum, whole saliva, and parotid saliva of patients with untreated celiac disease, normalized in relation to the comparable antibody levels present in healthy control subjects. Only whole saliva could be used in screening for celiac disease with a result matching that obtained with serum. (Based on data from Ref. 99 [study 1], Ref. 100 [study 2], Ref. 101 [study 3], and Ref. 102 [study 4].)

Allergy and Airway Infections

Most subjects with selective IgA deficiency in developed countries are clinically healthy, but some have increased tendency to allergy and infections, especially when compensatory production of SIgM is lacking.[103] It has been reported that recurrent respiratory infections and isolation of pathogens are rarely seen in IgA-deficient children when salivary IgM is increased.[104] This observation was subsequently supported by a study in adults with selective IgA deficiency; raised salivary levels of total IgM and of IgM antibodies to poliovirus and *E. coli* tended to be associated with relatively good resistance to infections of the respiratory tract.[105] Proneness to such infections in selective IgA deficiency has been shown, moreover, to be associated with replacement of nasal IgA$^+$ PCs mainly with cells producing IgG or IgD, whereas exclusive replacement with IgG$^+$ and IgM$^+$ PCs apparently confers satisfactory mucosal resistance.[106]

In addition, several studies have been performed on subjects with no overt immunodeficiency to see if the level of salivary IgA shows any negative relationship to contraction of allergy or infectious disease, mainly in the upper airways. Usually no clear-cut associations have appeared with infections,[107] although it was reported that infants born to atopic parents show significantly increased prevalence of salivary IgA deficiency, presumably mainly being transient.[108] Such a temporary IgA defect was not convincingly related to later development of allergy, in agreement with a subsequent study.[109] However, it was not excluded that variable compensation with SIgM (which was not studied) explained the apparent lack of association.

It has also been reported that total salivary IgA tends to be reduced in infection-prone children with no overt immunodeficiency.[110] This is in keeping with some other studies, but the possibility exists that the result could be explained by degradation of SIgA1 by microbial proteases. Salivary IgA may, moreover, be decreased in children with recurrent tonsillitis[111] or adenoid hyperplasia,[112] and also in asthmatic children when wheezing is precipitated mainly by recurrent respiratory tract infections.[111] A subsequent study reported a significant relationship of transient salivary IgA deficiency with bronchial hyperreactivity but not with asthma.[109] It has also been suggested that repeated antibiotic courses may lead to persistently low salivary IgA levels.[113]

Human Immunodeficiency Virus (HIV) Infection and AIDS

Saliva of HIV-infected subjects may variably carry this virus,[114] probably depending on the collection method and oral health, which both influence the number of HIV$^+$ lymphocytes in the samples.[115] Saliva from infected subjects contains IgA antibodies to HIV, which to some extent may neutralize the virus, and there are HIV-inhibiting factors in saliva unrelated to antibodies.[116]

Controversial results have been published concerning the effect of HIV infection on the salivary IgA level.[116] Such inconsistency may be ascribed both to the disease stage and the sample source. Because HIV infection is often associated with elevation of serum IgA (mainly IgA1), it is important to avoid whole saliva for conclusive IgA studies, especially when the patients have candidiasis or other oral health problems. Atkinson *et al.*[117] first reported significantly elevated IgA in submandibular but not in parotid secretion from such patients, and Mandel *et al.*[118] confirmed the lack of parotid elevation. These observations might reflect that the IgA response in the submandibular/sublingual glands is more closely related to GALT in terms of B cell homing than that in the parotid as discussed earlier, although the influence of a reduced flow rate on the salivary IgA level could not be excluded.[117] Untreated AIDS patients do indeed have a highly upregulated intestinal IgA system,[119] which would be expected to become disseminated to salivary glands with the possible exception of the parotid.

Nevertheless, Jackson[120] subsequently reported reduced whole saliva IgA in AIDS, which was ascribed to a selective decrease of IgA2, while our laboratory[26] showed that such patients had reduced output of both parotid IgA1 and IgA2—a result that might reflect the degree of secondary immunodeficiency.[116] This observation has subsequently been supported and extended to whole saliva.[121] Also salivary IgA antibodies to viral p24 and gp160 are reduced in symptomatic patients.[122] It is possible that in some studies, a leakage of monomeric IgA1 into whole saliva could have masked a decrease of locally produced salivary IgA1.

The level of IgG antibodies to p24 in whole saliva correlates with that in serum,[122] again suggesting paracellular leakage from blood,[116] either via gingival crevices (see earlier) or directly through the oral surface epithelium when touched by a sampling device, for example, a cotton swab. Tight junctions between epithelial cells are dynamic structures that rapidly open up under the influence of bacterial toxins and inflammatory mediators.[123] Thus, IgG immune complexes increase leakage of bystander protein through rabbit sublingual mucosa.[124] Moreover, the slightest irritation of human nasal or intestinal mucosa leads to bulk flow of serum proteins to the epithelial surface.[125] Also interestingly, salivary IgG has been reported to carry specific anti-gp160 activity 25-fold higher than that of serum IgG,[126] suggesting some local production—perhaps in tonsils, inflamed gingivae, or submandibular and labial salivary glands (which contain a higher density of IgG^+ PCs than the parotid).[46,49]

Adequately performed measurements of salivary IgG antibodies to HIV do show high specificity and sensitivity,[127] and commercial kits are available for this purpose. Sampling of whole saliva has many advantages in scientific field studies of HIV infection and may be preferable to blood for antibody screening also in clinical settings—and particularly for home testing—although issues concerning false-positive and false-negative results need to be resolved.[128]

CONCLUSIONS

Oral microorganisms and aerodigestive antigens are continuously influenced by the two major antibody classes in saliva: SIgA and IgG. The former is synthesized as pIgA by PCs in salivary glands and is exported by an epithelial receptor-mediated mechanism. Conversely, most IgG occurring in saliva represent systemic immunity because it is derived from serum by passive diffusion—preferentially through gingival crevices—although a minor fraction may originate from glandular, gingival, or tonsillar PCs. Along with the paracellular leakage of monomeric IgA and IgG antibodies, other serum-derived or locally produced factors and mediators will also appear in whole saliva, such as IL-6 in patients with ulcerative colitis.[129]

The secretory antibody system is subject to complex immunoregulation that influences distinctly the activity of the various cell types involved in SIgA generation. However, a number of biological phenomena induced by mucosal antigen exposure are poorly defined in experimental animals and still more obscure in humans. There is evidence to suggest that GALT and NALT do not contribute equally to the induction of secretory immunity in various regions of the body. Because of such compartmentalization, enteric immunization cannot be regarded as an adequate mode of stimulating salivary IgA antibodies, whereas NALT stimulation might be more effective. In fact, the various salivary glands may rely on different inductive sites for humoral immunity.

In addition to the apparent compartmentalization of the oral secretory immune system, there are several other important variables influencing the levels of total IgA and specific antibodies in salivary fluids. These include the impact of flow rate, protein loss during sample handling, difficulties with reproducibility and standardization of immunoassays, and uncontrolled admixture of serum-derived monomeric IgA and IgG to the samples. Despite these problems, saliva remains an interesting biological fluid with great scientific and clinical potentials.

ACKNOWLEDGMENTS

Studies in the author's laboratory have been supported by the Research Council of Norway, the University of Oslo, Anders Jahre's Foundation for the Promotion of Science, and Rikshospitalet-Radiumhospitalet Medical Center. Ms. Hege Eliassen and Mr. Erik Kulø Hagen are thanked for excellent assistance with the manuscript.

REFERENCES

1. ELLISON, S.A., P.A. MASHIMO & I.D. MANDEL. 1960. Immunochemical studies of human saliva. I. The demonstration of serum proteins in whole and parotid saliva. J. Dent. Res. **39:** 892–898.

2. HANSON, L.Å. & P. BRANDTZAEG. 1993. The discovery of secretory IgA and the mucosal immune system. Immunol. Today **14:** 416–417.
3. TOMASI, T.B., E.M. TAN, A. SOLOMON, *et al.* 1965. Characteristics of an immune system common to certain external secretions. J. Exp. Med. **121:** 101–124.
4. CRABBÉ, P.A., A.O. CARBONARA & J.F. HEREMANS. 1965. The normal human intestinal mucosa as a major source of plasma cells containing γA-immunoglobulin. Lab. Invest. **14:** 235–248.
5. BRANDTZAEG, P. 1973. Two types of IgA immunocytes in man. Nat. New Biol. **243:** 142–143.
6. BRANDTZAEG, P. 1974. Presence of J chain in human immunocytes containing various immunoglobulin classes. Nature **252:** 418–420.
7. BRANDTZAEG, P., I. FJELLANGER & S.T. GJERULDSEN. 1968. Immunoglobulin M: local synthesis and selective secretion in patients with immunoglobulin A deficiency. Science **160:** 789–791.
8. BRANDTZAEG, P., I. FJELLANGER & S.T. GJERULDSEN. 1970. Human secretory immunoglobulins. I. Salivary secretions from individuals with normal or low levels of serum immunoglobulins. Scand. J. Haematol. Suppl. **12:** 1–83.
9. BRANDTZAEG, P. 1975. Human secretory immunoglobulin M. An immunochemical and immunohistochemical study. Immunology **29:** 559–570.
10. BROWN, W.R., Y. ISOBE & P.K. NAKANE. 1976. Studies on translocation of immunoglobulins across intestinal epithelium. II. Immuno electron–microscopic localization of immunoglobulins and secretory component in human intestinal mucosa. Gastroenterology **71:** 985–995.
11. BRANDTZAEG, P. 1974. Mucosal and glandular distribution of immunoglobulin component. Immunohistochemistry with a cold ethanol-fixation technique. Immunology **26:** 1101–1114.
12. BRANDTZAEG, P. 1974. Mucosal and glandular distribution of immunoglobulin components. Differential localization of free and bound SC in secretory epithelial cells. J. Immunol. **112:** 1553–1559.
13. BRANDTZAEG, P. & H. PRYDZ. 1984. Direct evidence for an integrated function of J chain and secretory component in epithelial transport of immunoglobulins. Nature **311:** 71–73.
14. KRAJČI, P., K.H. GRZESCHIK, A.H.M. GEURTS VAN KESSEL, *et al.* 1991. The human transmembrane secretory component (poly-Ig receptor): molecular cloning, restriction fragment length polymorphism and chromosomal sublocalization. Hum. Genet. **87:** 642–648.
15. JOHANSEN, F.-E. & P. BRANDTZAEG. 2004. Transcriptional regulation of the mucosal IgA system. Trends Immunol. **25:** 150–157.
16. BRANDTZAEG, P. 2003. Role of secretory antibodies in the defence against infections. Int. J. Med. Microbiol. **293:** 3–15.
17. PHALIPON, A. & B. CORTHESY. 2003. Novel functions of the polymeric Ig receptor: well beyond transport of immunoglobulins. Trends Immunol. **24:** 55–58.
18. BRANDTZAEG, P. & R. PABST. 2004. Let's go mucosal: communication on slippery ground. Trends Immunol. **25:** 570–577.
19. NORDERHAUG, I.N., F.-E. JOHANSEN, H. SCHJERVEN, *et al.* 1999. Regulation of the formation and external transport of secretory immunoglobulins. Crit. Rev. Immunol. **19:** 481–508.
20. VAERMAN, J.-P., A.E. LANGENDRIES, D.A. GIFFROY, *et al.* 1998. Antibody against the human J chain inhibits polymeric Ig receptor-mediated biliary and epithelial transport of human polymeric IgA. Eur. J. Immunol. **28:** 171–182.

21. JOHANSEN, F.-E., R. BRAATHEN & P. BRANDTZAEG. 2001. The J chain is essential for polymeric Ig receptor-mediated epithelial transport of IgA. J. Immunol. **167:** 5185–5192.
22. BRAATHEN, R., V.S. HOHMAN, P. BRANDTZAEG, *et al.* 2007. Secretory antibody formation: conserved binding interactions between J chain and polymeric Ig receptor from humans and amphibians. J. Immunol. **178:** 1589–1597.
23. BRANDTZAEG, P. 1971. Human secretory immunoglobulins. VII. Concentrations of parotid IgA and other secretory proteins in relation to the rate of flow and duration of secretory stimulus. Arch. Oral Biol. **16:** 1295–1310.
24. BRANDTZAEG, P. 1998. Synthesis and secretion of human salivary immunoglobulins. *In* Glandular Mechanisms of Salivary Secretion. J.R. Garrett, J. Ekström & L.C. Anderson, Eds.: Frontiers of Oral Biology **10:** 167–199. Karger. London.
25. DELACROIX, D.L. & J.P. VAERMAN. 1983. Function of the human liver in IgA homeostasis in plasma. Ann. N. Y. Acad. Sci. **409:** 383–401.
26. MÜLLER, F., S.S. FRØLAND, M. HVATUM, *et al.* 1991. Both IgA subclasses are reduced in parotid saliva from patients with AIDS. Clin. Exp. Immunol. **83:** 203–209.
27. BOSCH, J.A. 2001. Stress and salivary defense systems. The effects of acute stressors on salivary composition and salivary function. Research thesis (Academisch Proefschrift), Vrije University, the Netherlands. ISBN 90-9015363-2 (Printpartners Ipskamp B.V.), pp. 213.
28. MANDEL, I.D. & H.S. KHURANA. 1969. The relation of human salivary γA globulin and albumin to flow rate. Arch. Oral Biol. **14:** 1433–1435.
29. GRÖNBLAD, E.A. 1982. Concentration of immunoglobulins in human whole saliva: effect of physiological stimulation. Acta Odontol. Scand. **40:** 87–95.
30. RUDNEY, J.D. & Q.T. SMITH. 1985. Relationships between levels of lysozyme, lactoferrin, salivary peroxidase, and secretory immunoglobulin A in stimulated parotid saliva. Infect. Immun. **49:** 469–475.
31. STUCHELL, R.N., I.D. MANDEL. 1978. Studies of secretory IgA in caries-resistant and caries-susceptible adults. Adv. Exp. Med. Biol. **107:** 341–348.
32. OON, C.H. & J. LEE. 1972. A controlled quantitative study of parotid salivary secretory IgA-globulin in normal adults. J. Immunol. Methods **2:** 45–48.
33. GAHNBERG, L. & B. KRASSE. 1981. Salivary immunoglobulin A antibodies reacting with antigens from oral streptococci: longitudinal study in humans. Infect. Immun. **33:** 697–703.
34. RUDNEY, J.D., K.C. KAJANDER & Q.T. SMITH. 1985. Correlations between human salivary levels of lysozyme, lactoferrin, salivary peroxidase and secretory immunoglobulin A with different stimulatory states and over time. Arch. Oral Biol. **30:** 765–771.
35. KORSRUD, F.R. & P. BRANDTZAEG. 1982. Characterization of epithelial elements in human major salivary glands by functional markers: localization of amylase, lactoferrin, lysozyme, secretory component and secretory immunoglobulins by paired immunofluorescence staining. J. Histochem. Cytochem. **20:** 657–666.
36. DESLAURIERS, N., M. OUDGHIRI, J. SEGUIN, *et al.* 1986. The oral immune system: dynamics of salivary immunoglobulin production in the inbred mouse. Immunol. Invest. **15:** 339–349.
37. WIDERSTRÖM, L. & D. BRATTHALL. 1984. Increased IgA levels in saliva during pregnancy. Scand. J. Dent. Res. **92:** 33–37.
38. GLEESON, M., A.W. CRIPPS & R.L. CLANCY. 1995. Modifiers of the human mucosal immune system. Immunol. Cell Biol. **73:** 397–404.

39. KERR, A.C. 1961. The physiological regulation of salivary secretions in man. *In* International Series of Monographs on Oral Biology, Vol. 1. R.C. Greulich, J.B. MacDonald & M.A. Rushton, Eds. Pergamon Press, Oxford.
40. TENOVUO, J., O.P. LEHTONEN, A.S. AALTONEN, *et al.* 1986. Antimicrobial factors in whole saliva of human infants. Infect. Immun. **51:** 49–53.
41. CRAWFORD, J.M., M.A. TAUBMAN & D.J. SMITH. 1975. Minor salivary glands as a major source of secretory immunoglobin A in the human oral cavity. Science **190:** 1206–1209.
42. BRANDTZAEG, P., I. FJELLANGER & S.T. GJERULDSEN. 1968. Adsorption of immunoglobulin A onto oral bacteria in vivo. J. Bacteriol. **96:** 242–249.
43. KRISTIANSEN, B.E. 1984. Collection of mucosal secretion by synthetic discs for quantitation of secretory IgA and bacteria. J. Immunol. Methods **73:** 251–257.
44. BERG, E. & J.C. TJELL. 1969. Paraffin chewing as a saliva-stimulating agent. J. Dent. Res. **48:** 325.
45. KORSRUD, F.R. & P. BRANDTZAEG. 1980. Quantitative immunohistochemistry of immuno globulin- and J-chain-producing cells in human parotid and submandibular salivary glands. Immunology **39:** 129–140.
46. BRANDTZAEG, P. 1983. The secretory immune system of lactating human mammary glands compared with other exocrine organs. Ann. N. Y. Acad. Sci. **409:** 353–381.
47. BRANDTZAEG, P. 1977. Immunohistochemical studies of various aspects of glandular immunoglobulin transport in man. Histochem. J. **9:** 553–572.
48. MORO, I., S. UMEMURA, S.S. CARGO, *et al.* 1984. Immunohistochemical distribution of immmunoglobulins, lactoferrin, and lysozyme in human minor salivary glands. J. Oral Pathol. **13:** 97–104.
49. MATTHEWS, J.B., A.J.C. POTTS & M.K. BASU. 1985. Immunoglobulin-containing cells in normal human labial salivary glands. Int. Arch. Allergy Appl. Immunol. **77:** 374–376.
50. KILIAN, M., J. REINHOLDT, H. LOMHOLT, *et al.* 1996. Biological significance of IgA1 proteases in bacterial colonization and pathogenesis: critical evaluation of experimental evidence. APMIS **104:** 321–338.
51. KETT, K., P. BRANDTZAEG, J. RADL, *et al.* 1986. Different subclass distribution of IgA-producing cells in human lymphoid organs and various secretory tissues. J. Immunol. **136:** 3631–3635.
52. BRANDTZAEG, P. & F.-E. JOHANSEN. 2005. Mucosal B cells: phenotypic characteristics, transcriptional regulation, and homing properties. Immunol. Rev. **206:** 32–63.
53. BRANDTZAEG, P. 1977. Human secretory component. VI. Immunoglobulin-binding properties. Immunochemistry **14:** 179–188.
54. NATVIG, I.B., F.-E. JOHANSEN, T.W. NORDENG, *et al.* 1997. Mechanism for enhanced external transfer of dimeric IgA over pentameric IgM. Studies of diffusion, binding to the human polymeric Ig receptor, and epithelial transcytosis. J. Immunol. **159:** 4330–4340.
55. MESTECKY, J., J.R. MCGHEE, R.R. ARNOLD, *et al.* 1978. Selective induction of an immune response in human external secretions by ingestion of bacterial antigen. J. Clin. Invest. **61:** 731–737.
56. WEISZ-CARRINGTON, P., M.E. ROUX, M. MCWILLIAMS, *et al.* 1979. Organ and isotype distribution of plasma cells producing specific antibody after oral immunization: evidence for a generalized secretory immune system. J. Immunol. **123:** 1705–1708.

57. JACKSON, D.E., E.T. LALLY, M.C. NAKAMURA, *et al.* 1981. Migration of IgA-bearing lymphocytes into salivary glands. Cell. Immunol. **63:** 203–209.
58. CZERKINSKY, C., A.M. SVENNERHOLM, M. QUIDING, *et al.* 1991. Antibody-producing cells in peripheral blood and salivary glands after oral cholera vaccination of humans. Infect. Immun. **59:** 996–1001.
59. JERTBORN, M., A.M. SVENNERHOLM & J. HOLMGREN. 1986. Saliva, breast milk, and serum antibody responses as indirect measures of intestinal immunity after oral cholera vaccination or natural disease. J. Clin. Microbiol. **24:** 203–209.
60. BRANDTZAEG, P. 1996. The B-cell development in tonsillar lymphoid follicles. Acta Otolaryngol. Suppl. **523:** 55–59.
61. BRANDTZAEG, P. 2003. Immunology of tonsils and adenoids: everything the ENT surgeon needs to know. International Congress Series (ICS) **1254:** 89–99, 2003 (Elsevier)/Int. J. Pediatr. Otorhinolaryngol. 67(Suppl. 1): S69–76.
62. FUKUIZUMI, T., H. INOUE, Y. ANZAI, *et al.* 1995. Sheep red blood cell instillation at palatine tonsil effectively induces specific IgA class antibody in saliva in rabbits. Microbiol. Immunol. **39:** 351–359.
63. FUKUIZUMI, T., H. INOUE, T. TSUJISAWA, *et al.* 1999. Tonsillar application of formalin-killed cells of *Streptococcus sobrinus* reduces experimental dental caries in rabbits. Infect. Immun. **67:** 426–428.
64. QUIDING-JÄRBRINK, M., G. GRANSTRÖM, I. NORDSTRÖM, *et al.* 1995. Induction of compartmentalized B-cell responses in human tonsils. Infect. Immun. **63:** 853–857.
65. STOLTENBERG, L., A. VEGE, O.D. SAUGSTAD, *et al.* 1995. Changes in the concentration and distribution of immunoglobulin–producing cells in SIDS palatine tonsils. Pediatr. Allergy Immunol. **6:** 48–55.
66. THRANE, P., T.O. ROGNUM & P. BRANDTZAEG. 1990. Increased immune response in upper respiratory and digestive tracts in SIDS. Lancet **335:** 229–230.
67. GLEESON, M., R.L. CLANCY & A.W. CRIPPS. 1993. Mucosal immune response in a case of sudden infant death syndrome. Pediatr. Res. **33:** 554–556.
68. BRANDTZAEG, P. 1999. Regionalized immune function of tonsils and adenoids. Immunol. Today **20:** 383–384.
69. JOHANSEN, F.-E., E.S. BAEKKEVOLD, H.S. CARLSEN, *et al.* 2005. Regional induction of adhesion molecules and chemokine receptors explains disparate homing of human B cells to systemic and mucosal effector sites: dispersion from tonsils. Blood **106:** 593–600.
70. MA, J.K., B.Y. HIKMAT, K. WYCOFF, *et al.* 1998. Characterization of a recombinant plant monoclonal secretory antibody and preventive immunotherapy in humans. Nat. Med. **4:** 601–606.
71. BROWN, T.A. & J. MESTECKY. 1985. Immunoglobulin A sublass distribution of naturally occurring salivary antibodies to microbial antigens. Infect. Immun. **49:** 459–462.
72. MANSA, B. & M. KILIAN. 1986. Retained antigen-binding activity of Fab fragments of human monoclonal immunoglobulin A1 (IgA1) cleaved by IgA1 protease. Infect. Immun. **52:** 171–174.
73. WEISER, J.N., D. BAE, C. FASCHING, *et al.* 2003. Antibody-enhanced pneumococcal adherence requires IgA1 protease. Proc. Natl. Acad. Sci. USA **100:** 4215–4220.
74. TAYLOR, H.P. & N.J. DIMMOCK. 1985. Mechanism of neutralization of influenzae virus by secretory IgA is different from that of monomeric IgA or IgG. J. Exp. Med. **161:** 198–209.

75. CHRISTENSEN, P. & V.-A. OXELIUS. 1975. A reaction between some streptococci and IgA myeloma proteins. Acta Pathol. Microbiol. Scand. [C] **83:** 184–188.
76. BIESBROCK, A.R., M.S. REDDY & M.J. LEVINE. 1991. Interaction of a salivary mucin–secretory immunoglobulin A complex with mucosal pathogens. Infect. Immun. **59:** 3492–3497.
77. QUAN, C., A. BERNEMAN, R. PIRES, *et al.* 1997. Natural polyreactive secretory immunoglobulin A autoantibodies as a possible immune barrier in humans. Infect. Immun. **65:** 3997–4004.
78. RUSSELL, M.W. & J. MESTECKY. 1986. Potential for immunological intervention against dental caries. J. Biol. Buccale **14:** 159–175.
79. TAUBMAN, M.A. & D.J. SMITH. 1993. Significance of salivary antibody in dental disease. Ann. N. Y. Acad. Sci. **694:** 202–215.
80. NOGUEIRA, R.D., A.C. ALVES, M.H. NAPIMOGA, *et al.* 2005. Characterization of salivary immunoglobulin A responses in children heavily exposed to the oral bacterium *Streptococcus mutans*: influence of specific antigen recognition in infection. Infect. Immun. **73:** 5675–5684.
81. GREGORY, R.L. & S.J. FILLER. 1987. Protective secretory immunoglobulin A antibodies in humans following oral immunization with *Streptococcus mutans*. Infect. Immun. **55:** 2409–2415.
82. GREGORY, R.L., S.M. MICHALEK, S.J. FILLER, *et al.* 1985. Prevention of *Streptococcus mutans* colonization by salivary IgA antibodies. J. Clin. Immunol. **5:** 55–62.
83. RUSSELL, M.W., N.K. CHILDERS, S.M. MICHALEK, *et al.* 2004. A caries vaccine? The state of the science of immunization against dental caries. Caries Res. **38:** 230–235.
84. SMITH, D.J., J.L. EBERSOLE, M.A. TAUBMAN, *et al.* 1985. Salivary IgA antibody to *Actinobacillus actinomyetemcomitans* in young adult population. J. Periodontal Res. **20:** 8–11.
85. MANDEL, M.A., K. DVORAK & J.J. DE COSSE. 1973. Salivary immunoglobulins in patients with oropharyngeal and bronchopulmonary carcinoma. Cancer **31:** 1408–1413.
86. BROWN, A.M., E.T. LALLY & A. FRANKEL. 1975. IgA and IgG content of the saliva and serum of oral cancer patents. Arch. Oral Biol. **20:** 395–398.
87. MANDEL, I.D. & H. BAURMASH. 1978. Salivary immunoglobulins in diseases affecting salivary glands. Adv. Exp. Med. Biol. **107:** 839–847.
88. THRANE, P.S., L.M. SOLLID, H.R. HAANES, *et al.* 1992. Clustering of IgA-producing immunocytes related to HLA-DR-positive ducts in normal and inflamed salivary glands. Scand. J. Immunol. **35:** 43–51.
89. ANDERSON, L.G., N.A. CUMMINGS, R. ASOFSKY, *et al.* 1972. Salivary gland immunoglobulin and rheumatoid factor synthesis in Sjögren's syndrome. Am. J. Med. **53:** 456–463.
90. DUNNE, J.V., D.A. CARSON, H.L. SPIEGELBERG, *et al.* 1979. IgA rheumatoid factor in the sera and saliva of patients with rheumatiod arthritis and Sjögren's syndrome. Ann. Rheum. Dis. **38:** 161–165.
91. TALAL, N., R. ASOFSKY & P. LIGHTBODY. 1970. Immunoglobulin synthesis by salivary gland lymphoid cells in Sjogren's syndrome. J. Clin. Invest. **49:** 49–54.
92. MATTHEWS, J.B., A.J.C. POTTS, J. HAMBURGER, *et al.* 1986. Immunoglobulin–producing cells in labial salivary glands of patients with rheumatoid arthritis and systemic lupus erythematosus. J. Oral Pathol. **15:** 520–523.

93. SCHMID, U., D. HELBRON & K. LENNERT. 1982. Development of malignant lymphoma in myoepithelial sialadenitis (Sjögren's syndrome). Virchows Arch. (Pathol. Anat.) **395:** 11–43.
94. JORDAN, R.C., E.W. ODELL & P.M. SPEIGHT. 1996. B-cell monoclonality in salivary lymphoepithelial lesions. Eur. J. Cancer B Oral Oncol. **32B:** 38–44.
95. JORDAN, R.C., Y. MASAKI, S. TAKESHITA, *et al.* 1996. High prevalence of B-cell monoclonality in labial gland biopsies of Japanese Sjogren's syndrome patients. Int. J. Hematol. **64:** 47–52.
96. BRANDTZAEG, P. 1971. Human secretory immunoglobulins. II. Salivary secretions from individuals with selectively excessive or defective synthesis of serum immunoglobulins. Clin. Exp. Immunol. **8:** 69–85.
97. BRANDTZAEG, P. 2002. The mucosal B cell and its functions. *In* Food Allergy and Intolerance. Second edition. J. Brostoff & S.J. Challacombe, Eds.: 127–171. Saunders: Elsevier Science. London.
98. HANSSON, T., A. DANNAEUS, W. KRAAZ, *et al.* 1997. Production of antibodies to gliadin by peripheral blood lymphocytes in children with celiac disease: the use of an enzyme-linked immunospot technique for screening and follow-up. Pediatr. Res. **41:** 554–559.
99. HAKEEM, V., R. FIFIELD, H.F. AL-BAYATY, *et al.* 1992. Salivary IgA antigliadin antibody as a marker for coeliac disease. Arch. Dis. Child. **67:** 724–727.
100. AL-BAYATY, H.F., M.J. ALDRED, D.M. WALKER, *et al.* 1989. Salivary and serum antibodies to gliadin in the diagnosis of celiac disease. J. Oral Pathol. Med. **18:** 578–581.
101. O'MAHONY, S., E. ARRANZ, J.R. BARTON, *et al.* 1991. Dissociation between systemic and mucosal humoral immune responses in coeliac disease. Gut **32:** 29–35.
102. GIBNEY, M.J. & C. BRADY. 1991. Systemic and salivary antibodies to dietary antigens (gliadin, casein and ovalbumin) in treated and untreated coeliac disease and in Crohn's disease. Eur. J. Int. Med. **2:** 115–119.
103. BRANDTZAEG, P. & D.E. NILSSEN. 1995. Mucosal aspects of primary B-cell deficiency and gastrointestinal infections. Curr. Opin. Gastroenterol. **11:** 532–540.
104. ÖSTERGAARD, P.A. 1980. Predictable clinical disorders related to serum and saliva Iglevels and the number of circulating T cells in asthmatic children. Clin. Allergy **10:** 277–284.
105. MELLANDER, L., J. BJÖRKANDER, B. CARLSSON, *et al.* 1986. Secretory antibodies in IgA-deficient and immunosuppressed individuals. J. Clin. Immunol. **6:** 284–291.
106. BRANDTZAEG, P., G. KARLSSON, G. HANSSON, *et al.* 1987. The clinical condition of IgA-deficient patients is related to the proportion of IgD- and IgM-producing cells in their nasal mucosa. Clin. Exp. Immunol. **67:** 626–636.
107. NAKAMURA, C., T. AKIMOTO, S. SUZUKI, *et al.* 2006. Daily changes of salivary secretory immunoglobulin A and appearance of upper respiratory symptoms during physical training. J. Sports Med. Phys. Fitness. **46:** 152–157.
108. VAN ASPEREN, P.P., M. GLEESON, A.S. KEMP, *et al.* 1985. The relationship between atopy and salivary IgA deficiency in infancy. Clin. Exp. Immunol. **62:** 753–757.
109. GLEESON, M., R.L. CLANCY, M.J. HENSLEY, *et al.* 1996. Development of bronchial hyperreactivity following transient absence of salivary IgA. Am. J. Respir. Crit. Care Med. **153:** 1785–1789.

110. LEHTONEN, O.-P.J., J. TENOVUO, A.S. AALTONEN, *et al.* 1987. Immunoglobulins and innate factors of immunity in saliva of children prone to respiratory infections. Acta Pathol. Microbiol. Scand. [C] **95:** 35–40.
111. ÖSTERGAARD, P.A. 1977. IgA levels and carrier rate of pathogenic bacteria in 27 children previously tonsillectomized. Acta Pathol. Microbiol. Scand. [C] **85:** 178–186.
112. HESS, M., J. KUGLER, D. HAAKE, *et al.* 1991. Reduced concentration of secretory IgA indicates changes of local immunity in children with adenoid hyperplasia and secretory otitis media. ORL J. Otorhinolaryngol. Relat. Spec. **53:** 339–341.
113. ÖSTERGAARD, P.A. 1983. Oral bacterial flora and secretory IgA in small children after repeated courses of antibiotics. Scand. J. Infect. Dis. **15:** 115–118.
114. GROOPMAN, J.E., S.Z. SALAHUDDIN, M.G. SARNGADHARAN, *et al.* 1984. HTLV-III in saliva of people with AIDS-related complex and healthy homosexual men at risk for AIDS. Science **226:** 447–449.
115. PEKOVIC, D., D. AJDUKOVIC, C. TSOUKAS, *et al.* 1987. Detection of human immunosuppressive virus in salivary lymphocytes from dental patients with AIDS. Am. J. Med. **52:** 188–189.
116. CHALLACOMBE, S.J. & J.R. NAGLIK. 2006. The effects of HIV infection on oral mucosal immunity. Adv. Dent. Res. **19:** 29–35.
117. ATKINSON, J.C., C. YEH, F.G. OPPENHEIM, *et al.* 1990. Elevation of salivary antimicrobial proteins following HIV-1 infection. J. Acquir. Immune Defic. Syndr. **3:** 41–48.
118. MANDEL, I.D., C.E. BARR & L. TURGEON. 1992. Longitudinal study of parotid saliva in HIV-1 infection. J. Oral Pathol. Med. **21:** 209–213.
119. NILSSEN, D.E., O. ØKTEDALEN & P. BRANDTZAEG. 2004. Intestinal B cell hyperactivity in AIDS is controlled by highly active antiretroviral therapy. Gut **53:** 487–493.
120. JACKSON, S. 1990. Secretory and serum IgA are inversely altered in AIDS patients. *In* Advances in Mucosal Immunology. T.T. MacDonald, S.J. Challacombe, P.W. Bland, C.R. Stokes, R.V. Heatly & A. McI. Mowat, Eds.: 665. Kluwer Academic Publishers. Lancaster.
121. SWEET, S.P., D. RAHMAN & S.J. CHALLACOMBE. 1995. IgA subclasses in HIV disease: dichotomy between raised levels in serum and decreased secretion rates in saliva. Immunology **86:** 556–559.
122. MATSUDA, S., S. OKA, M. HONDA, *et al.* 1993. Characteristics of IgA antibodies against HIV-1 in sera and saliva from HIV-seropositive individuals in different clinical stages. Scand. J. Immunol. **38:** 428–434.
123. FASANO, A. & T. SHEA-DONOHUE. 2005. Mechanisms of disease: the role of intestinal barrier function in the pathogenesis of gastrointestinal autoimmune diseases. Nat. Clin. Pract. Gastroenterol. Hepatol. **2:** 416–422.
124. BRANDTZAEG, P. & K. TOLO. 1977. Mucosal penetrability enhanced by serum-derived antibodies. Nature **266:** 262–263.
125. PERSSON, C.G., J.S. ERJEFALT, L. GREIFF, *et al.* 1998. Contribution of plasma-derived molecules to mucosal immune defence, disease and repair in the airways. Scand. J. Immunol. **47:** 302–313.
126. LU, X.S., J.F. DELFRAISSY, L. GRANGEOT-KEROS, *et al.* 1994. Rapid and constant detection of HIV antibody response in saliva of HIV-infected patients; selective distribution of anti-HIV activity in the IgG isotype. Res. Virol. **145:** 369–377.

127. Hunt, A.J., J. Connell, G. Christofinis, *et al.* 1993. The testing of saliva samples for HIV-1 antibodies: reliability in a non-clinic setting. Genitourin. Med. **69:** 29–30.
128. Wright, A.A. & I.T. Katz. 2006. Home testing for HIV. N. Engl. J. Med. **354:** 437–440.
129. Nielsen, A.A., J.N. Nielsen, A. Schmedes, *et al.* 2005. Saliva interleukin-6 in patients with inflammatory bowel disease. Scand. J. Gastroenterol. **40:** 144–148.

Oxytocin

Behavioral Associations and Potential as a Salivary Biomarker

C. SUE CARTER,[a] HOSSEIN POURNAJAFI-NAZARLOO,[a] KRISTIN M. KRAMER,[a,b] TONI E. ZIEGLER,[c] ROSEMARY WHITE-TRAUT,[d] DEBORAH BELLO,[d] AND DORIE SCHWERTZ[d]

[a]*Brain-Body Center, Department of Psychiatry, University of Illinois at Chicago, Chicago, Ilinois, USA*

[b]*Department of Biology, University of Memphis, Memphis, Tennessee 38152*

[c]*National Primate Research Center, University of Wisconsin, Madison, Wisconsin 53706, USA*

[d]*College of Nursing, University of Illinois at Chicago, Chicago, Illinois 60612, USA*

ABSTRACT: Oxytocin (OT) is a neuropeptide that is produced primarily in the hypothalamus and is best known for its role in mammalian birth and lactation. Recent evidence also implicates OT in social behaviors, including parental behavior, the formation of social bonds, and the management of stressful experiences. OT is reactive to stressors, and plays a role in the regulation of both the central and autonomic nervous system, including effects on immune and cardiovascular function. Knowledge of patterns of OT release would be of value in many fields of science and medicine. However, measurements of OT concentration in blood are infrequently performed, and previous attempts to measure OT in saliva have been unsuccessful. Using a sensitive enzyme immunoassay (EIA) and concentrated samples we were able to detect reproducible changes in salivary OT as a function of lactation and massage. These results indicate that measurements of biologically relevant changes in salivary OT are possible. These results confirm the biological relevance of changes in salivary OT with stressors and support saliva as a noninvasive source to monitor central neuroendocrine function.

KEYWORDS: oxytocin; stress; salivary neuropeptides; reproduction; social bonds; lactation; massage; enzyme immunoassay

Address for correspondence: C. Sue Carter, Ph.D., Brain-Body Center, Department of Psychiatry (mc 912), University of Illinois, 1601 West Taylor Street, Chicago, IL 60612. Voice: 312-355-1593; fax: 312-996-7658.
scarter@psych.uic.edu

Ann. N.Y. Acad. Sci. 1098: 312–322 (2007). © 2007 New York Academy of Sciences.
doi: 10.1196/annals.1384.006

INTRODUCTION AND BACKGROUND

Oxytocin (OT) is a mammalian nonapeptide (FIG. 1) with functional roles in birth, lactation, parenting, and forms of positive social interactions. Treatment with OT quickly facilitates positive social behaviors, including selective partner preferences and parental behavior.[1,2] OT receptors also correlate with positive social behavior. OT is made in and acts on the brain, primarily in the hypothalamus and areas of the nervous system that influence emotions and, as measured by mRNA levels, is the most abundantly expressed peptide in the hypothalamus.[5] OT is released during positive social interactions and may facilitate the ability to be vulnerable, feel safe, and relaxed by downregulating or buffering the response to stressors and the reactivity of the autonomic nervous system, including heart rate and blood pressure.[3,4] OT's role is well established in birth, lactation, and parenting, as well as sexual interactions and stress management. For example, in rats, touch and massage may release OT.[3] OT released in the brain may feed back on the nervous system to further enhance relaxation or other coping mechanisms, in part through direct effects on behavior, but also through actions on the hypothalamic-pituitary-adrenal axis and the autonomic nervous system. OT may have a protective role in the cardiovascular system, and can induce analgesia and facilitate wound healing, possibly through its anti-inflammatory actions, as well as effects on the autonomic nervous system.[3,4] Thus OT is an excellent candidate biomarker for the coordination of emotional states and feelings with physiological processes through which positive social interactions bestow health benefits.

OT is unusual in its ability to integrate multiple actions. Of particular relevance to the broad effect of OT is the fact that this hormone has only one known type of receptor.[6] Most neurochemicals, including the related neuropeptide arginine vasopressin (AVP), have several receptor subtypes that allow a single biologically active molecule to have diverse functions. However, OT has the potential to interact with AVP and its receptors, and many other systems implicated in stress and coping. Although AVP is structurally similar to OT in that its sequence differs by only two amino acids, several of the known effects of AVP are opposite to those of OT. OT and the neural systems that it helps to regulate also may counteract the tendency to be defensive, fearful, or anxious.[1,2] For example, during lactation, OT is released in pulses that both facilitate milk ejection and also may reduce anxiety and overreactivity to stressful stimuli.[7,8] Males also respond to OT: for example, a single treatment with an intranasal

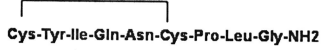

Cys-Tyr-Ile-Gln-Asn-Cys-Pro-Leu-Gly-NH2

FIGURE 1. OT is a 9 amino acid cyclic peptide. The PVN and SON are the primary hypothalamic sources of oxytocin in the brain.

OT spray reduces reactivity to stressors,[9] and also increases the willingness of men to "trust" others, as measured in a computerized game.[10] Intranasal OT also has direct effects on neural activation of the amygdala in response to fear-inducing stimuli and was associated with reductions in neural activation of brainstem regions implicated in fear and arousal.[11]

CENTRAL VERSUS PERIPHERAL PEPTIDES

Research in rats suggests that the major sources of OT in the peripheral circulation are hypothalamic nuclei, including the paraventricular nucleus (PVN) and the supraoptic nucleus (SON). OT from both the SON and PVN is carried via axoplasmic transport to the posterior pituitary for release into the blood stream. OT from the posterior pituitary is presumed to be the primary source of salivary OT. The PVN also is a major source for centrally released OT, although some OT of SON origin may also be released into the brain.[6] The relationship between central and peripherally produced peptides is controversial.[12] On the basis of central microdialysis data from rats, it appears that central and peripheral release of OT tends to be coordinated, while there is dissociation between central and peripheral AVP.[13] In voles, we have also found significant positive correlations between OT immunoreactivity in the PVN and SON and peripheral levels of OT measured by enzyme immunoassay (EIA) (Bales and Carter, unpublished data). Such experiments are not possible in humans, but support the usefulness of measures of peripheral levels of OT as an index of OT central synthesis. Additional evidence for sites of OT's action outside the central nervous system (CNS) includes spinal cord and nonspinal pathways via the vagus, and synergistic interactions between OT and CCK, mediated through the vagus.[3]

RATIONALE: METHODOLOGICAL ISSUES IN THE STUDY OF NEUROPEPTIDES

Compelling results from animal models and the emerging data on behavioral effects of peptides in humans implicate OT in both behavior and physiology,[1,2,6] although, measurements of neuropeptides are rare because of technical difficulties associated with collecting human samples. Such testing would be facilitated by noninvasive methods for measuring peptides associated with naturally occurring behavioral events. Commercial radioimmunoassays (RIA) for neuropeptides require large amounts of sample (~1 mL), and often require preanalytical chemical extraction that is associated with less than quantitative peptide recovery. Reported values for OT are frequently close to the level of detection for the assay and thus potentially result in a large experimental error.

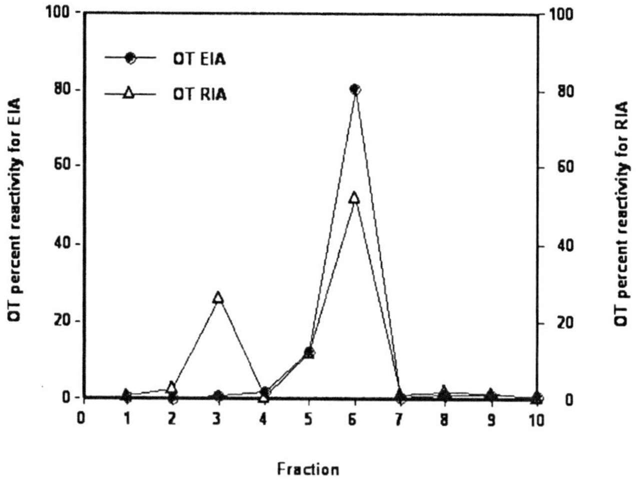

FIGURE 2. Oxytocin reactivity in prairie vole plasma samples. Comparison of oxytocin antibody specificities was made using HPLC separation and subsequent assay by either EIA (Assay Designs, Inc) or RIA (Peninsula, Bachem). OT elutes in the sixth fraction, while AVP elutes in the third fraction. Note that the EIA is more specific for OT levels since the RIA assay detects both AVP and OT.

Our research on the behavioral effects of OT in rodents necessitates dealing with amounts of sample that rendered the use of a RIA inadequate because of a lack of sensitivity. Consequently, we validated a commercial EIA[14] and used that with a preanalysis concentration step for human saliva. It was possible to measure OT above the detection limit of ~5 pg/mL with improved sensitivity and specificity (FIG. 2). However, there remains a need for improved sensitivity and assay technology.

METHODOLOGICAL ISSUES IN THE STUDY OF SALIVARY HORMONES

Salivary measures of biomarkers, such as steroid hormones, have over the last two decades become well accepted in human endocrinology. Most studies that relate salivary biomarkers to behavior have measured adrenal steroids (usually glucocorticoids) or gonadal hormones (testosterone, estrogen, or progesterone). This research has been productive and extensive databases are available describing the parameters of steroid hormones, including differences as a function of time of day of sampling, gender, age, and responses to stressors.[15] However, because these hormones originate primarily from tissue outside the nervous system they are at best only indirect measures of brain function. Neuropeptides such as OT that are primarily synthesized in the nervous system are more likely to be directly related to behavior.

Collection of saliva offers several advantages over blood, especially since this method is noninvasive, with few risks, and samples can be collected under both naturalistic and laboratory conditions. Steroid hormones in saliva are generally "free," while blood contains both free and protein-bound hormone. Collection can be made by the participant and, beyond the limitations of the participant's motivation and compliance, the frequency and time of the day of sampling are virtually unlimited. Samples can be frozen and briefly stored in a standard home freezer (without a thawing cycle), although a −80°C freezer is required for long-term storage.[16]

Several recent studies have identified methodological guidelines specific for the study of salivary hormones that suggest preferential collection from individuals who do not have mucosal sources of blood (bleeding gums, sores, etc). Recent brushing or flossing may release blood into saliva,[17] but microinjuries of the oral mucosa have little effect on the reliability of assays for cortisol.

Methods for collecting saliva that use either cotton or polyester may introduce variance into these assays. The impact of the collection method differs among hormones, but may be especially important when using an EIA.[16,18] In addition, methods used to increase salivation, including flavor crystals and chewing gum, may change the oral pH or dilute the analyte. On the basis of the research of Granger and colleagues,[16,17] saliva was collected by asking participants to "passively drool" into a cold collection tube and frozen immediately after collection.

It is possible that OT values from saliva may not be meaningfully related to other endocrine variables, including central peptides, which are of particular importance to behavior. As described above, most of the RIA commercial kits for OT require an extraction and large samples, and typical results are close to the limit of detection for the assay and consequently have not proven useful for salivary peptides.

VALIDATION OF THE EIA ASSAY FOR OXYTOCIN

Measurement and Validation of the EIA for Oxytocin

Oxytocin was measured using the EIA kit developed by Assay Designs, Inc. (Ann Arbor, MI, USA). Assay Designs reported a low cross-reactivity for similar neuropeptides found in mammalian sera at less than 0.001% (FIG. 2) and a minimum detection limit of 4.68 pg/mL. Manufacturer's instructions were followed without modification, although plasma was not extracted. Validation of the EIA was conducted using various methods[14]: parallelism was validated by measurements from a serial dilution of pooled vole plasma; accuracy of the assay was assessed by spiking samples of pooled vole plasma with varying amounts of standard; precision was determined from the variability surrounding multiple measurements of high and low controls (interassay CV) and variability in multiple measurements of unknowns (intra-assay CV); and *in vivo* determinations for biological validation were made in plasma from sexually naïve female prairie voles receiving a single subcutaneous injection of OT (5 μg OT/50 μL isotonic saline). Samples were collected prior to injection or at one of three time points after injection (5, 15, or 60 min). A dilution series of pooled vole plasma resulted in a displacement curve parallel to that of the standard curve ($y = 129.88–16.79\times; r^2 = 0.96$). Tests of accuracy resulted in a high correlation between expected and observed values ($y = -2.20 + 0.79\times; r^2 = 1.00$) and average recovery was 71%. Intra-assay CV averaged 2.68%; interassay CV averaged 14.45%; standard curves were similar across assays. Injection with OT resulted in a significant increase in plasma OT ($P < 0.0001$) with all pair-wise comparisons significant. Values were maximal 15 min following injection. Within 60 min, plasma OT was significantly lower ($P < 0.0001$), but still elevated over baseline levels ($P < 0.0001$).

Biological Validations in Blood: Species and Sex Differences

Prairie voles and rats of both sexes were used for comparisons to plasma OT. Blood samples for determination of baseline OT were collected from reproductively naïve male and female prairie voles (n =12 and 11, respectively) and male and female rats (n = 11 for each sex). There were significant differences in plasma OT by species and by sex, with voles having significantly higher OT than rats ($P < 0.0001$) and females having higher OT than males ($P < 0.001$). Female voles averaged 488.3 ± 87.6 pg/mL (n =11) and males, 264.4 ± 31.0 pg/mL (n =12). Female rats averaged 186.5 ± 53.4 pg/mL (n =11) and males, 78.9 ± 5.8 pg/mL (n =11).

On the basis of the parallelism data collected from initial measurements of plasma pools, plasma samples for both voles and rats were diluted 1:4 prior to assay (requiring 65 μL of plasma to assay samples in duplicate). Vole plasma collected after injection with OT was initially diluted 1:4, but in all cases OT concentrations that were measured exceeded the highest standard used to generate the standard curve. A second aliquot was available for each vole, allowing reassay at a higher dilution factor. A 1:20 dilution resulted in values within the range of the standards. Measured concentrations for the serial dilution of pooled plasma were parallel to the standard curve, indicating that extraction procedures are unnecessary and that a 1:4 or higher dilution eliminates significant matrix effects. The OT assay also was valid in terms of quantitative recovery, with good agreement between expected and observed OT concentrations for spiked samples.

OT concentrations observed in this study were higher than those reported using RIA and extracted plasma. Recent studies of nonextracted rat blood and CSF,[19] primate CSF and blood,[20] and human blood,[21,22] using the identical protocol produced OT values within the range of those measured in rodents in our lab. These laboratories also validated this assay (as described here) and found the same functional relationships. While it is difficult to quantitatively compare measurements between assays, especially using different assay systems, we assessed relative differences between species by comparison of split samples measured by both EIA and RIA (run courtesy of Dr. Janet Amico, University of Pittsburgh). The RIA had a range of 1–20 pg/mL. Results from this RIA showed that OT in rat plasma from our laboratory fell within the standard curve. However, all prairie vole plasma assayed by RIA far exceeded the standard curve, precluding comparisons of individual data. Results from both assays supported the hypothesis that prairie voles had significantly higher baseline OT levels and the usefulness of this assay.

Validation of EIA Antibody Specificity

To determine the specificity of the antibody used in the Assay Designs EIA, samples were chromatographed by high-performance liquid chromatography (HPLC) prior to immunoassay using previously described methods.[23] Samples were thawed and 200 μL added to acidified buffer (800 μL distilled water, 20 μL phosphoric acid) and purified by solid-phase extraction (SepPak, C18, Waters, Milford, MA, USA). The method consisted of conditioning the column with 1 mL methanol and then 1 mL distilled water prior to adding the acidified sample. The sample was washed on the column with 1 mL 1% trifluoroacetic acid (TFA) in 10% acetonitrile (ACN), and eluted with 80% ACN in H_2O. After drying, samples were reconstituted in 30 μL of ACN:H_2O (50:50) and 20 μL injected onto a Beckman HPLC system using the method reported by Fries *et al.*[23] Samples were eluted and 1-mL fractions collected for the 10-min runs. Standard of AVP and OT (Sigma Chemical Co., St. Louis, MO, USA) were eluted in the third and sixth fraction, respectively. All fractions were divided into two aliquots, dried and reconstituted in buffer for either EIA (Assay Design) or OT RIA (Peninsula, Bachem, CA, USA). For three independent prairie vole samples, the OT EIA showed no reactivity with other fractions, except the fraction containing OT, while in two of the samples the RIA did show cross-reactivity with the third fraction, where AVP elutes (FIG. 2). Therefore, the EIA was shown to be specific for OT in the prairie vole plasma, while the comparison RIA was less specific.

Salivary EIA Assay Methods

The same EIA used for the plasma validation in animal studies was used for the chemical validation of human salivary OT. The assay was validated for parallelism by assay of a serial dilution of pooled saliva samples. Quantitative recovery was examined to determine whether sample preparation methods adversely affected the results. Pooled salivary samples were spiked with varying amounts of standard of known concentration and then prepared for assay by concentration. Quantitative recovery was determined by comparing the assay results from the spiked samples to those for assay buffer spiked with the same volumes of standard. Precision was determined from the variability in multiple measurements of unknowns (intra-assay CV). A dilution series of pooled saliva resulted in a displacement curve parallel to that of the standard curve ($y = 134.75–17.04\times$; $r^2 = 0.94$). Tests of quantitative recovery resulted in a high correlation between expected and observed values ($y = 49.9 + 0.66\times$; $r^2 = 0.99$) and average recovery was 73%. Intra-assay CV averaged 2.1% and interassay CV values were 3.5% and 9.5% for high and low controls, respectively.

To allow measurement of salivary OT using a reliable portion of the standard curve, samples were concentrated prior to assay. After centrifugation, 1 mL of supernatant from each sample was dried by vacuum centrifugation and then reconstituted in 250 mL of assay buffer, resulting in a 4× sample.

Biological Validations of the Human Salivary OT Assay

Because of the well-established relationship between lactation and the synthesis of OT, we conducted validation studies for salivary OT in humans using saliva samples taken repeatedly from lactating women. In this study women ($n = 10$) provided 6 sets of 3 samples for a total of 180 samples. Samples were taken on two separate days at 30-min intervals before, during, and after breast feeding, and at three different times of day (morning, mid-day, and evening). Preliminary results indicated that OT concentrations in saliva samples (prior to concentration) were usually below the lowest standard. The results were

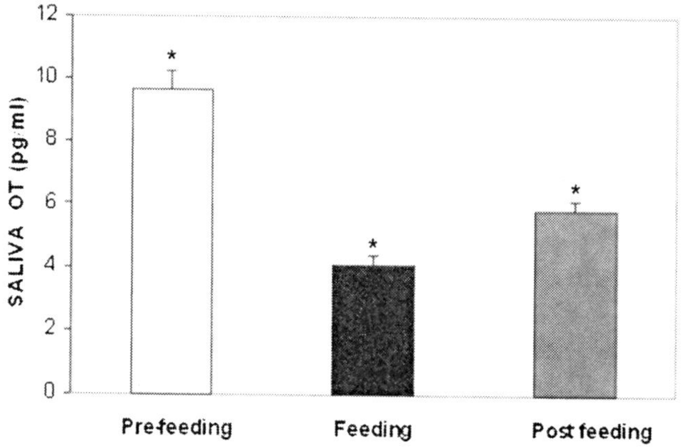

FIGURE 3. In lactating women, salivary oxytocin measured by EIA was elevated 30 min prior to breast feeding, declined during the feeding, and rebounded within 30 min after feeding. $^*P < 0.05$ or higher.

strikingly consistent (overall analysis of variance, $P < 0.001$; FIG. 3). In every set of samples OT was highest prior to breast feeding, declined by the time of the actual breast feeding of the infant and in most women began to rise again 30 min after the feeding ($P < 0.05$ or higher in all paired comparisons). Time of day of saliva collection or repeated collection on two different days did not have a significant effect on either levels of OT or the pattern of change in OT.

In a preliminary study, samples of both blood and saliva were collected before and 30 min after upper body massages in different male subjects ($n = 8$) with three separate replicates for each subject for a total of 48 samples (FIG. 4). In this study reliable, but relatively small increases in OT were detected following massage in both saliva and blood ($P < 0.05$).

One recent published study has reported that meaningful levels of OT could not be measured in saliva by EIA.[24] However, there are several methodological

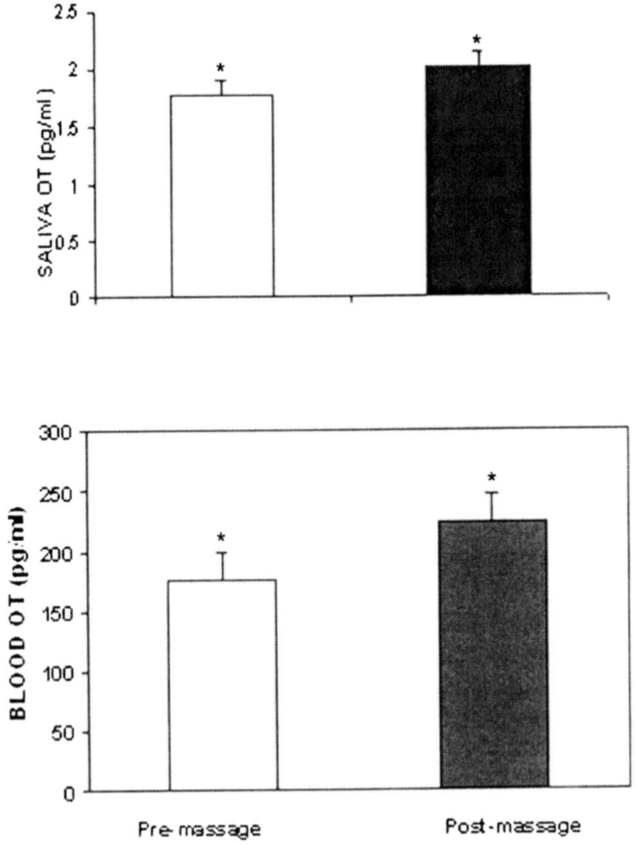

FIGURE 4. Oxytocin measured by EIA increased within 30 min following massage in saliva and blood in men. $^*P < 0.05$.

differences: the most important difference is that we quantitatively concentrate our samples fourfold to produce levels that are within the range of the EIA. Our study also avoided extractions, which can reduce the amount of measurable peptide.

In summary, this article supports the hypothesis that OT is present in human saliva. OT levels in saliva, measured by EIA, varied in a consistent pattern with relevant stimulation associated with lactation and massage. However, the amounts of OT present were low and the methods used here for concentrating samples are labor-intensive. A more sensitive assay is desirable, if salivary OT is to become a readily available "biomarker" for use in studies of human behavior or physiology.

ACKNOWLEDGMENTS

We are grateful to the staff of Assay Designs for consultation in the application of their EIA. This research was sponsored in part by NIH PO1 HD48390, MH072935, and MH 073022 and RR 000167.

REFERENCES

1. CARTER, C.S. 1998. Neuroendocrine perspectives on social attachment and love. Psychoneuroendocrinology **23:** 779–818.
2. CARTER, C.S. 2007. Sex differences in oxytocin and vasopressin: implications for autism spectrum disorders? Behav. Brain Res. **176:** 170–186.
3. UVNAS-MOBERG, K. 1998. Oxytocin may mediate the benefits of positive social interaction and emotions. Psychoneuroendocrinology **23:** 819–835.
4. PORGES, S.W. 2003. The polyvagal theory: phylogenetic contributions to social behavior. Physiol. Behav. **79:** 503–513.
5. GAUTVIK, K.M., L. DE LECEA, V.T. GAUTVIK, *et al.* 1996. Overview of the most prevalent hypothalamus-specific mRNAs, as identified by directional tag PCR subtractions. Proc. Natl. Acad. Sci. USA **93:** 8733–8738.
6. LANDGRAF, R. & I.D. NEUMANN. 2004. Vasopressin and oxytocin release within the brain: a dynamic concept of multiple and variable modes of neuropeptide communication. Front. Neuroendocrinol. **25:** 150–176.
7. ALTEMUS, M., P.A. DEUSTER, E. GALLIVAN, *et al.* 1995. Suppression of hypothalamic-pituitary-adrenal responses to exercise stress in lactating women. J. Clin. Endocrinol. Metabol. **80:** 2954–2959.
8. CARTER, C.S. & M. ALTEMUS. 1997. Integrative functions of lactational hormones in social behavior and stress management. Ann. N.Y. Acad. Sci. **807:** 164–174.
9. HEINRICHS, M., T. BAUMGARTNER, C. KIRSCHBAUM, *et al.* 2003. Social support and oxytocin interact to suppress cortisol and subjective responses to psychosocial stress. Biol. Psychiat. **54:** 1389–1398.
10. KOSFELD, M., M. HEINRICHS, P.J. ZAK, *et al.* 2005. Oxytocin increases trust in humans. Nature **435:** 673–676.

11. KIRSCH, P., C. ESSLINGER, Q. CHEN, *et al.* 2005. Oxytocin modulates neural circuitry for social cognition and fear in humans. J. Neurosci. **25:** 11489–11493.
12. AMICO, J.A., S.M. CHALLINOR & J.L. CAMERON. 1990. Pattern of oxytocin concentrations in the plasma and cerebrospinal fluid of lactating rhesus monkeys (*Macaca mulatta*): evidence for functionally independent oxytocinergic pathways in primates. J. Clin. Endocrinol. Metab. **71:** 1531–1535.
13. WOTJAK, C.T., J. GANSTER, G. KOHL, *et al.* 1998. Dissociated central and peripheral release of vasopressin, but not oxytocin, in response to repeated swim stress: new insights into the secretory capacities of peptidergic neurons. Neuroscience **85:** 1209–1222.
14. KRAMER, K.M., B.C. CUSHING, C.S. CARTER, *et al.* 2004. Sex and species differences in plasma oxytocin using an enzyme immunoassay. Can. J. Zool. **82:** 1194–1200.
15. KUDIELKA, B.M., A. BUSKE-KIRSCHBAUM, D.H. HELLHAMMER, *et al.* 2004. HPA axis responses to laboratory psychosocial stress in healthy elderly adults, younger adults, and children: Impact of age and gender. Psychoneuroendocrinology **29:** 83–98.
16. GRANGER, D.A., E.A. SHIRTCLIFF, A. BOOTH, *et al.* 2004. The "trouble" with salivary testosterone. Psychoneuroendocrinology **29:** 1229–1240.
17. KAVLIGHAN, K.T., D.A. GRANGER, E.B. SCHWARTZ, *et al.* 2004. Quantifying blood leakage into the oral mucosa and its effect on the measurement of cortisol, dehydroepiandrosterone and testosterone in saliva. Horm. Behav. **46:** 39–46.
18. SHIRTCLIFF, E.A., D.A. GRANGER, E. SCHWARTZ, *et al.* 2001. Use of salivary biomarkers in biobehavioral research: cotton-based sample collection methods can interfere with salivary immunoassay results. Psychoneuroendocrinology **26:** 165–173.
19. DEVARAJAN, K. & B. RUSAK. 2004. Oxytocin levels in the plasma and cerebrospinal fluid of male rats: effects of circadian phase, light and stress. Neurosci. Lett. **367:** 144–147.
20. WINSLOW, J.T., P.L. NOBLE, C.K. LYONS, *et al.* 2003. Rearing effects on cerebrospinal fluid oxytocin concentration and social buffering in rhesus monkeys. Neuropsychopharmacology **28:** 910–918.
21. ZAK, P.J., R. KURZBAN & W.T. MATZNER. 2005. Oxytocin is associated with human trustworthiness. Horm. Behav. **48:** 522–527.
22. TAYLOR, S.E., G.C. GONZAGA, L.C. KLEIN, *et al.* 2006. Relation of oxytocin to psychological stress responses and hypothalamic-pituitary-adrenocortical axis activity in older women. Psychosom. Med. **68:** 238–245.
23. FRIES, A.B., T.E. ZIEGLER, J.R. KURIAN, *et al.* 2005. Early experience in humans is associated with changes in neuropeptides critical for regulating social behavior. Proc. Natl. Acad. Sci. USA **102:** 17237–17240.
24. HORVAT-GORDON, M., D.A. GRANGER, E.B. SCHWARTZ, *et al.* 2005. Oxytocin is not a valid biomarker when measured in saliva by immunoassay. Physiol. Behav. **84:** 445–448.

Human Saliva Proteome Analysis

SHEN HU,[a] JOSEPH A. LOO,[d,e,f] AND DAVID T. WONG[a,b,c,d,e]

[a]*School of Dentistry and Dental Research Institute, University of California, Los Angeles, California, 90095, USA*

[b]*Division of Head and Neck Surgery/Otolaryngology, David Geffen School of Medicine, University of California, Los Angeles, California, 90095, USA*

[c]*Henry Samueli School of Engineering, University of California, Los Angeles, California, 90095, USA*

[d]*Jonsson Comprehensive Cancer Center, University of California, Los Angeles, California, 90095, USA*

[e]*Molecular Biology Institute, University of California, Los Angeles, California, 90095, USA*

[f]*Department of Chemistry and Biochemistry, University of California, Los Angeles, California, 90095, USA*

ABSTRACT: Human saliva contains proteins that can be informative for disease detection and surveillance of oral health. Comprehensive analysis and identification of the proteomic content in human whole and ductal saliva is a necessary first step toward the discovery of saliva protein markers for human disease detection. The article will review the recent advances in human saliva proteome analysis, including the efforts of the UCLA saliva proteome consortium funded by the National Institute of Dental and Craniofacial Research (NIDCR). We aim to summarize the proteomics technologies currently used for global analysis of saliva proteins and to elaborate on the application of saliva proteomics to discovery of disease biomarkers, in particular for oral cancer and Sjögren's syndrome, and discuss some of the critical challenges and perspectives for this emerging field. The impact of human saliva proteome analysis in the search for clinically relevant disease biomarkers will be realized through advances made using proteomics technologies.

KEYWORDS: human saliva proteome analysis; whole saliva; parotid; submandibular/sublingual; disease biomarkers; clinical proteomics

OVERVIEW

Human saliva proteome (HSP) analysis refers to the comprehensive identification and quantification of the total proteins and their posttranslational

Address for correspondence: Dr. David Wong, UCLA School of Dentistry, Los Angeles CA, 90095-1668. Voice: 310-206-3048; fax: 310-794-7109.
dtww@ucla.edu

Ann. N.Y. Acad. Sci. 1098: 323–329 (2007). © 2007 New York Academy of Sciences.
doi: 10.1196/annals.1384.015

modifications (PTMs) in human saliva, including their stratified origins from glandular/ductal secretions. Many proteins in human saliva have distinct biological functions, for example, antibacterial activity, lubrication, or digestion, and they help maintain the homeostasis of the oral cavity system. Traditional biochemical and molecular biological approaches, through the study of individual genes and proteins, have elucidated the structures and function of some major salivary proteins. However, since many salivary proteins and their functions remain unknown and uncharacterized, a more thorough investigation of the total protein composition of human saliva is required. A global analysis of saliva will enhance our understanding of oral health and disease pathogenesis and build a solid foundation for the characterization of saliva biomarkers for noninvasive diagnosis of human diseases, such as cancer, diabetes, and autoimmune disorders.

Mass spectrometry (MS) has become one of the core technologies for proteomics on account of its exquisite sensitivity for high-accuracy mass measurement of peptides and proteins that allows for identification of macromolecules. HSP analysis performed using 2D gel electrophoresis (2DGE) coupled with MS typically identifies approximately 100 salivary proteins.[1–9] A significant number of spots on a typical 2DGE of whole saliva (WS) arise as fragments from highly abundant salivary proteins, such as amylases, cystatins, and immunoglobulins.[6,8] It is critical to deplete highly abundant proteins prior to 2-DGE/MS in order to improve identification of lower-abundance salivary proteins. Shotgun proteome analysis based on advanced MS/MS techniques, such as quadruple time-of-flight (QqTOF), matrix-assisted laser desorption/ionization TOF/TOF (MALDI-TOF/TOF), linear ion trap (LIT), and LIT-Orbitrap provide significantly enhanced resolution for identification compared to 2DGE/MS.[7,10–17] In general, prefractionation of intact saliva proteins using high-resolution separation techniques, such as free-flow electrophoresis or capillary isoelectric focusing, is necessary to achieve significantly broad analytical coverage of the HSP.[14,15]

HUMAN SALIVA PROTEOME CONSORTIUM

A national consortium (John Yates, Ph.D., from Scripps Research Institute and Jim Melvin, Ph.D., D.D.S., from the University of Rochester; Susan Fisher, Ph.D., from UCSF; and David Wong, D.M.Sc., DDS, and Joseph Loo, Ph.D., from UCLA) was funded by the National Institute of Dental and Craniofacial Research to investigate the HSP, including cataloguing of the total proteins as well as their structurally modified forms (e.g., glycosylated and phosphorylated).[18] The focus for the three units of the consortium is to determine the proteomes of ductal fluids, parotid and submandibular/sublingual (SM/SL), and to centralize the acquired proteomic data from all three sites (www.hspp.ucla.edu).

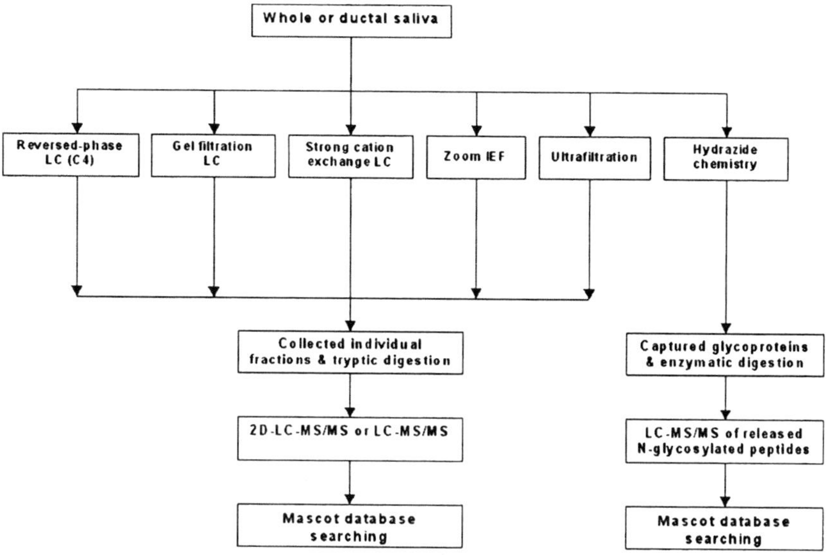

FIGURE 1. Analysis of the human saliva proteome using bottom-up proteomics. Whole or ductal saliva (parotid and SM/SL) proteins were initially fractionated with a variety of separation techniques including reversed-phase LC, strong cation exchange LC, gel filtration LC, ultrafiltration, and Zoom isoelectric focusing (Zoom IEF). The proteins in collected fractions were then digested and identified using 1D or 2D LC-MS/MS and Mascot database searching. Saliva glycoproteins were also identified using hydrazide resin-based pull-down and LC-MS/MS analysis of *N*-glycosylated peptides selectively released from the captured glycoproteins with PNGase F.[19]

The UCLA Saliva Proteome Consortium uses a multiplexed proteomics protocol as diagrammed in FIGURE 1. In general, protein from whole or ductal saliva (parotid and SM/SL) are initially fractionated using either reversed-phase LC, strong cation exchange LC, gel filtration LC, Zoom isoelectric focusing (Zoom IEF), or ultrafiltration separation techniques. Each protein fraction is then digested with trypsin and analyzed by 1D or 2D LC-MS/MS. The acquired MS data are processed and submitted to the Mascot database, which allows searching. We are also cataloguing saliva glycoproteins using a combined hydrazide precipitation release of *N*-glyocoloprotocol based on chemistry[19] and LC-MS/MS analysis of *N*-glycosylated peptides selectively released by PNGase F.

The multiplexed proteomic strategy has clearly enhanced HSP analysis. In total, we have catalogued more than 1,050 proteins in WS. Analysis of parotid and SM/SL saliva is ongoing. In parallel to the centralized database, we have also developed a saliva proteome knowledge base (SPKB) to annotate the acquired saliva proteomic data. The SPKB is accessible to the public for query of the identified proteins, each of which is also linked to public protein databases (e.g., GenBank).

Comparative analysis of HSP and human plasma proteome (HPP) suggests that extracellular proteins are predominant in HSP, whereas the membrane proteins are predominant in HPP. HSP proteins have significant binding and structural molecular activities, whereas the HPP proteins show significant activities of nucleotide/nucleic acid binding. In terms of biological process, a significant percentage of serum proteins is involved in cell cycle or signal transduction, whereas a significant percentage of saliva proteins is involved in physiological or other responses to stimulus processes.

Many salivary proteins (e.g., mucins and amylases) are also glycosylated. Our analysis profiles saliva glycoproteins using a glycoprotein pull-down strategy based on hydrazide chemistry. In this approach, glycoproteins are oxidized and coupled onto a hydrazide resin, and the proteins are digested with trypsin to remove non–glycopeptide-containing regions. Then the formerly *N*-glycosylated peptides are selectively cleaved from the resin by treatment with PNGase F, and are collected and analyzed by LC-MS/MS. This strategy combined with Zoom IEF prefractionation identified *84* *N*-glycosylated peptides from a total of 45 unique *N*-glycoproteins.[20]

Concurrent proteomic and transcriptomic profiles of WS samples from three healthy subjects were tested for coexistence of salivary proteins and their counterpart mRNA. Of the proteins with known function identified in WS, $\geq$ 60% contained their corresponding mRNA transcripts. For genes not detected at both protein and mRNA levels, further efforts were made to determine whether the counterpart was present. Of 19 selected genes detected only at the protein level, the mRNA of 13 (68%) genes was found in saliva after reverse transcriptase-polymerase chain reaction (RT-PCR). This study indicates that the saliva transcriptome may provide insights into the boundary of saliva proteome.[21]

SALIVARY PROTEIN MARKERS FOR ORAL CANCER AND SJÖGREN'S SYNDROME

Human saliva is an attractive medium for disease diagnosis because it is can be obtained using noninvasive techniques. As a clinical tool, saliva has many advantages over serum, including ease of collection, storing, and shipping. For patients, the noninvasive collection procedures dramatically reduce anxiety and discomfort and simplify procurement of repeated samples for monitoring over time.

Compared to blood, saliva may express more sensitive and specific markers for certain local oral diseases. For example, saliva contains locally expressed proteins distinct from serum that may be better indicators of oral disease. There are compelling reasons to use saliva as a diagnostic fluid to monitor the onset and progression of oral cancer. Saliva is the fluid that drains the lesions, and overexpressed RNA and proteins in oral cancer are elevated in saliva. Our studies have clearly demonstrated that saliva contains signature mRNAs and proteins that are diagnostic and predictive of the presence of the disease.[22,23] We have also compared the clinical accuracy of saliva with that of blood by using RNA biomarkers for oral cancer detection. With four serum mRNA markers, a sensitivity and specificity of 91% and 71% was obtained for oral cancer (ROC = 0.88). However, the four salivary mRNA markers had a collective ROC value of 0.95, which demonstrates that for oral cancer detection, salivary transcriptome diagnostics demonstrates a slight advantage over serum.[24]

Sjögren's syndrome (SS) is an autoimmune disease characterized by xerostomia (dry mouth) and xerophthalmia (dry eyes). This complex disease is poorly understood and its diagnosis is hampered by the lack of early diagnostic biomarkers. Salivary protein markers, such as anti-Ro/SS-A, anti-La/SS-B, and anti-α-fodrin, have been demonstrated for diagnosis of SS in a clinical setting.[25,26] However, it is expected that a comprehensive proteomic analysis can reveal a panel of markers that can collectively enhance the sensitivity and specificity for primary SS detection.[27] Using a combined proteomic and genomic approach, we have discovered a set of protein and mRNA signatures that can be informative for detection of primary SS. Although these candidates remain to be validated, the ideal salivary markers would be capable of differentiating SS from other autoimmune diseases. We have also demonstrated that WS is more informative than glandular/ductal saliva (parotid or SM/SL) for generating candidate biomarkers for primary SS detection (Unpublished results).

SUMMARY

HSP analysis is undoubtedly a valuable approach for identifying salivary protein markers for clinical diagnosis of human oral and systemic diseases. Tremendous efforts have been devoted to large-scale identification of proteins in human saliva. However, several issues remain to be addressed. First, we need to standardize the sample collection, preparation, and handling procedures, which are extremely important for subsequent saliva biomarker studies. This requires collaborative efforts and standardized protocols for sample acquisition, including the collection device, collection times, salivary flow rate, saliva protein preservation, and sample preparation.

An interesting dilemma in HSP analysis is how to monitor and compare protein compositions of individual secretions from stratified glands. Such studies will clarify what proteins are common and what are differentially expressed among individual glands. How to maintain the integrity of the stratified parotid, SM and SL saliva samples is crucial to the successful outcome of proteome analysis. Our preliminary analysis of separately collected SM and SL fluids suggested their proteomic profiles are different.[17] Whether we should segregate SM and SL fluid and how to segregate SM and SL fluids remains to be determined.

Secondly, HSP contains a large number of proteins whose concentration differs by an extraordinary dynamic range. Immunoaffinity depletion may be effective to unmask low-abundance proteins in saliva by removing highly abundant proteins, such as amylase and proline-rich proteins. Some salivary proteins are known to be heavily glycosylated and often form homotypic or heterotypic complexes, which requires a special analytical approach. For instance, heavily glycosylated mucins often selectively form heterotypic complexes with amylase, PRP, statherin, and histatins to concentrate those antimicrobial proteins at tissue interfaces for oral protection.[28,29]

Characterization of the normal saliva proteome is a crucial step toward the goal of biomarker discovery. Knowing the protein concentration, PTM state and dynamic range will illustrate how best to design a quantitative MS approach for proteomic profiling. Studying the variables affecting the production of salivary protein content will provide well-characterized clinical samples for analysis. It is important to point out that the robustness and the stability of a normal saliva proteome are unknown. We do not know whether the saliva proteomic patterns are constantly changing as a result of minor physiological events or if they are stable even in the face of major biological events. Despite these challenges, saliva proteomics remains one of the most promising approaches to human disease biomarker identification. We are enthusiastic about the new fast development of novel MS and proteomics technologies and expect good acceptance of saliva for clinical diagnostics.[30] The insights unveiled by HSP analysis will improve patient oral care and public health through better assessment of disease susceptibility, prevention of disease, and monitoring of therapeutics.[7,31]

ACKNOWLEDGMENTS

This work was supported by PHS Grant U01-DE016275.

REFERENCES

1. Ghafouri, B., C. Tagesson & M. Lindahl. 2003. Proteomics **3:** 1003–1015.
2. Yao, Y., E. A. Berg, C. E. Costello, *et al.* 2003. J. Biol. Chem. **278:** 5300–5308.
3. Vitorino, R., M. J. Lobo, A. J. Ferrer-Correira, *et al.* 2004. Proteomics **4:** 1109–1115.
4. Huang, C. M. 2004. Arch. Oral. Biol. **49:** 951–962.
5. Hardt, M., L. R. Thomas, S. E. Dixon, *et al.* 2005. Biochemistry **44:** 2885–2899.
6. Hirtz, C., F. Chevalier, D. Centeno, *et al.* 2005. J. Physiol. Biochem. **61:** 469–480.
7. Hu, S., Y. Xie, P. Ramachandran, *et al.* 2005. Proteomics **5:** 1714–1728.
8. Hirtz, C., F. Chevalier, D. Centeno, *et al.* 2005. Proteomics **5:** 4597–4607.
9. Walz, A., K. Stuhler, A. Wattenberg, *et al.* 2006. Proteomics **6:** 1631–1639.

10. MESSANA, I., T. CABRAS, R. INZITARI, *et al.* 2004. J. Proteome Res. **3:** 792–800.
11. WILMARTH, P. A., M. A. RIVIERE, D. L. RUSTVOLD, *et al.* 2004. J. Proteome Res. **3:** 1017–1023.
12. CASTAGNOLA, M., R. INZITARI, D. V. ROSSETTI, *et al.* 2004. J. Biol. Chem. **279:** 41436–41443.
13. YATES, J. R., D. COCIORVA, L. LIAO & V. ZABROUSKOV. 2006. Anal. Chem. **78:** 493–500.
14. XIE, H., N. L. RHODUS, R. J. GRIFFIN, *et al.* 2005. Mol. Cell. Proteomics **4:** 1826–1830.
15. GUO, T., P. A. RUDNICK, W. WANG, *et al.* 2006. J. Proteome Res. **5:** 1469–1478.
16. HARDT, M., H. E. WITKOWSKA, S. WEBB, *et al.* 2005. Anal. Chem. **77:** 4947–4954.
17. HU, S., P. DENNY, P. DENNY, *et al.* 2004. Int. J. Oncol. **25:** 1423–1430.
18. WONG, D.T. 2006. J. Am. Dent. Assoc. **137:** 313–321.
19. ZHANG, H., X. J. LI, D. B. MARTIN & R. AEBERSOLD. 2003. Nat. Biotechnol. **21:** 660–666.
20. RAMACHANDRAN, P., P. BOONTHEUNG, Y. XIE, *et al.* 2006. J. Proteome Res. **5:** 1493–1503.
21. HU, S., Y. LI, J. H. WANG, *et al.* 2006. J. Dent. Res. **85:** 1129–1133.
22. LI, Y., M. ST. JOHN, X. ZHOU, *et al.* 2004. Clin. Cancer Res. **10:** 8442–8450
23. HU, S., T. YU, Y. XIE, *et al.* 2006. Cancer Genom. Proteomics. In press.
24. LI, Y., D. ELASHOFF, M. OH, *et al.* 2005. J. Clin. Oncol.
25. BEN-CHETRIT, E. F. R., A. RUBINOW. 1993. Clin. Rheumatol. **12:** 471–474.
26. WITTE, T. 2005. Ann. NY Acad. Sci. **1051:** 235–239.
27. RYU, O. H., J. C. ATKINSON, G. T. HOEHN, *et al.* 2006. Rheumatology (Oxford) **45:** 1077–1086
28. IONTCHEVA, I., F. G. OPPENHEIM & R. F. TROXLER. 1997. J. Dent. Res. **76:** 734–743.
29. SOARES, R. V., C. C. SIQUEIRA, L. S. BRUNO, *et al.* 2003. J. Dent. Res. **82:** 471–475.
30. NGUYEN, S. & D. T. WONG. 2006. J. Calif. Dent. Assoc. **34:** 317–322.
31. WONG, D. T. 2006. J. Calif. Dent. Assoc. **34:** 283–285.

Genomics and Proteomics

The Potential Role of Oral Diagnostics

HANS J. TANKE

Department of Molecular Cell Biology, Leiden University Medical Center, Leiden, the Netherlands

ABSTRACT: Advances in genomics and proteomics increasingly contribute to the understanding of signal transduction pathways that control growth, differentiation, and death of cells. Since defects in these processes may result in the expression of inherited and or acquired disease, the identification of candidate disease genes and modifier genes by parallel use of genotyping together with an integrated study of gene expression and metabolite levels is instrumental for future health care. This approach, called systems biology, aims to recognize early onset of disease, institute preventive treatment, and identify new molecular targets for novel drugs in cancer, cardiovascular and metabolomic disease (e.g., diabetes), and neurodegenerative disorders. Gene interaction networks have recently been demonstrated, in which hub genes, that is, genes that show the highest level of interactions with other genes, play a special role. Hub genes, often chromatin regulators, may act as modifier genes (genes that modify the effect of other genes) in multiple mechanistically unrelated genetic diseases in humans. In addition, it has been shown that small metabolites such as hormones and cytokines, or proteins/enzymes such as C reactive protein (C-RP) and matrix metaloproteinase (MMP), reflect disease status in case of oral cancer, asthma, or periodontal and cardiac disease. Many of these molecular targets, as well as pathogen-specific DNA and RNA sequences, can be measured in oral fluids, providing a unique opportunity to develop novel noninvasive diagnostic tests. Efforts so far concentrate on the use of lab-on-a-chip technology in combination with novel reporters and microsensor arrays to measure multianalytes in oral fluids. Handheld devices that perform sensitive detection of multiple analytes in oral fluid will be obtainable in the near future.

KEYWORDS: oral fluid; diagnosis; genomics; proteomics; metabolomics; point-of-care testing; infectious diseases; cancer; diabetes; hub genes

Address for correspondence: Hans J. Tanke, Ph.D., Department of Molecular Cell Biology, Leiden University Medical Center, P.O. Box 9600-Postal Zone S1-P, 2300 RC Leiden, the Netherlands. Voice: 31-71-526-9201/9200; fax: 31-71-526-8270.
h.j.tanke@lumc.nl

Ann. N.Y. Acad. Sci. 1098: 330–334 (2007).
doi: 10.1196/annals.1384.042

GENOMICS, PROTEOMICS, AND METABOLOMICS

Achievements of genomics and the knowledge of the sequence of the human genome have inspired numerous *-omics* disciplines such as proteomics, glycomics, and metabolomics, mainly aiming to understand the signaling pathways that allow cells to divide, differentiate, and die in a controlled manner.[1] A new discipline has been coined called *systems biology* of which the central strategy is devoted to identify candidate disease genes and modifier genes in a unique set of population and clinical cohorts by parallel use of genotyping and an integrated study of gene expression and metabolite levels. Of the latest *-omics* disciplines, it is anticipated that metabolomics will provide essential information about various chronic and multifactorial diseases such as cardiovascular disease, diabetes, and diseases of the central nervous system. In the past decade the sensitivity of bioassays was spectacularly improved.[2] Often in the past, only high-concentration compounds that reflect homeostasis and housekeeping gene products could be measured in body fluids. However, now steroids, neurotransmitters, and trace elements can also be studied because of advances in analytical techniques such as mass spectroscopy,[3] chromatography, electrophoresis, or fluorescence-based detection technologies.

The sequencing of the human genome and many other species will identify major components of the proteome and result in a map of the major metabolites and their function. The key question is how this enormous amount of information will be amalgamated in order to understand the signal transduction pathways that allow cells to function normally, to divide and die.

TOWARD UNDERSTANDING SIGNALING PATHWAYS

Despite the large number of scientific papers published to date on signal transduction pathways, the major pathways are not well understood. This situation exists for several reasons: some of the published results are inevitably incorrect; it is almost certain that all the major regulatory mechanisms have not yet been described (e.g., the recent discovery of the regulatory role of micro RNAs); the majority of papers are mainly descriptive (e.g., "substance A helps B to get C") and lack detailed quantitative kinetic information (e.g., "how much of A and B is needed, how fast is the reaction, and how much C is produced"). In a noteworthy paper "Can a Biologist Fix a Radio? or, What I Learned While Studying Apoptosis," Lazebnik[4] makes a comparison between the views of a biologist and an electrical engineer, in particular with respect to signal transduction. An oversimplified statement of the different views is that cutting a wire in the radio stops the music, whereas that is seldom the case in biology. True quantitative analysis of pathways is essential and currently possible on account of the combined technical progress of advanced analytical instrumentation and computer power, so that analysis of the signaling systems in an integrated approach including extremely complex interconnectivity of the individual systems is possible. As an example, it has become clear that many genetic diseases are very complex in the sense that disease susceptibility is not necessarily caused by a single genetic lesion but often arises from the combined effect of mutation in multiple genes, suggesting genetic interactions.

THE CONCEPT OF HUB GENES

Recently, Lehner *et al.* used RNA interference techniques to systematically study 65,000 pairs of genes in *Caenorhabditis elegans* for their ability to interact and found about 350 genetic interactions between genes in signaling pathways that are mutated in human disease.[5] Note that a large part of the molecular machinery in *C. elegans* is conserved in humans. Interestingly, a subset of genes called "hub genes" showed a high degree of connectivity with other genes. They all encoded chromatin regulators and it was proposed by the authors that these hub genes could act as modifier genes in multiple genetic diseases that are mechanistically unrelated. It is unlikely that this observation and hypothesis are the last ones to occur. It took two decades to understand the role of micro RNAs.[6] Their presence and regulatory role implied the end of a central dogma, namely that noncoding sequences have no function; note that 98% of the transcriptional output in humans is noncoding RNA. Further studies of "junk" DNA, noncoding RNA sequences, and deeper insight in post-translational modifications likely will reveal new regulatory mechanisms. Thus, it leaves biologists with conflicting thoughts: on the one hand there is great excitement about any major discovery since completion of the sequencing of the human genome, while on the other hand there is the frustrating idea that "nothing is certain as yet" and the feeling that we only see the tip of the iceberg, to use an old cliché. In that dynamic uncertainty, it is not easy to identify the gene products and analytes that are diagnostically relevant in case of important diseases. To make the situation even more complex, it has become known that many key molecules have both promoting and repressing function, which implies that a single analyte measurement, without studying its effects on other analytes, may not be meaningful in many cases. Nevertheless, it remains a great challenge to identify the key modifier genes and their products, since that knowledge could drive new diagnostic platforms that recognize early disease onset, allow for the institution of preventive treatment, and identify new molecular mechanisms for novel drug development.

THE ROLE OF ORAL DIAGNOSTICS

The potential role of oral diagnostics has been clearly identified elsewhere. In short, there is convincing evidence that the various kinds of oral fluid contain many of the analytes that reflect normal function or disease status, as is the case in blood, cerebrospinal fluid, or urine. Second, technologies have emerged with high sensitivity that allow many important analytes to be routinely measured. Oral fluids are therefore an ideal target for a new generation of noninvasive diagnostic tools. This was recognized by the National Institutes of Health (NIH) and an NIH/NICDR grant program was started in 2002 called "Development of Technologies for Saliva/Oral Fluid Based Diagnostics." Seven groups were funded by that program and it is of interest to analyze their achievements in a bird's eye approach. In terms of technology, many groups use microfluidics and the latest lab-on-a-chip achievements. Furthermore, surface plasmon resonance (SPR) imaging platforms and novel nanosensors and microsensor arrays were developed, some allowing the measurements of hundreds of analytes. In terms of reporter molecules, electroluminescence and up-converting phosphor technology (UPT) are used next to conventional fluorescent reporters.

The analytes chosen by these seven groups basically cover the entire spectrum from viral/bacterial nucleic acids, including 16 S ribosomal genes, to mRNA transcripts, proteins and small metabolites, including hormones and cations. The targeted diseases are periodontal disease, viral and bacterial infections, renal disease, asthma, cardiac disease, and oral cancer. The analytes measured include C-reactive protein (C-RP), matrix metalloproteinases (MMPs), cytokines/chemokines, and tumor necrosis factor (TNF). Interestingly, the group of cytokines (IL-6, IL-8) and the MMPs are considered indicative of different diseases. For instance, cytokine levels are informative for periodontal disease, asthma, and oral cancer, while MMPs can process many bioactive mediators such as cytokines, growth factors, their receptors, and specific matrix protein anchors for these molecules.[7] MMPs also play a role in development, wound healing, learning, aging, and in diseases such as cancer. The processing of bioactive mediators can lead to profound alterations in cell behavior that result in shedding of cell surface molecules, activation, and inactivation of signaling molecules that occasionally convert agonists to antagonists.

From a conceptual point of view, mechanistically cytokines and MMP function in a manner parallel to the hub genes in that they reflect a complex "biostate" that is altered in case of periodontitis and asthma (both inflammatory diseases) and also in cancer.

STRATEGY AND APPROACH IN ORAL DIAGNOSTICS

With ongoing achievements in proteomics, metabolomics, and new classes of important biomolecules undoubtedly yet to be discovered, paralleled by further improvement of key analytical technologies, a main strategy for the development of oral diagnostic devices can be designed. The three key issues are: (1) performance of measurements of multiple analytes in order to profile the status of a (group) of diseases; (2) utilization of a generic approach in the design of detection platforms that can be easily adapted for the detection of different target molecules; note that this is achievable with easily substituted specific bioreagents such as antibodies and nucleic acid probes; and (3) the need to achieve the highest possible level of sensitivity since levels of target analytes may be orders of magnitude lower in saliva than in other body fluids, and newly discovered analytes are unlikely to occur in high concentration.

Lastly, diagnostics need to anticipate developments in future health care such as personalized medicine, aiming at early prediction and intervention.[8] It also needs to be noted that multinational electronics companies invest in the development of micro devices used for *in vivo* remote sensing. A noninvasive oral fluid test provides considerable advantages compared to a blood test. In Western societies the major target diseases are changing. Next to cancer there are increasing efforts to develop new diagnostic tests and therapeutics for cardiovascular diseases, metabolic disorders such as diabetes, and neurodegenerative disorders such as Alzheimer's and Parkinson's disease. Diagnostics for mental disorders are still in a relatively preliminary phase, despite their enormous socioeconomic impact. In contrast, in developing countries, infectious disease affecting more than a billion people remains the number one target disease and especially in this environment, hand-held oral diagnostic devices that are robust and easy to use at low operational costs can create a major step forward in basic health care.

REFERENCES

1. Collins, F.S. *et al.* 2003. A vision for the future of genomics research. Nature **422:** 835–847.
2. Greef, J. van der, P. Stroobant & R. van der Heijden. 2004. The role of analytical sciences in medical systems biology. Curr. Opin. Chem. Biol. **8:** 559–565.
3. Petricoin, E. *et al.* 2004. Clinical proteomics: revolutionizing disease detection and patient tailoring therapy. J. Proteome. Res. **3:** 209–217.
4. Lazebnik, Y. 2002. Can a biologist fix a radio? Or, what I learned while studying apoptosis. Cancer Cell **2:** 179–182.
5. Lehner, B. *et al.* 2006. Systematic mapping of genetic interactions in *Caenorhabditis elegans* identifies common modifiers of diverse signaling pathways. Nature Gen. **38:** 896–903.
6. Ruvkun, G. *et al.* 2004. The 20 years it took to recognize the importance of tiny RNAs. Cell **116:** 93–96.
7. Mott, J.D. & Werb, Z. 2004. Regulation of matrix biology by metalloproteinases. Curr. Opin. Cell Biol. **16:** 558–564.
8. Weston, A.D. & L. Hood. 2004. Systems biology, proteomics, and the future of health care: towards predictive, preventative, and personalized medicine. J. Proteome Res. **3:** 179–196.

SPR Imaging-Based Salivary Diagnostics System for the Detection of Small Molecule Analytes

ELAIN FU,[a] TIMOTHY CHINOWSKY,[b] KJELL NELSON,[a] KYLE JOHNSTON,[b] THAYNE EDWARDS,[a] KRISTEN HELTON,[a] MICHAEL GROW,[b] JOHN W. MILLER,[c] AND PAUL YAGER[a]

[a]*Department of Bioengineering, University of Washington, Seattle, Washington, USA*

[b]*Department of Electrical Engineering, University of Washington, Seattle, Washington, USA*

[c]*Regional Epilepsy Center and Department of Neurology, University of Washington, Seattle, Washington, USA*

ABSTRACT: Saliva is an underused fluid with considerable promise for biomedical testing. Its potential is particularly great for monitoring small-molecule analytes since these are often present in saliva at concentrations that correlate well with their free levels in blood. We describe the development of a prototype diagnostic device for the rapid detection of the antiepileptic drug (AED) phenytoin in saliva. The multicomponent system includes a hand-portable surface plasmon resonance (SPR) imaging instrument and a disposable microfluidic assay card.

KEYWORDS: SPR imaging; microfluidic assay; saliva; diagnostics

INTRODUCTION

Although the concentration of small molecules in saliva offers a wealth of information about the state of the body, saliva is not routinely used as a sample for biomedical testing. While blood samples are most often used in therapeutic drug monitoring (TDM), saliva is an ideal sample for TDM in outpatients since (i) saliva collection is noninvasive and painless, (ii) samples can be obtained more frequently than would be practical with blood, and (iii) small-molecule drugs in saliva often reflect the free drug concentration in blood.

For over 30 years, there has been significant interest in the use of saliva for TDM of antiepileptic drugs (AEDs).[1-3] Epilepsy affects an estimated 1%

Address for correspondence: Elain Fu, Department of Bioengineering, Box 355061, University of Washington, Seattle, WA 98195. Voice: 206-685-9891; fax: 206-616-3928.
efu@u.washington.edu

Ann. N.Y. Acad. Sci. 1098: 335–344 (2007). © 2007 New York Academy of Sciences.
doi: 10.1196/annals.1384.026

of the population worldwide,[4] and approximately 3 million people in the United States. The goal of pharmacological treatment of epilepsy is to control the patient's seizures without adverse effects. Monitoring of AEDs in blood is an important component of effective therapy based on the characteristics of many of the drugs including (i) toxicity above a narrow therapeutic concentration range, (ii) susceptibility to interactions with certain commonly prescribed drugs including other AEDs, and (iii) significant interindividual variability in pharmacokinetic properties.

Current determination of AED levels typically involves a visit to the physician's office or emergency room for a blood draw. Analysis requires 24 to 48 h when performed at an off-site reference lab, and may cost more than $100. A portable, rapid, and inexpensive therapeutic monitoring system that uses saliva as the sample has the potential to dramatically change clinical practice. The noninvasive nature of saliva collection is compatible with frequent sampling to monitor drug levels during critical time periods, such as the start of a new drug regimen. Rapid measure of salivary drug levels whenever seizures or adverse side effects occur would allow the physician to establish the optimal drug concentration level for an *individual* patient. Rapid determination of the optimal drug regimen for each patient would likely result in a decreased incidence of seizures and adverse side effects.

Phenytoin, the most widely used AED in the United States, is a particularly appropriate candidate drug, with salivary levels equivalent to serum levels.[5] It is a highly protein-bound drug that undergoes saturable hepatic metabolism, so that small variations in dose or compliance can result in larger changes in serum phenytoin concentration.[6] Thus, episodes of acute phenytoin toxicity with dizziness, ataxia, and nystagmus occur commonly. The complex, nonlinear pharmacokinetics of phenytoin makes measurement of phenytoin levels particularly important in clinical practice.

Surface plasmon resonance (SPR) imaging is well suited for use in a system for the TDM of AEDs for several reasons. Detection is based on measurement of changes in refractive index (RI) near the sensing surface, and so does not require labeling of the target substances. SPR imaging immunoassays are capable of rapidly detecting many small-molecule drugs at their therapeutic concentrations using simple optics that can be miniaturized for point-of-care diagnostics. Using spatially addressed capture arrays, SPR imaging can detect multiple types of binding events simultaneously, and incorporate multiple "reference" surfaces that can be used to correct measurements for interference from background signals (e.g., nonspecific binding to the SPR surface by interferants present in complex biological samples).

In this article, we describe progress toward the development of a prototype salivary diagnostic system for the detection of the therapeutic drug phenytoin. In order to achieve high performance with low per-test costs, our approach is to combine a permanent SPR imaging instrument (and fluid workstation) with a disposable card that incorporates the majority of the sample-contacting elements.

SPR Imaging-Based Reader

The SPR imaging-based reader[7] is designed for mechanical simplicity, compactness, ruggedness, and low-cost production. The instrument uses a near-infrared light-emitting diode (LED) source and stationary wide-field imaging optics. Instrument response can be optimized for a particular sample by small-range translation of the source. A compact liquid crystal polarizer (Bolder Vision Optik, Boulder, CO, USA) is used to electronically switch the source between transverse magnetic (TM, mode that supports surface plasmons) and transverse electric (TE, mode used for normalization) polarizations. Folding of the optical path is a key factor in the miniaturization of the optical module (approximately $25 \times 10 \times 5$ cm^3). A prototype hand-portable SPR imaging instrument is shown in FIGURE 1A.

The image detector is a 1/3″ CCD with 640×480 pixels chosen for a balance of cost and performance. The primary requirements for the detector are (i) fast readout to enable improvement of image statistics through averaging and (ii) low background noise when operating at high light levels. The CCD is integrated into a camera (Vision Components, Ettlingen, Germany) that contains a powerful digital signal processor. Images may be acquired and summed at 30 frames per second.

The custom-coded user software, created using LabWindows (National Instruments, Austin, TX, USA), allows for the integrated control of data acquisition functions (LED translation, polarization switching, image acquisition) and fluid motion (valve and pump actuation). The data analysis software was created using MATLAB (Mathworks, Natick, MA, USA). These high-level instrument functions are implemented on a tablet-style PC (Motion Computing, Austin, TX, USA) that provides ample processing power and a high-resolution display in a compact, lightweight form.

The performance of an SPR imaging instrument can be quantified in terms of the RI resolution and lateral resolution over the field of view. A RI resolution of 6×10^{-6} RIU[7] for a 300-pixel average was determined from noise measurements. Image resolution tests indicate a spatial resolution of less than 100 μm.[7] This spatial resolution is demonstrated in FIGURE 1B.

An additional factor that affects the lateral resolution is the quality of focus over the instrument field of view. The object (i.e., the sensing surface) of an SPR imaging instrument is tilted relative to the optical axis of the imaging optics, such that much of the object is out of focus when using standard imaging optics with the detector normal to the optical axis of the lenses. One potential solution is the use of a tilted image plane as described by the Scheimpflug

(A)

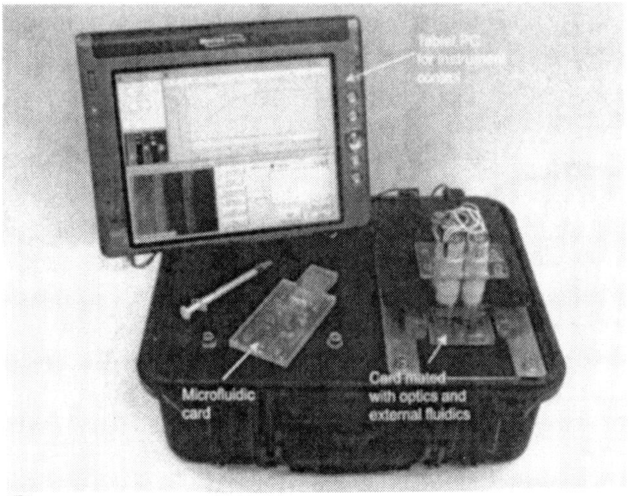

(B)

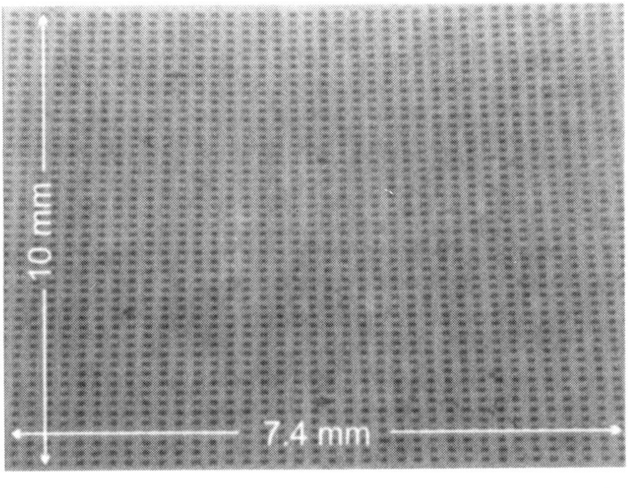

FIGURE 1. A prototype hand-portable SPR imaging instrument. (**A**) The instrument response is optimized using small-range translation of a near-infrared LED source. A compact liquid crystal polarizer is used to electronically switch the source polarization between TM and TE. Folding of the optical path is a key factor in the miniaturization of the optical module (approximately 25 × 10 × 5 cm^3) of the instrument. (**B**) SPR image of water-immersed PDMS stamps, fabricated with approximately 100-μm features, over the instrument field of view. (From Chinowsky *et al.*[7] Reproduced by permission.)

condition.[8] Implementation of this condition resulted in a significant improvement in the quality of focus over the 10 mm × 7.4 mm instrument field of view.

SPR Imaging Assay

The SPR imaging assay developed for the detection of the small-molecule drug phenytoin is a standard microfluidic indirect immunoassay.[7,9] To conduct this assay, the sample containing an unknown concentration of phenytoin is mixed with a known amount of antiphenytoin antibody before introduction to the detection zone containing surface-immobilized phenytoin. Only the antibody with free binding sites may bind to the surface-immobilized phenytoin. This surface binding produces a change in RI that is quantifiable by SPR imaging.

In the SPR imaging assay shown in FIGURE 2A, three samples of phenytoin in buffer (plus antiphenytoin) are measured simultaneously using controlled parallel flows.[10] The independence of the measurements is ensured by laminar flow (low Reynolds number) with negligible diffusion between the streams (high Peclet number). The single-channel/multiple-inlet design has the advantage of eliminating the potential for variable fluid resistances in separate channels (which would complicate comparisons between streams since flow rate affects the rate of antibody binding). Either on-board positive and negative controls or multipoint calibration may be integrated into the single-channel format by incorporating reference streams alongside the sample stream.

The SPR signals of FIGURE 2B and C (phenytoin in buffer and phenytoin in saliva, respectively) indicate that both the rate of binding and the total coverage of antiphenytoin are inversely correlated with the concentration of phenytoin,[10] as expected. Given that the low end of the therapeutic concentration of phenytoin in saliva is 200 nM,[11] the detection of 50 nM phenytoin provides a comfortable limit of detection margin for sample dilution due to potentially necessary preconditioning of saliva. The data indicate a rapid assay time of less than 5 min. Multiple analytes can be simultaneously detected by patterning different capture zones within a given fluid stream.[12]

Preconditioning of Complex Saliva Samples

Human saliva is a complex biological sample containing substances that can interfere with the detection of specific surface binding events and therefore complicate interpretation of SPR-based assays. For example, interactions between the analyte and substances in the sample may affect the transport or binding of the analyte to its antibody. Additionally, the nonspecific adsorption of substances in the sample to the detection zone may significantly interfere with targeted binding events and create a background signal. To mitigate these interfering effects, we use the following saliva collection and preconditioning method: Each donor is instructed to abstain from eating and drinking (other than water) for 1 h prior to saliva collection. The donor mouth is rinsed with water, paraffin wax is placed on the tongue, and saliva allowed to pool in the mouth. Approximately 2 mL of stimulated whole human saliva (WHS) per donor is collected in a screw-cap vial. The WHS samples are used the same day or are frozen for later use.

Following saliva collection, the sample is treated to remove substances that may interfere with the assay. Preconditioning of the WHS sample[10] is performed by two devices: (i) a 0.2-μm-pore conventional polymeric filter (Whatman, Middlesex, UK) and (ii) a microfluidic diffusion-based separations device, the flat H-filter.[13,14] TABLE 1 summarizes the effectiveness of each pretreatment step for an initial sample of WHS spiked with the small-molecule cortisol. The conventional filter removes debris, cells, and 73% of the glycoprotein/mucin content from the WHS sample, reducing the sample viscosity to nearly that of buffer. This reduction in sample viscosity is necessary for H-filter extraction. The combination of conventional filter and H-filter

(A)

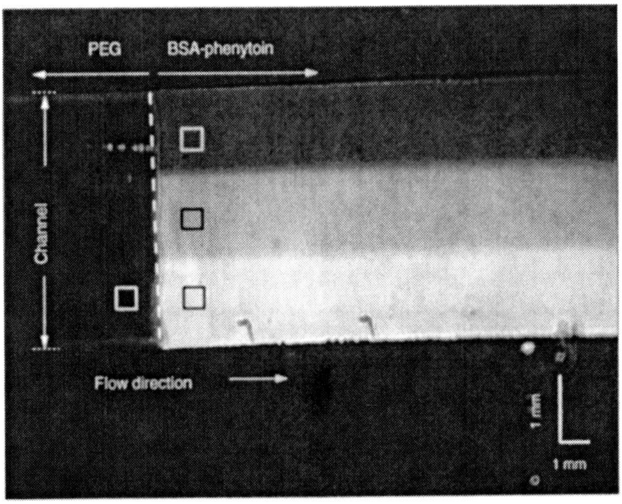

FIGURE 2. Parallel indirect immunoassays for the therapeutic drug phenytoin conducted using multiple flows. (From Yager *et al.*[10] Reproduced with permission.) (**A**) The SPR difference image shows the outcome of antiphenytoin binding to the surface from samples containing 0, 50, or 100 nM phenytoin in phosphate buffer premixed with 150 nM antiphenytoin after 5 min. The channel dimensions (62 μm deep, 3.6 mm wide) and flow rates (750 nL/sec total, 250 nL/sec/flow stream) used in this assay ensured laminar flow with negligible diffusion between the streams. (**B**) The averaged intensities calculated over the areas shown in FIGURE 2A demonstrate that both the rate of binding and the total coverage of the antibody anticorrelate with the amount of analyte in the sample. The detection zone consisted of bovine serum albumin (BSA)-phenytoin conjugate nonspecifically adsorbed to the gold layer and the region upstream of the detection zone was functionalized with a nonfouling layer of polyethyleneglycol-thiol. (**C**) Assay results for phenytoin spiked into preconditioned saliva. These results were obtained on a previous generation SPR imaging instrument.

(B)

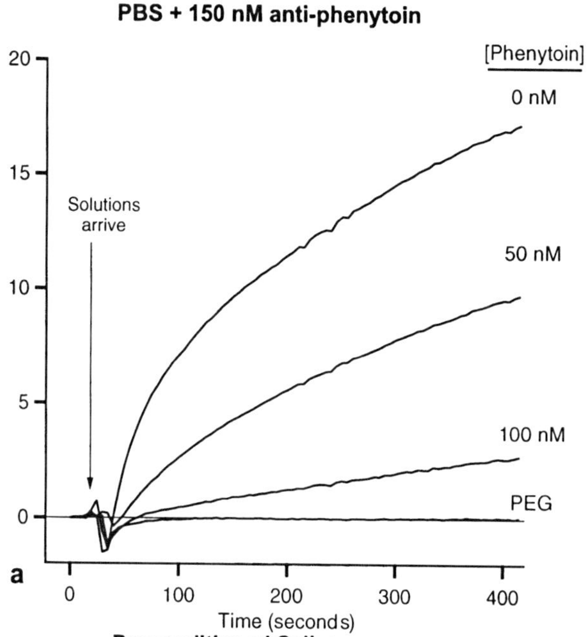

(C)

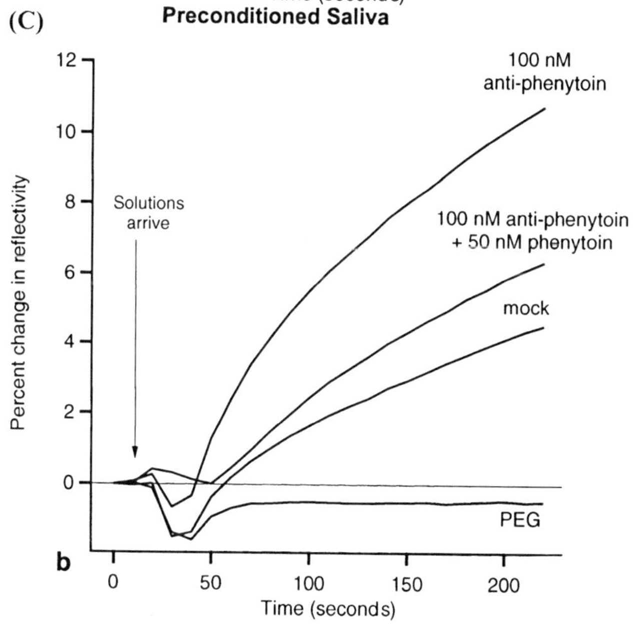

FIGURE 2. Continued.

removes 98% of the glycoprotein/mucin content and 92% of the protein content, while retaining 27% of the small-molecule analytes.[10]

TABLE 1. Composition of unprocessed and processed saliva

	Unprocessed WHS	WHS processed by conventional filter	WHS processed by conventional filter and H-filter
Glycoprotein/mucin	100%	27%	2%
Total protein	100%	53%	8%
Analyte (cortisol)	100%	92%	27%

NOTE: The H-filter separation was performed at a total flowrate of 2000 nL/s. The glycoprotein/mucin content was estimated using the Periodic acid-Schiff assay, the total protein content was quantified using the BCA Protein Assay Kit (Pierce, Rockford, IL, USA), and the cortisol content was quantified using an ELISA (Diagnostic Systems Laboratories, Webster, TX, USA). Reproduced with permission from Yager *et al.* NPG, 2006.

Disposable Diagnostic Card

In order to achieve high performance with low per-test cost, our approach for the salivary diagnostic system is to combine a permanent SPR imaging instrument (and fluid workstation) with a single-use disposable card. The disposable card contains many of the sample-contacting elements and eliminates the need for an extensive cleaning protocol between samples to prevent carry-over, thus simplifying routine use of the system.

Many of the steps necessary for analysis of a saliva sample (subsequent to mechanical filtration) have been combined onto the disposable diagnostic card[10] of FIGURE 3. Twelve layers of laser-cut Mylar™ are used to form the laminate that comprises the H-filter and most of the other structures for fluid manipulation on the card. The mixer, which requires components approximately 100 μm in size, is composed of polydimethylsiloxane (PDMS) and was created using standard SU-8 lithography and PDMS molding protocols. The immunoassay surface is a gold-coated glass substrate created using electron-beam evaporation. The gold-coated substrate is subsequently functionalized with the appropriate surface chemistry.

A set of six syringe pumps (Kloehn V6 Syringe Pumps, Las Vegas, NV, USA) drives fluids on the card and draws buffer from a common source. In the current design, the antibody solution sample loop and the reference sample loops are accessed through injection valves (Upchurch Scientific, Oak Harbor, WA, USA). The card is attached to a manifold, which also houses four 3-way valves (Lee Products Ltd., UK) for directing flow on the card. Testing of the card design and the development of on-card valves and miniature pumps are under way.

SUMMARY

We have described progress toward the development of a prototype diagnostic device for the rapid detection of the antiepileptic drug phenytoin in saliva.

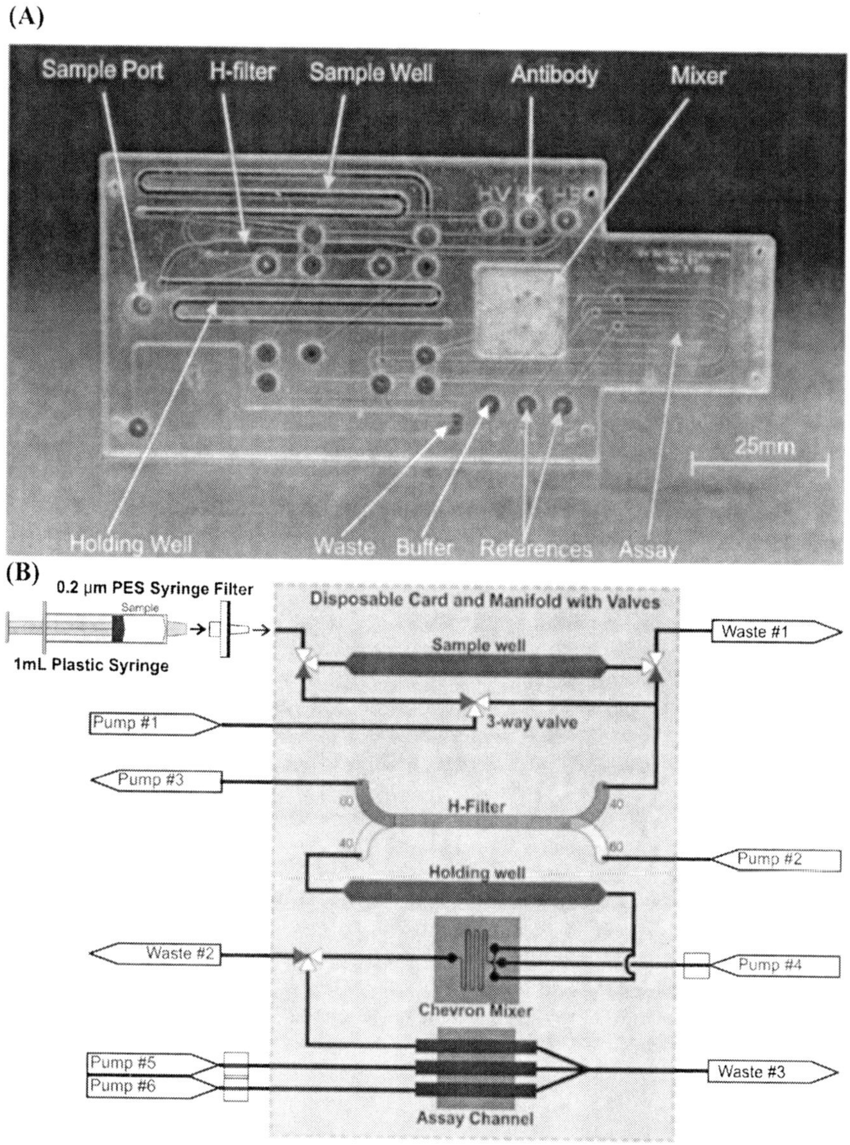

FIGURE 3. Disposable diagnostic card. (From Yager *et al.*[10] Reproduced with permission.) (**A**) Photograph of a card composed of polymeric laminate, PDMS, and gold-coated glass. (**B**) Schematic showing the fluidic processes on the disposable card as well as the off-card processes. The 3-way valves are located off-chip in this current design.

The system includes a hand-portable SPR imaging instrument and a disposable microfluidic assay card. This SPR imaging–based diagnostics platform will be generally useful for monitoring small-molecule analytes in saliva.

ACKNOWLEDGMENTS

We gratefully acknowledge the financial support of NIDCR Grant UO1-DE014971 and NSF Graduate Research Fellowship DGE0203031 (to K.H.).

REFERENCES

1. KNOTT, C. & F. REYNOLDS. 1984. The place of saliva in antiepileptic drug-monitoring. Ther. Drug Monit. **6:** 35–41.
2. HERKES, G.K. & M.J. EADIE. 1990. Possible roles for frequent salivary antiepileptic drug-monitoring in the management of epilepsy. Epilep. Res. **6:** 146–154.
3. DROBITCH, R.K. & C.K. SVENSSON. 1992. Therapeutic drug-monitoring in saliva—an update. Clin. Pharmacokinetics **23:** 365–379.
4. NEELS, H.M. *et al.* 2004. Therapeutic drug monitoring of old and newer anti-epileptic drugs. Clin. Chem. Lab. Med. **42:** 1228–1255.
5. TROUPIN, A.S. & P. FRIEL. 1975. Anticonvulsant level in saliva, serum, and cerebrospinal fluid. Epilepsia **16:** 223–277.
6. BROWNE, T. & B. LEDUC. 2002. Phenytoin and other hydantoins: chemistry and biotransformation. *In* Antiepileptic Drugs. R. Levy, *et al.*, Eds.: 565–580. Lippincott Williams & Williams. Philadelphia.
7. CHINOWSKY, T. *et al.* 2007. Compact surface plasmon resonance imaging system. Biosens. Bioelect. In press.
8. SMITH, W. 2000. Modern Optical Engineering. McGraw-Hill. New York.
9. NELSON, K.E. *et al.* 2006. Rapid, multiplexed competitive immunoassays using disposable microfluidic devices and SPR imaging. Micro Total Anal. Sys. **1:** 825–827.
10. YAGER, P. *et al.* 2006. Microfluidic diagnostic technologies for global public health. Nature **442:** 412–418.
11. LIU, H. & M.R. DELGADO. 1999. Therapeutic drug concentration monitoring using saliva samples—focus on anticonvulsants. Clin. Pharmacokinetics **36:** 453–470.
12. NELSON, K.E. *et al.* 2005. Rapid, parallel-throughput, multiple analyte immunoassays with on-board controls on an inexpensive, disposable microfluidic device. Micro Total Anal. Sys. **2:** 1000.
13. BRODY, J. & P. YAGER. 1997. Diffusion-based extraction in a microfabricated device. Sens. Act. A. **58:** 13–18.
14. BRODY, J. *et al.* 1996. A planar microfabricated fluid filter. Sens. Act. A. **54:** 704–708.

Saliva-Based Diagnostics Using 16S rRNA Microarrays and Microfluidics

E. MICHELLE STARKE,[a] JAMES C. SMOOT,[a] JER-HORNG WU,[b] WEN-TSO LIU,[b] DARRELL CHANDLER,[c,d] AND DAVID A. STAHL[a]

[a]*Civil and Environmental Engineering, University of Washington, Seattle, Washington 98195, USA*

[b]*Environmental Science and Engineering, National University of Singapore, 9 Engineering Drive 1, EA-07-23, Singapore 117576, Singapore*

[c]*Argonne National Laboratory, 9700 S. Cass Avenue, Argonne Illinois 60439, USA*

[d]*Akonni Biosystems, Inc., 9702 Woodfield Court, New Market, Maryland 21774, USA*

ABSTRACT: The development of a diagnostic system based on DNA microarrays for rapid identification and enumeration of microbial species in the oral cavity is described. This system uses gel-based microarrays with immobilized probes designed within a phylogenetic framework that provides for comprehensive microbial monitoring. Understanding the community structure in the oral cavity is a necessary foundation on which to understand the breadth and depth of different microbial communities in the oral cavity and their role in acute and systemic disease. Our ultimate goal is to develop a diagnostic device to identify individuals at high risk for oral disease, and thereby reduce its prevalence and therefore the economic burden associated with treatment. This article discusses recent improvements of our system in reducing diffusional constraints in order to provide more rapid and accurate measurements of the microbial composition of saliva.

KEYWORDS: microarrays; saliva; microbial ecology; microfluidics

MICROBIAL COMMUNITIES IN THE ORAL CAVITY

Oral health affects the quality of life of all individuals. In developed countries, such as the United States, dental caries affects 60–90% of school-aged

Address for correspondence: David A. Stahl, Ph.D., Civil and Environmental Engineering, 302 More Hall, Box 352700, University of Washington, Seattle, Washington 98195-2700. Voice: 206-685-3464; fax: 206-685-9185.
dastahl@u.washington.edu

**Ann. N.Y. Acad. Sci. 1098: 345–361 (2007). © 2007 New York Academy of Sciences.
doi: 10.1196/annals.1384.007**

children and a vast number of adults.[1] Dental caries is the most common chronic disease of childhood, five times greater than asthma and seven times greater than hay fever.[2] In adults aged 45 to 54 years and those aged 65 to 74 years, 14% and 23%, respectively, will experience severe periodontal disease.[2] Furthermore, caries onset and other oral diseases disproportionately affect those of lower economic status and minority groups. The far-reaching importance of oral health is also highlighted by the association of chronic oral infections with diabetes, heart and lung disease, and adverse pregnancy outcomes (recently reviewed by Scannapieco[3]).

Oral health is in large part the result of complex interactions between resident microbiota and host responses that, for the most part, go unnoticed. It is only when disease develops that the crucial role of bacteria and other microbiota is evident. Bacteria, such as *Streptococcus mutans* and *Lactobacillus* species, are associated with caries onset,[4,5] while *Actinobacillus actinomycetemcomitans* and *Porphyromonas gingivalis* are associated with periodontal disease.[6,7] However, the etiology of these diseases is complex and the result of many factors, including microbe–microbe interactions, diet, the immune status, and the genetic background of the individual. Given the importance of oral health, it is surprising that little is known about the dynamism of microbial communities that reside in the oral cavity.

Understanding the oral microbial community can be viewed as a microbial ecology problem whose questions are aimed at discovering which species are present, at what abundance, and how the community structure changes when various factors (e.g., diet, health, lifestyle) are altered. Most importantly, identifying the species present, and in what abundance, are the first steps toward initiating epidemiological studies to identify correlates and causal links to disease, both oral and systemic. Ribosomal sequences, such as the 16S ribosomal RNA (rRNA) gene in bacteria, are among the most useful markers of identification. They provide phylogenetic resolution that correlates with taxonomic rank. Current tools to identify all the 16S rRNA gene sequences present in a sample are informative, particularly sequencing 16S rRNA gene clone libraries. The results of these and other types of molecular studies estimate that the oral diversity is greater than 700 species, although the number of abundant species is much lower in a single individual.[8–10] However, clone library studies are laborious, expensive, and time-consuming. Alternatives, such as using terminal-restriction fragment length polymorphism (tRFLP)[11] or denaturing gradient gel electrophoresis (DGGE)[12,13] have greater throughput, but do not provide unambiguous species-level identification. Moreover, all of these techniques require enzymatic amplification of the 16S rRNA gene, a process that has been shown to skew measurements of abundance and lead to incorrect conclusions about the diversity of the bacterial populations.[14,15] An ideal method would combine the explicit identification of microbial species typically gained from clone library studies with direct detection methods and high-throughput capability.

HIGH-THROUGHPUT MICROBIAL IDENTIFICATION

Microarrays are well suited to bridge researchers' need for high-throughput assays and direct detection methods. Because a single microarray can comprise hundreds to hundreds of thousands of unique oligonucleotide probes, a single experiment can survey the 16S rRNA gene signatures present in a sample in a very short time (minutes to hours). To assist in the direct detection of microbes, microarrays can be designed to detect the naturally amplified 16S rRNA, which is estimated to have an average of about 20,000 copies in an actively growing bacterium. Because of its high abundance in the cell, no enzymatic amplification step is needed. In a typical experimental procedure, the rRNA is extracted and purified from a clinical sample and labeled with a fluorophore. The labeled rRNA, now referred to as target material, is subsequently applied to a microarray and allowed to hybridize at room temperature for several hours or overnight, after which the unbound target is washed away. The remaining target is detected using a CCD camera and the appropriate filters to detect the fluorophore. The measured signal intensity reflects the amount of bound target to each spot on the microarray and is interpreted to assess the microbial content in the original sample.

MICROARRAY TECHNOLOGY

To resolve the identity of microbial populations, the probes on the microarray are designed to take advantage of the conserved features of the 16S rRNA gene.[16,17] Probes complementary to the more conserved regions identify species in a large phylogenetic group, each group generally corresponding to a higher taxon (e.g., domain, phylum, class, order, or family). Probes complementary to more variable regions distinguish different genera and species. This design has several purposes: foremost, each species is identified by multiple probes, each identifying the organism at a different taxonomic (phylogenetic) rank, thus enabling explicit identification of the species present. Secondly, they provide redundancy that allows for greater confidence in determining which species are present. FIGURE 1 demonstrates the phylogenetic nesting of the probes and, in cartoon fashion, how a simple microarray pattern would be interpreted.

While most commercial phylogenetic microarrays are planar, that is, the oligonucleotide probes are fixed directly on a glass slide or silicon surface,[18] there are distinct advantages in using gel-based arrays. The gel-based arrays comprise three-dimensional pads or elements of polyacrylamide deposited on glass slides.[19,20] The oligonucleotide probes, approximately 20 nucleotides in length, are immobilized within the gel matrix. Therefore, the probe concentration can be several orders of magnitude greater than planar arrays,[21] permitting the detection of low-abundance targets—an essential feature for direct

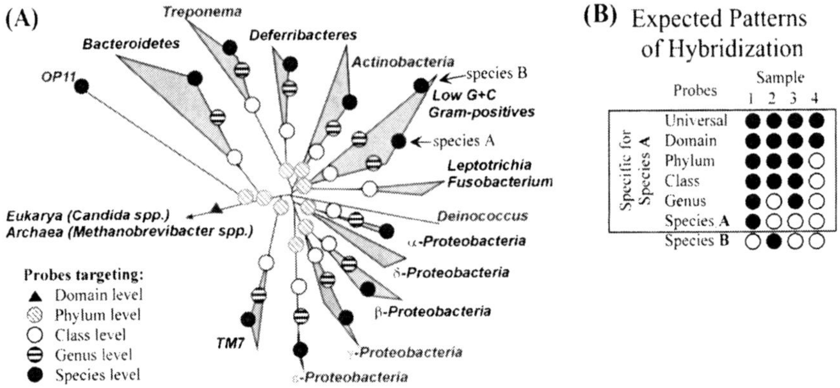

FIGURE 1. Probe design using 16S gene sequence. **(A)** A phylogenetic tree indicating the three domains (Archaea, Eukarya, and Bacteria [*black triangle*]), phylum (*diagonal hatch circles*), class (*open circles*), genus (*horizontal hatch circles*), and species-level probes (*filled circles*). Hypothetical species "A" and "B" are indicated. **(B)** The expected patterns of hybridizations when samples 1–4 are applied to a microarray. Sample 1 hybridizes to all probes specific for species A, indicating the presence of species A; sample 2, containing species B, hybridizes to the same universal, domain, phylum and class probes as species A because they share similar lineage. Sample 3 contains species C, which is in the same genera as species A, but there is no species-specific probe for species C. Sample 4 contains bacteria that is in a different phylum from species A, B, and C.

detection of low-abundance microorganisms. Moreover, as the gel matrix is mostly water, each element has the functional characteristics of an aqueous solution. As will be discussed in this review, this property facilitates melting analysis of each probe on the array. Another distinct advantage of gel-based arrays is their reusability, which reduces both cost and array-to-array variation.

All microarray experiments must resolve nonspecific hybridization from specific hybridization. One commonly used approach to estimate the amount of nonspecific hybridization employs mismatch probes that vary in sequence at a single nucleotide position relative to perfect-match probes. Because the two probes differ by a single nucleotide, the level of signal on the mismatch probe is thought to be comparable to the nonspecific binding of targets to the perfect-match probe. Another useful method to assist in distinguishing perfect-match hybridizations from closely related hybridizations is the use of thermal disassociation under nonequilibrium conditions.[22–27] Disassociation is controlled by increasing the temperature of the array incrementally using a Peltier-controlled thermal table and measuring the fluorescent signal at each temperature using a CCD camera until there is no signal remaining. The advantage of this approach is that it can independently measure the stability of the duplexes that form with each probe. Unstable duplexes, such as those with internal mismatches, will disassociate at lower temperatures than the more stable duplexes formed between perfectly matched probe and target. Because each element comprises mostly water, the melt analysis of duplexes is similar to what occurs in solution alone.[21] Melting profiles, graphs of signal intensity versus temperature, are sigmoid curves, from which metrics, such as T_d (the temperature at which 50% of the initial signal intensity remains) can be calculated and evaluated.

In this article we address one challenge of microarray experiments—diffusion-limited kinetics of target molecules—and present recent progress as part of ongoing efforts to develop a system for rapid identification and enumeration of the microbial species present in the oral cavity. To reduce the effects of diffusion in the microarray platform, we have explored several avenues, including introducing flow over the microarrays, modifying the target material, and modifying the gel elements themselves, all of which are described below.

THE DIFFUSION CHALLENGE

Hybridization to microarrays is diffusion-rate limited. Estimates of the diffusion coefficient range between 10^{-6} and 10^{-7} cm^2/sec for nucleic acid molecules, indicating that individual molecules move just a fraction of the volume in the hybridization chamber during a 24-h hybridization.[28–31] Attaining equilibrium with such diffusional constraints can take quite a long time, often beyond the feasible limits of the experiment and far beyond the requirements for a useful diagnostic device. Furthermore, the time to hybridization equilibrium is different for each probe. The higher the binding constant between the probe and target, the more time is required to reach equilibrium. Because mismatch targets have lower-binding constants, they reach equilibrium faster than perfect-match targets.[31] Thus, if insufficient time has elapsed for the perfect-match targets to reach equilibrium, the ratio of signal intensity values of perfect-match and mismatch will be skewed in favor of mismatch probes and may lead toward an underestimation of abundance.[32,33]

Another challenge with clinical samples is that the abundance of different targets will vary within a sample. A simulation study comparing the time to equilibrium for abundant, intermediate, and rare targets within a sample demonstrated that the abundant targets reached equilibrium within the typical time frame of an experimental protocol (overnight); whereas, rare targets never reached equilibrium, even after 24 h.[31] This phenomenon is particularly important in environmental and clinical samples where not only are the bacterial species unknown, but the abundance of each species is unknown. The probes most affected are the species-specific probes, which are specific for only a small fraction of the total microbial population. Because their cognate targets will be less abundant, and indeed rare relative to targets present in all bacteria, the microarray results will most likely underestimate their abundance.

An added challenge with gel-based microarrays is the effect of diffusion within the gel element itself. Described as "retarded diffusion," this process is the result of repeated duplex formation between the probe and the target nucleic acid molecules.[34] In order to penetrate the gel element, the target first binds to probe molecules on the surface of the element, then disassociates and diffuses inward to bind to a second probe, from which the target molecule must again disassociate and diffuse further inward to bind to a third probe molecule and so on. This process of association and disassociation reduces the diffusion rate (and the time to equilibrium) within the gel element and can present a challenge when interpreting signal intensity values. Diffusion, whether in solution or in gel elements, has been a major challenge in many microarray experiments. Any enhancements that hasten the diffusion of target are desirable to not only reduce the time to equilibrium and improve perfect-match/mismatch discrimination, but also to improve signal:noise ratio and reduce pad-to-pad variability.

ACCELERATE TARGET MOVEMENT

Several investigations have accelerated target movement and overcome the limits of diffusion in solution by various methods, such as applying electric fields,[35] acoustic waves,[36] inducing cavitation,[30] or forcing fluid motion.[28,29,37,38] Microfluidic devices are uniquely well suited for use in microarray technology, particularly because they can mix small volumes of solution and have the additional advantage that they can shorten analysis times and have the potential to lower operation and manufacturing costs. Several groups have reported marrying microarray and microfluidic technologies.[28,29,36,38,39] However, these previously reported technologies are not practical for thermal dissociation studies, which require real-time imaging of the fluorescent signal on the microarray to create melting profiles.

Initial adaptations of microfluidic devices for gel-based microarrays have been successful.[40] A prototype microfluidic card was manufactured using Mylar sandwiched between the microarray and a glass slide. Inlet and outlet ports were made at a 45° angle to the array and connected to a peristaltic pump (FIG. 2, panel A). Target material was then flowed over the surface of the microarray, and the increase in fluorescent signal was monitored in real time at room temperature using a custom-designed bench-top microarray imager. Using this system, the rate of hybridization is two times faster when compared to the static approach. In addition, the introduction of fluid recirculation also resulted in a 40% increase in signal intensity over static conditions at the end of the hybridization. Thus, not only does fluid movement (recirculation) result in shorter hybridization times, but it also results in higher signal intensity—an attribute that potentially may enhance our sensitivity for low-abundance species. An advantage of following hybridization in real time is that it allows

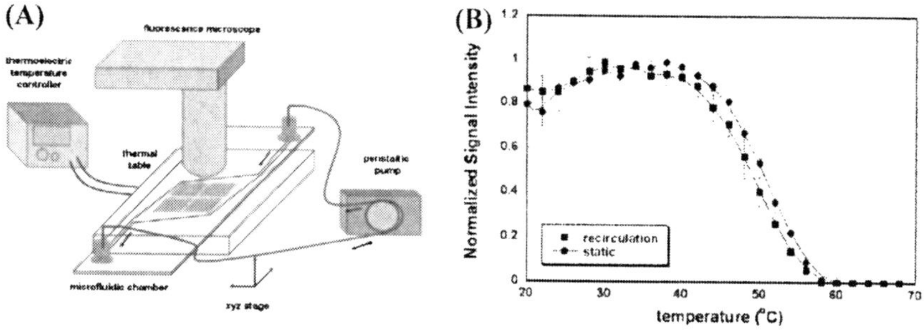

FIGURE 2. **(A)** Microfluidic device containing a gel-based microarray. The diagram demonstrates fluid recirculation over the surface of a microarray at a 45° angle. The microfluidic card is engineered to allow real-time imaging of the microarray and is manufactured to rest upon a thermal table to regulate temperature. **(B)** The melting profile of a perfect-match probe under conditions of recirculation using the microfluidic device and under static conditions without recirculation.[40] (From Lee *et al.*[40] Reproduced by permission of The Royal Society of Chemistry.)

the collection of information on the rate of hybridization for each probe. This information may be an additional tool to use in distinguishing closely related targets from perfect-match targets.

Because this microfluidic system reduces diffusion limitations, it is also valuable during thermal disassociation analyses. As the temperature increases on the array, the target molecules disassociate more readily from probes. However, in static experiments, it is difficult to distinguish recently unbound target from target that remains bound to probe because the rates of diffusion away from the gel elements are slow. By introducing active flow to remove unbound target from the vicinity of the element, we find that the resultant melting profiles shift to a lower melting temperature (FIG. 2, panel B).[40] We anticipate that continued technological improvements, such as the microfluidic device described by Lee *et al.* will continue to reduce the limitations imposed by diffusion and reduce the time for microarrays to reach equilibrium.

IMPROVING THE DIFFUSIVITY OF THE TARGET

The characteristics of the target molecule are also important factors that influence the diffusion rate, both in solution and in the gel elements. We have explored the effect of length of the target RNA on initial signal intensities and melting profiles using gel-based microarrays. In a comparison of *in vitro* transcribed RNA that was either left intact (362 bases) or fragmented, there is a marked difference in initial signal intensity when hybridized overnight using these gel-based microarrays (FIG. 3). The fragmented RNA resulted in a 10-fold increase in initial signal intensity over intact RNA when using the same amount (500 ng) of nucleic acid (data not shown). These results

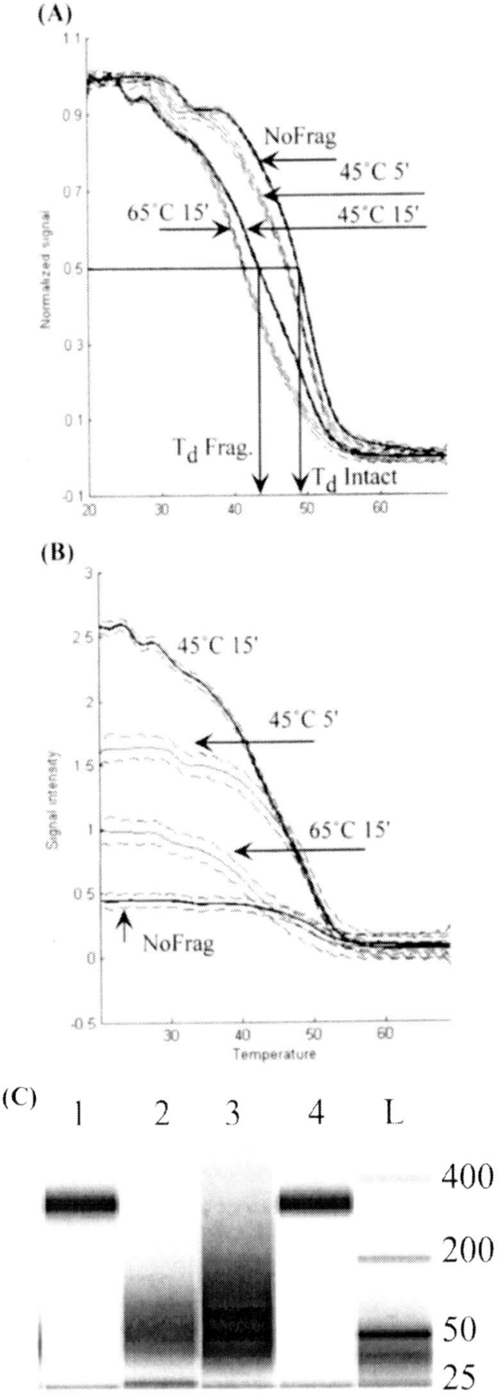
(A)
NoFrag
45°C 5'
65°C 15'
45°C 15'
Normalized signal
T_d Frag.
T_d Intact
(B)
45°C 15'
45°C 5'
65°C 15'
NoFrag
Signal intensity
Temperature
(C)
1 2 3 4 L
400
200
50
25

are not unique to gel-based microarrays, as they have been observed with other platforms[41,42] and indicate that shorter target lengths boost signal intensity with constant hybridization time, presumably because the diffusion constant is higher for smaller molecules. Similarly, melting profiles were dramatically different between the intact and fragmented RNA (FIG. 3, panel A). In triplicate experiments of quadruplicate probes, the mean T_d value was 48.5°C for the intact RNA (unfragmented) and 43.3°C for fragmented RNA (45°C, 15 min). If the fragmentation time is reduced (from 15 to 5 min), the T_d values are shifted only one degree; or if the temperature is raised during fragmentation (from 45°C to 65°C), the T_d values are more significantly reduced (seven degrees) relative to the unfragmented control. This clearly indicates that the degree of fragmentation influences the melting profiles of the target. Because the probe:target duplex is identical in all cases, we believe the apparent shift in melting dynamics and T_d is due to the reduced level of secondary structure and increased diffusion of the smaller-sized RNA molecules. The more fragmented RNA molecules diffuse more quickly within (and out of) the gel elements, resulting in lower disassociation temperatures.

Given the dramatic effect that fragmentation has on the melting profiles, it is important to develop reproducible and robust fragmentation methods as changes in the length of target rRNA may change useful metrics, such as

FIGURE 3. Fragmented RNA affects melt profiles. *In vitro* transcribed RNA (362-nt) was labeled with ULYSIS 594 (Molecular Probes, Carlsbad, CA, USA) following the manufacturer's instructions. Excess dye was removed. RNA fragmentation was carried out according to the modified protocol.[22] In this study, RNA was fragmented using 9 mM NaOH for 5 or 15 min at 45°C or for 15 min at 65°C and then neutralized with 8 mM HCl. Unfragmented and fragmented RNA was hybridized to the array overnight. These arrays contain acrylamide gel element matrices with dimensions of 300 by 300 by 10 μm, and a center-to-center spacing of 500 μm.[49] Each gel element contains 0.25 pmol immobilized probe. Afterward, it was washed and subjected to disassociation nonequilibrium analysis. The results shown here are triplicate experiments with unfragmented RNA, quadruplicate experiments with fragmented RNA for 15 min at 45°C, and single experiments with RNA fragmented for 45° for 5 min and 65° for 15 min. The curves shown represent the mean (*solid line*) and the upper and lower limits of the 95% confidence interval (*two dashed lines*) as calculated by functional analysis of variance (ANOVA) (Bugli *et al.* unpublished material).[50] Each experiment contained four elements immobilized with the probe of interest. All nine experiments used one of two arrays to reduce array-to-array variation. **(A)** Normalized signal intensity curves. **(B)** Signal intensity curves. **(C)** Bioanalyzer virtual gel profile of fragmented and unfragmented RNA. Lane 1: unfragmented *in vitro* transcribed RNA labeled with UYLSIS 594. Lane 2: ULYSIS 594-labeled *in vitro* transcribed RNA fragmented using NaOH. Lane 3: fragmented and labeled *in vitro* transcribed RNA using Cu^{2+} and O-Phenanthroline. Lane 4: a second unfragmented *in vitro* transcribed RNA labeled with ULYSIS 594. L: ladder with an additional 50 nucleotide oligo added. The bar at the bottom of each lane represents the 25-nt marker and lower cutoff of the assay.

T_d. In the past, we and others have used a radical-based method using Cu^{2+} and *o*-phenanthroline with success, despite the significant drawback that it is more laborious than other methods.[43,44] We are also exploring chemical agents that raise the pH sufficiently to fragment rRNA within 20–200 nucleotides but require fewer steps and are easier to use. We have success with 3-(cyclohexylamino)-1-propanesulfonic acid (CAPS), a zwitterionic buffer with a useful pH range between 9.7 and 11.1 (pKa = 10.4). After sufficient time and at sufficiently high temperature (to destabilize the secondary structure formed by rRNA), the solution is brought back to neutral pH using a weak solution of hydrochloric acid. Because the solution already contains Tris, the pH is stabilized upon neutralization.

IMPROVING THE DIFFUSIVITY WITHIN GEL ELEMENTS

Optimizing the chemistry and physical attributes of the gel elements themselves is another avenue we have explored to improve diffusivity within gel elements. The factors that influence diffusion within gel elements are distinctly different from those in solution. Gel porosity is determined by the chemical mixture of gel-forming monomers and the conditions of polymerization. Increased porosity allows the target material to move more quickly within the element. Another major influence on the diffusivity of target molecules within gel elements is the probe concentration immobilized within the element. Because probe:target associations are high-affinity interactions, reducing the probe density reduces the number of high-affinity interactions and permits the target to move more freely within the element. Altering the chemistry to increase the porosity and reducing the probe density in gel elements will promote target movement and diffusion within gel elements.

The first generation of gel-based microarrays (MAGIChips = MicroArray of Gel Immobilized Compounds on chip) consisted of regularly spaced three-dimensional gel pads (100 × 100 × 20 μm) that were loaded with oligonucleotide probes after the polymeric elements had been first polymerized in place on a glass slide through a photolithographic mask.[19,20,45] Copolymerized gel element arrays were invented to overcome some of the inherent performance and manufacturing limitations of preformed MAGIChips.[21] Because oligonucleotides are combined with the pre-polymer prior to printing, copolymerized arrays result in uniform probe distribution throughout the three-dimensional volume of the gel matrix[21,46] and are more amenable to large-scale, low-cost manufacture, an important feature for commercial production of low-cost diagnostic devices. Due to differences in manufacturing practice, copolymerization results in an approximately 10-fold reduction in immobilized probe relative to the preformed MAGIChip format (up to 0.25 mM immobilized probe concentration versus 2 mM probe concentration per element). Thus, target molecules have fewer chances to reinitiate the association–disassociation cycle.

Alternative gel compositions are under continued study in an effort to enhance the porosity of the gel matrix relative to the standard acrylamide compositions originally used in the production of preformed MAGIChip arrays. Together, the increase in porosity and reduction in probe concentration reduce the effects of retarded diffusion in a copolymerized gel element, approximating more closely the association–disassociation dynamics that occur in solution.

As the probe concentration is dramatically reduced in the new generation of gel-based microarrays relative to the MAGIChip platform, the sensitivity of the two platforms was compared to determine whether copolymerized arrays maintained the sensitivity that our clinical objectives and analyses require. For these analyses, we compared the signal intensity values of the same target material between the MAGIChip and the new copolymerized microarrays at the end of an overnight hybridization. In all comparisons, rRNA was fragmented[43] and labeled with ULYSIS Alexa 594 and pooled into a single stock, from which six 1-μg aliquots of target material were taken for triplicate hybridizations on each of the two platforms, for a total of six experiments per target material. Three unique target materials were prepared from pure cultures of *Streptococcus mutans, Streptococcus pyogenes,* and *Streptococcus sobrinus* (FIG. 4).

FIGURE 4 shows the initial signal intensity values of the MAGIChip and the copolymerized arrays using *S. mutans* rRNA genes. With this target, there was a significant increase in signal intensity with the copolymerized arrays in 6 out of 9 probes expected to have perfect-match hybridizations ($P < 0.05$). Similar to *S. mutans*, *S. pyogenes* signal intensity values were significantly higher on the copolymerized array for 7 out of 10 probes, including the species-specific probe for *S. pyogenes* (left-most column). Somewhat dissimilar results were seen with the *S. sobrinus* target, indicating that there is variation in platform performance depending on the target material. Of the 10 perfect-match probes, 3 had significantly greater in signal intensity on the MAGIChip array, while 1 was greater with the copolymerized array. Interestingly, the species-specific probe for *S. sobrinus* (left-most column) clearly shows that the copolymerized array has greater signal, indicative of greater sensitivity. As the species-level probes will have the fewest number of targets in complex mixtures, these probes require a high level of sensitivity.

It is clear with these results that the new copolymerized arrays are as sensitive, if not more so, than the first-generation gel-based microarrays. Comparison of the two microarray formats using rRNA for three target species showed that the copolymerized arrays offer better performance, having either greater (14 out of 29) or similar (12 out of 29) signal intensity values for specific targets relative to MAGIChip arrays (FIG. 4). This is likely the result of more uniform probe distribution and greater porosity of the gel matrix in the new arrays, both of which enhance the accessibility of the probe molecules for hybridization. Despite the reduction in immobilized probe concentration, probe density is still one or two orders of magnitude greater than can be achieved

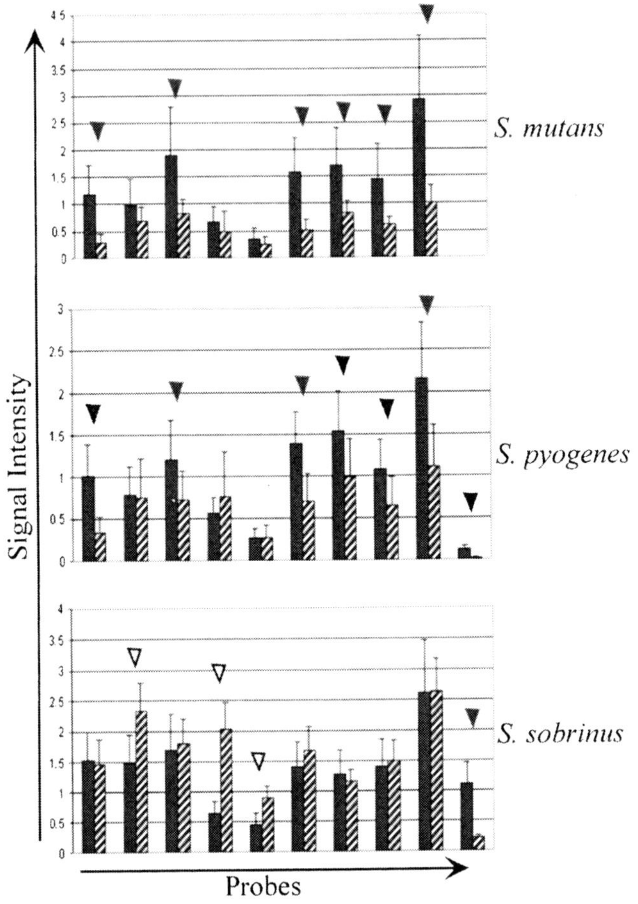

FIGURE 4. Comparison of initial signal intensity on the MAGIChip and copolymerized platform. The average signal intensity values for three reference targets: *S. mutans*, *S. sobrinus*, and *S. pyogenes* from copolymerized (*solid*) and MAGIChip (*diagonal hatch*) microarrays. The standard deviation is shown. Triangles indicate where the difference in mean intensity was significantly different on a Wilcoxon two-sample test ($P < 0.05$); results that favored copolymerized arrays or MAGIChip arrays are represented by *solid* or *open triangles*, respectively.

with conventional, planar arrays. Interestingly, the signal intensity values for *S. sobrinus* are much higher on MAGIChip arrays than for the other two targets for reasons that are not clear. It does not appear to be array-dependent as the same three arrays were used in all of the MAGIChip hybridizations and the standard deviation is similar to the values seen for the other two targets. However, these somewhat anomalous results also suggest that the copolymerized array offers more consistent probe-to-probe performance than the MAGIChip arrays.

In the near future, we aim to extend kinetic studies of both association and disassociation of targets on the new copolymerized arrays using the recirculation circuit in a microfluidic device. Ultimately, by increasing hybridization rates by modification of the device (recirculation, gel matrix chemistry) and sample preparation (optimizing target length), we anticipate enhanced discrimination between closely related target molecules and improved ability to identify low-abundance microbial populations.

FUTURE DIRECTIONS

Using high-throughput technologies, such as microarrays embedded in microfluidic cards, it is now possible to consider cohort studies to identify patterns of oral microbial community structure in subjects with varying states of oral and systemic health. Such studies are ongoing. Oral cavity samples, such as saliva, have been collected from healthy volunteers who vary in the levels of *Streptococcus mutans*, assessed by culture-based methods.[47,48] Because *S. mutans* is thought to be a major contributor to caries, these patients vary in their risk for caries development and represent a broad spectrum of oral health. Developing a microbial profile of these patients is a first step toward quantitatively evaluating the role of microbes in caries onset and periodontal disease and to evaluate intervention strategies.

The highly parallel nature of microarrays makes them uniquely well suited for such studies when large numbers of samples must be evaluated for more than just a few microorganisms. Other settings, such as water quality monitoring where water-borne pathogens are present or where an etiologic agent must be identified in biological and environmental samples are simple extensions of this technology into other areas. Future work is aimed toward extending the use of microfluidics in credit card–sized devices to extract, fragment, label, and hybridize rRNA from the microbial population present in saliva directly on a microarray without any intervention from human operators. The promise of microfluidic technology is immense; there are myriad potential applications for microfluidic devices, particularly in sample processing where small sample volumes are required. Given their potential for lower manufacturing costs on a large scale, microfluidic systems are also an ideal choice for diagnostic devices. We anticipate that this approach—where saliva is "read" for the microbiological content in the mouth—could be used in pediatricians' offices worldwide to identify children at high risk for caries onset or used by dental hygienists to evaluate the likelihood of tooth loss in older adults. Currently, oral diseases are the fourth most expensive disease to treat in industrialized countries,[1] thus highlighting the need for inexpensive and robust strategies to identify subjects at high risk or even moderate risk for oral disease. We propose that a device that incorporates high-throughput technology with reliable identification tools for microbial surveillance in the oral cavity would be a valuable resource for dentists and doctors alike. Eventually, as technology becomes less expensive and electronic devices more pervasive, we may even be able to assess our own oral health at home or while traveling abroad. Such information could be sent electronically to the dentist and visits scheduled as needed for appropriate dental care, thereby ensuring adequate dental care tailored for each individual.

ACKNOWLEDGMENTS

We would like to acknowledge our collaborators, P. Milgrom, M. Rothen, and M. Roberts, at the School of Dentistry, University of Washington. We gratefully acknowledge Maris Lemba and Kristina Hillesland for helpful discussions and editorial comments. This work is supported by grants from the National Institutes of Health (NIH)/NIDCR (U01 DE14955) to D.A.S.; NIH (5R01AI059517) to D.P.C.; NASA (MSMT-2004-0045-0066) to D.A.S.; US DARPA (DABT63-99-1-0009) to D.A.S.; and National University of Singapore (R-288-000-008-112) to W.T.L.

REFERENCES

1. PETERSEN, P.E. 2003. The World Oral Health Report 2003: continuous improvement of oral health in the 21st century—the approach of the WHO Global Oral Health Programme. Comm. Dent. Oral Epidemiol. **31**(Suppl 1): 3–23.
2. U.S.D.H.H.S. Oral Health in America: A Report of the Surgeon General—Executive Summary. 2000, U.S. Department of Health and Human Services, National Institutes of Dental and Craniofacial Research, National Institutes of Health: Rockville, MD.
3. SCANNAPIECO, F.A. 2005. Systemic effects of periodontal diseases. Dent. Clin. North Am. **49:** 533–550.
4. HAMDA, S. & H.D. SLADE. 1980. Biology, immunology, and cariogenicity of *Streptococcus mutans*. Microbiol. Rev. **44:** 331–384.
5. LOESCHE, W.J. 1986. Role of *Streptococcus mutans* in human dental decay. Microbiol. Rev. **50:** 353–380.
6. PIHLSTROM, B.L., B.S. MICHALOWICZ & N.W. JOHNSON. 2005. Periodontal diseases. Lancet **366:** 1809–1820.
7. HAFFAJEE, A.D. & S.S. SOCRANSKY. 1994. Microbial etiological agents of destructive periodontal diseases. Periodontol. 2000. **5:** 79–111.
8. PASTER, B.J., S.K. BOCHES, J.L. GALVIN, *et al.* 2001. Bacterial diversity in human subgingival plaque. J. Bacteriol. **183:** 3770–3783.
9. KOLENBRANDER, P.E. 2000. Oral microbial communities: biofilms, interactions, and genetic systems. Annu. Rev. Microbiol. **54:** 413–437.
10. MOORE, W.E. & L.V. MOORE. 1994. The bacteria of periodontal diseases. Periodontol. 2000 **5:** 66–77.
11. LIU, W., T. MARSH, H. CHENG, *et al.* 1997. Characterization of microbial diversity by determining terminal restriction fragment length polymorphisms of genes encoding 16S rRNA. Appl. Environ. Microbiol. **63:** 4516–4522.

12. Lessa, E. 1992. Rapid surveying of DNA sequence variation in natural populations. Mol. Biol. Evol. **9:** 323–330.
13. Muyzer, G., E.C. de Waal & A.G. Uitterlinden. 1993. Profiling of complex microbial populations by denaturing gradient gel electrophoresis analysis of polymerase chain reaction-amplified genes coding for 16S rRNA. Appl. Environ. Microbiol. **59:** 695–700.
14. de Lillo, A., F.P. Ashley, R.M. Palmer, *et al.* 2006. Novel subgingival bacterial phylotypes detected using multiple universal polymerase chain reaction primer sets. Oral Microbiol. Immunol. **21:** 61–68.
15. Acinas, S.G., R. Sarma-Rupavtarm, V. Klepac-Ceraj, *et al.* 2005. PCR-induced sequence artifacts and bias: insights from comparison of two 16S rRNA clone libraries constructed from the same sample. Appl. Environ. Microbiol. **71:** 8966–8969.
16. Woese, C.R. 1987. Bacterial evolution. Microbiol. Rev. **51:** 221–271.
17. Stahl, D.A., B. Flesher, H.R. Mansfield, *et al.* 1988. Use of phylogenetically based hybridization probes for studies of ruminal microbial ecology. Appl. Environ. Microbiol. **54:** 1079–1084.
18. Wilson, K.H., W.J. Wilson, J.L. Radosevich, *et al.* 2002. High-density microarray of small-subunit ribosomal DNA probes. Appl. Environ. Microbiol. **68:** 2535–2541.
19. Yershov, G., V. Barsky, A. Belgovskiy, *et al.* 1996. DNA analysis and diagnostics on oligonucleotide microchips. PNAS **93:** 4913–4918.
20. Guschin, D., B. Mobarry, D. Proudnikov, *et al.* 1997. Oligonucleotide microchips as genosensors for determinative and environmental studies in microbiology. Appl. Environ. Microbiol. **63:** 2397–2402.
21. Rubina, A.Y., S.V. Pan'kov, E.I. Dementieva, *et al.* 2004. Hydrogel drop microchips with immobilized DNA: properties and methods for large-scale production. Anal. Biochem. **325:** 92–106.
22. Liu, W.T., A.D. Mirzabekov & D.A. Stahl. 2001. Optimization of an oligonucleotide microchip for microbial identification studies: a non-equilibrium dissociation approach. Environ. Microbiol. **3:** 619–629.
23. Li, E.S.Y., J.K.K. Ng, J.-H. Wu, *et al.* 2004. Evaluating single-base-pair discriminating capability of planar oligonucleotide microchips using a non-equilibrium dissociation approach. Environ. Microbiol. **6:** 1197–1202.
24. Fotin, A., A. Drobyshev, D. Proudnikov, *et al.* 1998. Parallel thermodynamic analysis of duplexes on oligodeoxyribonucleotide microchips. Nucl. Acids Res. **26:** 1515–1521.
25. El Fantroussi, S., H. Urakawa, A.E. Bernhard, *et al.* 2003. Direct profiling of environmental microbial populations by thermal dissociation analysis of native rRNAs hybridized to oligonucleotide microarrays. Appl. Environ. Microbiol. **69:** 2377–2382.
26. Kelly, J.J., S. Siripong, J. McCormack, *et al.* 2005. DNA microarray detection of nitrifying bacterial 16S rRNA in wastewater treatment plant samples. Water Res. **39:** 3229–3238.
27. Urakawa, H., P.A. Noble, S. El Fantroussi, *et al.* 2002. Single-base-pair discrimination of terminal mismatches by using oligonucleotide microarrays and neural network analyses. Appl. Environ. Microbiol. **68:** 235–244.
28. McQuain, M.K., K. Seale, J. Peek, *et al.* 2004. Chaotic mixer improves microarray hybridization. Anal. Biochem. **325:** 215–226.

29. ADEY, N.B., M. LEI, M.T. HOWARD, *et al.* 2002. Gains in sensitivity with a device that mixes microarray hybridization solution in a 25-μm-thick chamber. Anal. Chem. **74:** 6413–6417.
30. LIU, R.H., R. LENIGK, R.L. DRUYOR-SANCHEZ, *et al.* 2003. Hybridization enhancement using cavitation microstreaming. Anal. Chem. **75:** 1911–1917.
31. GADGIL, C., A. YECKEL, J.J. DERBY, *et al.* 2004. A diffusion–reaction model for DNA microarray assays. J. Biotechnol. **114:** 31–45.
32. SOROKIN, N.V., V.R. CHECHETKIN, M.A. LIVSHITS, *et al.* 2005. Discrimination between perfect and mismatched duplexes with oligonucleotide gel microchips: role of thermodynamic and kinetic effects during hybridization. J. Biomol. Struct. Dyn. **22:** 725–734.
33. LIVSHITS, M.A. & A.D. MIRZABEKOV. 1996. Theoretical analysis of the kinetics of DNA hybridization with gel-immobilized oligonucleotides. Biophys. J. **71:** 2795–2801.
34. LIVSHITS, M.A., V.L. FLORENT'EV & A.D. MIRZABEKOV. 1994. Dissociation of duplexes formed by hybridization of DNA with gel-immobilized oligonucleotides. J. Biomol. Struct. Dyn. **11:** 783–795.
35. ODDY, M.H., J.G. SANTIAGO & J.C. MIKKELSEN. 2001. Electrokinetic instability micromixing. Anal. Chem. **73:** 5822–5832.
36. TOEGL, A., R. KIRCHNER, C. GAUER, *et al.* 2003. Enhancing results of microarray hybridizations through microagitation. J. Biomol. Tech. **14:** 197–204.
37. KI, Y.P., L. GUANGSHAN, B. YIJIA, *et al.* 2003. Microfluidic devices for fluidic circulation and mixing improve hybridization signal intensity on DNA arrays. Lab Chip. **3:** 46–50.
38. NOERHOLM, M., H. BRUUS, M.H. JAKOBSEN, *et al.* 2004. Polymer microfluidic chip for online monitoring of microarray hybridizations. Lab Chip. **4:** 28–37.
39. BYNUM, M.A. & G.B. GORDON. 2004. Hybridization enhancement using microfluidic planetary centrifugal mixing. Anal. Chem. **76:** 7039–7044.
40. LEE, H.H., J. SMOOT, Z. MCMURRAY, *et al.* 2006. Recirculating flow accelerates DNA microarray hybridization in a microfluidic device. Lab Chip. **6:** 1163–1170.
41. LIU, W.-T., H. GUO & J.-H. WU. 2006. Effects of target length on the hybridization efficiency and specificity of ribosomal RNA-based oligonucleotide microarrays. Appl. Environ. Microbiol. **73:** 73–82.
42. MEHLMANN, M., M.B. TOWNSEND, R.L. STEARS, *et al.* 2005. Optimization of fragmentation conditions for microarray analysis of viral RNA. Anal. Biochem. **347:** 316–323.
43. KELLY, J.J., B.K. CHERNOV, I. TOVSTANOVSKY, *et al.* 2002. Radical-generating coordination complexes as tools for rapid and effective fragmentation and fluorescent labeling of nucleic acids for microchip hybridization. Anal. Biochem. **311:** 103–118.
44. BAVYKIN, S.G., Y.P. LYSOV, V. ZAKHARIEV, *et al.* 2004. Use of 16S rRNA, 23S rRNA, and gyrB gene sequence analysis to determine phylogenetic relationships of *Bacillus cereus* group microorganisms. J. Clin. Microbiol. **42:** 3711–3730.
45. GUSCHIN, D., G. YERSHOV, A. ZASLAVSKY, *et al.* 1997. Manual manufacturing of oligonucleotide, DNA, and protein microchips. Anal. Biochem. **250:** 203–211.
46. STARKE, E.M.L., J.C. SMOOT, L.M. SMOOT, *et al.* 2006. Technology development to explore the relationship between oral health and the oral microbial community. BMC Oral Health **6**(Suppl 1): S10.

47. Ly, K.A., P. Milgrom, M.C. Roberts, *et al.* 2006. Linear response of mutans streptococci to increasing frequency of xylitol chewing gum use: a randomized controlled trial [ISRCTN43479664]. BMC Oral Health **6:** 6.
48. Milgrom, P., K.A. Ly, M.C. Roberts, *et al.* 2006. *Mutans streptococci* dose response to xylitol chewing gum. J. Dent. Res. **85:** 177–181.
49. Liu, W.-T., J.-H. Wu, E. S.-Y. Li, *et al.* 2005. Emission characteristics of fluorescent labels with respect to temperature changes and subsequent effects on DNA microchip studies. Appl. Environ. Microbiol. **71:** 6453–6457.
50. Eyers, L., J.C. Smoot, L.M. Smoot, *et al.* 2006. Discrimination of shifts in a soil microbial community associated with TNT-contamination using a functional ANOVA of 16S rRNA hybridized to oligonucleotide microassays. Environ. Sci. Technol. **40:** 5867–5873.

Integrated Microfluidic Platform for Oral Diagnostics

AMY E. HERR,[a] ANSON V. HATCH,[a] WILLIAM V. GIANNOBILE,[b] DANIEL J. THROCKMORTON,[a] HUU M. TRAN,[a] JAMES S. BRENNAN,[a] AND ANUP K. SINGH[a]

[a]*Biosystems Research Department, Sandia National Laboratories, Livermore, California 94550, USA*

[b]*School of Dentistry, University of Michigan, Ann Arbor, Michigan 84106, USA*

ABSTRACT: While many point-of-care (POC) diagnostic methods have been developed for blood-borne analytes, development of saliva-based POC diagnostics is in its infancy. We have developed a portable microfluidic device for detection of potential biomarkers of periodontal disease in saliva. The device performs rapid microfluidic chip-based immunoassays (<3–10 min) with low sample volume requirements (10 μL) and appreciable sensitivity (nM–pM). Our microfluidic method facilitates hands-free saliva analysis by integrating sample pretreatment (filtering, enrichment, mixing) with electrophoretic immunoassays to quickly measure analyte concentrations in minimally pretreated saliva samples. The microfluidic chip has been integrated with miniaturized electronics, optical elements, such as diode lasers, fluid-handling components, and data acquisition software to develop a portable, self-contained device. The device and methods are being tested by detecting potential biomarkers in saliva samples from patients diagnosed with periodontal disease. Our microchip-based analysis can readily be extended to detection of biomarkers of other diseases, both oral and systemic, in saliva and other oral fluids.

KEYWORDS: microfluidics; periodontal disease; diagnostics; point-of-care; POC; immunoassay; lab-on-a-chip; saliva

Throughout the last decade, research studies using saliva as a diagnostic fluid have increased exponentially.[1] The primary benefits of saliva-based tests, over more common blood tests, include easier, noninvasive saliva collection, and the lower costs associated with saliva testing.[2–4] The increased interest in

Address for correspondence: Anup K. Singh, Biosystems Research Department, Sandia National Laboratories, P.O. Box 969, MS 9292, Livermore, CA 94551-0969. Voice: 925-294-1260; fax: 925-294-1260.
aksingh@sandia.gov

Ann. N.Y. Acad. Sci. 1098: 362–374 (2007).
doi: 10.1196/annals.1384.004

saliva diagnostics has also been spurred by rapidly accumulating evidence of correlation between saliva analyte levels and those in serum. Thus, saliva testing is not only a means to monitor oral health, but is now viewed as a potential window into the overall systemic health of an individual. Saliva diagnosis of autoimmune disorders (i.e., Sjögren's syndrome), cardiovascular diseases, abnormal endocrine function, presence of infection (viral, bacterial), renal disease, cancer, and abuse of drugs are all areas that have benefited from using saliva as a diagnostic fluid.[1] Saliva testing is also being explored for directing and monitoring treatment options. For example, drug doses can be monitored without inconvenient and costly visits to blood-drawing facilities.

Agencies such as the Office of the Surgeon General[5] and the NIDCR[6] have recognized the potential of saliva as a diagnostic fluid and have thus called for increased research and development of saliva-based testing. While the potential value of saliva as a diagnostic fluid has become more apparent, the adoption of saliva as a routine diagnostic fluid has met with three general types of barriers. As outlined by the 1999 NIDCR workshop for saliva diagnostics development, barriers to acceptance and use of saliva can be broadly characterized as follows: need for innovation in accurate analyte measurements in small volumes and development of standardized saliva collection methods; lack of investment in product development; and low technology adoption rates by clinicians and medical insurance companies.

Since 1999, substantial technological barriers to widespread use of saliva in diagnostics have begun to diminish. Recent development of techniques that combine the power of miniaturization with cutting-edge discoveries in fields once as distinct as biology and engineering is leading to rapid, high-throughput, automated, portable, low-cost, and efficient biochemical analyses with small sample volumes.

MICROFLUIDIC METHODS CONFER ADVANTAGES TO DIAGNOSTIC DESIGN AND USAGE PARADIGMS

With recent advances in proteomics and systems biology, it is evident that multi-parameter diagnostic approaches are needed for clinical use.[7] Traditional single-marker approaches are based on the expectation that a change in the concentration of a single protein can unambiguously specify disease. In reality, diseases exhibit great heterogeneity between individuals; the same disease can be initiated by numerous factors and can cause a range of molecular changes. Thus, single-marker tests generally suffer from lack of sensitivity and specificity.[8] Diagnostics developed for complex diseases require the analysis of multiple components to effectively (1) predict the onset of disease, (2) stratify disease (e.g., prostate cancer could be subcategorized as three or four distinct diseases), (3) indicate the progression of the disease, (4) direct treatment (identify resistance to drugs or potential adverse reactions, etc.), and (5) monitor treatment. For each type of disease, the informative set of markers will be different. To accomplish this goal, low-cost high-throughput methods that safely make use of small sample volumes are necessary.

Researchers have envisioned miniaturized diagnostic technologies that could accurately ascertain disease states using droplets (tens of microliters) of human body fluid. Recent reports describe lab-on-a-chip instruments that perform multiple operations in parallel in extralaboratory settings (i.e., field deployment, near-patient environments, and resource-poor settings).[9–11] Current technologies not only provide paths toward such implementation, but are also presently being exploited and demonstrated. While recent reports show promise for microfluidic detection of analytes in various body fluids,[11–14] few, if any, reports on detection of endogenous biomarkers in saliva using a semiportable instrument have been made. We believe that operating specifications for such instrumentation can be described by the following characteristics:

(1) Saliva-based: Ready collection of samples by trained or untrained personnel.
(2) Microfluidic: Requires small volumes of saliva.
(3) Multiplexed detection: Simultaneous analysis of multiple analytes for accurate assessment of complex diseases.
(4) Portable and easy to use: Point-of-care (POC) testing with simple user interface.
(5) Rapid: Simultaneous protein measurements within a routine clinical visit.
(6) Low cost: Feasible widespread screening, diagnosis, and monitoring.

Key advances in bioanalytical instrument design, both at Sandia and elsewhere, have relied upon microfluidic technologies to enable sensitive and fast analyses, especially for manipulation of sub-microliter fluid volumes. Such significant technological advances in miniaturization (at the micro- and nanoscales) have led to the advent of valuable new diagnostic formats, known as "lab-on-a-chip" devices.

Realizing the potential miniaturization and automation made possible through the lab-on-a-chip paradigm requires integration of various functionalities. Incorporation of unit functionalities (fluid containment and fluid handling in channels similar in dimensions to a human hair; hands-free operation of sample preparation steps; automated bioanalytical assays; integrated optical systems for fluorescence-based detection; and subsequent reporting of assay results) is essential to conduct sophisticated analyses. Sandia has recently reported one of the first lab-on-a-chip systems to demonstrate full integration and hand-portable operation.[9] Innovation regarding portable, rapid, specific bioanalytical methods and associated hardware has been undertaken at Sandia for applications ranging from biotoxin detection[9,15,16] to proteomics[17–19] to cell sorting[20,21] and clinical diagnostics.[22]

Building on technologies introduced above, our group is actively developing an integrated microfluidic platform for oral diagnostics (IMPOD). An image of an early hardware version of the portable diagnostic is based on that developed at Sandia previously[9] and is shown in FIGURE 1. A descriptive summary of the technology is given in TABLE 1. IMPOD, as well as quantitative, portable instrumentation like it, has the potential to be translated to clinical settings for use in rapid, near-patient analysis of human saliva and oral fluids. The methods and technologies presented here also have applicability to nonoral diagnostic fluids, as well as other local and systemic diseases.

MICROFLUIDIC ELECTROPHORETIC IMMUNOASSAYS AS CORE ANALYTICAL METHOD

Use of microfluidic technologies naturally confers assay speed, low-reagent volume consumption, and streamlined sample preparation and assay integration to the IMPOD diagnostic platform. By incorporating photopolymerized cross-linked polyacrylamide gels within a microfluidic device and using immunoassays, we have further conferred high-resolution sieving-based performance and fine specificity to the microanalytical platform.

Combining the specificity and selectivity typical of conventional immunoassays with high-efficiency separations achieved by microfluidic methods, our demonstrated technology efficiently separates receptor-bound and unbound species, resulting in quantitation of only those analytes of interest. A schematic description of the assay protocol is in FIGURE 2. Such an approach avoids

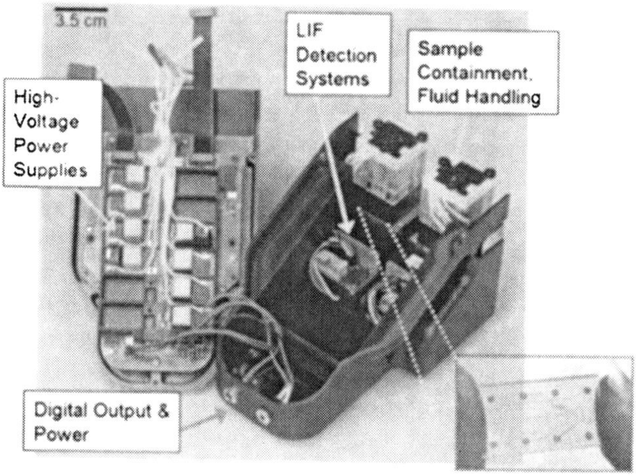

FIGURE 1. Image of early generation IMPOD with hardware components labeled. *Inset* shows glass microdevice used in performing immunoassays.

TABLE 1. Summary of IMPOD characteristics

Automated operation
Manual volume-independent sample introduction *via* syringe or pipette (no volume dependence, if > 50 μL)
Automated sample preparation and associated timing
Integrated high-voltage power supply (drives mixing, preconcentration, and electrophoresis)
Automated assay and replicate runs (timing)
Electronic data collection via software interface
Sample introduction and pre-processing
Off-chip: centrifugation and dilution
On-chip: debris removal (filtering); sample preconcentration; rapid mixing
Microfluidic immunoassay specificity and selectivity
Multiplexing: serial format demonstrated; parallel format in testing stage
"Capture chemistry": high-affinity, fluorescent receptors provide specificity
Tunable sieving matrix provides assay selectivity and resolution
Analysis of saliva from subjects in both healthy and periodontal-diseased categories
Gold-standard results comparison *via* validation assays (ELISA, protein microarray)
Miniaturized detection system
Fluorescent receptor molecules provide specific detection signal
Diode laser induced fluorescence excitation
Photomultiplier tube (PMT) fluorescence detection
Data analysis
Software-based data analysis (research stage)

problems associated with open-channel protein electrophoresis. Polyacrylamide gel electrophoresis (PAGE) has been used for decades for protein analysis and does not exhibit nonspecific adsorption of proteins. Channel surfaces are coated with linear polyacrylamide[23,24] using photoinitiated polymerization. Cross-linked polyacrylamide gels are made *in situ* by photopolymerization using methods reported by our group.[25,26]

Electrophoretic immunoassays offer a number of advantages over conventional radioimmunoassays or enzyme-linked immunosorbent assays (ELISA).[27] Capillary electrophoresis is a highly effective separation method allowing separation of bound from unbound species in a single step, thus eliminating the multiple wash steps required in conventional immunoassays. Immobilization of reporter molecules (antibodies) on sensor surfaces is unnecessary. Electrophoretic immunoassays are typically performed in an open or surface-modified microfluidic channel.[28–31] In contrast to conventional capillary zone electrophoresis approaches, we have relied upon sieving–gel-based electrophoretic immunoassays, as open channel (or free solution) immunoassays suffer from a number of disadvantages:

(1) Difficulty in attaining adequate species discrimination with electrophoresis-based immunoassays on account of the small differences in charge-to-mass characteristics between large antibodies and immune complexes.

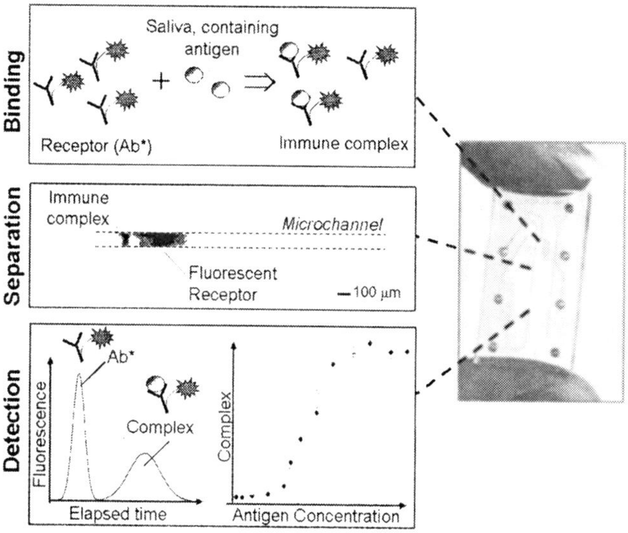

FIGURE 2. Microchip-based electrophoretic immunoassays. (left) Protocol schematic: Binding/Preconcentration: The sample is mixed with a known amount of fluorescently labeled antibody. The mixing is done off-chip or on-chip. During initial methods development, we carry out the mixing outside the chip (requires ~15 min). Upon optimization of immunoassay conditions, the sample and the labeled antibody are added to different reservoirs in the sample manifold and are mixed (in <3 min) electrokinetically on-chip. Separation: Mixed sample is electrophoretically injected into a separation channel. The antibody-bound antigen and unbound antibody are separated based on size and charge by native polyacrylamide gel electrophoresis. Detection & Analysis: A laser-induced fluorescence detector detects the unbound antibody and bound antibody (i.e., immune complex) peaks. Migration times are recorded and peak areas are calculated. The immunoassay is repeated with samples containing known amounts of analyte to generate a multi-point calibration curve. The calibration curve is used to estimate the concentration of analyte in the unknown sample. (right) Image of glass microdevice.

(2) Vulnerability to analyte dispersion arising from difficult-to-control hydrodynamic flow.
(3) Difficulty associated with integration of open channel components with complex chip-based systems (e.g., sample preprocessing functionalities).

The microfluidic components that form the core of IMPOD integrate multiple functions (e.g., sample injection, mixing and incubation with antibodies, sample enrichment, and separation of bound and unbound immune complexes). The immunoassays themselves take minutes to complete (compared to hours for conventional quantitative ELISA) and are capable of analyzing sample volumes of less than 10 μL. Integration of parallel analyses using multichannel chips and spectral multiplexing for simultaneous multi-analyte analysis is currently under way. Integrated sample enrichment improves the detection limit to the low picomolar range.[32]

PROTEIN BIOMARKERS FOR CLINICAL DIAGNOSIS OF PERIODONTAL DISEASE

Saliva is a fluid that can be easily collected, contains locally derived and systemically derived biomarkers of periodontal disease, and hence may offer the basis for a patient-specific biomarker assessment for periodontitis and other systemic diseases. Periodontitis is a group of inflammatory diseases affecting the connective tissue attachment and supporting bone around the teeth. Once initiated, the disease progresses with the loss of tooth-supporting collagen fibers and attachment to the cemental surface, apical migration of the junctional epithelium, formation of deepened periodontal pockets, and resorption of alveolar bone. If left untreated, the disease continues with progressive bone destruction, leading to tooth mobility and subsequent tooth loss. Periodontal disease afflicts over 50% of the adult population in the United States, with approximately 10% displaying severe disease concomitant with early tooth loss. Recent evidence suggests a strong association among periodontal disease, cardiovascular disease, and pulmonary and other serious systemic diseases.

Because of the noninvasive and undemanding nature of saliva collection, analysis of saliva may be especially beneficial in the determination of current periodontal status and a means of monitoring the response to treatment. Studies report that the determination of inflammatory mediator levels in biological fluids is a good indicator of inflammatory activity.[33] Accordingly, studies related to the pathogenesis of periodontal disease examine the link between biochemical and immunological markers in saliva and the extent of periodontal destruction—with potential for predicting future disease progression. Oral fluid biomarkers studied for periodontal diagnosis include proteins of host origin (i.e., enzymes and immunoglobulins), phenotypic markers, host cells, hormones, bacteria and bacterial products, ions and volatile compounds. There are a variety of other biomarkers of skeletal homeostasis. These biomarkers may have significant potential in other bone metabolic diseases, such as osteoporosis, rheumatoid arthritis, and osteolytic bone metastases.

Risk factors are considered to be modifiers of the nature of disease. Associated with host susceptibility and a variety of local and systemic conditions, risk factors influence the initiation and progression of periodontitis and alter biomarker levels. Longitudinal studies of biomarkers play an important role in life sciences and have begun to assume a greater role in diagnosis, monitoring of therapy outcomes, and drug discovery. The challenge for biomarkers is to allow earlier detection of the disease evolution and more robust assessment of therapeutic efficacy. For biomarkers to assume their rightful role, a greater understanding of the mechanism of disease progression and therapeutic intervention is essential. Consequently, there is a need for the development of new diagnostic tests that can detect the presence of active disease, forecast future disease progression, and evaluate the response to periodontal therapy, thereby improving the clinical management of periodontal patients. The diagnosis of active phases of periodontal disease and the identification of patients at risk for active disease represent challenges for both clinical investigators and practitioners.

ONGOING CLINICAL MEASUREMENTS USE IMPOD

Our group has recently demonstrated the ability to measure putative protein biomarkers in whole saliva, gingival crevicular fluid (GCF), and other oral fluids. We have completed a pilot analysis in a recent cross-sectional investigation of periodontal disease and health, as well as recently published evaluation of protein biomarkers in oral fluids during periodontal reconstructive therapy using local[34,35] or systemic drug delivery.[36] Our studies employing the IMPOD diagnostic have shown the reproducible assessment of putative periodontal disease biomarkers.[37] FIGURE 3 shows IMPOD analysis of saliva for two cytokines, tumor necrosis factor-α (TNF-α) and interleukin-6 (IL-6). While these results demonstrate IMPOD-based measurement of spiked cytokines in whole saliva, the data show how rapid the analyses are (completed in less than 250 sec) and that the assays are quantitative—allowing generation of calibration curves for measurement of endogenous saliva in pilot samples.

Low nanomolar detection limits are not always adequate for screening of low-abundance disease markers.[38] Our group has demonstrated experience with development of on-chip functional components (e.g., dialysis membranes,[39] buffer exchange membranes,[40] and pressure-actuated valving[18,41]) using photolithographically controlled polymerization. We employ these fabrication methods to create size-exclusion membranes within microfluidic channels for sample concentration or enrichment. The pore size of the miniaturized size-exclusion membrane is controlled via the formulation of the precursor solution (monomer and cross-linker).

Our group has shown that the incorporation of photolithographically fabricated size-exclusion membranes has extended the sensitivity of protein-sizing assays to the femtomolar level.[32] When implemented with our electrokinetic immunoassays, online sample enrichment allows rapid identification (<10 min) of protein markers in human saliva at clinically relevant concentrations (10^{-12} M) without the need for additional off-chip sample preparation. Trapping of analytes in the volume near the membrane obviates the need for off-chip sample incubation and reporter-binding steps. FIGURE 4 shows results from the high-sensitivity IMPOD immunoassay technique, compared to immunoassays not incorporating sample enrichment, for analysis of C-reactive protein (CRP). An appreciable increase in assay sensitivity is observed when IMPOD operates using sample enrichment. Further, the duration of sample (as well as antibody) enrichment can be used to "tune" the assay sensitivity and dynamic range in real time, thus making a single assay applicable to measurement of both high- and low-abundance protein species in a given diagnostic sample.

A key, defining feature of our microfluidic approach is development of an instrument that readily measures multiple analytes simultaneously. Such an approach is essential for early and accurate diagnosis of a chronic inflammatory disease, such as periodontitis, where measuring overall "composite profile" or "signature" of a set of biomarkers may have significantly higher diagnostic value than measurement of individual analytes. In addition to measuring endogenous protein content in saliva using the IMPOD approach, our group is assessing a large panel of putative biomarkers for relevance to periodontal disease. Data currently being generated from these proteomic measurements

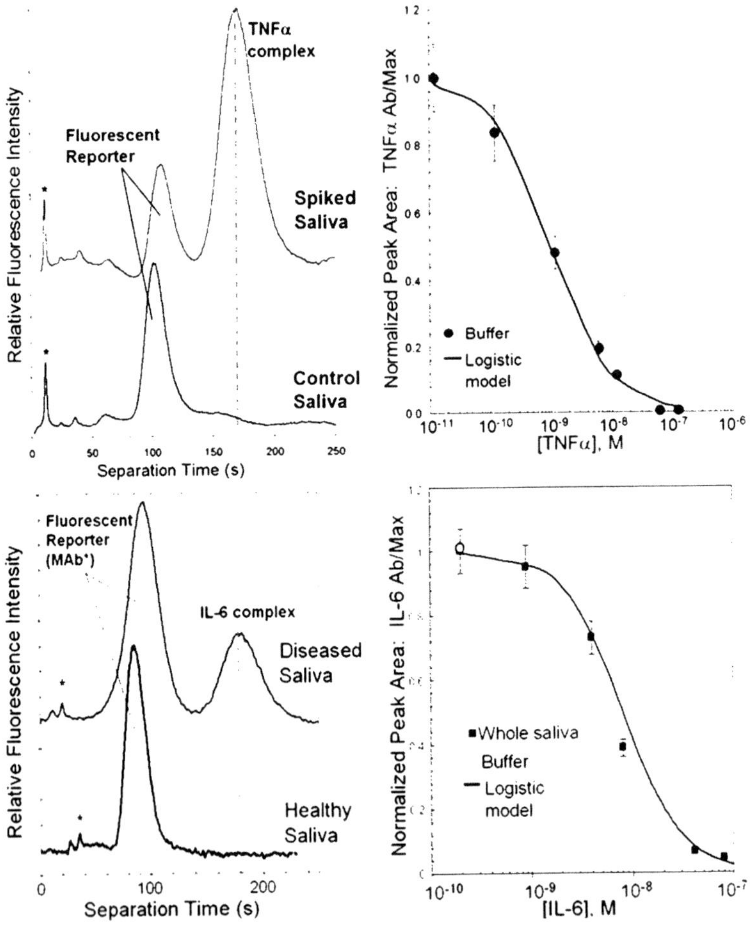

FIGURE 3. IMPOD analysis is quantitative and completes in less than four minutes. (*Top*) Electropherograms for exogenous TNF-α analysis in whole saliva. Companion dose-response curve acquired in model buffer system. (*Bottom*) IL-6 measurements from IMPOD analysis of spiked whole saliva. Companion dose-response curve from spiked saliva.

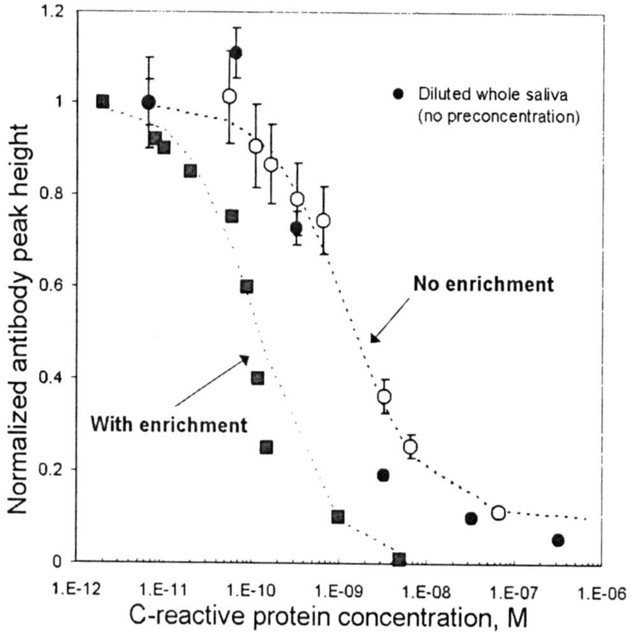

FIGURE 4. On-chip sample enrichment extends the C-reactive protein dynamic range and increases the sensitivity of IMPOD. Buffer samples spiked with known concentrations of C-RP were analyzed with (*solid squares*) and without (*hollow circles*) on-chip sample enrichment. Measurement of C-RP in spiked whole saliva, without enrichment, is shown for comparison (*solid circles*).

allow us to gauge the predictive ability of putative disease biomarkers (cytokines, enzymes, and bone collagen fragments). Currently, the use of digital subtraction radiography and longitudinal clinical parameter assessment is being compared to oral fluid samples (saliva, GCF) to develop the appropriate metrics in evaluating the interrelationship of disease biomarkers to various fluid components.

Analyte classes, shown here in order of progression of periodontal disease, are currently being evaluated for their predictive value in diagnosis of periodontal disease by our group and others: (1) inflammatory mediators and host-response modifiers (Th1/Th2 cytokines and CRP), (2) host-derived enzymes (collagenases, gelatinases), (3) connective and bone breakdown proteins (osteocalcin, osteonectin, laminin, C-telopeptide pyridinoline cross-links [ICTP]). Our preliminary proteomic studies suggest that at least three analytes (IL-1B, MMP-8, and ICTP) have significant correlation with the progression of periodontal disease. Each of these three analytes is implicated in a different stage of periodontal disease progression. Further work regarding identification and validation of these protein biomarkers is currently under way by our group.

CONCLUSIONS

Over the past 3 years, our collaborative project has focused on the development of a novel diagnostic strategy for the evaluation of human periodontal disease. A goal of periodontal diagnostic procedures is to provide the clinician with useful information regarding the present periodontal disease type, location, and severity. These findings serve as a basis upon which treatment planning is formulated and provide useful information during periodontal maintenance and disease monitoring phases of treatment. Traditional periodontal diagnostic parameters used by clinicians include several semisubjective assessment parameters, many of which develop during advanced disease. Advances in oral and diagnostic research are moving toward methods by which periodontal risk can be identified and quantified by objective measures, such as biomarkers that can be determined in a rapid fashion, such as is the case with our IMPOD instrument.

Ongoing work centers on identification and validation of panels of protein biomarkers, with particular emphasis on identifying groups of proteins that have relevance to all stages of periodontal disease progression—both as a means to enable early diagnosis of periodontal disease and as a means to assess the activity of disease in a particular patient or site. Our group is working to incorporate the capability for multiplex analyses (i.e., assaying a single sample for multiple biomarkers in parallel). Lastly, the research reported here lays the groundwork for extension of these methods to other diagnostic fluids and illnesses. Clinical and engineering approaches, such as those described in this work, present compelling advantages for furthering personalized medicine in the 21st century.

ACKNOWLEDGMENTS

The authors thank V. VanderNoot, R. Renzi and J. Stamps at Sandia National Laboratories. The authors also thank M. McReynolds, K. Ghandi, and D. Degrasse at Caliper Life Sciences. At the University of Michigan, the authors gratefully acknowledge J. Kinney, C. Ramseier, L. Rayburn, and J. Sugai.

This work was supported by the U.S. National Institute of Dental and Craniofacial Research (Grant NIDCR U01-DE014961). Sandia is a multiprogram laboratory operated by Sandia Corp., a Lockheed Martin Co., for the United States Department of Energy under Contract DE-AC04-94AL85000.

REFERENCES

1. Streckfus, C.F. & L.R. Bigler. 2002. Saliva as a diagnostic fluid. Oral Dis. **8:** 69–76.

2. FERGUSON, D.B. 1987. Current diagnostic uses of saliva. J. Dent. Res. **66:** 420–424.
3. MANDEL, I.D. 1990. The diagnostic uses of saliva. J. Oral Path. **19:** 119–125.
4. MALAMUD, D. 1992. Saliva as a diagnostic fluid. Br. Med. J. **305:** 207–208.
5. HHS. 2000. Oral health in America. A report of the Surgeon General. U.S. Department of Health and Human Services, National Institute of Dental and Craniofacial Research, National Institutes of Health. Rockville, MD.
6. NIDCR. 1999. Workshop on Development of New Technologies for Saliva and Other Oral Fluid-Based Diagnostics. Airlie House Conference Center, Virginia, September 12–14, 1999.
7. WESTON, A.D. & L. HOOD. 2004. Systems biology, proteomics, and the future of health care: toward predictive, preventative, and personalized medicine. J. Proteome Res. **3:** 179.
8. ANDERSON, N.L. & N.G. ANDERSON. 2002. The human plasma proteome—History, character, and diagnostic prospects. Mol. Cell Proteomics **1:** 845–867.
9. RENZI, R.F. *et al.* 2005. Hand-held microanalytical instrument for chip-based electrophoretic separations of proteins. Anal. Chem. **77:** 435–441.
10. SIA, S.K. *et al.* 2004. An integrated approach to a portable and low-cost immunoassay for resource-poor settings. Angewandte Chemie Int. Ed. **43:** 498.
11. SRINIVASAN, V., V.K. PAMULA & R.B. FAIR. 2004. An integrated digital microfluidic lab-on-a-chip for clinical diagnostics on human physiological fluids. Lab Chip. **4:** 310.
12. HATCH, A. *et al.* 2001. A rapid diffusion immunoassay in a T-sensor. Nat. Biotechnol. **19:** 461–465.
13. YANG, C.Y. *et al.* 2005. Detection of picomolar levels of interleukin-8 in human saliva by SPR. Lab Chip. **5:** 1017.
14. CHRISTODOULIDES, N. *et al.* 2005. Application of microchip assay system for the measurement of C-reactive protein in human saliva. Lab Chip. **5:** 261–269.
15. BAILEY, C.G. *et al.* 2000. Chip-based, multiple-channel, total-analysis system. Abstr. Papers Am. Chem. Soc. **219:** U85.
16. FRUETEL, J.A. *et al.* 2005. Microchip separations of protein biotoxins using an integrated hand-held device. Electrophoresis **26:** 1144.
17. HERR, A.E. & A.K. SINGH. 2004. Photopolymerized cross-linked polyacrylamide gels for on-chip protein sizing. Anal. Chem. **76:** 4727–4733.
18. REICHMUTH, D.S., T.J. SHEPODD & B.J. KIRBY. 2005. Microchip HPLC of peptides and proteins. Anal. Chem. **77:** 2997.
19. THROCKMORTON, D.J., T.J. SHEPODD & A.K. SINGH. 2002. Electrochromatography in microchips: reversed-phase separation of peptides and amino acids using photopatterned rigid polymer monoliths. Anal. Chem. **74:** 784–789.
20. CUMMINGS, E.B. & A.K. SINGH. 2003. Dielectrophoresis in microchips containing arrays of insulating posts: theoretical and experimental results. Anal. Chem. **75:** 4724.
21. BARRETT, L.M. *et al.* 2005. Dielectrophoretic manipulation of particles and cells using insulating ridges in faceted prism microchannels. Anal. Chem. **77:** 6798–6804.
22. HERR, A.E., A.A. DAVENPORT & A.K. SINGH. 2005. Microchip immunoassays using photodefined polyacrylamide for rapid native gel electrophoresis of immune complexes. Anal. Chem. **77:** 585–590.
23. HJERTEN, S. *et al.* 1993. Reversed-phase chromatography of proteins and peptides on compressed continuous beds. Chromatographia **37:** 287.

24. HJERTEN, S. & M.D. ZHU. 1985. Adaptation of the equipment for high-performance electrophoresis to isoelectric-focusing. J. Chromatogr. **346:** 265–270.
25. HERR, A.E. & A.K. SINGH. 2004. Photopolymerized cross-linked polyacrylamide gels for on-chip protein sizing. Anal. Chem. **76:** 4727–4733.
26. HAN, J. & A.K. SINGH. 2004. Rapid protein separations in ultra-short microchannels: microchip sodium dodecyl sulfate-polyacrylamide gel electrophoresis and isoelectric focusing. J. Chromatogr. A. **1049:** 205.
27. SCHMALZING, D. & W. NASHABEH. 1997. Capillary electrophoresis based immunoassays: a critical review. Electrophoresis 2184–2193.
28. SHIMURA, K. & B.L. KARGER. 1994. Affinity probe capillary electrophoresis: analysis of recombinant human growth-hormone with a fluorescent-labeled antibody fragment. Anal. Chem. **66:** 9–15.
29. SCHULTZ, N.M. & R.T. KENNEDY. 1993. Rapid immunoassays using capillary electrophoresis with fluorescence detection. Anal. Chem. **1:** 3161–3165.
30. CHENG, S.B. *et al.* 2001. Development of a multichannel microfluidic analysis system employing affinity capillary electrophoresis for immunoassay. Anal. Chem. **73:** 1472–1479.
31. WANG, Y.C. *et al.* 2001. Enhancement of the sensitivity of a capillary electrophoresis immunoassay for estradiol with laser-induced fluorescence based on a fluorescein-labeled secondary antibody. Anal. Chem. **15:** 5616–5619.
32. HATCH, A.V. *et al.* 2006. Integrated preconcentration-sizing of proteins in microchips using photopatterned polyacrylamide gels. Anal. Chem. **78:** 4976–4984.
33. KAUFMAN, E. & I.B. LAMSTER. 2000. Analysis of saliva for periodontal diagnosis—A review. J. Clin. Periodontol. **27:** 453–465.
34. COOKE, J.A. *et al.* 2006. Effect of rhPDGF-BB delivery on mediators of periodontal wound repair. Tissue Eng. **12:** 1441–1450.
35. SARMENT, D.P. *et al.* 2006. Effects of rhPDGF-BB on bone marrow turnover during periodontal repair. J. Clin. Periodontol. **33:** 135–140.
36. GAPSKI, R. *et al.* 2004. Effect of systemic matrix metalloproteinase inhibition on periodontal wound repair: a proof of concept trial. J. Periodontol. **75:** 441–452.
37. HERR, A.E. *et al.* 2007. Microfluidic immunoassays as rapid saliva-based clinical diagnostics. Under review.
38. OZMERIC, N. 2004. Advances in periodontal disease markers. Clin. Chim. Acta **343:** 1–16.
39. SONG, S. *et al.* 2004. Microchip dialysis of proteins using in situ photopatterned nanoporous polymer membranes. Anal. Chem. **76:** 2367–2373.
40. SONG, S., A.K. SINGH & B.J. KIRBY. 2004. Electrophoretic concentration of proteins at laser-patterned nanoporous membranes in microchips. Anal. Chem. **76:** 4589.
41. KIRBY, B.J. *et al.* 2005. Microfluidic routing of aqueous and organic flows at high pressures: fabrication and characterization of integrated polymer microvalve elements. Lab Chip. **5:** 184.

Development of a Microfluidic Device for Detection of Pathogens in Oral Samples Using Upconverting Phosphor Technology (UPT)

WILLIAM R. ABRAMS,[a] CHERYL A. BARBER,[a] KURT McCANN,[a] GARY TONG,[a] ZONGYUAN CHEN,[b] MICHAEL G. MAUK,[b] JING WANG,[b] ALEX VOLKOV,[c] PETE BOURDELLE,[c] PAUL L. A. M. CORSTJENS,[d] MICHEL ZUIDERWIJK,[d] KEITH KARDOS,[e] SHANG LI,[e] HANS J. TANKE,[d] R. SAM NIEDBALA,[c] DANIEL MALAMUD,[a] AND HAIM BAU[b]

[a]*Department of Basic Sciences, New York University College of Dentistry, New York, New York 10010, USA*

[b]*School of Engineering and Applied Science, University of Pennsylvania, Philadelphia, Pennsylvania 19104, USA*

[c]*Chemistry Department, Lehigh University, Lehigh, Pennsylvania 18015, USA*

[d]*Leiden University Medical Center, Leiden, the Netherlands*

[e]*OraSure Technologies, Inc., Bethlehem, Pennsylvania 18015, USA*

ABSTRACT: Confirmatory detection of diseases, such as HIV and HIV-associated pathogens in a rapid point-of-care (POC) diagnostic remains a goal for disease control, prevention, and therapy. If a sample could be analyzed onsite with a verified result, the individual could be counseled immediately and appropriate therapy initiated. Our group is focused on developing a microfluidic "lab-on-a-chip" that will simultaneously identify antigens, antibodies, RNA, and DNA using a single oral sample. The approach has been to design individual modules for each assay that uses similar components (e.g., valves, heaters, metering chambers, mixers) installed on a polycarbonate base with a common reporter system. Assay miniaturization reduces the overall analysis time, increases accuracy by simultaneously identifying multiple targets, and enhances detector sensitivity by upconverting phosphor technology (UPT). Our microfluidic approach employs four interrelated components: (1) sample acquisition–OraSure UP*link*™ collectors that pick-up and release bacteria, soluble analytes, and viruses from an oral sample; (2) microfluidic processing–movement of microliter volumes of analyte, target analyte extraction and

Address for correspondence: William R. Abrams, New York University College of Dentistry, Department of Basic Sciences, 345 East 24th Street, New York, NY 10010. Voice: 212-998-9241; fax: 212-995-4087.
william.abrams@NYU.edu

Ann. N.Y. Acad. Sci. 1098: 375–388 (2007).
doi: 10.1196/annals.1384.020

amplification; (3) detection of analytes using UPT particles in a lateral flow system; and (4) software for processing the results. Ultimately, the oral-based microscale diagnostic system will detect viruses and bacteria, associated pathogen antigens and nucleic acids, and antibodies to these pathogens.

KEYWORDS: microfluidic; HIV; pathogen; point-of-care; diagnostic; saliva; oral fluid; multiplex analysis; lateral flow; confirmatory

BACKGROUND

Scope of the HIV Diagnostic Problem

HIV is a major pathogen challenging the global community and it is well recognized that HIV/AIDS is a leading cause of illness and death in the United States. In 2006, more than 1 million people living in the United States have been diagnosed with HIV/AIDS and more than 500,000 individuals have died over the last 25 years.[1] It was estimated that globally in 2005, 34 to 46 million people were infected with HIV, approximately 4.1 million new infections occurred, and 3 million deaths resulted. Dr. Kevin M. De Cock, Director of the WHO AIDS program, reported at the 16th International AIDS Conference that approximately 2.3 million children up to the age of 15 years around the world are infected with HIV.[2] Of the 38.6 million people worldwide with HIV, about 6.8 million are in low- and middle-income communities and can be expected to die within 2 years without antiretroviral therapy. Only about 4% of this infected population are now receiving therapy.[2] The rate of HIV infection in the USA is estimated to be 0.6% annually compared to a level of ~19% in South Africa for populations aged between 15 and 49 years. In South Africa alone ~320,000 individuals are reported to die annually on account of an AIDS-related illness.[3–7] Even in a cosmopolitan city like New York, it is estimated that there are 100,000 HIV-positive individuals and of these ~20% have not been diagnosed.[8–10] In Africa the percentage of undiagnosed HIV-positive individuals is higher than in North America.[11,12] As a result of this large undiagnosed population, the CDC recently modified its position and now recommends HIV testing of individuals aged 13 to 64 years as part of an annual medical examination.[13]

Early identification of infection is associated with increased initial cost, but intervention based on a diagnostic result allows an appropriate therapeutic response (FIG. 1),[14] ultimately reducing costs to society. Great Britain's National Health Service recently asked "How much is it worth spending to prevent one new HIV infection or transmission?" The Department's health economists estimated that the monetary value of preventing a new single transmission of HIV is between pound 0.5 and pound 1 million in individual health benefits and treatment costs.[15] In South Africa, SABMiller, one of the world's largest

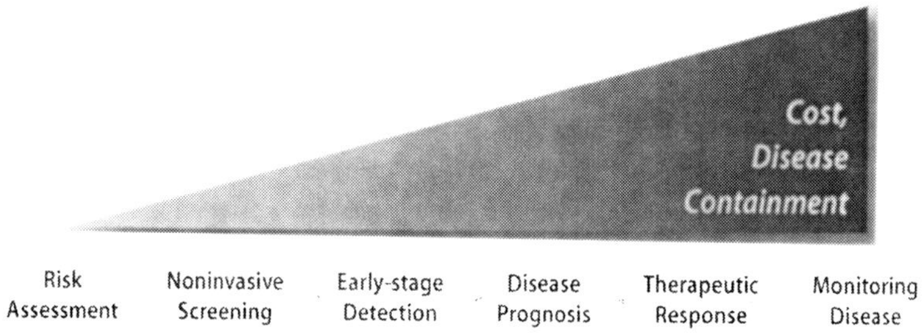

FIGURE 1. Schematic of the link between diagnostic testing and disease assessment. (Adapted from Hartwell *et al.*[14])

brewers, was reported by the Wall Street Journal to have initiated an employee testing and treatment program that provides counseling and therapy in order to decrease the loss of highly trained workers.[16] The economic consequences of NOT testing for HIV infection are staggering. Early diagnosis through testing ultimately saves money by breaking the chain of transmission and has been responsible for the dramatic decrease in the spread of the disease in the United States. Closing the "window" from the time of infection to seroconversion has the potential to further reduce the spread of HIV disease and increase therapeutic effectiveness.[17,18] Because people do not die from an HIV infection, but rather from other opportunistic infections, such as tuberculosis, intestinal tract infections, and progressive multifocal leukoencephalopathy,[19] we envision developing modules for multiple HIV-associated pathogens. Existing screening tests for HIV require the subject to return for results of a confirmatory test, which has a low compliance rate and is associated with a number of social dynamic issues.[9] A point-of-care (POC) confirmatory test would alleviate this problem.

Device Concept

Our focus from the outset was to address the lack of diagnostic testing for infectious diseases. We wanted to build a simple to use, rapid, POC device for diagnosis of HIV and other bacterial–viral infections because of the pressing need. HIV is clearly an important disease, but it is also a good model for a proof-of-concept of a confirmatory test by employing a multiplex analysis paradigm: parallel processing of multiple analytes for antigen, antibody, and/or nucleic acid. However, because of opportunistic infections associated with HIV, it is important to simultaneously detect other bacterial and viral pathogens. Other infectious agents have become apparent in patients receiving antiretroviral drug treatment, including many bacteria,[20] fungi, and protozoan and viral pathogens that can effectively be treated with drugs if diagnosed.[19]

Challenge

We are using a modular design strategy to build a device that ultimately will provide testing for multiple pathogens using saliva as a noninvasive source of test material, detecting pathogens by antigen, antibody, and/or nucleic acid, by use of multiplex analysis. Any one of these tests is available today although not packaged together and the impact of simultaneous, multiple analyte testing is to increase the overall test accuracy and its diagnostic usefulness.[17]

A schematic was constructed of the tests the device needed to include. All the techniques were available as bench-top procedures; including nucleic acid isolation and amplification (reverse transcriptase-polymerase chain reaction [RT-PCR] and/or PCR) and identification of antigen or antibody by an enzyme-linked immunosorbent assay (ELISA) or Western blot approach. Modifications of these techniques to microscale presented engineering challenges that included controlling the surface area of channels and chambers relative to the volume of sample, dealing with the movement of small volumes of liquid, small amounts of target, and the required sensitivity of the reporter. Also, to be of maximum use, results are best obtained in less than 1 h.

Sample Collection

The standardized collection of oral fluid was addressed in a study that compared eight different collectors for their ability to pick up and release bacteria, protein, and virus.[21] The OraSure UP*link*™ was chosen because of its overall performance characteristics and in addition it was already an FDA-approved device. The collector is a pencil-like device with a compressed cellulose-like sponge that expands and can collect ~300 μL of fluid when inserted into the mouth between the gum and the cheek for 1 min (Fig. 2A). The schematic (FIG. 2B) shows the compressed sponge releasing oral fluid into a channel as it is inserted into the assay cassette.

Microchip Components

Several components are required for a disposable microfluidic chip construction and operation. These include the chip substrate, mechanism for fluid movement, mixing, measurement of volume, PCR chamber configuration, passivation of surfaces, valves, connectors, reporter, and reporter detection. FIGURE 3 shows a number of different valves that were developed to evaluate which was the most efficient and effective one. The phase change (FIG. 3A) uses a thermoelectric unit to maintain a channel at −20°C so that the flow stops on freezing when liquid enters the cooled area, and subsequent warming allows the flow to continue;[22] the hydrogel valve[23] (FIG. 3B) uses a hydroscopic matrix

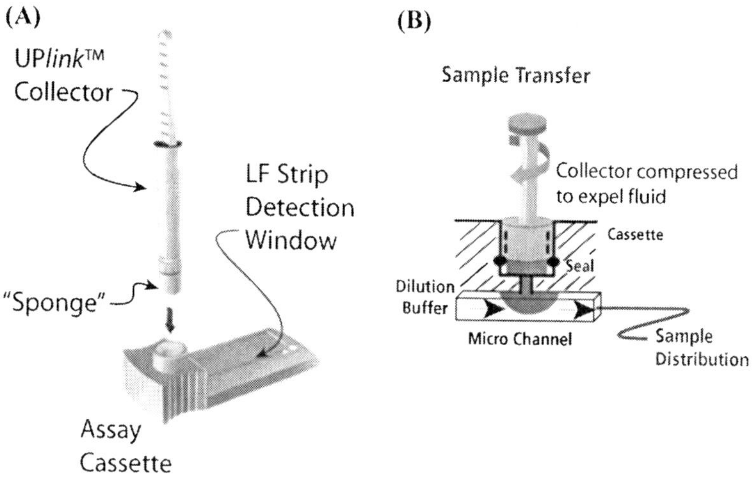

FIGURE 2. (**A**) UP*link*™ collector and cassette, which houses the microfluidic chip. (**B**) Schematic showing transfer of fluid from pick-up sponge to a channel on the chip.

prepared from acrylamide that swells when in contact with an aqueous solution at ambient temperatures and shrinks again when heated; a passive valve (FIG. 3C) that takes advantage of surface tension to stop flow under low pressures; a check valve (FIG. 3D) that prevents back flow of reagent; and a deformable valve (FIG. 3E) that is fabricated with a thin-film deformable membrane over a channel combined with an external actuator pin that blocks flow.

PCR Chambers

PCR has been performed by shuttling a bolus of fluid containing nucleic acid with appropriate reagents between zones of different temperatures using

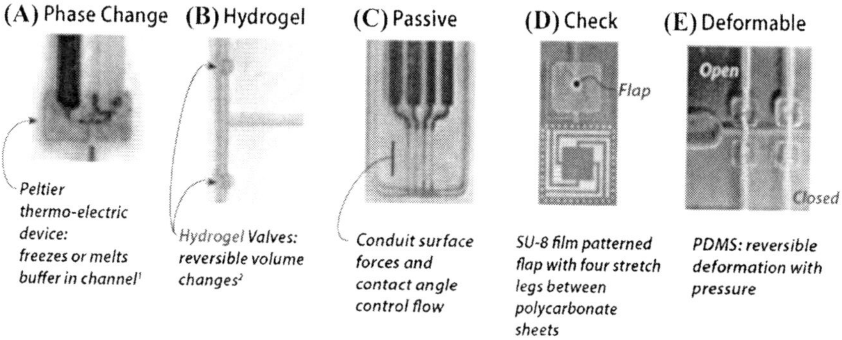

FIGURE 3. Panel of valves for use in a microfluidic chip.

a syringe pump under computer control.[24] We also developed a buoyancy or thermosiphon-based PCR (FIG. 4), a unique design that automatically generates a circular flow through changes in thermally induced density changes within the loop that includes all the reagents necessary for PCR.[22,25] Amplification of target occurs as reagents cycle through the temperature zones. However, the operating plane of the loop must be set at an angle of 90° to gravity, which is currently not compatible with the cassette's horizontal format. The current chip will use a static PCR chamber consisting of a 10-μL well fitted with a thermoelectric heater and thermister.

Microfluidic Chips

An example of a microfluidic chip for PCR-based detection is shown in FIGURE 5. This hydraulically driven chip is capable of receiving a fluid sample, metering its volume, lysing any pathogen present, isolating nucleic acid, performing PCR, incubating with reporter conjugate, lateral flow immunochromatography, and detecting the amplified product. FIGURE 5B shows the chip incorporated into a cassette. Another chip module has been constructed to perform consecutive flow upconverting phosphor technology (UPT) simultaneous detection of TB, HCV, and HIV antibodies.[26]

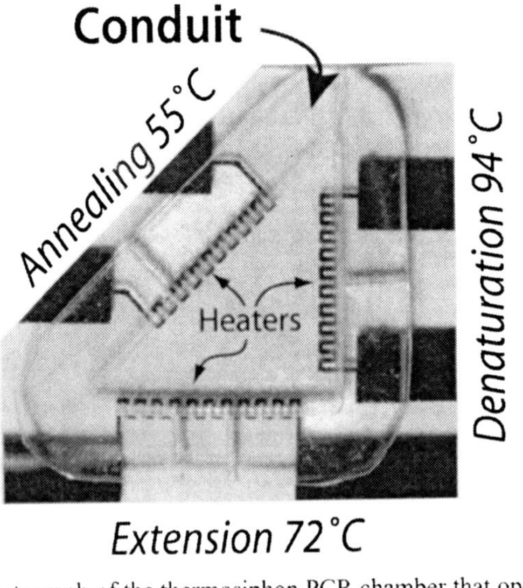

FIGURE 4. Photograph of the thermosiphon PCR chamber that operates in a vertical orientation by the density changes induced by heating and cooling of the PCR reaction mix.

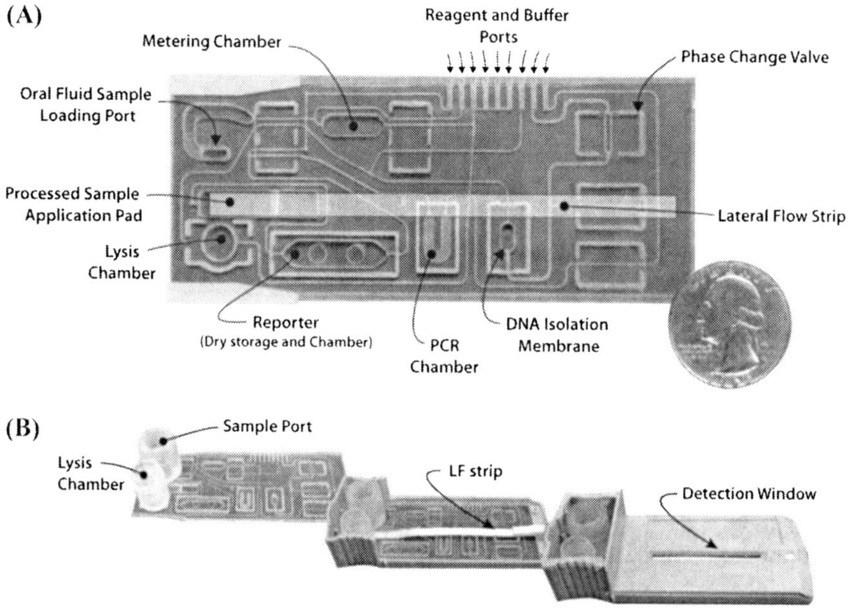

FIGURE 5. Photograph of DNA chip. (**A**) Top view. (**B**) Sequential images showing the fit of the chip into the cassette.

Common Reporter

UPT was selected as the detection reporter because of its high signal-to-noise ratio for biological samples in a miniaturized format that is compatible with lateral flow.[27] UPT particles are ceramic spheres containing rare earth molecules that can absorb multiple photons of infrared light (980 nm) and emit visible light in the green part of the spectrum (550 nm).[27] Different combinations of rare earths can be used to produce different visible colors. This process converts lower energy to higher energy, a process that does not normally exist in nature and consequently has no natural background (FIG. 6). The particles are silica coated to both passivate the ceramic surface and allow covalent conjugation of biomolecules, which "functionalizes" the UPT particles as a reporter. The UPT particles have been used in a number of different diagnostic assays.[27–30]

Lateral Flow Detection

A nitrocellulose-based lateral flow strip is a simple and inexpensive technique for the chromatographic separation of target bound to reporter from excess reporter.[31] Typically, a lateral flow strip (FIG. 7) has at least two capture lines, a test and a control line. The test line binds the target reporter complex and the control line verifies that flow has reached the end of the strip by binding

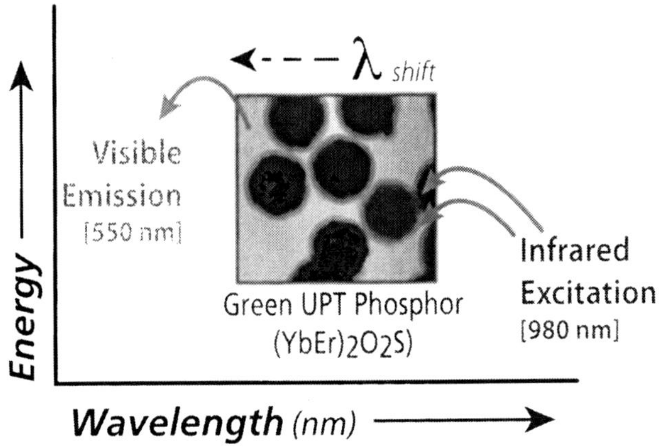

FIGURE 6. Schematic of up-phosphor conversion. *Inset* is an SEM of 400 nm UPT particles.

free reporter. The waste pad functions as a sink that ensures that reagents do not flow backward toward the test line area. The capture lines take advantage of high-affinity binding pairs, such as biotin to avidin or between an antigen and its antibody to form a sandwich. The schematic (FIG. 7B) illustrates a simplified view of how the strip functions for recognition of a PCR amplicon (the DNA target) that was labeled with biotin and digoxigenin. The target

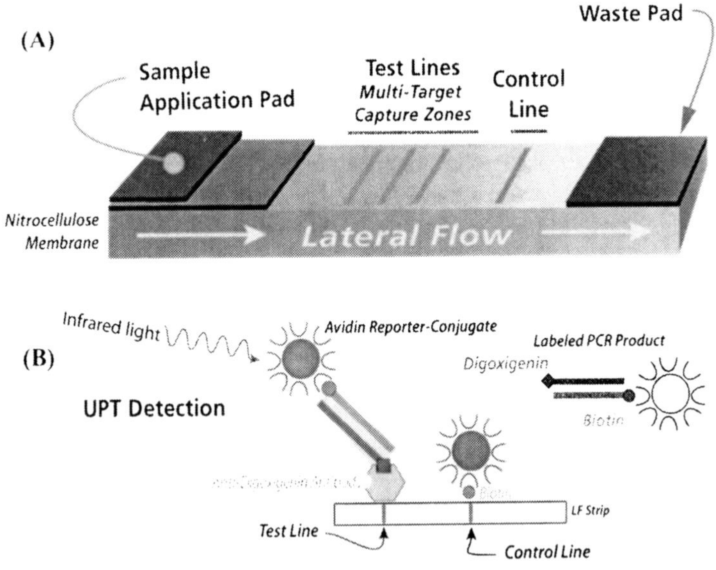

FIGURE 7. (**A**) Lateral flow strip anatomy. (**B**) Schematic showing the binding of target complex to the test line. DNA amplicon with a biotin tag binds to the UPT-avidin reporter. Antidigoxigenin antibody on the test line captures the digoxigenin tag on the amplicon during immunochromatography. Free avidin-UPT reporter binds to the control line.

reporter complex (biotinylated DNA target bound to avidin-coated UPT reporter particles) and free reporter (avidin-coated UPT reporter particles) are trapped at the test and control lines, respectively, as they flow past their binding partner. The capture lines are then interrogated with IR light (980 nm) and the UPT-emitted light (550 nm) is recorded.

METHODS AND RESULTS

Comparison of Nucleic Acid Fluorescence Staining with UPT Detection

PCR was performed on *S. pyogenes* DNA isolated from ~10^7 cells in a twofold dilution series (FIG. 8). Each dilution was then amplified using biotinylated and digoxigenin-labeled primers. Panel A shows the PCR reaction product (10 μL) after 25 cycles of PCR, electrophoresed on a 2% agarose gel. Panel B shows the relative fluorescence ratios (test line/control line) obtained with UPT detection using 1 μL of PCR product as described in the methods section. Note the increased sensitivity with UPT detection.

Chip Docking

Mating the chip to a buffer or reagent reservoir requires a quick connection to occur as the disposable chip is inserted into the reader. Fabrication from polyethylene or aluminum of a simple conical male fitting allows for a

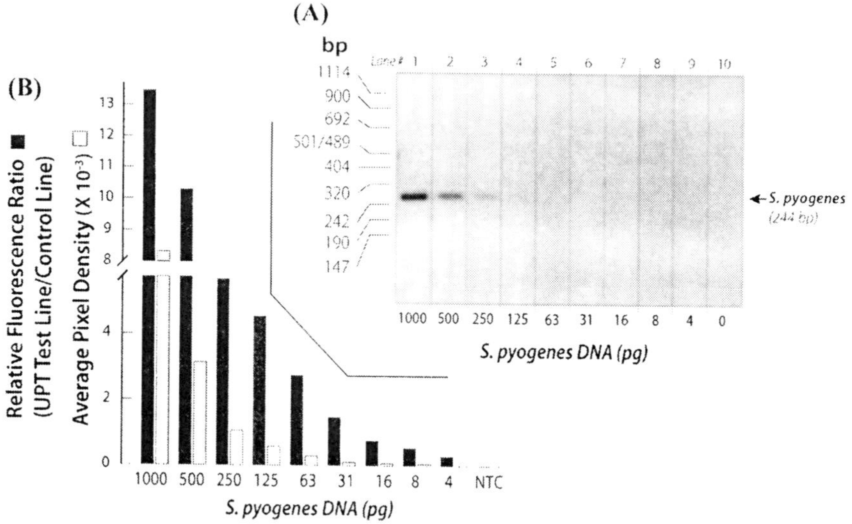

FIGURE 8. Comparison of UPT detection sensitivity with ethidium bromide staining.

leak-free seal to 29 psi when mated with a polycarbonate seat using minimal joining pressure, which is well below the chip's operating pressures.

Multiplex Detection

S. pyogenes *and* B. cereus

Multiplexing was carried out with an optimized PCR protocol using Eppendorf Taq DNA polymerase and primers designed for each organism (FIG. 9). These results confirm that multiple bacteria can be simultaneously amplified.

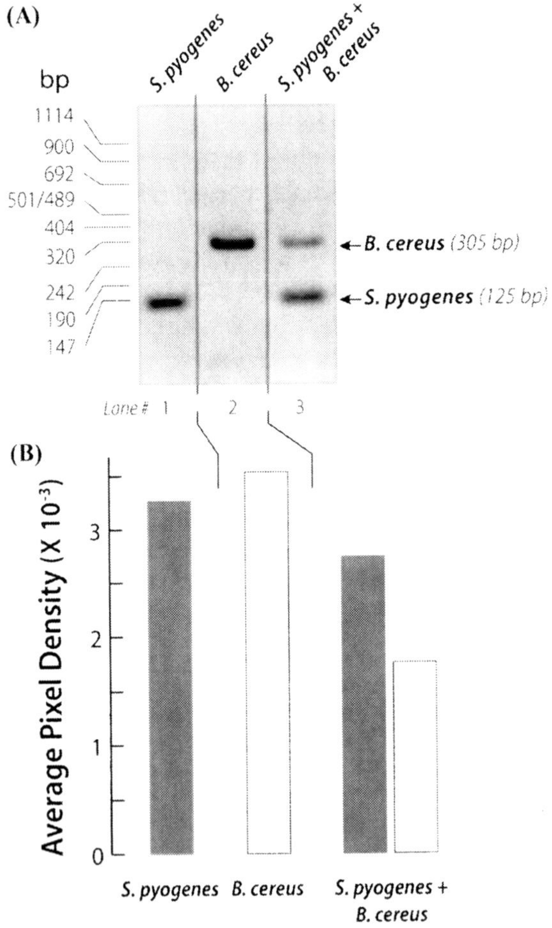

FIGURE 9. Simultaneous amplification of *B. cereus* and *S. pyogenes* nucleic acids. (**A**) Ethidium bromide-stained gel showing the amplicons produced from *B. cereus* and *S. pyogenes*. (**B**) Digitalization of the agarose gel result. Enzymatic lysis was performed with lysozyme (0.5 mg/mL), proteinase K (0.5 mg/mL), and mutanolysin (100 U/mL) for 15 min at 37°C and then for 15 min at 67°C. Nucleic acid purification used the DNeasy Tissue Kit (Qiagen, Germany) as per manufacturer's recommendation.

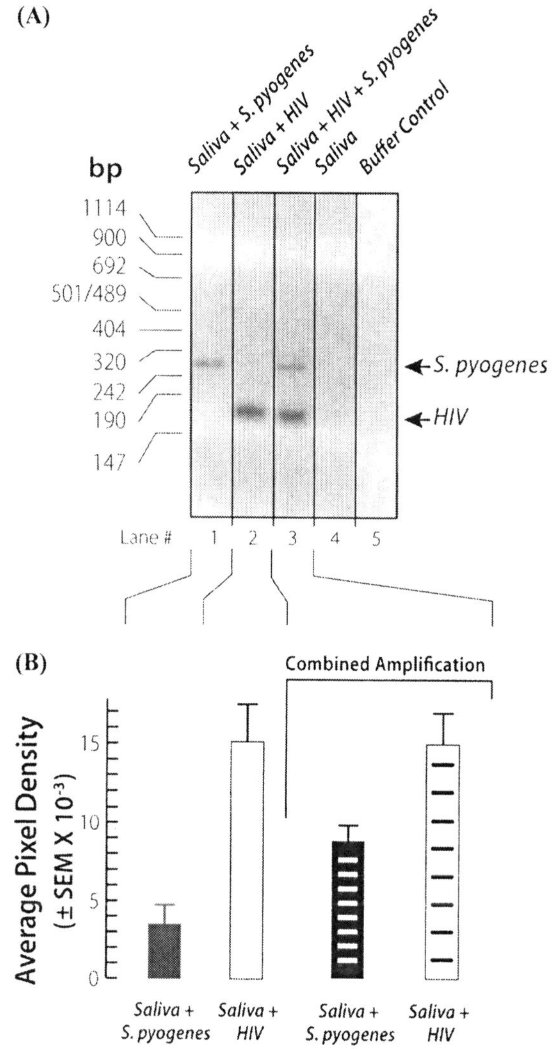

FIGURE 10. Simultaneous detection of *S. pyogenes* and HIV in saliva. (**A**) Representative lanes of an ethidium bromide–stained agarose gel showing the amplicons produced from nucleic acid isolated from virus and bacteria and amplified together. (**B**) Digitization of analysis of amplification from four different saliva samples.

Multiplex Detection of S. pyogenes *or* B. cereus *and HIV*

To demonstrate the ability to simultaneously detect virus and bacteria in an oral fluid sample, we isolated these together in saliva or plasma and performed simultaneous PCR and RT-PCR. Virus and bacteria were added to whole saliva and nucleic acid isolated using a Qiagen RNAeazy RNA extraction kit. The capability to simultaneously detect amplicon products from both HIV and

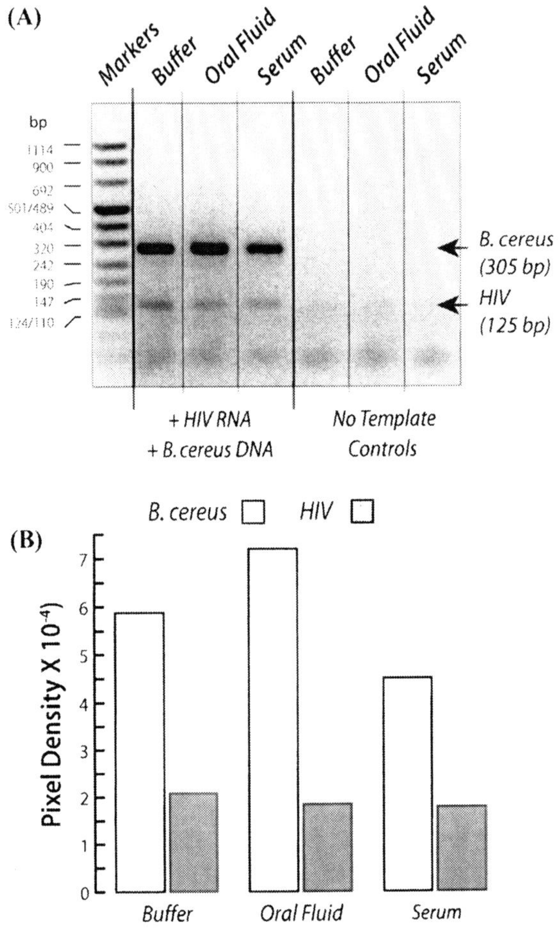

FIGURE 11. Simultaneous detection of HIV and *B. cereus* in buffer, oral fluid, and serum. (**A**) Representative lanes of an ethidium bromide–stained agarose gel showing the amplicons produced from nucleic acid isolated from HIV and *B. cereus*. (**B**) Digitization of panel A.

S. pyogenes in saliva is documented in FIGURE 10, which shows representative lanes from an ethidium bromide–stained agarose gel electrophoresis separation together with the results of digital quantitation. FIGURE 11 shows a similar experiment for the simultaneous detection of *B. cereus* and HIV comparing the ability to amplify in the presence of buffer, saliva, or serum.

SUMMARY

A modular design strategy has been used to build a diagnostic device for infectious disease detection by performing multiple tests for targets in parallel and employing a standardized collection device, the UP*link*™ collector, and a common sensitive reporter (UPT). We have carried out simultaneous determinations of *S. pyogenes* and *B. cereus*, HIV and *B. cereus*, and HIV and *S. pyogenes* in oral fluid. The use of UPT for a reporter provides a 3-order-of-magnitude enhancement of sensitivity over conventional ethidium bromide–stained agarose gels.

ACKNOWLEDGMENTS

This work was supported by grants to Dan Malamud from the NIH (UO1DE0114964 and UO1DE017855) and the New York State Office of Science Technology and Academic Research (NYSTAR).

REFERENCES

1. MMWR. 2006. Twenty-five years of HIV/AIDS–United States, 1981–2006. MMWR Morb. Mortal. Wkly. Rep. **55:** 585–589.
2. Altman, L.K. 2006. Children slip through cracks of AIDS efforts. The New York Times, August 17. New York.
3. Sanchez, T. *et al.* 2006. Human immunodeficiency virus (HIV) risk, prevention, and testing behaviors—United States, National HIV Behavioral Surveillance System: men who have sex with men, November 2003-April 2005. MMWR Surveill. Summ. **55:** 1–16.
4. Eaton, D.K. *et al.* 2006. Youth risk behavior surveillance–United States, 2005. MMWR Surveill. Summ. **55:** 1–108.
5. MMWR. 2006. Trends in HIV-related risk behaviors among high school students—United States, 1991–2005. MMWR Morb. Mortal. Wkly. Rep. **55:** 851 854.
6. MMWR. 2006. HIV prevalence among populations of men who have sex with men—Thailand, 2003 and 2005. MMWR Morb. Mortal. Wkly. Rep. **55:** 844–848.
7. MMWR. 2006. The Global HIV/AIDS pandemic, 2006. MMWR Morb. Mortal. Wkly. Rep. **55:** 841–844.
8. Frieden, T.R. 2006. Stopping the HIV/AIDS epidemic in NYC. PRN Notebook **11:** 3–7.
9. Frieden, T.R. *et al.* 2005. Applying public health principles to the HIV epidemic. N. Engl. J. Med. **353:** 2397–2402.
10. Centers for Disease Control and Prevention. 2005. HIV/AIDS Surveillance Report, 2004, Vol. 16 US Department of Health and Human Services, CDC. Atlanta, GA. 1–46.
11. De Cock, K.M., E. Marum & D. Mbori-Ngacha. 2003. A serostatus-based approach to HIV/AIDS prevention and care in Africa. Lancet **362:** 1847–1849.
12. Rosenberg, T. 2006. When a pill is not enough. The New York Times, August 6. New York.
13. Branson, B.M. *et al.* 2006. Revised recommendations for HIV testing of adults, adolescents, and pregnant women in health-care settings. MMWR Recomm. Rep. **55:** 1–17.

14. HARTWELL, L. *et al.* 2006. Cancer biomarkers: a systems approach. Nat. Biotechnol. **24:** 905–908.
15. THE UK COLLABORATIVE GROUP FOR HIV AND STI SURVEILLANCE. 2004. Focus on Prevention. HIV and Other Sexually Transmitted Infections in the United Kingdom in 2003. Health Protection Agency Centre for Infections. London.
16. ECHIKSON, W. & A. COHEN. 2006. SABMiller's AIDS test program gest results. The Wall Street Journal, August 19. New York.
17. PILCHER, C.D. *et al.* 2005. Detection of acute infections during HIV testing in North Carolina. N. Engl. J. Med. **352:** 1873–1883.
18. PILCHER, C.D. *et al.* 2004. Acute HIV revisited: new opportunities for treatment and prevention. J. Clin. Invest. **113:** 937–945.
19. ABERG, J.A. 2000. Reconstitution of immunity against opportunistic infections in the era of potent antiretroviral therapy. AIDS Clin. Rev. 115–138.
20. MCNEIL, D.G. JR. 2006. Worrisome new link: AIDS drugs and leprosy. The New York Times, October 24. New York.
21. HOLM-HANSEN, C. *et al.* 2004. Comparison of oral fluid collectors for use in a rapid point-of-care diagnostic device. Clin. Diagn. Lab. Immunol. **11:** 909–912.
22. CHEN, Z. *et al.* 2005. Thermally-actuated, phase change flow control for microfluidic systems. Lab Chip. **5:** 1277–1285.
23. WANG, J. *et al.* 2005. Self-actuated, thermo-responsive hydrogel valves for lab on a chip. Biomed. Microdev. **7:** 313–322.
24. MALAMUD, D. *et al.* 2005. Point detection of pathogens in oral samples. Adv. Dent. Res. **18:** 12–16.
25. CHEN, Z. *et al.* 2004. Thermosiphon-based PCR reactor: experiment and modeling. Anal. Chem. **76:** 3707–3715.
26. CORSTJENS, P. 2006. Consecutive flow microfluidic chip for detection of TB, HCV and HIV antigens. NYAS Oral-Based Diagnostics. Ann. N. Y. Acad. Sci. October 10–13, 2006.
27. CORSTJENS, P. *et al.* 2005. Infrared up-converting phosphors for bioassays. IEEE Proc. Nanobiotechnology **152:** 64–72.
28. CORSTJENS, P. *et al.* 2001. Use of up-converting phosphor reporters in lateral-flow assays to detect specific nucleic acid sequences: a rapid, sensitive DNA test to identify human papillomavirus type 16 infection. Clin. Chem. **47:** 1885–1893.
29. ZUIDERWIJK, M. *et al.* 2003. An amplification-free hybridization-based DNA assay to detect *Streptococcus pneumoniae* utilizing the up-converting phosphor technology. Clin. Biochem. **36:** 401–403.
30. CORSTJENS, P.L. *et al.* 2003. Lateral-flow and up-converting phosphor reporters to detect single-stranded nucleic acids in a sandwich-hybridization assay. Anal. Biochem. **312:** 191–200.
31. WEISS, A. 1999. Concurrent engineering for lateral-flow diagnostics. IVD Technology Magazine. Canon Communications. Los Angeles, CA.

Microsensor Arrays for Saliva Diagnostics

DAVID R. WALT,[a] TIMOTHY M. BLICHARZ,[a] RYAN B. HAYMAN,[a] DAVID M. RISSIN,[a] MICHAELA BOWDEN,[a] WALTER L. SIQUEIRA,[b] EVA J. HELMERHORST,[b] NERLINE GRAND-PIERRE,[b] FRANK G. OPPENHEIM,[b] JASVINDER S. BHATIA,[c] FRÉDÉRIC F. LITTLE,[c] AND JEROME S. BRODY[c]

[a]*Department of Chemistry, Tufts University, Medford, Massachusetts 02155, USA*

[b]*Department of Periodontology and Oral Biology, Boston University School of Dental Medicine, Boston, Massachusetts 02118, USA*

[c]*Department of Medicine, Boston University School of Medicine, Boston, Massachusetts 02118, USA*

ABSTRACT: Optical fiber microarrays have been used to screen saliva from patients with end-stage renal disease (ESRD) to ascertain the efficacy of dialysis. We have successfully identified markers in saliva that correlate with kidney disease. Standard assay chemistries for these markers have been converted to disposable test strips such that patients may one day be able to monitor their clinical status at home. Details of these developments are described. In addition, saliva from asthma and chronic obstructive pulmonary disease (COPD) patients is being screened for useful diagnostic markers. Our goal is to develop a multiplexed assay for these protein and nucleic acid biomarkers for diagnosing the cause and severity of pulmonary exacerbations, enabling more effective treatment to be administered. These results are reported in the second part of this article.

KEYWORDS: noninvasive diagnostics; end-stage renal disease; asthma; COPD; antibody array; DNA array

INTRODUCTION

Saliva has been used increasingly as a sample matrix for systemic disease diagnosis, based on the premise that saliva reflects the composition of blood and is a window to an individual's general health.[1,3,4] Our interdisciplinary project team is focused on developing point-of-care diagnostic systems for common disease states. We initially screen for potentially useful biomarkers

Address for correspondence: David R. Walt, Department of Chemistry, Tufts University, 62 Talbot Avenue, Medford, MA 02155. Voice: 1-617-627-3470; fax: 1-617-627-3443. david.walt@tufts.edu

Ann. N.Y. Acad. Sci. 1098: 389–400 (2007).
doi: 10.1196/annals.1384.031

using standard assays and then transition to microsphere-based assays for any potentially useful analytes. These microsphere-based sensors and probes are then integrated into a multiplexed detection platform. This article describes the implementation of these strategies for two disease case studies to produce tests that may be used in point-of-care diagnostics.

Salivary Analysis of End-Stage Renal Disease (ESRD)

ESRD is a condition in which kidney functions are severely compromised. Patients with ESRD require kidney transplantation or frequent hemodialysis to prevent clinical complications or death due to the buildup of waste products in the blood.[2] It is critical to monitor kidney function in pre-ESRD patients to diagnose conversion to the acute disease state. We examined numerous potential renal function biomarkers in the saliva of ESRD patients. These patients should be an ideal study cohort because the concentration of some blood analytes decreases dramatically during dialysis. A noninvasive, self-administered, and rapid method for monitoring kidney function could reduce the need for periodic hospital visits and blood testing for pre-ESRD patients and could potentially be used to evaluate dialysis efficacy for ESRD patients.

Initial ESRD Biomarker Screening Study

ESRD patients in various states of disease progression were enrolled at the dialysis clinic of Boston University Medical Center (BUMC) and were asked to donate saliva before and after undergoing dialysis on a weekly basis for a 2-month period. A panel of candidate analytes was screened for consistent trends between the pre- and post-dialysis saliva samples. Preliminary tests were performed for sodium (Na^+), potassium (K^+), magnesium (Mg^{2+}), calcium (Ca^{2+}), chloride (Cl^-), phosphate (PO_4^{3-}), and nitrite (NO_2^-) ions, pH, thiols, uric acid (UA), amylase, lactoferrin, esterase, total protein, nucleic acids, and glucose levels (TABLE 1). Analytes shown in italics exhibited differences between pre- and post-dialysis saliva composition in initial screening. Analytes listed in bold type (NO_2^-, Cl^-, Na^+, and UA) showed the best correlations and were monitored in a more extensive study by collecting saliva samples at regular time intervals throughout dialysis.

ESRD Salivary Biomarker Monitoring During Dialysis

To determine whether these analytes were good indicators for monitoring the efficacy of dialysis, a study was conducted where saliva samples were collected from ESRD patients immediately before and after dialysis, as well

TABLE 1. Salivary analytes initially screened for dialysis correlation

Analytes deemed not useful	Analytes deemed potentially useful
PO_4^{3-}	***NO_2^-***
Glucose	***UA***
Thiols	***Na^+***
Esterase	***Cl^-***
Nucleic acids	*Total protein*
Ca^{2+}	*pH*
Mg^{2+}	*Amylase*
K^+	*Lactoferrin*

NOTE: Analytes in italics showed differences between pre- and post-dialysis saliva composition in initial screening. Analytes in boldface showed the best correlations; for more detail see text.

as at hourly intervals throughout treatment. Saliva levels of NO_2^- and UA consistently tracked dialysis, exhibiting decreasing concentrations throughout the process; the rate of decrease, however, varied by individual. These two analytes were selected for further evaluation.

Salivary Test Strips for Monitoring Renal Disease

A simple colorimetric test strip was developed to semiquantitatively determine concentrations of NO_2^- and UA in saliva. This approach offers the potential for a low-tech and low-cost method for monitoring renal status. Chromatography paper was impregnated with the NO_2^- and UA detection chemistries were followed by adhesion of the test papers onto a vinyl support material. The colorimetric test paper for salivary NO_2^-determination is based on the Griess reaction, a common method for nitrite quantification,[5,6] while the colorimetric test paper for UA uses a sodium bicinchoninate chelate method.[7,8] Brief immersion in solution produced test pad color intensities proportional to the concentrations of NO_2^- and UA in the sample. We developed a calibration color chart to visually determine the concentrations of these analytes (FIG. 1). FIGURE 2 demonstrates the color change of the test strips after immersion in archived saliva supernatant samples collected from pre- and post-dialysis ESRD patients.

Following this proof-of-principle study, test strips were employed at the BUMC Dialysis Clinic for point-of-care salivary NO_2^- and UA determinations. Stimulated, whole saliva was collected from 19 ESRD patients both before and after dialysis and was tested immediately using the NO_2^-/UA strips. Similarly, time-matched samples were donated by 10 healthy controls and were similarly analyzed using the test strips. The outcome confirmed our earlier results; the test strips could be used to follow NO_2^- and UA concentrations during dialysis (FIG. 3).

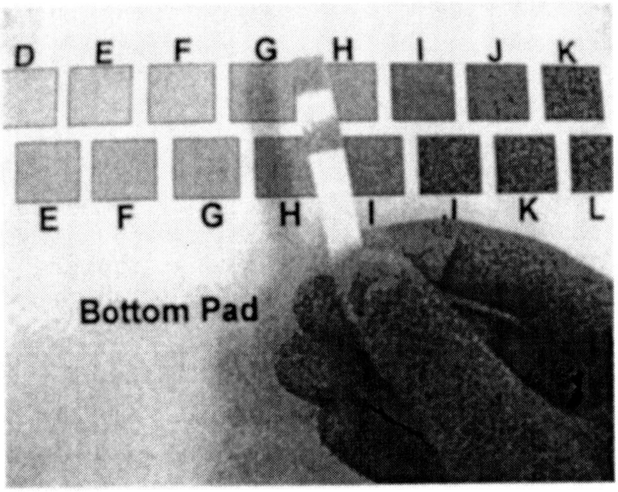

FIGURE 1. Photograph of a NO_2^-/UA test strip used to determine the concentrations of NO_2^- (*top pad*) and UA (*bottom pad*) in a saliva sample.

Salivary Analysis of Pulmonary Inflammatory Diseases

Pulmonary inflammatory diseases, such as asthma and chronic obstructive pulmonary disease (COPD), are becoming increasingly prevalent. Asthma

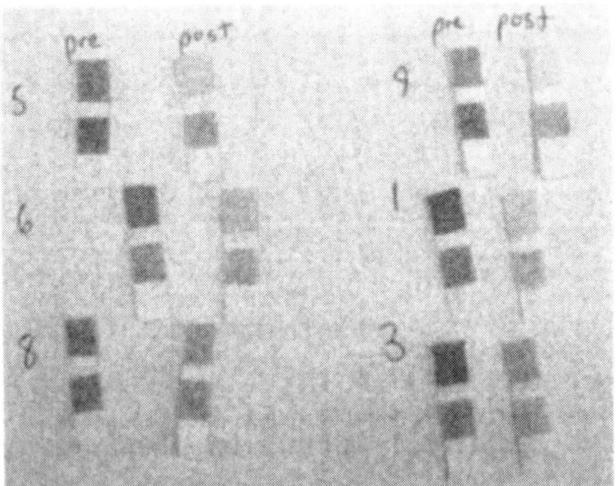

FIGURE 2. Test strips following immersion in archived saliva supernatant samples from six ESRD patients collected prior to ("pre") and immediately after ("post") undergoing dialysis treatment. Note the comparative difference in test strip color intensity between pre- and post-dialysis saliva samples.

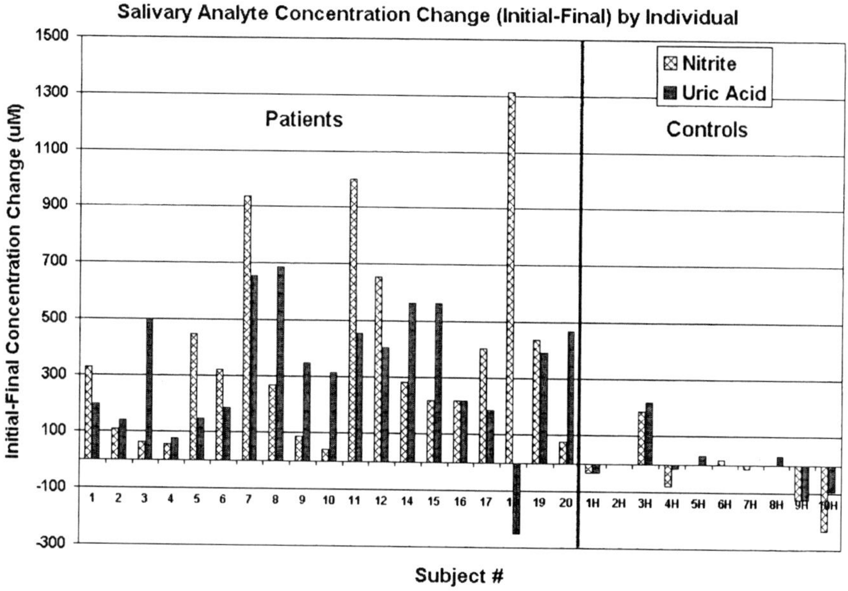

FIGURE 3. Test strip results compiled by examining stimulated whole saliva samples from ESRD patients in the BUMC Dialysis Clinic and healthy controls (not dialyzed). Each strip was evaluated by two analysts and the two concentration readings were averaged.

affects nearly 20 million Americans and costs $11.5 billion in direct expenditures in 2004.[9] COPD affects nearly 16 million Americans, with another 14 million estimated as living with undiagnosed disease.[10]

Our goal is to develop a portable, point-of-care device that can rapidly monitor multiple biomarkers in the saliva of patients suffering from obstructive pulmonary inflammatory diseases. The pathogenesis of pulmonary inflammation in obstructive diseases is the result of a complex network of specialized immune cells and their protein products.[11,12] Causes of exacerbation include: allergens, irritants, heat/cold/humidity, and bacterial or viral infection.[13,14] A platform capable of simultaneously monitoring both the causes of exacerbation as well as the levels of numerous biomarkers resulting from the pathogenic response would be a powerful tool for elucidating the differences associated with the different causes of exacerbation (extrinsic vs. intrinsic). By monitoring many analytes simultaneously, salivary protein and pathogen "fingerprints" could be created. These individual profiles could be regularly monitored to elucidate the causes of exacerbation and to evaluate the effectiveness of treatment.

Initial Screening of Salivary Cytokines and Chemokines

To determine the endogenous cytokines and chemokines present at detectable concentrations in saliva, we initially screened a small number of

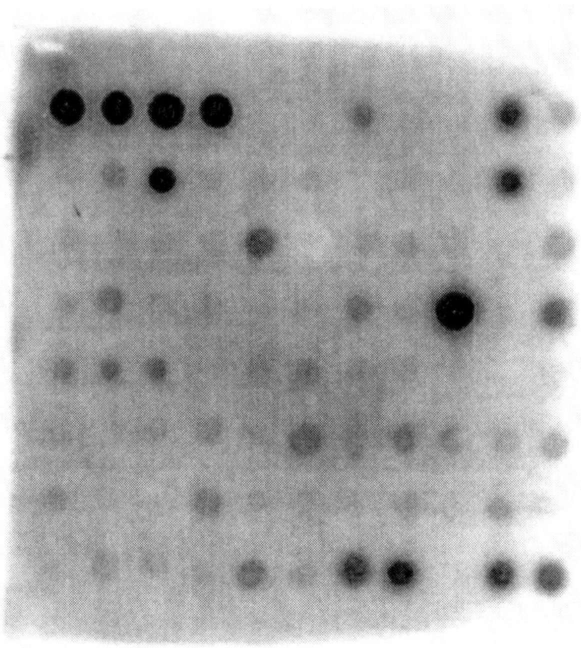

FIGURE 4. Representative salivary cytokine and chemokine screening results using the RayBio Cytokine Array V. Each dark spot visible on the array corresponds to an analyte present in saliva above the detection limit of the kit. Cytokines and chemokines detected on the array that are associated with pulmonary inflammatory diseases could be examined with secondary screening studies.

archived saliva supernatant samples from asthmatics and healthy controls using Human Cytokine Array V Kits from RayBiotech (Norcross, GA, USA). These commercially available tests are based on a multiplexed enzyme-linked immunosorbent assay (ELISA) with chemiluminescent detection and they provide qualitative results identifying the relative levels of 79 cytokines and chemokines (FIG. 4). Using this method, we were able to identify a number of salivary cytokines and chemokines that were examined in greater detail using secondary quantitative assays.

Secondary Screening of Salivary Cytokines and Chemokines

A number of cytokines and chemokines showing elevated levels with pulmonary inflammation on the RayBio Cytokine Array V were examined with quantitative ELISA screening studies. Representative preliminary quantitative salivary screening results of 12 asthmatic patients and 12 healthy controls are shown in FIGURE 5. When examined closely, no single analyte correlates

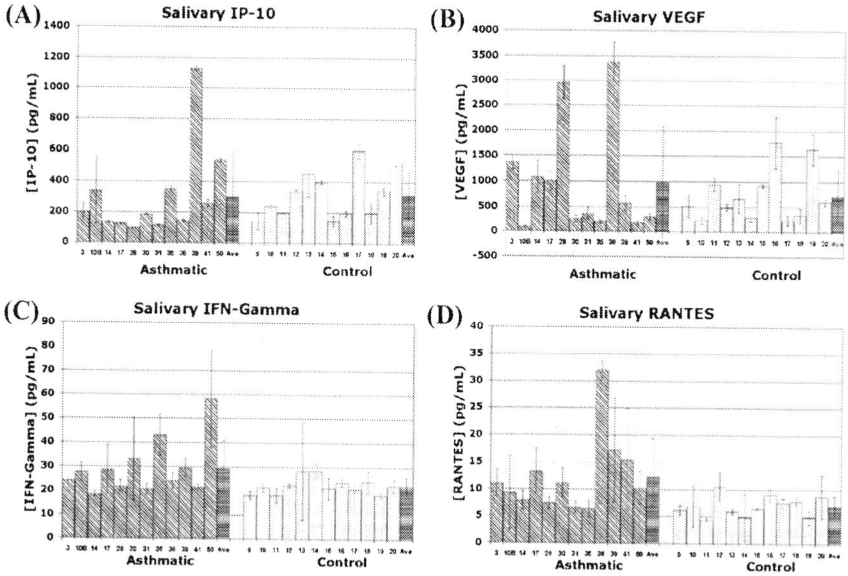

FIGURE 5. Representative quantitative screening results for (**A**) IP-10, (**B**) VEGF, (**C**) IFN-γ, and (**D**) RANTES for 12 asthmatic patients and 12 healthy controls determined using microtiter plate-based ELISA. The average for each population (asthmatic or healthy control) is represented by the *dark gray bar* on the right of each data set.

with pulmonary inflammatory state, but elevated levels of multiple analytes are present in most of the patients. Cytokines and chemokines showing potential correlations with pulmonary inflammation will be further investigated to confirm their utility as asthma biomarkers. Finally, by examining numerous inflammatory proteins simultaneously using a multiplexed assay, we hope to develop a better understanding of the different ways in which asthma can manifest itself in different patients.

Development of a Multiplexed Fiber Optic Microsphere-Based Cytokine Array

Assays for cytokines or chemokines shown to have potential correlation with pulmonary inflammatory disease or exacerbation could be converted to microsphere-based probes and pooled for multiplexed screening studies. To perform multiplexed microsphere-based fiber optic measurements, amine-functionalized 3.1-μm diameter polymer microspheres were first encoded with distinct concentrations of a fluorescent europium dye. Microspheres were then converted into cytokine–chemokine probes by the covalent attachment of monoclonal antibodies via glutaraldehyde chemistry. Probes recognizing different analytes were pooled and deposited into the wells of a fiber optic array to produce a multiplexed immunoassay. Additional cytokine and chemokine probes can be included on the multiplexed array by modifying the composition of the microsphere bead pool. The current iteration of the multiplexed cytokine array includes probes specific for IFN-γ, IP-10, RANTES, eotaxin-3, and VEGF.[15]

Salivary Analysis of Pulmonary Pathogens

We hypothesize that there is a natural exchange of bacteria and viruses associated with pulmonary exacerbations between upper respiratory tract fluids and saliva. We have identified a variety of organisms and are developing multiplexed bead-based fiber optic sensor arrays to screen saliva samples for these pathogens. Probe sequences specific to polymerase chain reaction (PCR) amplicons for the pathogens listed in TABLE 2 have been incorporated into Sentrix and BeadChip arrays (Illumina, Inc., San Diego, CA, USA), containing 96 and 16 bundles of 50,000 bead sensors, respectively. Oral control microorganisms have also been included to verify the validity of our detection strategy.

Our present nucleic acid detection approach involves PCR amplification of the pathogen sequences followed by direct hybridization to oligonucleotide microarrays. Primers were tested using culture samples or commercially available extracted bacterial DNA. Whole saliva samples were first centrifuged to separate cells and particulate matter. Nucleic acids were then isolated and purified from the resulting pellet using a commercial kit (QIAmp DNA Mini Kit, Qiagen Inc.) and amplified using asymmetric PCR. Hybridization of the PCR amplicons to the oligonucleotide arrays was detected by staining the biotinylated primers with streptavidin-Cy3. Fluorescence intensities of PCR products from asthmatic patients versus healthy control samples for *Actinomyces naeslundii* are presented in FIGURE 6.

TABLE 2. Pathogens included in Illumina direct hybridization arrays

Bacteria	Viruses	Oral controls
Haemophilus influenzae	Metapneumovirus (hMPV)	*Actinomyces gerencseriae*
Streptococcus pneumoniae	Respiratory syncytial virus	*Actinomyces naeslundii*
Moraxella catarrhalis	Parainfluenza 1, 2, 3, 4 a/b	*Streptococcus oralis*
(*Branhamella catarrhalis*)	Influenza A virus	*Candida albicans*
Chlamydophila/Chlamydia pneumoniae	Influenza B virus	*Fusobacterium nucleatum*
Mycoplasma pneumoniae	Coronavirus	*Prevotella melaninogenica*
Legionella pneumophila	Adenovirus A,B,C,D,E,F	*Capnocytophaga gingivalis*
	Rhinovirus A, B	*Clostridium difficile*
		Streptococcus pyogenes

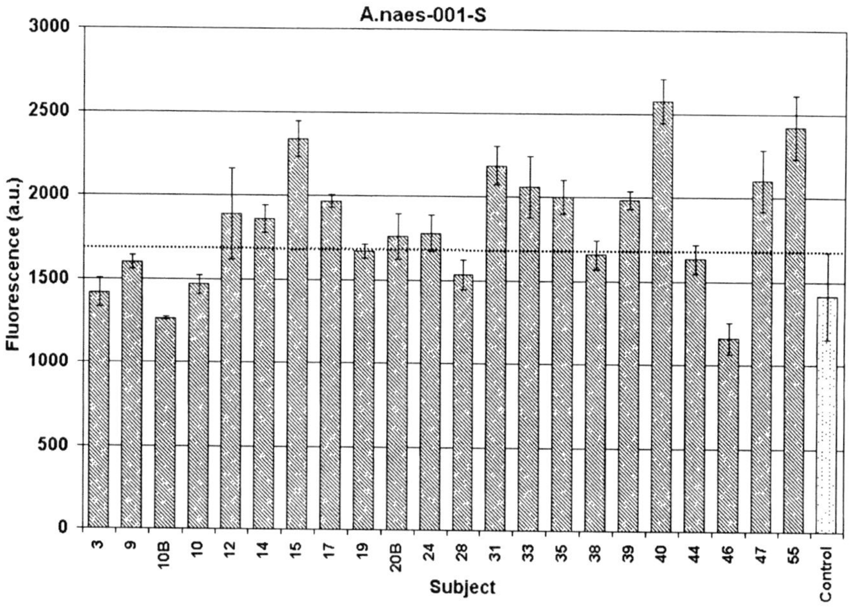

FIGURE 6. Fluorescence intensities of *Actinomyces naeslundii*-001 probes for 22 asthmatics (*dark gray*, left) versus the average of 20 healthy controls (*light gray*, right) determined using multiplexed direct hybridization experiments. The limit of detection of this assay was determined to be 568 a.u.

A threshold was set to the mean of the controls plus one standard deviation. Interestingly, higher intensities were observed for this oral control organism in asthmatic saliva samples (13/22, 59%) than in control patient samples (1/20, 5%). Oligonucleotide probes complementary to *H. influenzae* amplicons tended to show higher fluorescence intensity for asthmatic samples (17/22, 77% above threshold) than for controls (2/20, 10%), as seen in FIGURE 7. When the patient population was limited to asthmatics with COPD, the results were similar (data not shown).

Our oligonucleotide microarray approach to detecting respiratory pathogens has several advantages over current Taqman or immunoassay-based methods. Hybridizing PCR amplification products to the arrays incorporates an added level of specificity. The built-in redundancy of ~30 beads per fiber bundle ensures that these hybridization events are statistically significant. Conventional culture-based methods are typically slow and unable to differentiate between strains or serotypes. Direct detection of nucleic acids from pulmonary pathogens should provide a more accurate profile of current infection than clinical diagnostic kits that detect antibodies to respiratory pathogens because of the inherent delay in immune system response and continued antibody production following infection clearance.

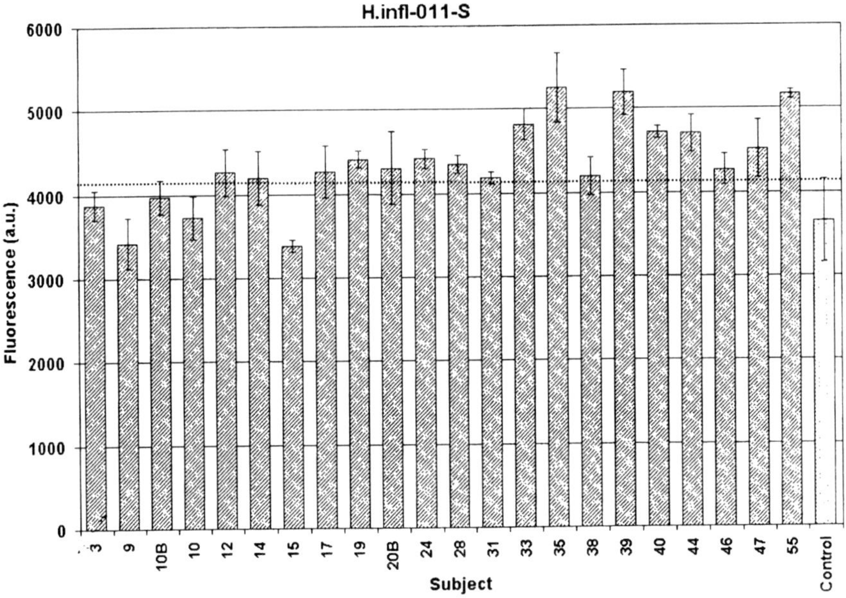

FIGURE 7. *Haemophilus influenzae*-011 probe signals are higher for a majority of the asthmatics (*dark gray*, left) relative to controls (*light gray*, right).

Incorporation of Microfluidics

Sample collection, handling, and pretreatment encompass a wide range of challenges for a point-of-care device, especially in saliva diagnostics. All interfaces between the device and saliva samples must be cleaned between sampling events to eliminate the possibility of interpatient sample contamination. We are developing disposable microfluidics cassettes that will incorporate all necessary extraction, concentration, amplification, and detection chemistries. Embedded arrays will be included in the cassettes and will contain bead-based sensors analogous to the BeadChip design used for analyte screening. On the basis of our experience that hybridization kinetics and limits of detection are improved by the agitation of samples across sensors,[16] the microfluidics platform will also employ thermopneumatic flow oscillation in the detection chamber.

CONCLUSION

In the studies reported here, we have analyzed saliva from ESRD patients and asthmatics to determine whether this sample matrix could be used for systemic disease diagnosis and monitoring. For ESRD, we were able to identify two salivary analytes (NO_2^- and UA) that were elevated in predialysis patients and were shown to be reduced following dialysis. Detection chemistries for these two analytes have been converted to a colorimetric test strip format for the rapid and facile semiquantitative determination of NO_2^- and UA in saliva. The test strips have notable advantages over solution-based screening methods, namely the ability to provide instantaneous measurements for two analytes simultaneously in undiluted saliva without expensive instrumentation. We foresee the NO_2^-/UA test strips potentially improving the quality of life for ESRD and especially for pre-ESRD patients, as these individuals could monitor their salivary analyte levels at home, thereby eliminating periodic visits to the clinic and/or invasive blood testing. For asthma, our goal is to elucidate the complex network of proteins and pathogens implicated in pulmonary exacerbations using whole saliva as a diagnostic fluid. By incorporating assays for promising pulmonary inflammation biomarkers into a multiplexed point-of-care platform for saliva, physicians would be able to make better-informed decisions about the cause of exacerbation and appropriate treatment options.

ACKNOWLEDGMENT

This work was supported by Grant No. U01 DE14950 from the National Institute of Dental and Craniofacial Research (NIDCR).

REFERENCES

1. MUKHOPADHYAY, R. 2006. Devices to drool for. Anal. Chem. **78:** 4255–4259.
2. NATIONAL KIDNEY FOUNDATION. 2006. Chronic Kidney Disease (CKD). http://www.kidney.org/kidneydisease/ckd/index.cfm. Accessed on November 15, 2006.
3. FERGUSON, D.B. 1987. Current diagnostic uses of saliva. J. Dent. Res. **66:** 420–424.
4. MALAMUD, D. 1992. Saliva as a diagnostic fluid. Br. Med. J. **305:** 207–208.
5. FEIN, H., M. BRODERICK, X. ZHANG, *et al.* 2003. Measurement of nitric oxide production in biological systems by using Griess reaction assay. Sensors **3:** 276–284.
6. TAKINO, K., H. ASAHI & H. WADA, INVENTORS; EIKEN KAGAKU KABUSHIKI KAISHA, ASSIGNEE. 1986. U.S. patent 4,631,255. Date of application: December 23.
7. GINDLER, E.M. 1970. Automated determination of uric acid via reductive formation of lavender Cu(I)-2,2′-bicinchoninate chelate. Clin. Chem. **16:** 536.
8. LEE, T.Y., Y.C. LEI, S.-Y. SHEU, *et al.* INVENTORS; DEVELOPMENT CENTER FOR BIOTECHNOLOGY, ASSIGNEE. 2004. U.S. patent 6,699,720. Date of application: March 2.
9. AMERICAN LUNG ASSOCIATION. July 2006. Trends in asthma morbidity and mortality, 1–40.
10. COPD INTERNATIONAL. COPD information and support. http://www.copd-international.com/. Accessed on October 19, 2006

11. Kips, J.C. 2001. Cytokines in asthma. Eur. Resp. J. **34**(Suppl.): 24S–33S.
12. Barnes, P.J. 2001. Th2 cytokines and asthma: an introduction. Resp. Res. **2:** 64–65.
13. Sethi, S. 2004. Bacteria in exacerbations of chronic obstructive pulmonary disease. Proc. Am. Thorac. Soc. **1:** 109–114.
14. Wedzichha, J. 2004. Role of viruses in exacerbations of chronic obstructive pulmonary disease. Proc. Am. Thorac. Soc. **1:** 115–120.
15. Blicharz, T.M. & D.R. Walt. 2006. Detection of inflammatory cytokines using a fiber optic microsphere immunoassay array. Proc. SPIE- Int. Soc. Optic. Engng. **6380:** 638010/1–638010/6.
16. Bowden, M., L. Song & D.R. Walt. 2005. Development of a microfluidic platform with an optical imaging microarray capable of attomolar target DNA detection. Anal. Chem. **77:** 5583–5588.

Oral Fluid Nanosensor Test (OFNASET) with Advanced Electrochemical-Based Molecular Analysis Platform

VINCENT GAU[a] AND DAVID WONG[b]

[a]*GeneFluidics Inc., Monterey Park, California, USA*

[b]*Division of Oral Biology and Medicine, School of Dentistry at UCLA, Los Angeles, California, USA*

ABSTRACT: High-impact diseases, including cancer, cardiovascular disease, and neurological disease, are challenging to diagnose without supplementing clinical evaluation with laboratory testing. Even with laboratory tools, definitive diagnosis often remains elusive. The lack of three crucial elements presents a road block to achieving the potential of clinical diagnostic tests: (1) definitive disease-associated protein and genetic markers, (2) easy and inexpensive sampling methods with minimal discomfort for the subject, and (3) an accurate and quantitative diagnostic platform. Our aim is to develop and validate a solution for requirement (3) and also to develop a portable system. Requirements (1) and (2) will be addressed through the utilization of novel and highly specific oral cancer saliva proteomic and genomic biomarkers and the use of saliva as the biofluid of choice, respectively. The Oral Fluid NanoSensor Test (OFNASET) technology platform combines cutting-edge technologies, such as self-assembled monolayers (SAM), bionanotechnology, cyclic enzymatic amplification, and microfluidics, with several well-established techniques including microinjection molding, hybridization-based detection, and molecular purification. The intended use of the OFNASET is for the point of care multiplex detection of salivary biomarkers for oral cancer. We have demonstrated that the combination of two salivary proteomic biomarkers (thioredoxin and IL-8) and four salivary mRNA biomarkers (SAT, ODZ, IL-8, and IL-1b) can detect oral cancer with high specificity and sensitivity. Our preliminary studies have shown compelling results. We sequentially delivered a serial dilution of IL-8 antigen, probe solution, wash, enzyme solution, wash, and mediator solution to sensor reaction chambers housed in a prototype cartridge and demonstrated strong signal separation at 50 pg/mL above a negative control.

Address for correspondence: Vincent J. Gau, GeneFluidics, Inc., 2540 Corporate Place, Suite B101, Monterey Park, CA 91754. Voice: 323-269-0900; fax: 323-269-0988.
vgau@genefluidics.com

**Ann. N.Y. Acad. Sci. 1098: 401–410 (2007). © 2007 New York Academy of Sciences.
doi: 10.1196/annals.1384.005**

KEYWORDS: salivary diagnostics; electrochemical detection; cyclic enzymatic reaction; bionanotechnology; genetic assay; immunoassay; simultaneous multichannel detection; point-of-care; PCR-less detection; ultrasensitive detection

INTRODUCTION

Microfabrication technology has enabled the development of electrochemical biosensors with the capacity for sensitive and marker-specific detection of nucleic acids and proteins.[1–5] The ability of electrochemical sensors to directly identify nucleic acids or proteins in complex mixtures is a significant advantage over approaches, such as PCR, that require target purification and amplification. Application of universal molecular analysis to cancer screening has the potential for recognition of cancer-specific signature sequences and protein markers in biological fluids. The most challenging body fluid submitted to clinical microbiology laboratories is saliva on account of its complexity and the low concentration of analytes. PCR-based methods for purification and identification of cancer-related genetic targets are time-consuming and labor-intensive. Oral cancer is the sixth most common cancer in the United States, affecting 38,000 Americans annually and killing 7,200. Worldwide, oral cancer annually affects 350,000 individuals (http://www.oralcancerfoundation.org/). Over 90% of these cancers are squamous cell carcinoma. Despite treatment advances that have resulted in reductions in patient morbidity, the overall 5-year survival rate for oral squamous cell carcinoma (OSCC) remains among the worst of all cancer death rates (approximately 30–40%), considerably lower than survival rates for colorectal, cervical, and breast cancer.[6,7]

A rapid, automated, point-of-care system using microfluidic and micro-electro-mechanical systems (MEMS) technologies for measuring DNA, gene transcripts (mRNA), proteins (cardiac and periodontal disease markers), electrolytes, and small molecules in saliva for cardiovascular disease would have a significant impact on the future development of salivary diagnostics.

A general approach for cancer-specific identification of genetic material (DNA, RNA) or protein markers using an electrochemical sensor involves hybridization of single-stranded oligonucleotide capture and detector probes to target 16S rRNA[2,8] or affinity binding of antibodies to cancer-related antigens. The capture probe anchors the target to the sensor, while the detector probe signals the presence of the target through a reporter molecule (FIG. 1). Both oligonucleotide probes will be replaced by antibody pairs for protein marker detection, such as thioredoxin or IL-8, as shown in FIGURE 2. Binding of the capture and detector probes to the nucleic acid target creates a three-component "sandwich" complex on the sensor surface.[9–12] The fluorescein-modified detector probe or antibody enables binding of an antifluorescein-conjugated horseradish peroxidase (HRP) reporter enzyme to the target–probe

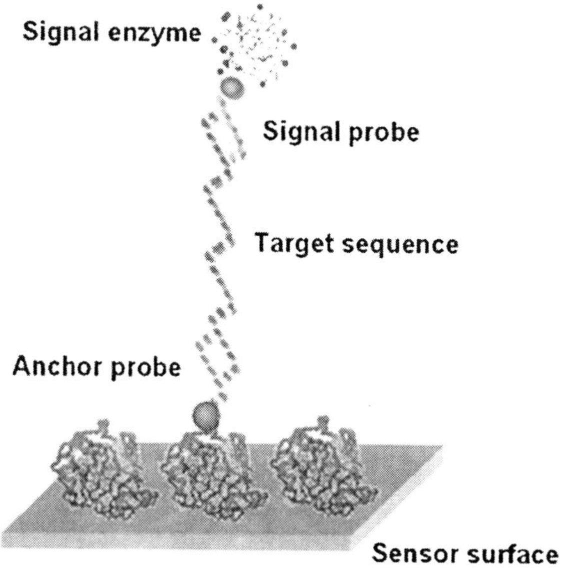

FIGURE 1. The detection of a salivary genetic marker after lysing the cells by mixing saliva with lysis buffer.

complex.[8] Addition of a redox substrate and application of a fixed potential between the working and reference sensor electrodes creates a HRP-mediated redox cycle that is detected by the electrochemical sensor as current.[13,14] In this way, the amplitude of the electroreduction current reflects the concentration of the target–probe complexes on the sensor surface. Both genetic and immunoassays share the same reagents except for the binding solutions (capture and detector), and it is possible to multiplex different assays onto the sensor array chip for simultaneous multianalyte detection.

Initial proof-of-concept result has indicated that this electrochemical sensor approach has the potential for detection of cancer-specific protein markers in saliva while preliminary clinical studies have identified markers for infectious disease in urine.[15,16]

MATERIALS AND METHODS

Sensor Preparation for IL-8 Immunoassay

Microfabricated electrochemical sensor arrays with an alkanethiolate self-assembled monolayer (SAM) were obtained from GeneFluidics (Monterey Park, CA, USA). SAM integrity was confirmed by cyclic voltammetry (CV)[17] using a 16-channel potentiostat (GeneFluidics). After CV characterization,

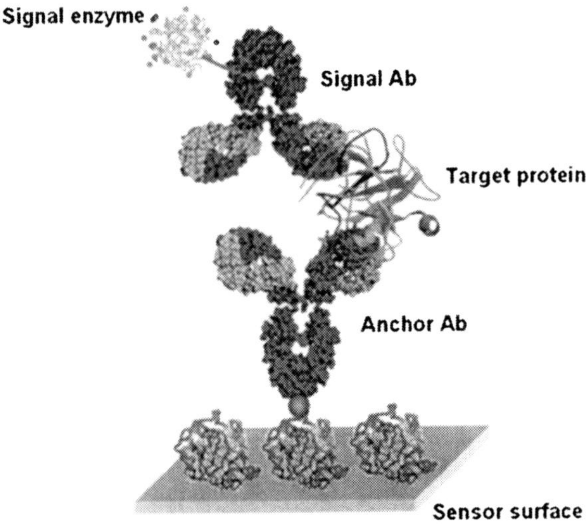

FIGURE 2. The detection of a salivary proteomic marker, such as thioredoxin of IL-8 directly from saliva without pretreatment.

sensor arrays were washed and dried. Washing steps were carried out by applying a stream of deionized H_2O to the sensor surface for approximately 2–3 sec followed by 5 sec of drying under a stream of nitrogen. To functionalize the sensor surface, 4 μL of 1.04 mg/mL polyclonal human anti-IL-8 capture antibodies in 1 × phosphate-buffered saline (PBS) with 0.1% sodium azide was added to the alkanethiol-activated sensors. After 30 min of incubation at room temperature, the sensor array was washed and dried, completing the surface preparation.

Amperometric Detection of IL-8 Protein Marker

For each individual sensor of a 16-sensor array chip, 4 μL of the biotinylated IL-8 antibody (12.5 μg/mL) in 1 × PBS, was mixed with the same volume of saliva. The antibody–saliva lysate mixture was applied onto each sensor and incubated for 30 min to allow affinity binding of antibody to IL-8 analyte in saliva. After washing and drying, 4 μL of streptavidin-conjugated POD 1.25 μg/mL or 13.4 nM; Zymed, CA, USA) in 0.5% BSA in 1 × PBS was deposited on each of the working electrodes for 15 min. After washing and drying, a prefabricated plastic well manifold (GeneFluidics) was bonded to the sensor array. Eighty microliters of HRP substrate solution (K-Blue Aqueous TMB; Neogen, Lexington, KY, USA) was placed on each of the sensors in the array so as to cover all three of the electrodes. Measurements were immediately and simultaneously taken for all 16 sensors. The entire assay protocol was completed within 45 min from the initiation of saliva mixing. Amperometric current versus time was measured using a multichannel potentiostat (GeneFluidics). The voltage was fixed at –200 mV (vs. reference), and the electroreduction current was measured at 60 sec after the HRP redox reaction reached steady state.

Sensor Preparation for IL-8 RNA Genetic Assay

The integrity of the SAM chip was confirmed by CV using a 16-channel potentiostat (GeneFluidics). To functionalize the sensor surface, 4 μL of 0.5 mg/mL streptavidin (Calbiochem, San Diego, CA, USA) in H_2O was added to the alkanethiol-activated sensors, incubated for 10 min at room temperature and washed. Biotinylated capture probes (4 μL, 1 μM in 1 M phosphate buffer, pH 7.4) were added to the streptavidin-coated sensors. Phosphate buffer (1 M), pH 7.4, was prepared by mixing 1 M NaH_2PO_4 and 1 M K_2HPO_4 in a 19:81 (vol/vol) ratio, respectively, and adjusting the pH to 7.4. After 30 min of incubation at room temperature, the sensor array was washed and dried, completing the surface preparation.

Amperometric Detection of IL-8 RNA Molecule

Salivary RNA samples were obtained from Dr. David Wong's laboratory at the University of California, Los Angeles. Fifty microliters of the detector probe (0.25 μM) in 2.5% bovine serum albumin (Sigma)–1 M phosphate buffer, pH 7.4, was added to the salivary RNA sample. The detector probe/sample mixture was incubated for 10 min at 65°C to allow hybridization of the detector probe to target rRNA. Four microliters of the sample/detector probe mixture was deposited on each of the working electrodes in the sensor array. The sensor array was incubated for 15 min at 65°C in a humidified chamber. After washing and drying, 4 μL of 0.5 U/mL anti-fluorescein HRP Fab conjugate (Roche), diluted in 0.5% casein in 1 M phosphate buffer, pH 7.4, was deposited on each of the working electrodes for 15 min. After washing and drying, a prefabricated plastic well manifold (GeneFluidics) was bonded to the sensor array. Eighty microliters of HRP substrate solution (K-Blue Aqueous TMB) was placed on each of the sensors in the array so as to cover all three of the electrodes. Measurements were immediately and simultaneously taken for all 16 sensors. Amperometric current versus time was measured using a multichannel potentiostat (GeneFluidics). The voltage was fixed at –200 mV (vs. reference), and the electroreduction current was measured at 60 sec after the HRP redox reaction reached steady state.

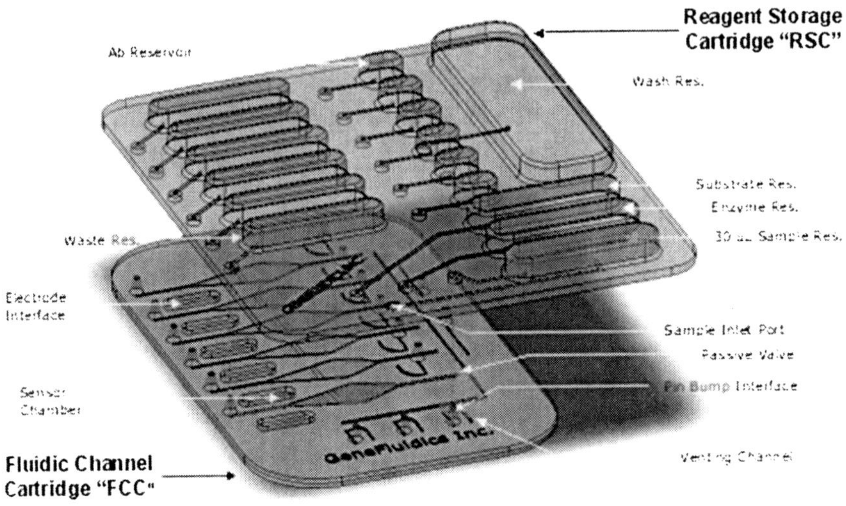

FIGURE 3. Six-channel immunoassay schematic of RSC and FCC.

The Oral Fluid NanoSensor Test (OFNASET) Cartridge

The microfluidics cartridge was designed to enable all steps in a manual assay protocol to be conducted in a "hands free" format. When the cartridge is loaded with reagents specific for a panel of assays, such as those for oral or breast cancer detection, the cartridge can serve as a disposable device for saliva-based diagnostics.

The current prototype has a microfludic cartridge with two main parts—a top piece or reagent storage cartridge (RSC), which would house all of the reagent and sample reservoirs, and a bottom piece or fluidic channel cartridge (FCC), which would house all of the fluidic channels, reaction chambers, and valve structures. FIGURE 3 shows an interface schematic of the two pieces and FIGURE 4 shows an FCC prototype cartridge with RSC unit in the back.

While still in development, RSC and FCC can be integrated to form the OFNASET–Cartridge to be inserted into the reader Instrument and initiate the experiment through an instrument interface where actuators depress the flexible sample, reservoir. Simultaneously, the sample passive valve (described below) would be opened for each of the sensor chambers. The resulting pressure would drive the sample from the RSC to the FCC through the pin bump structure and onward to each sensor chamber with an open valve. Note that without the opening of the valves, the sample would remain in the sample reservoir. Note further that air in the microfluidic channels would be vented through venting channels.

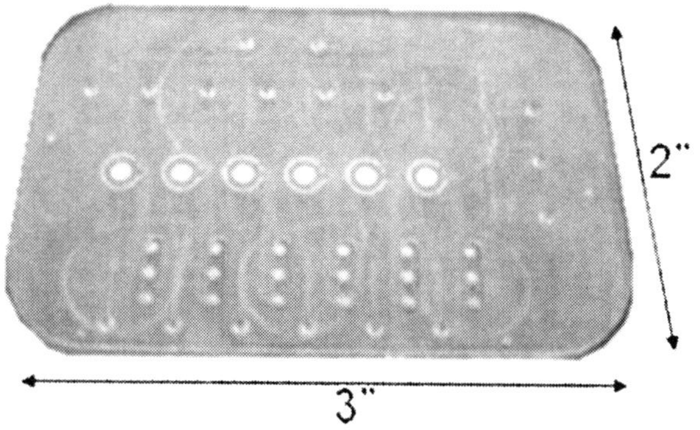

FIGURE 4. Prototype cartridge of FCC and RSC.

For protein or cell-surface antigen detection, the detecting antibody solution would be introduced to each sensor chamber and mixed with the sample. The capture antibodies would immobilize the target–signal antibody complexes onto the sensor surface before the wash solution was delivered to remove nonbound material. A sensor chamber "ramp structure" would enable a consistent back-pressure to make the sheer force wash more efficient. Enzyme solution would then be delivered to each chamber, which would bind to the signal antibodies. A second wash would remove nonbound enzyme. During the last step, the substrate solution would be delivered and the bias potential applied through the electrode interface. Amperometric current would be measured at 60 sec. A postassay CV would provide a baseline signal for calibration of the specific sample, sensor, and reagents. All waste solutions would be routed back to the RSC.

RESULTS

IL-8 Protein Detection in Saliva Using FCC

To date, most of our assays have been conducted with the sensor chip and reader (manual sample preparation). We have also developed the use of FCC for automated sample preparation. These studies have shown preliminary but compelling results. In one example, we have demonstrated the successful sequentially delivered serial dilution of IL-8 antigen, probe solution, wash, enzyme solution, wash, and mediator solution to sensor reaction chambers housed in a prototype cartridge. As shown in FIGURE 5, the experiment demonstrates strong signal separation at 50 pg/mL above a negative control. Note the smaller error bars due to the minimization of error from manual handling.

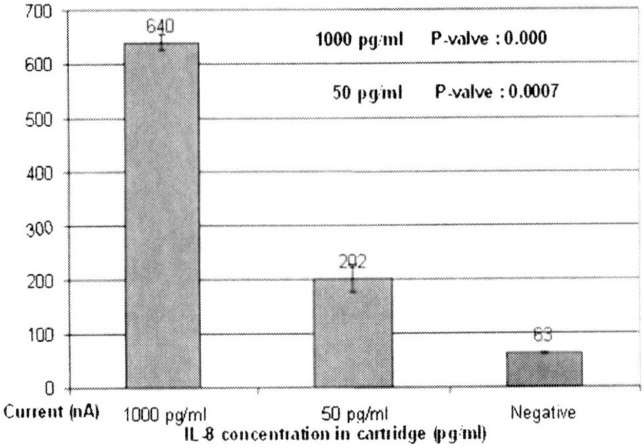

FIGURE 5. Detection of IL-8 protein using FCC.

Use of the OFNASET Sensor Chip for Salivary IL-8 Protein and RNA Detection

Saliva samples from patients with oral cancer and matched controls were tested with GeneFluidics's electrochemical-based detection platform; IL-8 protein and IL-8 RNA were the targets. The results shown in FIGURE 6A to C demonstrate a clear separation between the signal from the cancer patients and the controls. Each sample was tested in duplicate and no pretreatment was conducted on the saliva prior to testing, although RNA was isolated prior to detection.

DISCUSSION

We have developed a sensor array chip for direct electrochemical detection of the cancer markers (RNA and protein) from saliva associated with oral cancer. The sensor assay system relies on efficient binding of target RNA molecules or proteins onto the sensor surface. This sensor array chip is part of the developmental work for production of the final microfluidics cartridge OFSANET. The detection system involves oligonucleotide capture and detector probes or antibodies that bind to RNA or antigens present in saliva. Preliminary results show validation of rapid salivary diagnostics for point-of-care applications. The initial study was conducted using IL-8 RNA and protein as the cancer marker targets. Additional cancer markers, such as thioredoxin-1 (protein), OAZ1 (RNA), SAT (RNA), S100P (RNA), DUSP1 (RNA), and HIST3H3 (RNA), will be tested and validated with clinical specimens to further expand the oral cancer test panel using this electrochemical-based detection platform.

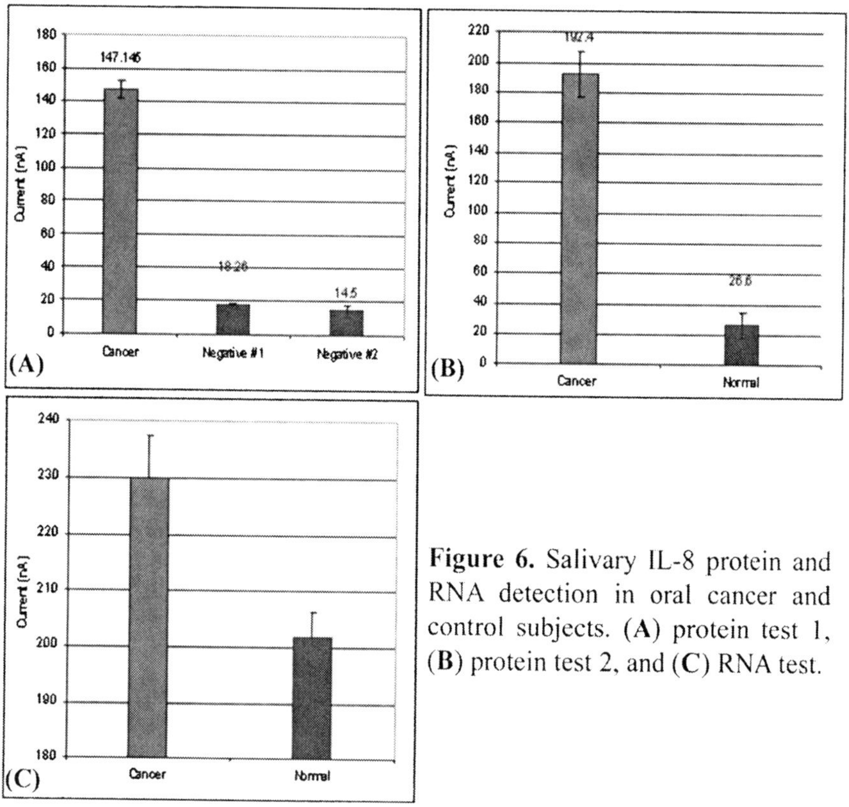

Figure 6. Salivary IL-8 protein and RNA detection in oral cancer and control subjects. (**A**) protein test 1, (**B**) protein test 2, and (**C**) RNA test.

ACKNOWLEDGMENTS

We thank other members of the UCLA OFNASET team—David Wong, Chih-Ming Ho, and Jim Dow at Becton Dickson—for joint cooperation. We also thank David Haake and Bernard Churchill for helpful discussions and long-term collaboration on assay development. The development of this study was supported by U01 Grant DE17790 from the National Institute of Dental and Craniofacial Research (NIDCR) and Bioengineering Research Partnership Grant EB00127 (to B.M.C.) from the National Institute of Biomedical Imaging and Bioengineering. The primary author is the founder and majority shareholder of GeneFluidics.

REFERENCES

1. DRUMMOND, T.G., M.G. HILL & J.K. BARTON. 2003. Electrochemical DNA sensors. Nat. Biotechnol. **21:** 1192–1199.
2. GAU, J.J., E.H. LAN, B. DUNN, *et al.* 2001. A MEMS based amperometric detector for *E. coli* bacteria using self-assembled monolayers. Biosens. Bioelectron. **16:** 745–755.

3. GOODING, J.J. 2002. Electrochemical DNA hybridization biosensor. Electroanalysis **14:** 1149–1156.
4. PALECEK, E. & F. JELEN. 2002. Electrochemistry of nucleic acids and development of DNA sensors. Crit. Rev. Anal. Chem. **3:** 261–270.
5. WANG, J. 2002. Electrochemical nucleic acid biosensors. Anal. Chim. Acta **469:** 63–71.
6. SCHANTZ, S.P. 1993. Carcinogenesis, markers, staging, and prognosis of head and neck cancer. Curr. Opin. Oncol. **5:** 483–490.
7. SCHANTZ, S.P. 1993. Biologic markers, cellular differentiation, and metastatic head and neck cancer. Eur. Arch. Otorhinolaryngol. **250:** 424–428.
8. GAU, V., S.C. MA, H. WANG, *et al.* 2005. Electrochemical molecular analysis without nucleic acid amplification. Methods **37:** 73–83.
9. CAMPBELL, C.N., D. GAL, N. CRISTLER, *et al.* 2002. Enzyme-amplified amperometric sandwich test for RNA and DNA. Anal. Chem. **74:** 158–162.
10. DEQUAIRE, M. & A. HELLER. 2002. Screen printing of nucleic acid detecting carbon electrodes. Anal. Chem. **74:** 4370–4377.
11. UMEK, R.M., S.W. LIN, J. VIELMETTER, *et al.* 2001. Electronic detection of nucleic acids: a versatile platform for molecular diagnostics. J. Mol. Diagn. **3:** 74–84.
12. WILLIAMS, E., M.I. PIVIDORI, A. MERKOCI, *et al.* 2003. Rapid electrochemical genosensor assay using a streptavidin carbon-polymer biocomposite electrode. Biosens. Bioelectron. **19:** 165–175.
13. FANJUL-BOLADO, P., M.B. GONZALEZ-GARCIA & A. COSTA-GARCIA. 2005. Amperometric detection in TMB/HRP-based assays. Anal. Bioanal. Chem. **382:** 297–302.
14. MECHERI, B., L. PIRAS, L. CIOTTI & G. CAMINATI. 2004. Electrode coating with ultrathin films containing electroactive molecules for biosensor applications. IEEE Sens. J. **4:** 171–179.
15. SUN, C.P., J.C. LIAO, Y.H. ZHANG, *et al.* 2005. Rapid, species-specific detection of uropathogen 16S rDNA and rRNA at ambient temperature by dot-blot hybridization and an electrochemical sensor array. Mol. Genet. Metab. **84:** 90–99.
16. LIAO, J.C., M. MASTALI, V. GAU, *et al.* 2006. Use of electrochemical DNA biosensors for rapid molecular identification of uropathogens in clinical urine specimens. J. Clin. Microbiol. **44:** 561–570.
17. BARD, A.J. & L.R. FAULKNER. 2001. Potential Sweep Methods.: 226–260. John Wiley & Sons. Hoboken, New Jersey.

Lab-on-a-Chip Methods for Point-of-Care Measurements of Salivary Biomarkers of Periodontitis

NICOLAOS CHRISTODOULIDES,[a] PIERRE N. FLORIANO,[a] CRAIG S. MILLER,[b,c] JEFFREY L. EBERSOLE,[b,c] SANGHAMITRA MOHANTY,[a] PRIYA DHARSHAN,[a] MICHAEL GRIFFIN,[a] ALEXIS LENNART,[a] KARRI L. MICHAEL BALLARD,[a] CHARLES P. KING, JR.,[b] M. CHRIS LANGUB,[d,e] RICHARD J. KRYSCIO,[f] MARK V. THOMAS,[b] AND JOHN T. McDEVITT[a,g,h]

[a]*Department of Chemistry & Biochemistry, The University of Texas at Austin, Austin, Texas, USA*

[b]*Department of Oral Health Practice and Center for Oral Health Research, College of Dentistry, University of Kentucky, Lexington, Kentucky, USA*

[c]*Department of Microbiology, Immunology & Molecular Genetics, and Department of Internal Medicine, University of Kentucky, Lexington, Kentucky, USA*

[d]*Office of Extramural Programs, National Institute for Occupational Safety and Health, Centers for Disease Control and Prevention, Atlanta, Georgia, USA*[i]

[e]*Department of Internal Medicine, College of Medicine, University of Kentucky, Lexington, Kentucky, USA*[j]

[f]*Department of Biostatistics, College of Public Health, University of Kentucky, Lexington, Kentucky, USA*[j]

[g]*Center for Nano and Molecular Science and Technology, The University of Texas at Austin, Austin, Texas, USA*

[h]*Texas Materials Institute, The University of Texas at Austin, Austin, Texas, USA*

ABSTRACT: Salivary secretions contain a variety of molecules that reflect important pathophysiological activities. Quantitative changes of specific salivary biomarkers could have significance in the diagnosis and management of both oral and systemic diseases. Modern point-of-care technologies with enhanced detection capabilities are needed to implement

Address for correspondence: Dr. John T. McDevitt, University of Texas at Austin, Department of Chemistry and Biochemistry, 1 University Station - A5300, Austin, TX 78712. Voice: 512-471-0046; fax: 512-232-7052.

mcdevitt@mail.utexas.edu

[i]Current affiliation.

[j]Affiliation at the time the work was performed.

Ann. N.Y. Acad. Sci. 1098: 411–428 (2007).
doi: 10.1196/annals.1384.035

a significant advancement in salivary diagnostics. One such promising technology is the recently described lab-on-a-chip (LOC) assay system, in which assays are performed on chemically sensitized beads populated into etched silicon wafers with embedded fluid handling and optical detection capabilities. Using this LOC system, complex assays can be performed with small sample volumes, short analysis times, and markedly reduced reagent costs. This report describes the use of LOC methodologies to assess the levels of interleukin-1β (IL-1β), C-reactive protein (CRP), and matrix metalloproteinase-8 (MMP-8) in whole saliva, and the potential use of these biomarkers for diagnosing and categorizing the severity and extent of periodontitis. This study demonstrates that the results achieved by the LOC approach are in agreement with those acquired with standard enzyme-linked immunosorbent assay (ELISA), with significant IL-1β and MMP-8 elevations in whole saliva of periodontitis patients. Furthermore, because of the superior detection capacities associated with the LOC approach, unlike those with ELISA, significant differences in CRP levels between periodontitis patients and normal subjects are observed. Finally, principal component analysis (PCA) is performed to yield an efficient method to discriminate between periodontally healthy and unhealthy patients, thus increasing the diagnostic value of these biomarkers for periodontitis when examined with the integrated LOC sensor system.

KEYWORDS: lab-on-a-chip; salivary diagnostics; inflammation; biomarkers; periodontitis

INTRODUCTION

Periodontitis is a chronic inflammatory disorder of the tissues supporting the teeth and represents one of the most widely distributed and prevalent microbial diseases of humans. Current figures estimate that 10–15% of adults worldwide have advanced periodontal disease.[1] Although implementation of preventive measures during the past 30 years has significantly improved the oral health of many individuals, periodontal disease still contributes to widespread oral health dysfunction and increased susceptibility to other systemic health risks.[2–4]

The World Health Organization (WHO) recently published a global overview of oral health, and described its approach to promoting further improvement in oral health during the 21st century.[1] The report emphasized that despite great improvements in the oral health status of populations across the world, major problems still persist for historically under-served groups in both developed and developing communities. The WHO recognizes oral health as an integral part of general health, as recent data support that chronic infection of the periodontium with chronic stimulation of inflammatory responses contributes to systemic sequelae, such as preterm delivery of low-birth-weight babies, cardiovascular disease (CVD), and diabetes.[5–7] Indeed, numerous case–control and cohort studies have indicated that patients with periodontitis have an increased risk of CVD, that is, acute myocardial infarction (AMI), stroke, and peripheral arterial disease, when compared with subjects with healthy periodontium.[8–11]

Although the diagnosis and treatment of periodontitis has historically focused on mechanical approaches, research in molecular mechanisms and the outcomes of the Human Genome Project have provided increasing information demonstrating the specific biological pathways and biomolecules that could be used as biomarkers for risk assessment, diagnosis, and prognosis. Obviously, saliva, because of its locality, represents a key and very relevant diagnostic fluid for periodontal disease. Many important biological substances including electrolytes[12,13] drugs,[14–20] proteins (e.g., cytokines, hormones, enzymes),[21–24] antibodies,[25–27] microbes,[28–30] and RNAs[31–34] have been identified in saliva. However, significant correlates that could be used as adjunctive diagnostic/prognostic information by clinicians remain elusive. In contrast to medicine's rapid use of point-of-care (POC) diagnostic devices with focus on biomarkers, or risk factors, in serum, urine, and cerebrospinal fluid, progress in oral health studies targeting the use of saliva as a diagnostic fluid for local and/or systemic diseases has been slow. Impediments to the use of oral fluids have been the relatively low concentration of various important biomolecules in saliva, compared with serum or plasma, and the lack of sufficiently sensitive and simple assays and equipment that could be used in the dentist's office.

The development and implementation of modern technologies that use specific biomarkers at the "chair-side" are likely to increase clinical diagnostic and prognostic insights. Their application in broad health-care settings and in populations that lack access to necessary medical and dental infrastructure should be a substantial benefit to public health. In particular, biomarkers used for the diagnosis of periodontitis should lead to early identification with increased potential for referral, early intervention, and prevention that could contribute to improved overall oral and systemic health.

Tremendous advances have been made recently in the area of lab-on-a-chip (LOC) devices exploiting the advantages offered by miniaturization, such as reduced reagent and sample volume requirements, rapid analysis times, and cost-effective assays that can be operated with fewer technological constraints, making them amenable to POC testing. Most importantly, these characteristics, when fully developed into a functional system, have the potential to lead to a significant reduction in the time that is needed for an accurate diagnosis and treatment.

Over the past five decades, the microelectronics industry has sustained tremendous growth and has become what is arguably the most dominant industrial sector for our society. The availability of powerful microfabrication tools based on photolithographic methods that can be used to process these devices in highly parallel manner has led to this explosive growth. Our group has combined and adapted the tools of nano-materials and microelectronics for the practical implementation of miniaturized sensors that are suitable for application to a variety of health-care areas. Importantly, the performance metrics of these miniaturized sensor systems have been shown to correlate closely with established macroscopic gold-standard methods, making them suitable for use as subcomponents of highly functional detection systems for analysis of complex fluid samples. These efforts remain unique in terms of functional LOC methods having a demonstrated capacity to meet or exceed the analytical characteristics of mature macroscopic instrumentation for a variety of analyte measures, including pH, DNA oligonucleotides, metal cation, biological cofactors, and inflammatory mediators.[35–47]

These LOC methods offer the ability to perform multiplex assays in small sample volumes, with enhanced sensitivity, thus making them amenable to applications involving a variety of bodily fluids, including saliva. Salivary biomarkers that were previously undetectable by standard methods can now be targeted to assess periodontal disease in a noninvasive fashion. Here, we report the application of the LOC system for the concomitant measurement of salivary biomarkers C-reactive protein (CRP), matrix metalloproteinase-8 (MMP-8), and interleukin-1β (IL-1β) as related to the clinical expression of periodontitis. Our results suggest the LOC approach is suitable for the detection of three important biomarkers in saliva, which makes this technology relevant to the diagnosis, staging, and management of periodontal disease.

METHODS AND MATERIALS

Patient Populations and Collection of Salivary Fluids

Orally healthy and periodontitis patients were recruited from the population at the University of Kentucky College of Dentistry. The protocol and consent forms were approved by the Institutional Review Boards of the University of Texas at Austin and the University of Kentucky. Subjects included were $\geq$18 years of age, had a minimum of 20 teeth, and no medical history of chronic illnesses (e.g., diabetes, rheumatoid arthritis).

Two groups of adult patients matched by race and age participated. The first group consisted of 29 normal healthy volunteers without any clinically detectable periodontal lesions. Specifically, this group of patients had at least 20 teeth with fewer than 10% of gingival sites with bleeding on probing (BOP), no probing pocket depths (PD) $\geq$5 mm, $<$1% of interproximal sites with clinical attachment loss $>$2 mm, and no evidence of radiographic bone loss as determined by posterior vertical bitewings films, and no PD sites greater than 5 mm. The second group was derived from 28 patients with moderate-to-severe periodontal disease. This group of patients had noticeable loss of connective tissue attachment and bone around the teeth in conjunction with the formation of periodontal pockets due to the apical migration of the junctional epithelium. Specifically, this group comprised individuals with at least 20 teeth and $>$30% of gingival sites with BOP, $>$20% probing depths $\geq$4 mm, and

>5% of interproximal sites with clinical attachment loss (CAL) of >2 mm, and evidence of radiographic bone loss as determined by posterior vertical bitewings films.

Unstimulated whole saliva was collected from each of the subjects as described previously.[48] In brief, after the mouth was rinsed with water, saliva was allowed to accumulate in the floor of the mouth for approximately 2 min and was repeatedly expectorated into a test tube to collect 5 mL. Following collection, the samples were aliquoted and immediately stored at –80°C either until evaluated by enzyme-linked immunosorbent assay (ELISA) at the University of Kentucky or transported on dry ice to the University of Texas at Austin for LOC testing.

ELISA Analyses

ELISA analyses of IL-1β, MMP-8, and CRP were performed using commercial kits (Human IL-1β Quantikine kit and Human Quantikine MMP-8 ELISA kit, R&D Systems, Minneapolis, MN, USA and ALPCO, Windham, NH, USA, respectively) and following the manufacturers' instructions. The whole saliva was evaluated in duplicate using a microQuant plate reader using KC4 Kineticalc software for curve fitting and calculation of the levels.

LOC Procedures and Analyses

Bead Preparation, Sieving, Activation, and Antibody Conjugation

Agarose beads used in the LOC system in past studies were purchased from a commercial source.[46] More recently, we have acquired the expertise and an in-house capacity to produce large batches of agarose beads. All the experiments described in this study utilized exclusively beads developed in our laboratories. As the precision of the LOC assays is highly dependent on the size homogeneity of its component sensor beads, the beads are exposed to a sieving process that produces a consistent population of microspheres 280 ± 10 μm. Reactive aldehyde groups within the agarose matrix are generated by mixing gently overnight 1 mL of the sieved beads in a 10-mL solution of 1-M sodium hydroxide containing 20 mg of sodium borohydride and 3 mL of glycidol. The beads are then washed copiously with water and exposed to 0.16-M sodium periodate solution followed with successive water washes. Analyte-specific and control (analyte-irrelevant) antibodies are coupled to the beads by reductive amination, as described previously.[35]

Reagents for LOC Assays

CRP, tumor necrosis factor (TNF)-α, IL-6, myeloperoxidase (MPO), MMP-8, and IL-1β protein standards were obtained from Cortex Biochemicals

(San Leandro, CA, USA), Biomol International (Plymouth, PA, USA), eBioscience (San Diego, CA, USA), Biodesign International (Saco, ME, USA), Chemicon International (Temecula, CA, USA) and BD Biosciences Pharmingen (Chicago, IL, USA), respectively. Capturing and detecting antibodies for CRP assay were from Accurate Chemical (Westbury, NY, USA). Capture and detection antibodies for the TNF-α assay were from R&D Systems (Minneapolis, MN) and Sigma Aldrich (St. Louis, MO, USA), respectively. Capture and detection antibodies for the MPO assay were from Biodesign International and Abcam (Cambridge, UK), respectively. Capture and detection antibodies for both the MMP-8 and IL-1β assays were from R&D Systems.

Description of Assay Run on LOC System

Prior to each assay, beads are placed in addressable regions within the array. The bead-loaded chip is encased into the flow cell, which is located at a fixed position with regards to the optical station and the charge-coupled device (CCD). Wash buffer (i.e., phosphate-buffered saline [PBS]), detecting antibody, blocking agent (3% bovine serum albumin [BSA] in PBS), and sample are then primed into the system to minimize the introduction of air bubbles into the analysis area.

During each assay, bead components are first blocked with 3% BSA/PBS, to eliminate nonspecific binding. Protein standards, or the unknown sample, are then delivered into the array for analysis. Following a brief rinse with PBS, the presence of captured analyte on the beads is achieved with an analyte-specific detecting antibody conjugated to AlexaFluor-488® (Molecular Probes). Excess antibody reagent is removed from the flow cell by washing with PBS, and final image of the bead array is acquired with a CCD camera and stored digitally for analysis.

Image and Data Analysis

Digital information from each array/run is obtained using Image Pro Plus software and analyzed with SigmaPlot® (Systat Software Inc [SSI], San Jose, CA, USA). The concentration of the unknown sample is extrapolated from the generated standard curve, as described below.

First, an area of interest is drawn on the periphery of each bead of the array, for each run. The intensity in the green channel for each bead, for standards and unknowns, is measured and recorded as density of green (average intensity per pixel). Data from all beads are then exported to a SigmaPlot® datasheet. The signal intensity obtained from redundant beads of the array is averaged for each assay run. A *Q*-test is then applied to identify and discard outliers. The remaining data are analyzed using a four-parameter logistic equation process within the SigmaPlot® environment to generate a standard, dose-response curve and to predict concentrations of the unknowns.

Statistical Analyses

Evaluation of the analyte levels in saliva emphasized the use of nonparametric Wilcoxon-Mann-Whitney *U* rank sum analyses based upon the small number of samples and the nonnormal distribution of the data. As such, correlations were determined using a nonparametric Spearman rank correlation analysis. Logistic and multiple regressions were performed on the analyte levels in the saliva. All statistical analyses were performed using SigmaStat 3.0 (Chicago, IL, USA).

Principal component analysis (PCA) was performed using Statistica 5.5 (Tulsa, OK, USA). Two principal components were extracted from a total of six measured variables across 48 patients. Patients with incomplete measurements across all six variables were excluded from this study. The raw data were normalized to zero mean and unit variance. Only factors with eigenvalues greater than 1.0 were utilized.

RESULTS AND DISCUSSION

Previously, we have described studies of the design, fabrication, and testing of nano-biochip structures whereby immunoassays are performed on chemically sensitized beads that are arranged in an array of wells etched on silicon wafers with integrated fluid handling and optical detection capabilities (FIG. 1).[35,46] Each bead within the array serves as its own independent self-contained microreactor sensor, with specificity determined by the antibody element that it hosts. The bead-loaded nano-biochip is sandwiched between two optically transparent poly-methyl-methacrylate (PMMA) inserts, packaged within a casing described here as the "flow cell." The flow cell allows for

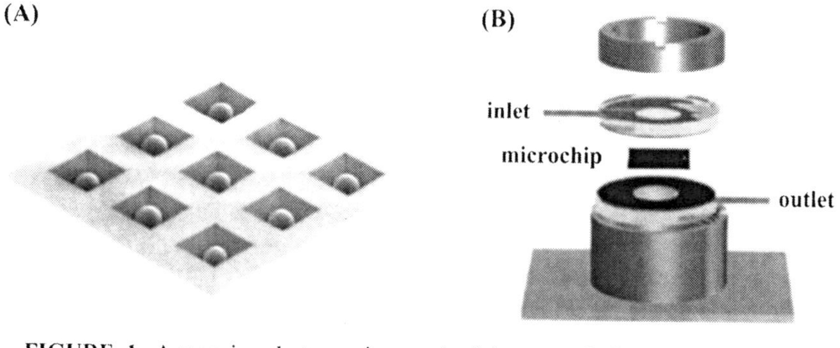

FIGURE 1. A scanning electron micrograph of the assay platform of the LOC system composed of a silicon microchip with a microetched array of addressable wells that host beads sensitized to analytes of interest (**A**). The bead-loaded microchip is sandwiched between two PMMA inserts and packaged within a casing to create the analysis flow cell of the LOC system (**B**).

delivery of sample and detecting reagents to the nano-biochip and the associated beads. Fluids are delivered to the beads via the inlet of the top PMMA insert, soaking evenly the beads located therein, while unspent reagents are directed to a waste reservoir through the bottom drain element of the bottom PMMA element.

In this study, sandwich-type immunoassays are used for the measurement of CRP, IL-1β, and MMP-8 biomarkers on the LOC system, as shown in the immunoschematic in FIGURE 2A. Here, the capture antibody-coated beads are sequentially exposed to the analyte of interest (protein standard or sample) and to a detection antibody conjugated to Alexafluor-488 to produce an analyte/dose-dependent fluorescent signal within and around the bead. The top insert of the flow cell allows for the microscopic evaluation of signals generated within the array, which are subsequently captured by a CCD video chip along with the use of transfer optics (FIG. 2B). Here, after each assay run, the final image of the bead array is captured with the CCD (FIG. 2C), digitally processed and analyzed, and the signal intensity converted for each bead into a quantitative measurement based on the generated standard curve.

Typical dose-response (standard) curves for LOC-based assays for MMP-8, IL-1β, and CRP are shown in FIGURE 3. Highly sensitive immunoassays targeting these analytes were developed to accommodate efficient measurement of diluted saliva samples (1:20, 1:100, and 1:1,000 for MMP-8, IL-1β, and CRP, respectively). With these dilutions, possible interference problems associated with the viscous nature of saliva are thus avoided. Likewise, these assays were designed to provide a useful quantitation range, consistent with the reported pathophysiological levels of these analytes in saliva.[49] Here, the quantitation ranges of the MMP-8 and IL-1β assays (after dilution) were at 20–20,000 ng/mL and 10–10,000 pg/mL, respectively.

As described in a previous report, the detection limit of the LOC-based assay for CRP is significantly lower than that of ELISA, which is the current gold standard for high-sensitivity (hs) measurements of CRP.[46] From a comparison of the two methods, it is clear that the LOC approach yields a five orders of magnitude lower limit of detection than that exhibited by the hsCRP ELISA method. Furthermore, the LOC assay was linear for three orders of magnitude, whereas ELISA was only linear for two orders of magnitude. Here, the LOC assay procedure demonstrates a detection limit at 5 fg/mL and a useful range between 10 fg/mL to 10 pg/mL. Therefore, even with 1:1,000 dilution of saliva samples, the LOC assay affords a useful quantitation range of 10–10,000 pg/mL. In contrast, the hsCRP ELISA method yields a detection limit of 2 ng/mL and a useful detection range between of 2–100 ng/mL CRP.

It is well known that the utility of an assay is directly associated with its capacity to detect low concentrations of the analyte it targets. From this perspective, the diagnostic utility of the LOC assay system was recently demonstrated with the application of ultrasensitive CRP measurements in human saliva from both healthy and periodontal disease patients.[46]

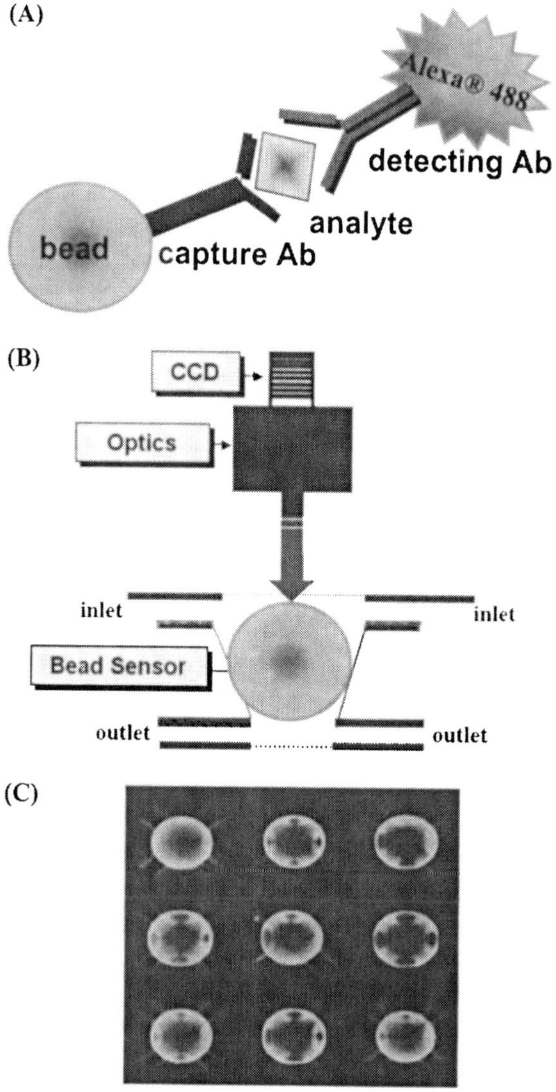

FIGURE 2. The relevant immunocomplexes of the bead-based, sandwich-type of immunoassays on the LOC system are shown in (**A**). Here, an analyte-specific capturing antibody sequesters the antigen on the beads. Detection of the captured antigen is achieved in fluorescence mode with an Alexafluor-488®-conjugated detecting antibody. Completion of assays on the LOC system relies on the integrated function of its microfluidic and optical components (**B**). Here, during each assay, sample and detecting reagents are delivered to the beads via the top inlet of the flow cell, while the bottom drain provides an outlet for the direction of unspent reagents to a waste reservoir. The flow cell allows for the microscopic analysis and capture of signals generated on the array in conjunction with an optical station equipped with a CCD camera. Shown in (**C**) is a typical image of an array of beads captured in the last step of an LOC assay run.

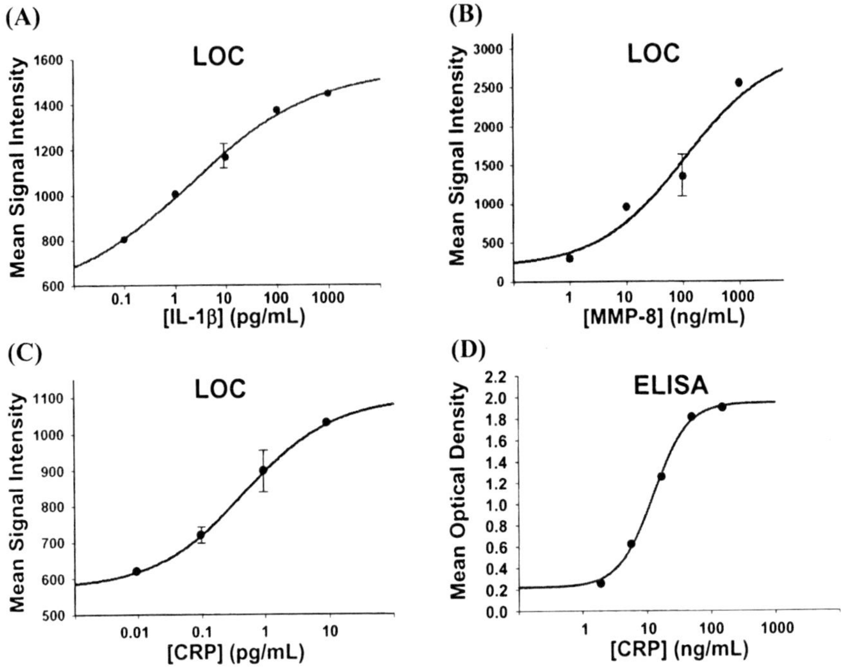

FIGURE 3. Dose-dependent curves obtained for IL-1β (**A**), MMP-8 (**B**), and CRP (**C**), as achieved by the LOC method. Shown in (**D**) is a CRP dose-response curve achieved with a commercial high-sensitivity ELISA method.

Having demonstrated the capacity of the LOC system to measure CRP, and use it to discriminate between healthy and periodontitis patients, we next evaluated the capacity of the LOC method to assess multiple proteins as biomarkers of periodontal disease. Here, both LOC and ELISA methodologies were applied in parallel for the measurement of salivary levels of biomarkers CRP, IL1-β, and MMP-8 in healthy and periodontitis patients. Ten patients who had at least 20 erupted teeth provided whole expectorated saliva and were clinically examined for oral disease. Subjects ranged in age from 29 to 58 years. The study group with oral disease was homogeneous for type II (moderate) periodontitis and was demographically similar to the nine healthy controls. Concentrations of IL1-β and MMP-8 in whole expectorated saliva were successfully measured by both systems. Salivary levels of IL-1β and MMP-8 were higher in subjects who had periodontitis compared with healthy controls, using both LOC and ELISA methods (FIG. 4). Here, salivary levels of IL-1β in the periodontal disease group were 2.6 times higher (as measured by LOC) and 1.5 times higher (as measured by ELISA) than the controls. Salivary MMP-8 levels were 2.0 times higher (as measured by LOC) and 2.6 times higher (as measured by ELISA) in the periodontal disease group compared with the healthy

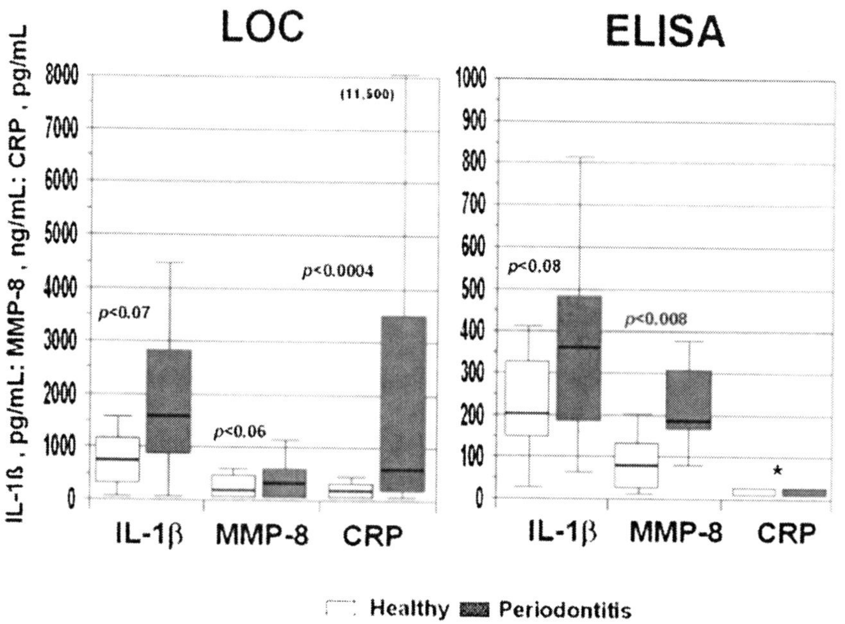

FIGURE 4. Comparison of (**A**) IL-1β (pg/mL), (**B**) MMP-8 (ng/mL), and (**C**) CRP (pg/mL) levels between healthy and periodontitis groups, as achieved by LOC and ELISA methods. The *bars* denote the 95% confidence interval, the *horizontal line* denotes the median of the data, and the *vertical brackets* provide the range of values for each mediator. *Salivary CRP levels were below the level of detection of ELISA.

controls. Inasmuch as salivary CRP levels were below the level of detection of high-sensitivity ELISA, only the LOC approach measured CRP successfully in both groups. Consistent with our previous report,[46] whole saliva from patients with periodontitis contained 18.2 times higher CRP levels than those of healthy patients.

The diagnostic value of strategic biomarkers of periodontitis is expected to increase significantly when these biomarkers are evaluated in a multiplexed manner. Multiplexing is consistent with miniaturization efforts and a multimarker screening strategy, which target a reduction in sample and reagent volumes, and assay time. From a diagnostic point of view, multiplexing offers the opportunity to eliminate the main disadvantage of parallel testing in lowering costs, while also improving turnaround time, usually an issue for sequential testing. Further, this approach promotes monitoring of multiple parameters in a simultaneous manner with the generation of large amounts of complex data that no longer necessarily behave in a linear way. The unique opportunity to probe for several markers as a multivariate observation improves diagnostic efficiency by minimizing the tradeoff between high specificity and sensitivity characteristic of combinations of parallel or sequential testing strategies.[50]

The nonlinear relationships in the data can be deciphered and thus generate patterns, or fingerprint, of disease, which can be expanded to include both biomarker and physical data.

Previous work from Diehl *et al.*[51] has shown that it is possible to apply PCA and discriminant function analysis (DFA) to classify quantitative measurements of periodontitis based on the evaluation of physical parameters. PCA is an unsupervised mathematical method used to simplify a multivariate dataset by choosing a new coordinate system that maximizes the variance represented by the initial dataset. Rather than being limited to the graphical representation of two or three different variables, PCA can be used to effectively represent the information from multiple variables in a two-dimensional plot.

In order to take full advantage of the multianalyte testing capacity of the LOC method, PCA was used here as a multivariate analysis method to discriminate between healthy and periodontitis patients using data aggregated from both physical parameters and biomarker measurements. PCA was applied to a large dataset that included clinical measures of percent BOP, percent CAL >2 mm, percent PD >4 mm, and the recently described biomarkers MMP-8, IL-1β,[49] and CRP.[35,46] For this study, MMP-8 and IL-1β were measured with ELISA, while CRP was measured by means of the more sensitive LOC system. FIGURE 5 shows a principal component plot using all six variables with complete separation between healthy ($N = 24$) and periodontitis patients ($N = 22$). The tight

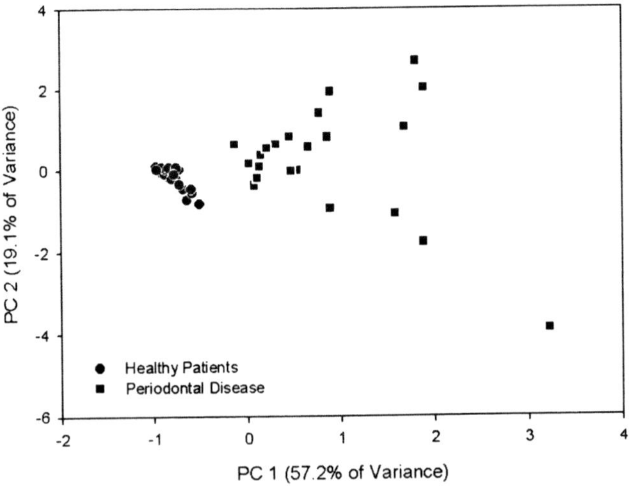

FIGURE 5. A principal component plot of 46 patients showing distinct clustering of healthy patients from the periodontal disease population. Two factors were extracted to account for 76.3% of the total variance. For the principal component analysis, a total of six variables was used including percent of teeth with probe depth greater than 4 mm (% > 4 mm), percent of teeth with blood on probe (% BOP), percent of teeth with CAL >2 mm (% CAL > 2 mm), IL-1β, MMP-8, and CRP.

grouping of healthy patients is indicative of minimal distribution of measured values, while the scatter in the unhealthy patients may be indicative of additional, unclassified subgroups. Work is currently under way to see whether additional clusters can be identified in the diseased population to characterize systemic versus localized disease and to identify the different stages of periodontitis.

In addition to being able to discriminate between different patient populations, it is possible to assign a loading score to each marker that is indicative of its contribution to the magnitude of variance in the dataset. For two well-separated groups, the loading value can be used as one potential indicator of the diagnostic efficacy of that biomarker. Aided by this information, a subset of biomarkers can be chosen for effective diagnosis. Here, CRP, IL-1β, MMP-8, as measured by LOC and ELISA, and oral physical parameters were analyzed with PCA to rapidly identify the biomarkers with near-perfect discrimination between healthy and unhealthy patients. These initial successes bode well for the future use of such methods for the classification of patient samples.

Having demonstrated the capacity of the LOC approach to measure and differentiate three biomarkers of periodontal disease in saliva, the expansion to a larger panel of analytes was initially attempted with a multiplexed assay for six biomarkers, CRP, IL-1β, IL-6, MMP-8, MPO, and TNF-α, in phosphate buffer. Prior to attempting simultaneous detection of all six analytes, each analyte is first captured by its specific bead and then detected individually with a cocktail of antibodies specific for all six antigens (FIG. 6A). Here noted is the absence of signal on control (CTL) beads and of any "cross talk" between irrelevant bead sensors and detecting reagents. In a separate assay run, requiring the same amount of time and reagents as the individual test, all analytes are detected concurrently from the same solution without an apparent loss in signal or a significant increase in nonspecific background level (FIG. 6B). These results demonstrate the capacity of the bead array system to detect simultaneously six biomarkers, all of which have been shown to be associated with periodontal disease.[10,46,49,52–54] Our team is currently working on augmenting further the LOC salivary analyte panel with inclusion of additional biomarkers for both local and systemic disease.

CONCLUSIONS

In conclusion, the selection of salivary biomarkers for the diagnostic assessment of chronic inflammatory diseases, including periodontal disease, from the array of potential markers, has benefited from use of evidence-based information describing biochemical, physiological, and immunological phases of inflammation and tissue destruction. To be useful, salivary biomarkers must be accurate, biologically relevant, discriminatory, and at measurable concentrations. The development and implementation of modern technologies that

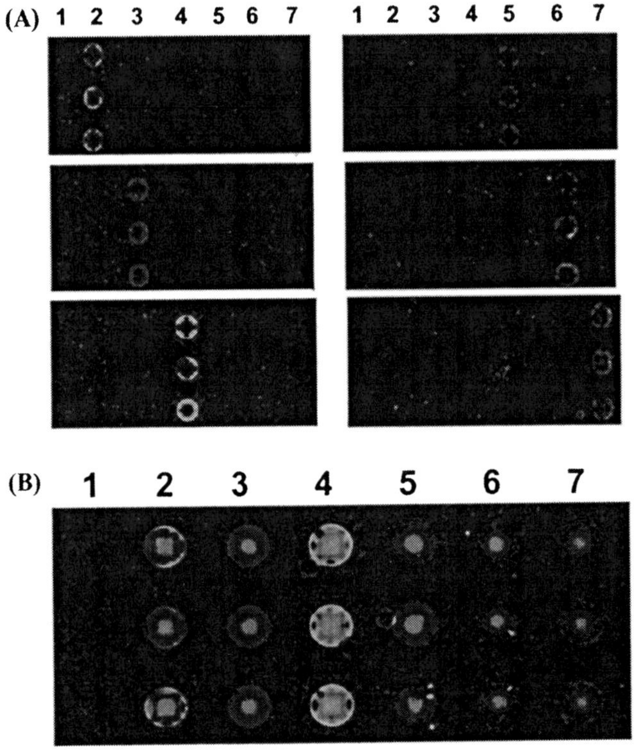

FIGURE 6. Expansion of the multiplexed panel of LOC system to six analytes. Detection of CRP, TNF-α, IL-6, MPO, MMP-8, and IL-1β protein standards is shown first in a single-analyte assay format (**A**). Here, an array of beads sensitized to the six analytes, as indicated below, is exposed to six assay runs. A different analyte is introduced to a fresh array of beads during each assay run. The bead-captured analyte is then probed with a cocktail of detection antibodies. Note the absence of signal on control (CTL) beads loaded with an irrelevant antibody to the analyte, as well as the absence of "cross talk" between irrelevant bead sensors and detecting reagents. The simultaneous detection of all six analytes accomplished in one assay run is shown in (**B**). Here, the bead array is sequentially exposed to a solution containing a mixture of the six analytes and to a cocktail solution of detecting antibody probes specific for all six analytes. The positioning of each bead type (with a threefold redundancy per column) in the array is as follows: column 1, CTL beads; column 2, CRP; column 3, TNFα; column 4, IL-6; column 5, MPO; column 6, MMP-8; and column 7, IL-1β.

use specific biomarkers at the chair-side appear to be on the horizon and are likely to increase clinical diagnostic and prognostic insight. In this study, the LOC methodology is shown to serve as a useful analytical tool that promises to extend the potential diagnostic value of salivary biomarkers of disease. Future opportunities that are afforded through the marriage of the bead-based LOC system with salivary diagnostics are promising. Indeed, attractive goals for future research include the technological abilities to evaluate both local and systemic diseases from their onset and progression and the ability to influence treatment outcomes through noninvasive means.

ACKNOWLEDGMENTS

Funding for this project was provided by the National Institute of Dental and Craniofacial Research (UO1 DE15017), the Welch Foundation (F-1193), Philip Morris USA Inc. and Philip Morris International (UTA 02-400, AMD3), the General Clinical Research Center at the University of Kentucky (M01-RR02602), and from the National Center for Research Resources (P20 RR 020145). The authors thank Ken Westberry and Jason Stevens for providing technical assistance. This article was written in a personal capacity and does not necessarily represent the opinions or reflect the views of the National Institutes of Health, the U.S. Department of Health and Human Services, or the federal government (M.C.L.).

REFERENCES

1. PETERSEN, P.E. & H. OGAWA. 2005. Strengthening the prevention of periodontal disease: the WHO approach. J. Periodontol. **76:** 2187–2193.
2. HERRERA, D., S. ROLDAN & M. SANZ. 2000. The periodontal abscess: a review. J. Clin. Periodontol. **27:** 377–386.
3. PIHLSTROM, B.L., B.S. MICHALOWICZ & N.W. JOHNSON. 2005. Periodontal diseases. Lancet **366:** 1809–1820.
4. SOUTHERLAND, J.H., G.W. TAYLOR, K. MOSS, *et al.* 2006. Commonality in chronic inflammatory diseases: periodontitis, diabetes, and coronary artery disease. Periodontology 2000 **40:** 130–143.
5. KHADER, Y.S. & Q. TA'ANI. 2005. Periodontal diseases and the risk of preterm birth and low birth weight: a meta-analysis. J. Periodontol. **76:** 161–165.
6. MATTILA, K.J., P.J. PUSSINEN & S. PAJU. 2005. Dental infections and cardiovascular diseases: a review. J. Periodontol. **76:** 2085–2088.
7. OFFENBACHER, S. 2004. Maternal periodontal infections, prematurity, and growth restriction. Clin. Obstet. Gynecol. **47:** 808–821.
8. AJWANI, S., K.J. MATTILA, T.O. NARHI, *et al.* 2003. Oral health status, C-reactive protein and mortality—a 10 year follow-up study. Gerodontology **20:** 32–40.
9. MEURMAN, J.H., S.J. JANKET, M. QVARNSTROM & P. NUUTINEN. 2003. Dental infections and serum inflammatory markers in patients with and without severe heart disease. Oral Surg. Oral Med. Oral Pathol. Oral Radiol. Endod. **96:** 695–700.
10. NOACK, B., R.J. GENCO, M. TREVISAN, *et al.* 2001. Periodontal infections contribute to elevated systemic C-reactive protein level. J. Periodontol. **72:** 1221–1227.
11. WEHRMACHER, W.H. 2001. Periodontal disease predicts and possibly contributes to acute myocardial infarction. Dent. Today **20:** 80–81.
12. LEVY, G. & R. LAMPMAN. 1975. Relationship between pH of saliva and pH of urine. J. Pharm. Sci. **64:** 890–891.

13. THAYSEN, J.H., N.A. THORN & I.L. SCHWARTZ. 1954. Excretion of sodium, potassium, chloride and carbon dioxide in human parotid saliva. Am. J. Physiol. **178:** 155–159.
14. BREIMER, D.D. & M. DANHOF. 1980. Saliva—fluid for measuring drug concentrations. Pharm. Int. **1:** 9–11.
15. CADDY, B. 1984. *In* Advances in Analytical Toxicology, volume 1. R.C. Baselt, Ed.: 198–254. Biomedical Publications, Foster City, CA.
16. DANHOF, M. & D.D. BREIMER. 1978. Therapeutic drug monitoring in saliva. Clin. Pharmacokin. **3:** 39–57.
17. DROBITCH, R.K. & C.K. SVENSSON. 1992. Therapeutic drug-monitoring in saliva—an update. Clin. Pharmacokinet. **23:** 365–379.
18. GORODETZKY, C.W. & M.P. KULLBERG. 1974. Validity of screening methods for drugs of abuse in biological fluids. II. Heroin in plasma and saliva. Clin. Pharmacol. Ther. **15:** 579–587.
19. IDOWU, O.R. & B. CADDY. 1982. A review of the use of saliva in the forensic detection of drugs and other chemicals. J. Forensic Sci. Soc. **22:** 123–135.
20. SCHRAMM, W., R.H. SMITH, P.A. CRAIG & D.A. KIDWELL. 1992. Drugs of abuse in saliva—a review. J. Anal. Toxicol. **16:** 1–9.
21. CHU, F.W. & R.P. EKINS. 1988. Detection of corticosteroid binding globulin in parotid fluids—evidence for the presence of both protein-bound and non-protein-bound (free) steroids in uncontaminated saliva. Acta Endocrinol. **119:** 56–60.
22. DEBOEVER, J., F. KOHEN, J. BOUVE, *et al.* 1990. Direct chemiluminescence immunoassay of estradiol in saliva. Clin. Chem. **36:** 2036–2041.
23. REY, F., G. CHIODONI, K. BRAILLARD, *et al.* 1990. Free testosterone levels in plasma and saliva as determined by a direct solid-phase radioimmunoassay—a critical evaluation. Clin. Chim. Acta **191:** 21–30.
24. SELBY, C., P.A. LOBB & W.J. JEFFCOATE. 1988. Sex-hormone binding globulin in saliva. Clin. Endocrinol. **28:** 19–24.
25. ARCHIBALD, D.W., L. ZON, J.E. GROOPMAN, *et al.* 1986. Antibodies to human T-lymphotropic virus type-Iii (Htlv-Iii) in saliva of acquired-immunodeficiency-syndrome (AIDS) patients and in persons at risk for AIDS. Blood **67:** 831–834.
26. GRAU, A.J., A.W. BODDY, D.A. DUKOVIC, *et al.* 2004. Leukocyte count as an independent predictor of recurrent ischemic events. Stroke **35:** 1147–1152.
27. SKIDMORE, S.J. & C.A. MORRIS. 1992. Salivary testing for HIV infection. Br. Med. J. **305:** 1094.
28. KAUFMAN, E. & I.B. LAMSTER. 2000. Analysis of saliva for periodontal diagnosis—a review. J. Clin. Periodontol. **27:** 453–465.
29. KLOCK, B., M. SVANBERG & L.G. PETERSSON. 1990. Dental-caries, mutans streptococci, lactobacilli, and saliva secretion rate in adults. Commun. Dent. Oral Epidemiol. **18:** 249–252.
30. KOHLER, B. & S. BJARNASON. 1992. Mutans streptococci, lactobacilli and caries prevalence in 15- to 16-year olds in Goteborg. 2. Swed. Dent. J. **16:** 253–259.
31. EL-NAGGAR, A.K., L. MAO, G. STAERKEL, *et al.* 2001. Genetic heterogeneity in saliva from patients with oral squamous carcinomas—implications in molecular diagnosis and screening. J. Mol. Diagn. **3:** 164–170.
32. LIAO, P.H., Y.C. CHANG, M.F. HUANG, *et al.* 2000. Mutation of p53 gene codon 63 in saliva as a molecular marker for oral squamous cell carcinomas. Oral Oncol. **36:** 272–276.

33. MERCER, D.K., K.P. SCOTT, C.M. MELVILLE, *et al.* 2001. Transformation of an oral bacterium via chromosomal integration of free DNA in the presence of human saliva. FEMS Microbiol. Lett. **200:** 163–167.
34. STAMEY, F.R., M. DELEON-CARNES, M.M. PATEL, *et al.* 2003. Comparison of a microtiter plate system to Southern blot for detection of human herpesvirus 8 DNA amplified from blood and saliva. J. Virol. Methods **108:** 189–193.
35. CHRISTODOULIDES, N., M. TRAN, P.N. FLORIANO, *et al.* 2002. A microchip-based multianalyte assay system for the assessment of cardiac risk. Anal. Chem. **74:** 3030–3036.
36. CUREY, T.E., A. GOODEY, A. TSAO, *et al.* 2001. Characterization of multicomponent monosaccharide solutions using an enzyme-based sensor array. Anal. Biochem. **293:** 178–184.
37. GOODEY, A., J.J. LAVIGNE, S.M. SAVOY, *et al.* 2001. Development of multianalyte sensor arrays composed of chemically derivatized polymeric microspheres localized in micromachined cavities. J. Am. Chem. Soc. **123:** 2559–2570.
38. GOODEY, A.P. & J.T. MCDEVITT. 2003. Multishell microspheres with integrated chromatographic and detection layers for use in array sensors. J. Am. Chem. Soc. **125:** 2870–2871.
39. LAVIGNE, J.J., S. SAVOY, M.B. CLEVENGER, *et al.* 1998. Solution-based analysis of multiple analytes by a sensor array: toward the development of an "electronic tongue." J. Am. Chem. Soc. **120:** 6429–6430.
40. MCCLESKEY, S.C., M.J. GRIFFIN, S.E. SCHNEIDER, *et al.* 2003. Differential receptors create patterns diagnostic for ATP and GTP. J. Am. Chem. Soc. **125:** 1114–1115.
41. MCCLESKEY, S.C., P.N. FLORIANO, S.L. WISKUR, *et al.* 2003. Citrate and calcium determination in flavored vodkas using artificial neural networks. Tetrahedron **59:** 10089–10092.
42. WISKUR, S.L., P.N. FLORIANO, E.V. ANSLYN & J.T. MCDEVITT, 2003. A multicomponent sensing ensemble in solution: differentiation between structurally similar analytes. Angewandte Chemie International Edition **42:** 2070–2072.
43. ALI, M.F., R. KIRBY, A.P. GOODEY, *et al.* 2003. DNA hybridization and discrimination of single-nucleotide mismatches using chip-based microbead arrays. Anal. Chem. **75:** 4732–4739.
44. CHRISTODOULIDES, N., P.N. FLORIANO, S.A. ACOSTA, *et al.* 2005. Towards the development of a lab-on-a-chip dual function white blood cell and C-reactive protein analysis method for the assessment of inflammation and cardiac risk. Clin. Chem. **51:** 2391–2395.
45. LI, S., P.N. FLORIANO, N. CHRISTODOULIDES, *et al.* 2005. Disposable polydimethylsiloxane/silicon hybrid chips for protein detection. Biosens. Bioelectron. **21:** 574–580.
46. CHRISTODOULIDES, N., S. MOHANTY, C.S. MILLER, *et al.* 2005. Application of microchip assay system for the measurement of C-reactive protein in human saliva. Lab Chip **5:** 261–269.
47. LI, S., D. FOZDAR, M.F. ALI, *et al.* 2005. A continuous flow polymerasse chain reaction microchip with regional velocity control. J. Microelectromech. Syst. **15:** 223–236.
48. NAVAZESH, M. 1993. Methods for collecting saliva. Ann. N. Y. Acad. Sci. **694:** 72–77.
49. MILLER, C.S., C.P. KING, C. LANGUB, *et al.* 2006. Salivary biomarkers of existing periodontal disease—A cross-sectional study. J. Am. Dent. Assoc. **137:** 322–329.

50. VITZTHUM, F., F. BEHRENS, N.L. ANDERSON & J.H. SHAW. 2005. Proteomics: from basic research to diagnostic application. A review of requirements and needs. J. Proteome Res. **4:** 1086–1097.
51. DIEHL, S.R., T.X. WU, B.S. MICHALOWICZ, *et al.* 2005. Quantitative measures of aggressive periodontitis show substantial heritability and consistency with traditional diagnoses. J. Periodontol. **76:** 279–288.
52. ASSUMA, R., T. OATES, D. COCHRAN, *et al.* 1998. IL-1 and TNF antagonists inhibit the inflammatory response and bone loss in experimental periodontitis. J. Immunol. **160:** 403–409.
53. DELIMA, A.J., S. KARATZAS, S. AMAR & D.T. GRAVES. 2002. Inflammation and tissue loss caused by periodontal pathogens is reduced by interleukin-1 antagonists. J. Infect. Dis. **186:** 511–516.
54. GRAVES, D.T., A.J. DELIMA, R. ASSUMA, *et al.* 1998. Interleukin-1 and tumor necrosis factor antagonists inhibit the progression of inflammatory cell infiltration toward alveolar bone in experimental periodontitis. J. Periodontol. **69:** 1419–1425.

A Microfluidic System for Saliva-Based Detection of Infectious Diseases

ZONGYUAN CHEN,[a] MICHAEL G. MAUK,[a] JING WANG,[a] WILLIAM R. ABRAMS,[b] PAUL L. A. M. CORSTJENS,[c] R. SAM NIEDBALA,[d] DANIEL MALAMUD,[b] AND HAIM H. BAU[a]

[a]*University of Pennsylvania School of Engineering and Applied Science, Philadelphia, Pennsylvania, USA*

[b]*New York University College of Dentistry, New York, New York, USA*

[c]*Leiden University Medical Center, Leiden, the Netherlands*

[d]*Department of Chemistry, Lehigh University, Bethlehem, Pennsylvania, USA*

ABSTRACT: A "lab-on-a-chip" system for detecting bacterial pathogens in oral fluid samples is described. The system comprises : (1) an oral fluid sample collector; (2) a disposable, plastic microfluidic cassette ("chip") for sample processing including immunochromatographic assay with a nitrocellulose lateral flow strip; (3) a platform that controls the cassette operation by providing metered quantities of reagents, temperature regulation, valve actuation; and (4) a laser scanner to interrogate the lateral flow strip. The microfluidic chip hosts a fluidic network for cell lysis, nucleic acid extraction and isolation, PCR, and labeling of the PCR product with bioconjugated, upconverting phosphor particles for detection on the lateral flow strip.

KEYWORDS: microfluidics; lab-on-a-chip; infectious diseases; diagnostics; immunochromatography

INTRODUCTION AND BACKGROUND

Currently, point-of-care (POC) testing for infectious diseases is routinely performed using immunochromatographic assays of serum, urine, or saliva. These assays are typically implemented as easy-to-use nitrocellulose lateral flow test strips. Similar test strips are widely employed for a variety of applications, such as home pregnancy testing and detecting drugs of abuse. Serologic lateral flow strips that detect HIV antibodies in oral fluid are completely self-contained, requiring no addition of reagents or accessory instrumentation.[1]

Address for correspondence: Haim H. Bau, 216 Towne Building, Mechanical Engineering and Applied Mechanics, University of Pennsylvania, Philadelphia, PA 19104-6315. Voice: 215-898-8363; fax: 215-573-6114.
bau@seas.upenn.edu

Ann. N.Y. Acad. Sci. 1098: 429–436 (2007).
doi: 10.1196/annals.1384.024

The strips include a control test function for quality assurance, and can provide test results in less than 10 min. The lateral flow test strip device provides convenient, fast, and low-cost POC and home testing. However, there is considerable incentive for improving sensitivity and increasing functionality. For example, a more quantitative assay could be used to determine viral loads. Additionally, parallel testing of nucleic acid (DNA and RNA) and antigen targets, from both viral and bacterial pathogens, as well as detection of antibodies, would enable simultaneous robust, multipurpose, and confirmatory testing from the same sample with the same test device. However, multiplexed nucleic acid testing and immunoassays require elaborate sample processing, which complicates implementation as a simple-to-use POC device. Microfluidic "lab-on-a-chip" devices in the form of miniaturized networks of conduits, chambers, and valves formed in a plastic substrate "chip" facilitate sample metering, cell and virus lysis, isolation of nucleic acids and proteins, amplification of nucleic acids using PCR or analogous techniques, and labeling of analytes for detection. Thus, a lateral flow strip combined with a microfluidic cassette for sample processing would significantly expand the range of applications and tasks that could be performed by immunochromatographic methods.

Most lateral flow test strips use colloidal gold, dye, or latex beads as reporters to allow visual inspection of test results.[1] Lateral flow strip tests that incorporate upconverting phosphors as reporters have also been developed.[2–5] Upconverting phosphor technology (UPT) is based on sub-micron-sized ceramic particles coated with lanthanides that absorb infrared light (excitation) and emit visible light (response signal). The particles can be functionalized with antigens and antibodies for use as labels on lateral flow strips. UPT particles avoid interference from background fluorescence and do not exhibit photobleaching effects. UPT reporters have been shown to improve the limit of detection in various bioassays by 10-fold or more.[3–6] In one particular application, a POC diagnostics system (OraSure Up*link*™) includes an FDA-approved oral fluid sample collector, a molded plastic cartridge that houses the lateral flow strip and features a loading port for the sample collector, and a tabletop reader instrument with laser scanner and photomultiplier tube detector to interrogate the lateral flow strip.[7] Also, UPT-based immunochromatography flow strips have been adapted for detecting nucleic acid targets,[3–5] such as UPT-labeled PCR products, thus opening the way for applying this technology for nucleic acid testing. This UPT lateral flow system thus provides an excellent platform for developing a microfluidics system for multiplexing nucleic acid–based diagnostics and immunoassays. The approach is to develop a disposable credit card–sized plastic cassette that hosts a microfluidic circuit integrating components for (1) sample metering; (2) enzymatic and chemical lysis using lysozyme, proteinase K, chaotropic salts, and detergents; (3) solid-phase extraction using a porous silica membrane embedded in the cassette; (4) PCR amplification; (5) labeling of the PCR product with UPT phosphor; and

(6) immunochromatography assay of the PCR amplicon with a nitrocellulose lateral flow strip mounted in the cassette.

Efforts to add microfluidic capabilities to lateral flow strip assays have already met with success. Chen *et al.*[8] developed a consecutive lateral flow assay microfluidic chip with UPT reporters for use with the UP*link*™ laser scanner. The chip includes functions for sample introduction, buffer distribution, metering, mixing, and thermopneumatic pumping. After a serum sample is loaded on the chip, the chip is connected to an instrumentation platform that provides fluidic power, temperature control, and valve actuation. Automated processing on the chip comprises a sequence of steps including sample aliquoting and metering, dilution with buffer, and blotting on a nitrocellulose strip followed by a separate wash step of the strip with buffer, followed by addition of buffer with UPT label. The UPT labels are dry-stored and preloaded on the chip in a lyosphere. The chip is then inserted into the UP*link*™ reader for interrogation of the strip by IR (980 nm) and signal readout. Along similar lines, Wang *et al.*[9] reported a disposable microfluidic cassette for DNA amplification and detection combining a microfluidic PCR chamber and an incubation chamber to label the PCR product with UPT particles, integrated with a lateral flow strip to capture UPT-labeled PCR amplicon for detection in the laser scanner. Here, we report a microfluidic cassette for PCR-based detection of Gram-positive bacteria using a lateral flow strip with UPT reporters. The cassette provides complete processing including sample introduction and metering, enzymatic cell lysis, nucleic acid isolation, PCR amplification with pathogen-specific primers conjugated for labeling and capture on a lateral flow strip, and blotting of labeled PCR product on the lateral flow strip affixed to the cassette. The microfluidic system is tested with diluted *B. cereus*, a safe surrogate for anthrax, but the processing steps are sufficiently generic and the chip design is sufficiently flexible so that it can be adapted for detection of different viral and bacterial pathogens.

CHIP DESIGN, FABRICATION, AND OPERATION

FIGURE 1A is a photograph of a microfluidic chip for PCR-based detection of bacteria targets. The chip is made as a thermally bonded three-layer laminate of polycarbonate sheets with channels and chambers defined in two of the layers by CNC machining. Cross-section dimensions of the conduits range from 250 to 500 μm. The microfluidic cassette is mounted in a molded plastic cartridge compatible with insertion into the UP*link*™ laser scanner/reader unit (FIG. 1B). Phase-change "ice" valves actuated by small, electrically controlled thermoelectric elements on the processing platform are used for flow control and sealing the 10-μL PCR chamber.[10] Flow is halted when the fluid in a channel freezes at the section of conduit in contact with a cooled thermoelectric element, and flow resumes when the thermoelectric element is heated to thaw

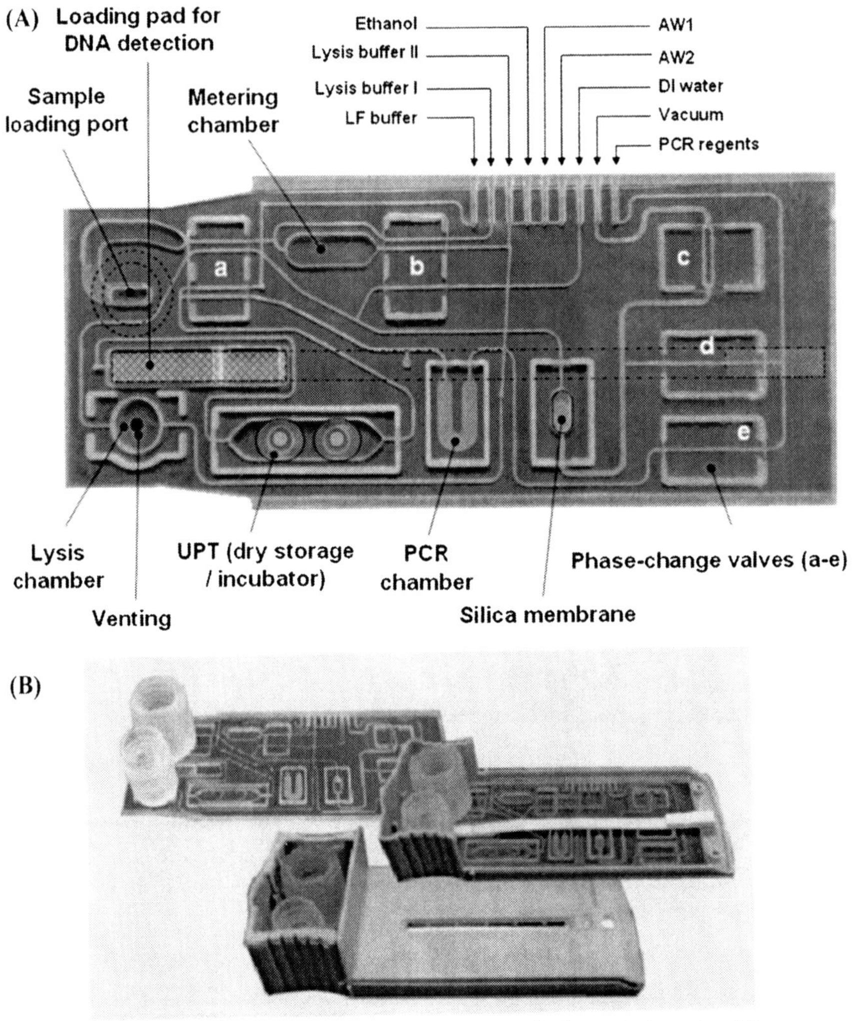

FIGURE 1. (**A**) Polycarbonate microfluidic cassette for bacteria detection. The cassette features conduits, chambers, and phase-change valve sites for sample metering, lysis, solid-phase extraction, and isolation of nucleic acids with a porous silica membrane-binding phase, PCR, labeling, and nitrocellulose lateral flow test strip. The cassette measures 85 × 35 mm and is 3 mm thick. (**B**) Microfluidic cassettes with lateral flow strips are mounted in a molded plastic cartridge for the UP*link*™ laser scanner/reader unit.

the fluid. The PCR chamber is also heated and cooled by a thermoelectric element. The chip is interfaced with programmable syringe pumps, reservoirs for buffers and reagents, and electronic controllers for the thermoelectric elements, controlled by a PC with LabVIEW™ software (FIG. 2).

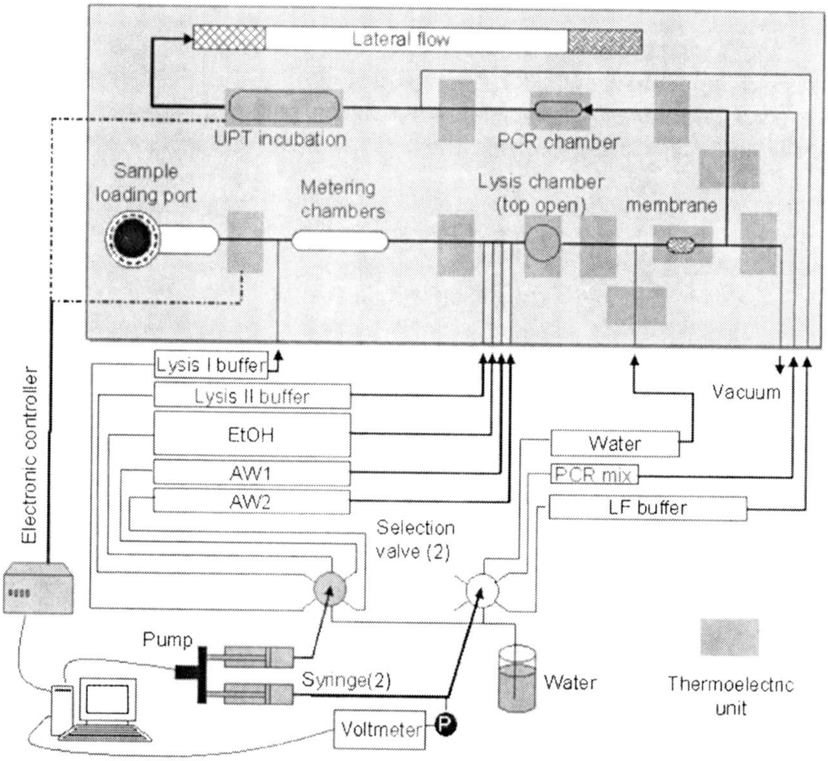

FIGURE 2. Schematic of supporting instrumentation for microfluidic cassette including buffer and reagent reservoirs, programmable syringe pumps, thermoelectric elements for actuating phase-change "ice" valves and heating/cooling PCR chamber, controlled power supplies, and PC.

The sample is introduced through a loading port and 50 μL of sample aliquoted in a metering chamber. The sample is then subjected to a two-step lysis process. The sample is first enzymatically digested for 30 min at 37°C with lysozyme (20 mg/mL lysozyme, 1.2% Triton X-100, 2mM EDTA, 20 mM Tris-Cl), and then incubated with SDS detergent (0.1%), proteinase K (10 mg/mL) and 6M guanidinium chloride at 60°C for 15 min. The lysate is forced through a porous silica membrane embedded in the cassette. The chaotropic salts induce nucleic acids binding to the silica. The silica-bound nucleic acids are washed with ethanol-based solutions to remove debris. Next, the nucleic acids are desorbed from the silica by elution with low-salt, pH-neutral buffer, and a 5-μL elution fraction is mixed with PCR reagents (0.3 U Taq polymerase, 200 μM dNTP, 0.1 μg/uL BSA, 50 mM Tris-Cl, 1.5–3.5 mM $MgCl_2$) and primers (forward: 5′-TCT CGT TTC ACT ATT CCC AAG T-3′ conjugated to dioxygenin; reverse: 5′ AAG GTT CAA AAG ATG GTA TTC AGG-3′ conjugated to biotin) and sealed in a 10-μL PCR chamber, where it is thermally cycled 25 times (95°C, 15 sec; 55°C, 25 sec; 72°C, 20 sec). The 305-bp amplicon is labeled by incubation at room temperature for 15 min with 150 ng of avidin-conjugated UPT particles. The UPT-labeled PCR product is applied to the sample pad of a lateral flow strip and captured by immobilized anti-DIG antibodies at the test line stripe. The strip is then scanned with the UP*link*™ reader.

RESULTS

To assess microfluidic lysis and nucleic acid isolation for producing a DNA template for PCR from cell culture samples, nucleic acid eluted from the chip silica membrane was PCR-amplified and compared to PCR-amplified nucleic acid isolated by a standard benchtop protocol (QIAGEN DNeasy™ Kit). In the benchtop protocol, a 50-μL sample of *B. cereus* (~10^6 cells/mL) was subjected to a two-step enzymatic lysis process with lysozyme, chaotropic salts, and detergents. The nucleic acid was then isolated from the lysate in successive

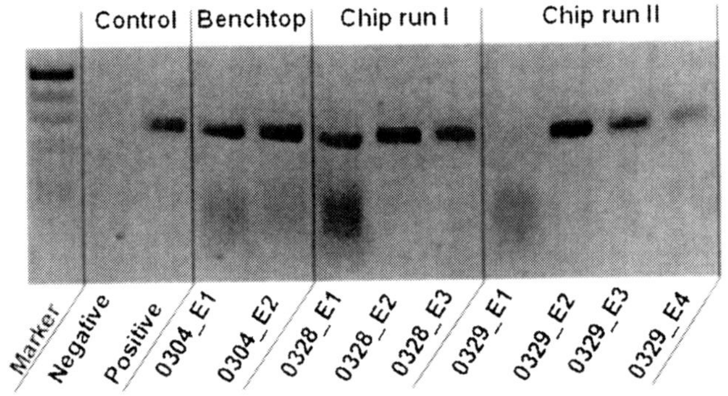

FIGURE 3. Comparison of on-chip and benchtop lysis and nucleic acid isolation. To determine the efficiency of on-chip lysis and nucleic acid isolation relative to benchtop lysis and nucleic acid isolation, elutions from the silica membrane on the chip and from a spin column were amplified by PCR and analyzed by gel electrophoresis. Control: PCR of pure *B. cereus* genomic DNA. Benchtop: *B. cereus* (50 μL, ~10^6 cells/mL) was subjected to two-step lysis (lysozyme digestion followed by treatment with chaotropic salt and proteinase K according to QIAGEN DNeasy™ protocol). Nucleic acid was isolated from lysate using QIAGEN DNeasy™ spin column. Aliquots of 5 μL from two successive 50-μL elutions (E1 and E2) from the spin column were mixed with 5 μL of PCR mix (Taq, dNTP, buffer, primers, BSA), amplified for 25 cycles, and run on an agarose gel (1.5%, ethidium bromide staining). Chip runs (two replicates, I and II): Similar *B. cereus* samples (50 μL, ~10^6 cells/mL) were loaded into the chip and subjected to two-step lysis and nucleic acid isolation using the silica-membrane component of the chip. Aliquots (5 μL) from successive 50-μL elutions (E1, E2, E3...) were mixed with 5 μL of PCR mix (Taq, dNTP, buffer, primers, BSA), amplified for 25 cycles on a benchtop thermal cycler, and run on an agarose gel (1.5%, ethidium bromide staining). Weak or absent PCR band for first elution (E1) for chip runs suggests a PCR inhibitor is contained in the initial elution. Subsequent elutions from the chip membrane, however, showed strong PCR amplification.

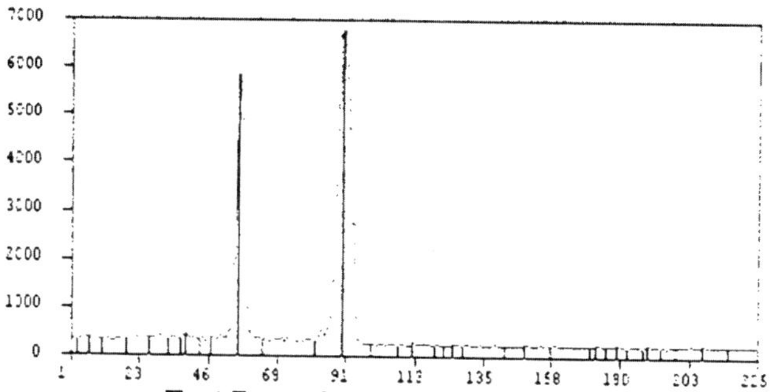

FIGURE 4. A representative UP*link*^{TM} laser scan of the lateral flow strip for measurement of UPT-labeled PCR amplicon from bacteria (50 μL of $\sim 10^6$ cells/mL) processed on a microfluidic cassette.

50-μL elutions from a QIAGEN DNeasy^{TM} spin column (chaotrope–silica method), and amplified by benchtop PCR. Similarly, a 50-μL sample of *B. cereus* was loaded into the microfluidic cassette for on-chip lysis and nucleic acid isolation. Successive 50-μL elutions from the silica membrane embedded in the chip were collected and amplified by benchtop PCR. In each case, 5-μL aliquots from successive 50-μL elution of the spin column and from the chip silica membrane were mixed with 5 μL of PCR mix (Taq, dNTP, buffer, primers, BSA), amplified for 25 cycles, loaded and run on a 1.5% agarose gel with ethidium bromide stain (FIG. 3). The weak or absent PCR product from the initial elution of the chip indicates the presence of an inhibitor, probably residual ethanol from the silica membrane wash step. This initial elution fraction should be avoided for providing a PCR template, but otherwise the subsequent elutions from the chip yielded easily detected PCR product. There appears to be some benefit to removing inhibitors from the silica membrane by a vacuum drying step prior to elution (data not shown). To test the chip for complete sample processing (i.e., on-chip lysis, nucleic acid isolation, PCR amplification, labeling, and lateral flow assay), the chip was loaded with a 50-μL aliquot of *B. cereus* ($\sim 10^6$ cells/mL), and operated according to the sequence detailed above. A typical UP*link*^{TM} scan of the chip flow strip is shown in FIGURE 4, indicating a detectable *B. cereus*–specific PCR product produced–from the sample.

DISCUSSION AND CONCLUSIONS

We have demonstrated a microfluidic cassette that can prepare UPT-labeled microfluidic PCR product for detection on a lateral flow strip assay interfaced with the cassette, thus showing the feasibility of microfluidic PCR-based nucleic acid testing of raw samples. The current development effort is focused on

Gram-positive bacterial targets (*B. cereus*, a safe surrogate for anthrax), but the system can be readily adapted for detection of viral and bacterial pathogens, multiplexed analysis of several pathogens, and serologic testing for antibodies. The device will serve as a module in a comprehensive, POC saliva-based diagnostic system that will also include modules for antigen and antibody assays.

ACKNOWLEDGMENTS

The work was supported by NIH Grants UO1DE0114964 and UO1DE017855, and in collaboration with OraSure Technologies, Inc. (Bethlehem, PA, USA).

REFERENCES

1. KETEMA, F., H.L. ZINK, K.M. KREISEL, *et al.* 2005. A 10-minute, US Food and Drug Administration-approved HIV test. Expert Rev. Mol. Diagn. **5:** 135–143.
2. NIEDBALA, R.S., H. FEINDT, K. KARDOS, *et al.* 2001. Detection of analytes by immunoassay using up-converting phosphor technology. Anal. Biochem. **293:** 22–30.
3. CORSTJENS, P.L.A.M., M. ZUIDERWIJK, M. NILSSON, *et al.* 2003. Lateral-flow and up-converting phosphor reporters to detect single-stranded nucleic acids in a sandwich-hybridization assay. Anal. Biochem. **312:** 191–200.
4. CORSTJENS, P.L.A.M., M. ZUIDERWIJK, A. BRINK, *et al.* 2001. Use of up-converting phosphor reporters in lateral flow assays to detect specific nucleic acid sequences: a rapid, sensitive DNA test to identify human papilloma type 16 infection. Clin. Chem. **47:** 1885–1893.
5. HAMPL, J., M. HALL, N.A. MUFTI, *et al.* 2001. Upconverting phosphor reporters in immunochromatographic assays. Anal. Biochem. **288:** 176–187.
6. CORSTJENS, P.L.A.M., S. LI, M. ZUIDERWIJK, *et al.* 2005. Infrared up-converting phosphors for bioassays. IEEE Proc. Nanobiotechnol. **152:** 64–72.
7. MALAMUD, D., H. BAU, S. NIEDBALA & P. CORSTJENS. 2005. Point detection of pathogens in oral samples. Adv. Dent. Res. **18:** 12–16.
8. CHEN, Z., P.L.A.M. CORSTEJENS, M. ZUIDERWIJK, *et al.* 2005. A disposable microfluidic point-of-care device for detection of HIV: a new upconverting phosphor technology application. Proceedings of the 9th Int. Conf. Miniaturized Sys. Chem. Life Sci. 791–793.
9. WANG, J., Z. CHEN, P.L. CORSTJENS, *et al.* 2006. A disposable microfluidic cassette for DNA amplification and detection. Lab. Chip **6:** 46–53.
10. WANG, J., Z. CHEN, M. MAUK, *et al.* 2005. Self-actuated, thermo-responsive hydrogel valves for lab on a chip. Biomed. Microdevices **7:** 313–322.

Rapid Assay Format for Multiplex Detection of Humoral Immune Responses to Infectious Disease Pathogens (HIV, HCV, and TB)

PAUL L. A. M. CORSTJENS,[a] ZONGYUANG CHEN,[b] MICHEL ZUIDERWIJK,[a] HAIM H. BAU,[b] WILLIAM R. ABRAMS,[c] DANIEL MALAMUD,[c] R. SAM NIEDBALA,[d] AND HANS J. TANKE[a]

[a]*Department of Molecular Cell Biology, Leiden University Medical Center, Leiden, the Netherlands*

[b]*Department of Mechanical Engineering and Applied Mechanics, University of Pennsylvania, Philadelphia, Pennsylvania, USA*

[c]*Department of Basic Sciences, New York University College of Dentistry, New York, New York, USA*

[d]*Department of Chemistry, Lehigh University, Bethlehem, Pennsylvania, USA*

ABSTRACT: A novel assay is described for multiplex detection of antibodies against different pathogens from a single sample. The assay employs a modified lateral flow format (consecutive flow, CF) together with a sensitive reporter particle technology (up-converting phosphor technology, UPT) that allows for fully instrumented assay analysis. Lateral flow (LF) strips developed for the detection of human antibodies against human immunodeficiency virus type-1 and -2 (HIV-1 and -2) with additional capture zones to detect antibodies against *Myobacterium tuberculosis* (TB) and hepatitis C Virus (HCV) provided the strips to test multiplexing. Data are presented that show the performance of the TB and HCV test, as well as two multiplex assays, TB with HIV and HCV with HIV. The TB/HCV assays demonstrate excellent detection capability, and HIV multiplexing does not affect the qualitative test result. The bench-top CF format was converted to a microfluidic platform and a first prototype semiautomated chip capable of performing CF is presented here.

KEYWORDS: infectious disease; multiplex assay; up-converting phosphor; lateral flow; TB; HCV; HIV

Address for correspondence: Paul L. A. M. Corstjens, Department of Molecular Cell Biology, Leiden University Medical Center, PO Box 9600, 2300 RC Leiden, the Netherlands. Voice: +31-71-5269209; fax: +31-71-5268270.
Corstjens@LUMC.NL

Ann. N.Y. Acad. Sci. 1098: 437–445 (2007). © 2007 New York Academy of Sciences.
doi: 10.1196/annals.1384.016

INTRODUCTION

The presented antibody test platform allows convenient, rapid, sensitive, and cost-effective detection of infectious disease pathogens. It was initially developed for the detection of antibodies to HIV. In this report we explore the potential of the platform for multiplex detection of antibodies associated with other infectious diseases, for example, simultaneous detection of HIV, TB, and HCV. TB often causes opportunistic infection in HIV/AIDS patients, and HCV has a high prevalence in HIV-infected individuals and is suspected to co-infect with HIV.[1,2] The antibody test platform was designed to become part of a modular microfluidic device, a point-of-care (POC) device to analyze oral fluid for simultaneous detection of pathogenic antigens, nucleic acids, and the host antibodies to the pathogen.[3]

The antibody assay uses a modified LF immunochromatography format with three consecutive flow (CF) steps: (1) the first flow with diluted specimen (e.g., plasma, oral fluid, or urine); (2) the second flow with wash buffer; and (3) the third flow with protein A–coated reporter particles. The multiple flow system (CF) accommodates a 10-fold higher sample volume compared to a single flow format, and demonstrates a better signal-to-noise value. Previously described up-converting phosphor technology (UPT) reporter particles are applied for ultrasensitive and instrumented assay analysis.[4,5] These UPT reporters generate a visible light emission signal upon excitation with low-energy infrared (IR) light.[6] Interrogation of LF strips exposed to UPT reporters can be performed with a portable UPT reader (UP*link*).[7] In the UP*link* system LF strips are integrated in disposable plastic cassettes. We developed and constructed a prototype semiautomatic microfluidic module to perform CF that fits into existing UP*link*-compatible cassettes.[8]

RESULTS

Consecutive Flow

FIGURE 1 shows an illustration of UPT–CF designed for detection of human antibodies against HIV. Antibodies to HIV-1 and -2 are captured at the test line which consists of HIV-specific antigens and the remaining human IgGs bind downstream to an anti-human IgG flow-control line. Nitrocellulose sheets for CF with proprietary HIV and HCV test lines were provided by OraSure Technologies. A third test line with a proprietary TB antigen mixture (Courtesy of H.J. Houthoff and G.J. van Dam) was added to the CF strips in-house using a Camag Linomat IV striper.[9] Nitrocellulose sheets with only a TB-specific test line were prepared in-house in a format similar to the sheets with the HIV-specific test line. Nitrocellulose sheets were assembled into LF strips as described earlier.[10] The assays described here used plasma, but can be easily

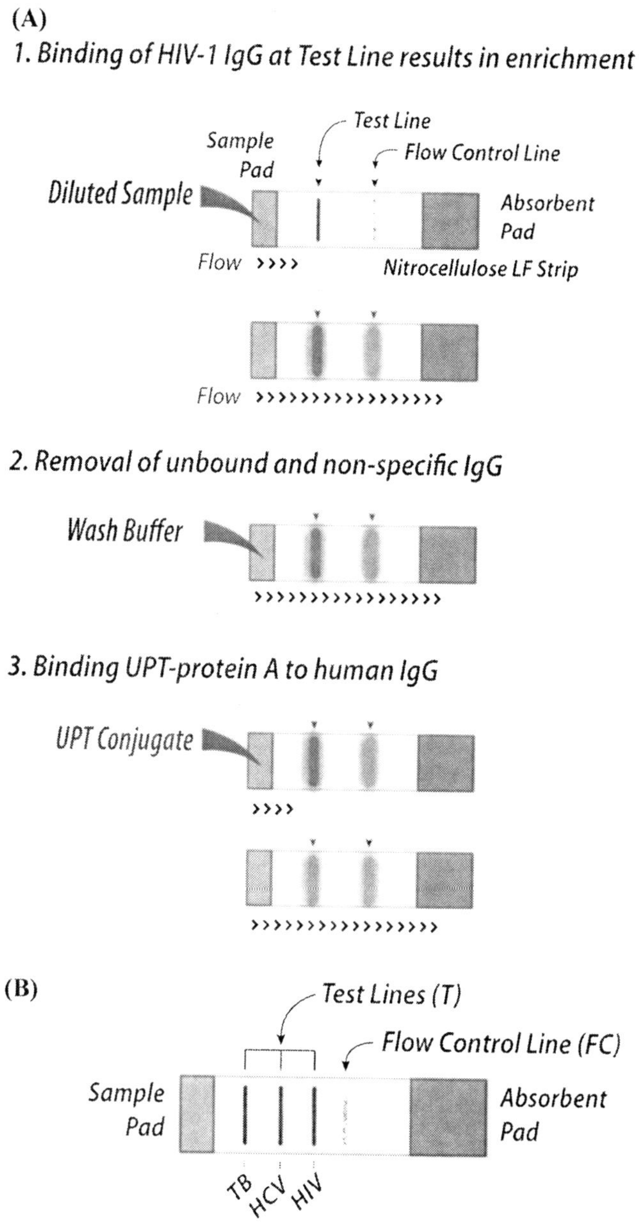

FIGURE 1. Consecutive flow format and antibody multiplex strips. (**A**) Consecutive flow for plasma samples comprises three successive flow steps: (1) diluted specimen; (2) wash buffer; and (3) protein A–coated UPT reporter particles. Bold capture lines (T and FC) indicate the deposition of the UPT particles that excite green light upon excitation with infrared light. (**B**) Schematic illustration of an antibody multiplex strip for the detection of antibodies against TB, HCV, and HIV.

modified to allow testing with oral fluid or urine. For bench-top assays, plasma samples are diluted 100-fold in assay buffer (100 mM Hepes, pH 7.2, 270 mM NaCl, 0.5 % Tween-20, 1% w/v BSA). The dilution factor will vary, depending on the type of specimen. After dilution, 10 μL is mixed in a microtiter plate well containing 40 μL assay buffer and flow initiated by inserting the LF strips into the well, followed by a 20 μL assay buffer wash immediately after the sample has migrated into the LF strip. These two flow steps take approximately 2 min, after which LF strips are added to a new well containing 70 μL of assay buffer with 100 ng UPT^{protA} conjugate. An illustration with a schematic presentation of a microfluidic device capable of mimicking the above bench-top CF format is shown in FIGURE 2. The microfluidic device differs from the bench-top assay in that the sample input is approximately 2 μL of either diluted sample or undiluted sample. Oral fluid specimens do not require dilution, whereas plasma needs to be diluted 10- or 100-fold prior to applying the sample to the loading well.

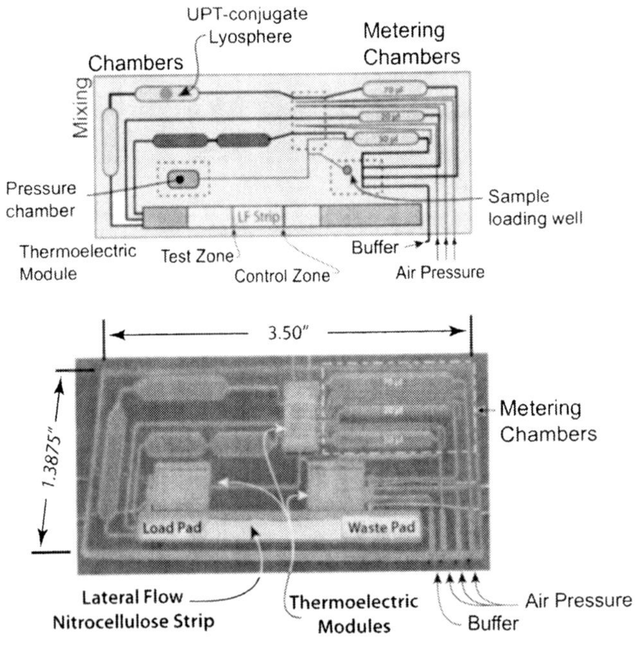

FIGURE 2. Semiautomatic prototype microfluidic device for consecutive flow. Pneumatic pressure is used to prefill the three buffer metering chambers. The specimen (around 5 μL) is applied to sample loading well, and cooling of the pressure chamber creates a void that pulls an approximately 2-μL specimen into the capillary system. Pneumatic pressure and ice valves operated by thermoelectric modules control the liquid flow.

HIV/TB Multiplexing

A collection of 300 banked plasma specimens (provided by H.J. Houthoff and G.J. van Dam) were analyzed with UPT–CF. The specimens consisted of 100 healthy control samples (HIV–/TB–), 100 samples from HIV-negative but TB-infected patients (HIV–/TB+), and 100 samples from patients infected with HIV and TB (HIV+/TB+). The UPT assay threshold for plasma samples as determined from an earlier study was verified by analyzing samples on LF strips prepared with an HIV-specific test line only. In this experiment, samples generating a ratio signal (test signal divided by flow-control signal) >0.04 were considered antibody-reactive. A ratio value rather than an actual UPT signal value (measured in relative fluorescent units, RFU) is preferred as this provides a convenient method to normalize test results; however, normalization is not a requirement. TB infection status was then tested on LF strips carrying a TB-specific test line only. The resulting clinical parameters regarding TB testing are presented in TABLE 1 together with lab-based enzyme-linked immunosorbent assay (ELISA) results that were provided with the specimens. From these results we concluded that the accuracy of the rapid UPT assay is superior to the lab-based ELISA assay; it is important to note that both assays used the same antigen mixture to capture TB-specific antibodies, so that the increased performance is related to the difference in assay platform (ELISA vs. LF) as well as the difference in reporter technology (fluorescence vs. UPT).

Compared with testing for other infectious diseases, the actual specificity and sensitivity value for TB is relatively low, which is a known problem.[11] Attempts to develop a low-complexity, rapid, immunologic-based TB assay with sufficient accuracy have failed so far. In developing countries TB diagnosis is especially problematic because children and HIV-compromised patients are low responders, which increases the number of false negatives. TABLE 1

TABLE 1. UPT consecutive flow analysis of plasma specimen from 200 TB-diagnosed patients (50% being HIV compromised) and 100 healthy controls

TB/HIV Status[a,b]	Mixed ($n = 300$) HIV–/TB– ($n = 100$) HIV–/TB+ ($n = 100$) HIV+/TB+ ($n = 100$)		Without HIV+/TB+ ($n = 200$) HIV–/TB– ($n = 100$) HIV–/TB+ ($n = 100$)		Without HIV–/TB+ ($n = 200$) HIV+/TB+ ($n = 100$) HIV–/TB– ($n = 100$)	
Assay	ELISA[b]	UPT	ELISA[b]	UPT	ELISA[b]	UPT
Specificity	96.0	95.0	96.0	95.0	96.0	95.0
Sensitivity	52.5	62.5	66.0	74.0	39.0	51.0
Accuracy	50.4	59.4	63.4	70.3	37.4	48.5

[a]Plasma specimens of 200TB- diagnosed patients (TB+) and 100 healthy controls (HIV–/TB–) were analyzed. Half ($n = 100$) of the TB-diagnosed patients were also diagnosed as HIV+.

[b]Plasma specimens and corresponding ELISA data were provided by H.J. Houthoff and G.J. van Dam.

indicates that the UPT assay sensitivity improves from 62.5% to 74.0% when the HIV-compromised patient group (HIV+/TB+ samples) is omitted. This is also evident when examining the ELISA data.

Analysis of selected samples on multiplex LF strips with TB and HIV test lines did not affect the qualitative result of the assay (results not shown). A detailed study is ongoing to examine the potential effect on signal value of "preceding capture line interference," including differences in the location/distance of the capture line from the sample application pad.

HIV/HCV Multiplexing

The UPT–CF format was also used for the detection of HCV infection. Analysis of three seroconversion panels demonstrated excellent performance of the UPT assay compared to an EIA (TABLE 2), which demonstrates the applicability as a model in developing UPT–CF multiplex assays. To further explore the extent of potential "preceding capture line interference," a "dilution checkerboard matrix" of a high-reactive HCV panel member mixed with a high-reactive HIV plasma sample was prepared in normal human plasma (NHP) and analyzed on multiplex LF strips. Each multiplex LF strip produced an HIV as well as HCV test signal, and a flow-control signal. HIV and HCV ratio signals were calculated by dividing their individual test signal by the joint flow-control signal. In FIGURE 3 the ratio value is indicated on the *y* axis of three-dimensional histograms; panel A shows the results for HIV and panel B shows the results for HCV. In both histograms the *x*- and *z* axis represents the

TABLE 2. Relative sensitivity of UPT consecutive flow for the analysis of HCV plasma seroconversion panels

BBI Panel PHV905[a]			BBI Panel PHV907[a]			BBI Panel PHV914[a]		
Days[b]	EIA[c]	UPT[d]	Days[b]	EIA[c]	UPT[d]	Days[b]	EIA[c]	UPT[d]
0	0.0	0.02	0	0.0	0.02	0	0.0	0.01
4	0.0	0.01	4	0.0	0.02	5	0.0	0.02
7	0.0	0.03	7	0.0	0.02	9	0.0	0.01
11	0.3	0.02	13	0.1	**0.09**	12	0.0	0.04
14	0.7	**0.06**	18	0.7	**0.24**	16	0.2	**0.16**
18	0.7	**0.07**	21	**1.5**	**0.42**	19	0.3	**0.21**
21	**2.5**	**0.10**	164	**5.0**	**1.48**	24	**3.2**	**0.37**
25	**>5.0**	**1.10**				30	**>4.7**	**0.71**
28	**>5.0**	**2.00**				33	**>4.7**	**0.95**

[a]Detailed panel information at www.seracare.com/bbidx/hcv_panels.htm.
[b]Days after first bleed.
[c]BBI-provided data; test values ≥ 1 were considered positive (indicated in bold).
[d]UPT ratio values above > 0.04 were considered positive (indicated in bold).

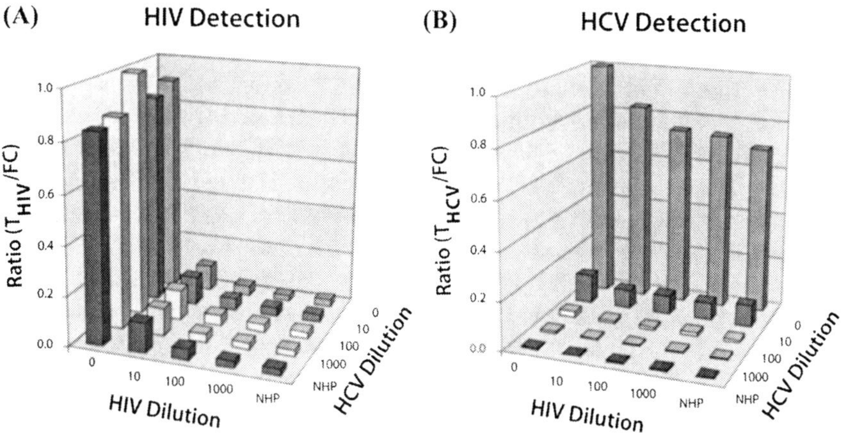

FIGURE 3. Result of a multiplex analysis; the simultaneous detection of antibodies against HIV and HCV in single plasma specimens. A dilution series of plasma specimen from an HIV-compromised patient and an HCV-compromised patient were mixed in equal amounts. In the histograms "0" indicates the highest amount of HIV and/or HCV-compromised plasma specimen; the indicated dilutions were made in NHP. NHP is also used as the (no signal) control sample. The dilution of HIV plasma is indicated on the x axis, the dilution of HCV plasma (z axis) (**A**) the HIV ratio signal, (**B**) the HCV ratio signal. All data points are the average value of three individual experiments.

test matrix with the HIV dilution on the x axis and the HCV dilution on the z axis. The experiment was performed in triplicate and the average signal is presented in the histograms. For purpose of comparison the maximum ratio value was normalized to 1.

The actual ratio value as determined for undiluted HIV and HCV samples was 0.96 and 6.0, respectively (in FIGURE 3 these values were normalized to 1). In patients infected with both viruses, HIV and HCV ratio signals expectedly are different because HIV and HCV test lines use different capture antigens to detect their respective antibodies; the amount of antibodies specific against HIV and HCV is variable, as is the binding affinity of the applied antigen–antibody pairs. In the experiments performed here the HCV test line was closer to the sample application pad than the HIV test line (FIG. 1). As a consequence, the HCV test line signals are higher in comparison to assays where the same HCV test line is localized at the (further downstream) position of the HIV test line. In theory this implies an assay cut-off threshold value that is dependent on the distance of the test line from the sample application pad. For the experiment described here, the actual test line ratio signal obtained on multiplex LF strips with NHP (x_5, z_5 in FIG. 3) generated approximately the same value for HCV and HIV (respectively 0.032 and 0.027 [$n = 3$]) and the maintained cut-off threshold value was > 0.04.

DISCUSSION

We describe a modified lateral flow format (CF) for antibody detection and explore its potential for multiplexing. The CF format was originally developed to detect human antibodies against HIV-1 and -2. Multiplexing is achieved by providing the HIV-specific LF strips with additional capture lines specific for antibodies against other pathogens. Similar to the HIV-specific test line, the additional capture lines comprise pathogen-specific antigens. The different test lines are placed transversely across the LF strip such that the tested specimen will sequentially pass individual test lines. Antibodies present in the specimen can bind to these antigens. This procedure demands careful development and optimization of the various capture lines in order to avoid nonspecific binding and to prevent undesired cross-reactivity of the antibodies with preceding test lines. Furthermore, when applying an antibody-generic reporter, a reporter flow completely disconnected from the antibody flow is desired. CF applies an initial flow of specimen, followed by a wash step and a final flow with antibody-generic UPT reporter particles. The initial flow allows a free flow of the antibodies present in the specimen, so that the antibodies against various pathogens can enrich in the spatially separated capture zones. A wash step then removes non-specific-bound antibodies from the LF strip, and in the final flow UPT^{protA} reporters bind to the antibodies in the pathogen-specific capture zones.

The results presented in this article indicate that multiple antibody test lines are feasible. Multiplex analysis of combined specimens with high loads of antibodies against HIV and HCV did not show relevant interference. Further studies are necessary to evaluate the maximum number of test lines that can be applied to a LF strip without disturbing the UPT reporter particle flow. In this context the effect of high signals in the test lines closest to the sample pad needs to be carefully analyzed. But as most specimens will only show antibody reactivity for a limited number of pathogens, disturbance of the reporter flow to further downstream test lines as a consequence of multiple coinciding high test signals may only be a theoretical problem.

In earlier studies using LF strips with a single test line and a flow-control line, ratio calculations of the test (T) divided by the flow control (FC) were applied. This normalization allows for interassay comparison of results obtained with different LF strips. Increasing the number of test lines will probably affect the normalization algorithm. When the number of test lines is increased, differences in distance and signal may become more evident. In the multiplex assays presented here for TB and HCV these effects were minor and did not affect the outcome of the multiplex tests. Note also that qualitative assays do not necessarily demand a ratio calculation (see, e.g., Mokkapati *et al.*[7]).

The CF format described here allows multiplex detection of various antibodies from a single specimen with an antibody-generic UPT reporter. As such, the CF format suits its objective and was therefore selected as the antibody test module in a (modular) microfluidic device that permits simultaneous detection of viral and bacterial antigens, nucleic acids, and antibodies to these pathogens. The bench-top CF format was therefore converted to a microfluidic module and a first semiautomated prototype is presented.

ACKNOWLEDGMENT

Shang Li and Geraldine Guillon are acknowledged for providing nitrocellulose striped with HIV and HCV antigens and the UPT reporters. Hendrik-Jan Houthoff and Govert J. van Dam are acknowledged for providing the TB antigen mixture and 200 plasma samples from TB-diagnosed patients together with 100 healthy control samples. Claudia J. de Dood and Dieuwke Kornelis are acknowledged for technical assistance. Part of this work was supported by NIH Grant UO1-DE-017855.

REFERENCES

1. Rockstroh, J.K. & U. Spengler. 2004. HIV and hepatitis C virus co-infection. Lancet Infect. Dis. **4:** 437–444.
2. Vernet, G. 2004. Molecular diagnostics in virology. J. Clin. Virol. **31:** 239–247.
3. Malamud, D. *et al.* 2005. Point detection of pathogens in oral samples. Adv. Dent. Res. **18:** 12–16.
4. Li, S. *et al.* 2002. Preparation, characterization and fabrication of uniform coated $Y_2O_2S{:}Er^{3+}$ up-converting phosphor particles for biological detection applications. Proc. SPIE-Int. Soc. Opt. Eng. **4809:** 100–109.
5. Corstjens, P.L.A.M. *et al.* 2005. Infrared up-converting phosphors for bioassays. IEE Proc. Nanobiotechnol. **152:** 64–72.
6. Zarling, D.A. *et al.* 1997. Up-converting reporters for biological and other assays using laser excitation techniques. US Patent **5:** 674–698.
7. Mokkapati, V.K. *et al.* 2007. Evaluation of UP*link*-RSV: a prototype rapid antigen test for detection of respiratory syncytial virus infection. Ann. N.Y. Acad. Sci. This volume.
8. Chen, Z. *et al.* 2005. A disposable microfluidic point-of-care device for detection of HIV: a new up-converting phosphor technology application. Presented at the 9th International Conference on Miniaturized Systems for Chemistry and Life Sciences (μTAS).: 791–793.
9. Niedbala, R.S. *et al.* 2001. Detection of analytes by immunoassay using up-converting phosphor technology. Anal. Biochem. **293:** 22–30.
10. Corstjens, P.L.A.M. *et al.* 2001. Use of up-converting phosphor reporters in lateral-flow assays to detect specific nucleic acid sequences: a rapid, sensitive DNA test to identify human papillomavirus type 16. Infection **47:** 1885–1893.
11. Charles, M. & J.W. Pape. 2006. Turberculosis and HIV: implications in the developing world. Curr. HIV/AIDS Rep. **3:** 139–144.

Patterns of Salivary Estradiol and Progesterone across the Menstrual Cycle

BEATRICE K. GANDARA,[a] LINDA LERESCHE,[a] AND LLOYD MANCL[b]

[a]*Department of Oral Medicine, School of Dentistry, University of Washington, Seattle, Washington 98195, USA*

[b]*Department of Dental Public Health Sciences, School of Dentistry, University of Washington, Seattle, Washington 98195, USA*

ABSTRACT: The aim of this study was to characterize the normality of menstrual cycles on the basis of progesterone and estradiol levels in self-collected saliva samples. Twenty-two women, ages 19–40 years, self-collected whole unstimulated saliva specimens each morning for two consecutive menstrual cycles. On the basis of presence/timing of hormone peaks, two investigators classified 24 cycles as normal, 10 as likely normal, and 10 as clearly not normal with respect to expected profiles. Our results show that whole saliva samples collected at home on a daily basis provide a noninvasive, feasible method of determining menstrual cycle profiles.

KEYWORDS: saliva; estradiol; progesterone; menstrual cycle; hormone; women; pain; temporomandibular dysfunction

INTRODUCTION

Historically, salivary analyses of female sex hormones were used for fertility and pregnancy monitoring.[1–3] However, recent findings indicate that these assays may be useful beyond the study of reproductive concerns. There is emerging evidence that females are at greater risk for various diseases and experience diseases differently than males. For example, autoimmune diseases, such as Sjögren's syndrome, systemic lupus erythematosus, rheumatoid arthritis, multiple sclerosis, and autoimmune thyroiditis, are more prevalent in women than in men. Progesterone and estradiol, which are at high levels in women during their reproductive years, may cause increased immunoreactivity responsible for this phenomenon.[4]

Other sex-related differences include differences in pain perception, which have been demonstrated at both experimental and clinical levels. For example,

Address for correspondence: Beatrice K. Gandara, Department of Oral Medicine, Box 356370, School of Dentistry, University of Washington, Seattle, Washington 98195. Voice: 206-616-6010; fax: 206-685-8412.

bgandara@u.washington.edu

Ann. N.Y. Acad. Sci. 1098: 446–450 (2007).
doi: 10.1196/annals.1384.022

chronic pain conditions, such as migraine headaches and temporomandibular muscle and joint disorders, are more common in women than men, and occur at peak prevalence during the reproductive years.[5,6] The effects of female sex hormones are also evident in the study of changes in disease susceptibility, severity, and drug pharmacokinetics in relation to the menstrual cycle.[7]

Frequent serum sampling for hormone analysis is invasive, inconvenient, and requires skilled personnel to draw samples. However, whole saliva provides an excellent specimen for monitoring estradiol and progesterone levels across the menstrual cycle. It can be self-collected at home on a daily basis and stored in a home freezer for a month or longer before delivery to a laboratory for analysis, and it can be subjected to repeated freezing and thawing without adverse effects on assay results.[8]

There are large differences in the concentration of hormones in each phase of the menstrual cycle and in the timing of menstrual cycle events from woman to woman. In addition, variations exist across multiple cycles for an individual woman, including occurrence of some nonovulatory cycles.[9] The aim of this study was to characterize the use of salivary progesterone and estradiol in self-collected whole saliva samples to monitor menstrual cycles.

METHODS

Twenty-two healthy women, ages 19–40 years, who were recruited for a study of hormones and orofacial pain, self-collected unstimulated whole saliva specimens each morning for two consecutive menstrual cycles. Subjects were asked to spit the saliva into a 10-mL tube until a 2.5-mL sample was collected. The specimen was dated and stored in the subject's home freezer until the end of the second cycle, when it was picked up by research personnel and stored at –20°C in the laboratory for up to 6 months until analysis.

On the day of analysis, each specimen was thawed and heated at 57°C for 2 h and centrifuged at 9,000×*g* for 4 min at 10°C. Estradiol and progesterone concentrations were determined by enzyme immunoassay by Saliva Testing and Reference Laboratory (Seattle, WA, USA) with commercial kits (EIA 537 and EIA 574 kits, Pantex, Santa Monica, CA, USA). Laboratory personnel were blinded to the day of the cycle.

Two investigators independently examined plots of the daily salivary hormone levels for each subject over two menstrual cycles ($n = 44$ cycles). On the basis of specified criteria, which included presence/timing of hormone peaks and missing data, cycles were classified as normal, likely normal, or clearly not normal with respect to expected profiles (TABLE 1).

Median hormone levels were calculated for each subject and averaged across subjects. Agreement between the two investigators regarding classification of cycles was assessed using the kappa statistic.

TABLE 1. Criteria for menstrual cycle assessment

Normal cycle:

1. Two estradiol peaks are present.
2. Primary (earlier) estradiol peak precedes the second estradiol peak by 5 days or more.
3. Primary peak is higher than the second peak.
4. Progesterone rising over the course of the cycle (a little variation/ a few small dips are acceptable).
5. Progesterone peaks within 2 days (+ or –) of the second estradiol peak. (e.g., if estradiol peaks on day 27, progesterone peaks on day 25, 26, 27, 28, or 29)

Likely normal cycle:

One or more of the criteria cannot be judged on account of missing data, but the criteria that can be judged are fulfilled.

Clearly abnormal cycle:

Sufficient data are available and one or more of the criteria for normal cycle are not fulfilled.

RESULTS AND DISCUSSION

Menstrual cycles ranged from 22–44 days long. The median concentrations for estradiol and progesterone were 2.81 pg/mL and 139.1 pg/mL, respectively. There was a good agreement between the two investigators in categorizing the menstrual cycles ($\kappa = 0.69$). After resolution of disagreements through discussion, 24 cycles were designated as normal, 10 as likely normal, and 10 as clearly abnormal (FIGS. 1–3). If "likely normal" cycles are combined with

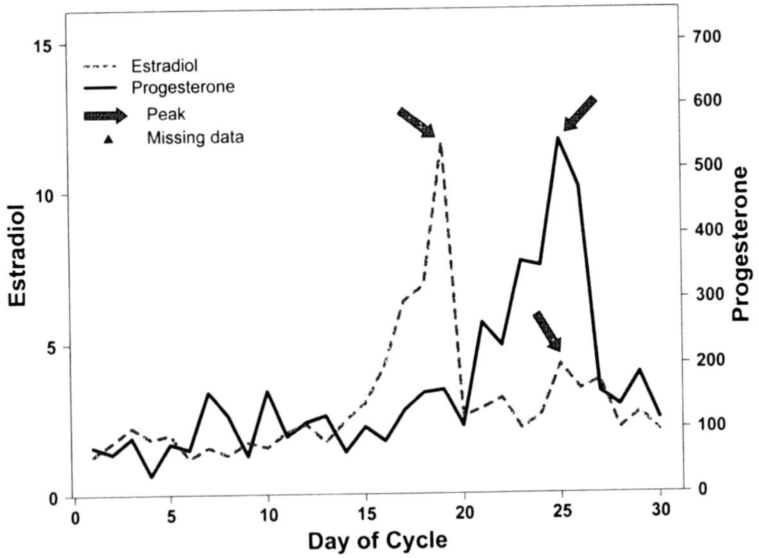

FIGURE 1. Representative patterns of salivary estradiol and progesterone levels across the normal menstrual cycle of one subject.

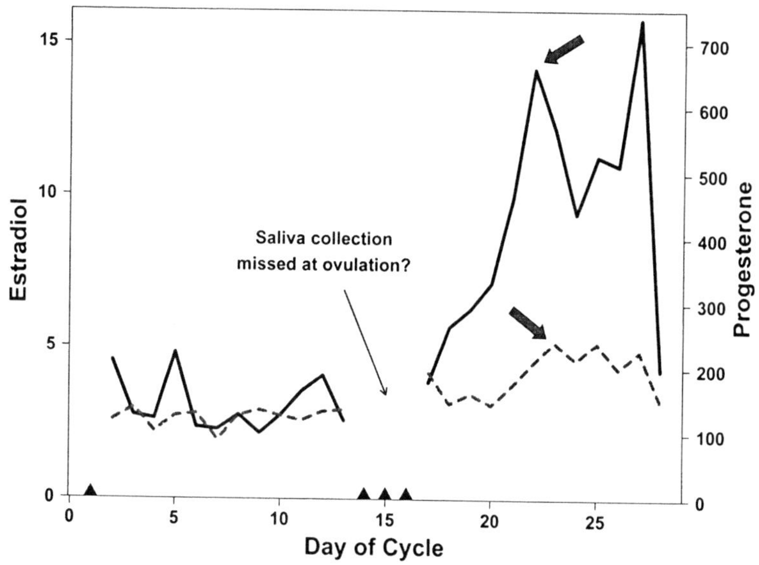

FIGURE 2. Representative patterns of salivary estradiol and progesterone levels across the likely normal menstrual cycle of one subject. Symbols as in FIGURE 1.

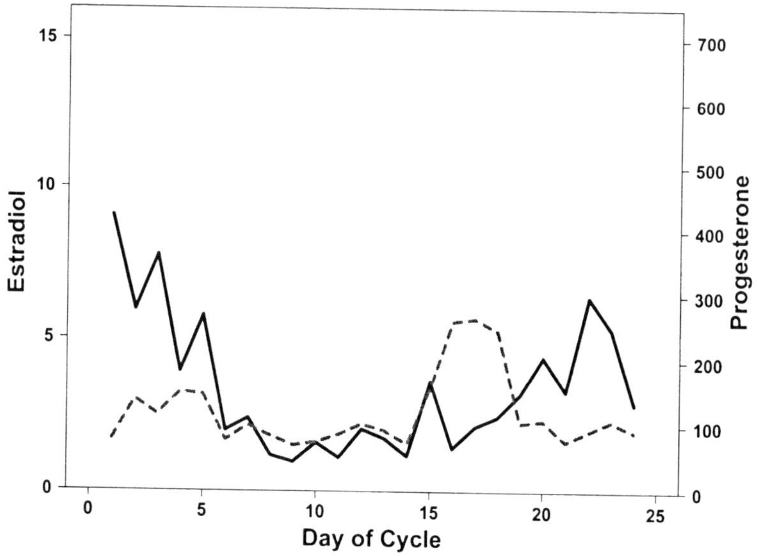

FIGURE 3. Representative patterns of salivary estradiol and progesterone levels across the clearly abnormal cycle of one subject. Symbols as in FIGURE 1.

"normal" cycles, then 22.7% of the samples are abnormal (possibly anovulatory) cycles, which is comparable with serum values.[10]

SUMMARY

Our results show that whole salivary samples collected at home by the subject on a daily basis provide a noninvasive, feasible method of determining menstrual cycle profiles. These biomarkers have great potential for studying hormone levels in research on gender differences in health and disease.

ACKNOWLEDGMENTS

This work was supported by NIH/NIDCR Grant DE016212.

REFERENCES

1. Ellison, P.T. 1993. Measurements of salivary progesterone. Ann. N.Y. Acad. Sci. **694:** 161–176.
2. Read, G.F. 1993. Status report on measurement of salivary estrogens and androgens. Ann. N.Y. Acad. Sci. **694:** 146–160.
3. Hofman, L. 2001. Human saliva as a diagnostic specimen. J. Nutr. **131:** 1621S–1625S.
4. Markovic, Nina. 2001. Women's oral health across the lifespan. Dent. Clin. N Am. **45:** 513–521.
5. Unruh, A.M. 1996. Gender variations in clinical pain experience. Pain **65:** 123–167.
6. LeResche, L. 2000. Epidemiologic perspectives on sex differences in pain. *In* Sex, Gender and Pain. R.B. Fillingim, Ed.: 233–249. IASP Press. Seattle, WA
7. Anthony, M. & M.J. Berg. 2002. Biologic and molecular mechanisms for sex differences in pharmacokinetics, pharmacodynamics and pharmacogenetics: Part II. J. Women's Health Gender-based Med. **11:** 617–629.
8. Gandara, B., L. LeResche & L. Mancl. 2006. Effects of repeated freeze-thaw in self-collected salivary hormone specimens [abstract]. J. Dent. Res. **85:** 1045. (www.dentalresearch.org).
9. Becker, J.B., A.P. Arnold, K.J. Berkley, *et al.* 2006. Strategies and methods for research on sex differences in brain and behavior. Endocrinology **146:** 1650–1673.
10. Metcalf, M.G., D.S. Skidmore, G.F. Lowry & J.A. Mackenzie. 1983. Incidence of ovulation in the years after the menarche. J. Endocrinol. **97:** 213–219.

Layered Peptide Arrays

A Diverse Technique for Antibody Screening of Clinical Samples

GALLYA GANNOT,[a] MICHAEL A. TANGREA,[b] RODRIGO F. CHUAQUI,[a] JOHN W. GILLESPIE,[c] AND MICHAEL R. EMMERT-BUCK[a]

[a]*Pathogenetics Unit, Laboratory of Pathology and Urologic Oncology Branch, Surgery Branch, Center for Cancer Research, National Cancer Institute, National Institutes of Health, Bethesda, Maryland 20892-4605, USA*

[b]*Tumor Angiogenesis Section, Surgery Branch, Center for Cancer Research, National Cancer Institute, National Institutes of Health, Bethesda, Maryland 20892-4605, USA*

[c]*SAIC Frederick Inc. Frederick, Maryland, USA*

ABSTRACT: The layered peptide array (LPA) is a recently developed technique designed to measure antibody levels in a multiplex, high-throughput manner. LPAs can assess antibody presence either in fluid samples or from tissues while maintaining the two-dimensional orientation of the life science platform. In this manuscript, we evaluated and assessed the performance of the LPA platform, focusing on throughput capability, sensitivity, and specificity of the assay in several different systems.

KEYWORDS: layered peptide array; multiplex; high-throughput; ELISA; immunohistochemistry

INTRODUCTION

Layered expression scanning (LES) is a molecular analysis technique that integrates two-dimensional life science platforms with a third, molecular array dimension.[1–3] We describe here a novel application of the method that permits multiplex antibody measurements to be performed on a variety of life science platforms, including multiwell arrays of sera samples, tissue sections, and immunoblots. Similar to the parent technique, the new approach, termed layered

Address for correspondence: Michael Emmert-Buck, Advanced Technology Center, Center for Cancer Research, National Cancer Institute, 8717 Grovemont Circle, Bethesda, MD 20892-4605. Voice: (301)-496-2912; fax: (301)-594-7582.
buckm@mail.nih.gov

Ann. N.Y. Acad. Sci. 1098: 451–453 (2007). © 2007 New York Academy of Sciences.
doi: 10.1196/annals.1384.041

peptide array (LPA), can be described in terms of three dimensions, the *x–y* plane of the life science platform and the *z*-dimension representing the analysis layers. Membranes are individually coated with peptides specific to antibodies of interest. Biological samples are then placed adjacent to the membrane set, and the antibodies passed through the layers while maintaining their original two-dimensional *x–y* positions. If present in a sample, antibodies are specifically captured by their target peptide as they pass through the layers, and subsequently detected using standard secondary antibody-based methods. Thus, each membrane measures one specific antibody, and the overall membrane set can be designed for multiplex analysis to suit the needs of the investigator. The core LPA technique can be utilized in one of two configurations, either as a direct assay system (dLPA[4]) to measure antibodies in sera or other fluid samples, or as an indirect assay (iLPA[5]) to measure antigens on solid surfaces such as tissue sections. The simplest application is direct measurement of antibodies in a multiwell plate of liquid samples, such as sera or other patient fluids, conditioned media from cells in culture, or hybridoma supernatants. Quantification of antibodies is important for several clinical and laboratory studies, including those associated with cancer surveillance and detection, microbial exposures, and autoimmune diseases.[6–13] The performance of the dLPA method was evaluated by analyzing sera and saliva from patients with Sjögren's syndrome, an autoimmune connective tissue disorder with characteristic autoantibodies.[14,15] The data demonstrated that the prototype dLPA device is capable of producing up to, but not limited to, 5,000 antibody measurements per experiment, and appears to be scalable to higher throughput levels. Detection of the Sjögren's syndrome antigen B in patient sera samples produced results similar to those with standard ELISA. The detection sensitivity of LPAs and ELISAs were nearly identical based on the optimal sample dilutions that provided readable signals.[4]

Alternatively, the iLPA application of the technique permits measurement of antibodies that are prebound to target antigens on a solid surface such as a tissue section or immunoblot. In this approach, the antibodies serve as reporters for the amount of antigen present in the specimen under study. Since the two-dimensional architecture of the specimen is maintained, all of the subelements in the samples are simultaneously measured. For example, different histological regions of a tissue section, or protein bands on a blot, could be evaluated in this manner. In this study, minor salivary glands from Sjögren's patients were studied using a prototype iLPA system. Quantitative, multiplex proteomic analysis of histological sections was achieved with up to 20 different membranes, and the data compared to sections stained using immunohistochemistry with an experiment variability of 18%.[5]

Overall, the evaluation of the two LPA configurations suggests that the method is simple, versatile, and relatively inexpensive for multiplex molecular measurements from biological samples.

ACKNOWLEDGMENT

This research was supported by the Intramural Research Program of the NIH, National Cancer Institute, Center for Cancer Research, Bethesda, MD.

REFERENCES

1. ENGLERT, C.R., G.V. BAIBAKOV & M.R. EMMERT-BUCK. 2000. Layered expression scanning: rapid molecular profiling of tumor samples. Cancer Res. **60:** 1526–1530.
2. CHUAQUI, R.F. *et al.* 2002. Post-analysis follow-up and validation of microarray experiments. Nat. Genet. **32:** 509–514.
3. TANGREA, M.T. *et al.* 2003. Layered expression scanning: multiplex analysis of RNA and protein gels. Biotechniques **35:** 1280–1285.
4. GANNOT, G. *et al.* 2005. Layered peptide arrays—high-throughput antibody screening of clinical samples. J. Mol. Diagn. **7:** 427–436.
5. GANNOT, G. *et al.* 2007. Layered peptide array for multiplex immunohistochemistry. J. Mol. Diagn. **9:** in press.
6. ZHANG, J.Y. *et al.* 2003. Enhancement of antibody detection in cancer using panel of recombinant tumor-associated antigens. Cancer Epidemiol. Biomarkers Prev. **12:** 136–143.
7. MACBEATH, G. & S.L. SCHREIBER. 2000. Printing proteins as microarrays for high-throughput function determination. Science **289:** 1760–1763.
8. ROBINSON, W.H. *et al.* 2002. Autoantigen microarrays for multiplex characterization of autoantibody responses. Nat. Med. **8:** 295–301.
9. NAM, M.J. *et al.* 2003. Molecular profiling of the immune response in colon cancer using protein microarrays: occurrence of autoantibodies to ubiquitin C-terminal hydrolase L3. Proteomics **3:** 2108–2115.
10. QIU, J. *et al.* 2004. Development of natural protein microarrays for diagnosing cancer based on an antibody response to tumor antigens. J. Proteome. Res. **3:** 261–267.
11. NEUMAN DE VEGVAR, H.E. & W.H. ROBINSON. 2004. Microarray profiling of antiviral antibodies for the development of diagnostics, vaccines, and therapeutics. Clin. Immunol. **111:** 196–201.
12. PICKERING, J.W. *et al.* 2002. Comparison of a multiplex flow cytometric assay with enzyme-linked immunosorbent assay for quantitation of antibodies to tetanus, diphtheria, and *Haemophilus influenzae* Type b. Clin. Diagn. Lab. Immunol. **9:** 872–876.
13. LIN, Y. *et al.* 2002. Profiling of human cytokines in healthy individuals with vitamin E supplementation by antibody array. Cancer Lett. **187:** 17–24.
14. GILBURD, B. *et al.* 2004. Autoantibodies profile in the sera of patients with Sjogren's syndrome: the ANA evaluation—a homogeneous, multiplexed system. Clin. Dev. Immunol. **11:** 53–56.
15. GANNOT, G., H.E. LANCASTER & P.C. FOX. 2000. Clinical course of primary Sjogren's syndrome: salivary, oral, and serologic aspects. J. Rheumatol. **27:** 1905–1909.

Whole Saliva Proteolysis

Wealth of Information for Diagnostic Exploitation

EVA J. HELMERHORST

Department of Periodontology and Oral Biology, Boston University, Goldman School of Dental Medicine, Boston, Massachusetts 02118, USA

ABSTRACT: Whole saliva (WS) protein profiles differ significantly from those of glandular salivary secretions. Rapid proteolysis of the prominent members of the salivary protein families by WS resident proteases appears to be a major cause for the observed differences. We propose that the rate and mode of glandular salivary protein degradation in the oral cavity contains information that could be of unique value in the diagnosis of oral disease.

KEYWORDS: glandular salivary proteins; whole saliva; oral disease; salivary proteolysis

INTRODUCTION

A major impetus for salivary research originates from the recognition that all molecular surface interactions on oral soft and hard tissues occur in a medium dictated in large part by the constituents of whole saliva (WS). The predominant contributors to WS are exocrine in nature and consist of secretions from the major and minor salivary glands. In view of the importance of the parotid (PS) and submandibular, sublingual (SMSL) secretions as major contributors to oral fluid, efforts in the past decades have been geared toward defining their composition and function. Through traditional biochemical approaches the structural characteristics of the salivary proteins and polymorphic isoforms of these proteins have been identified, generating the fundamental basis of the salivary secretome.[1,2] More in-depth detail is being achieved through the rapid advances in mass spectrometry, allowing the analysis of proteins down to femtomole levels.[3–7] In earlier work it has been noted that the protein composition of WS is not simply representing the mixture of proteins in glandular secretions

Address for correspondence: Eva J. Helmerhorst, Ph.D., Department of Periodontology and Oral Biology, Boston University, Goldman School of Dental Medicine, 700 Albany Street, W-201, Boston, MA 02118. Voice: 617-414-1119; fax: 617-638-4924.
helmer@bu.edu

Ann. N.Y. Acad. Sci. 1098: 454–460 (2007).
doi: 10.1196/annals.1384.013

as might be expected from their large volumetric contributions to WS.[8–10] As a matter of fact, many proteins that make up a major fraction of the glandular salivary protein pool appear in significantly lower levels in WS. Knowledge of the biochemical basis for the differences in protein concentrations and patterns between glandular secretions and WS is important in understanding the variability of the WS proteome, and for the study of biomarkers that could be of value in monitoring oral and systemic health and disease.[11,12]

WS PROTEOLYSIS

Using one- and two-dimensional gel electrophoretic systems, differences have been reported between the protein composition of WS and glandular secretions.[3,8,9,13] Some differences can be explained by nonexocrine contributions to WS. For example, albumin is present in higher levels in WS than in glandular secretions because of the serum-like contributions of crevicular fluid that enters the oral cavity through the gingival crevice.[8] In contrast, a number of small glandular salivary proteins show much lower concentrations in WS than in PS or SMSL.[10,14–16] Various quantification studies indicate that while proline-rich proteins (PRPs) and histatins form a major portion of the glandular proteins, they disappear rapidly as soon as they are mixed with WS.[17,18] The origin and fate of histatins in the oral cavity is diagrammatically depicted in FIGURE 1. Comparison of the electrophoretograms of WS with PS reflects a dramatic reduction of histatin levels in WS, and similar observations have been made with statherins and PRPs.[14–16]

The presence of proteolytic activity in the oral cavity has long been known[19–21] and has been the major focus of investigations directed toward explaining protein concentration differences in glandular secretions and WS.[17,18] Multiple fragments of histatins, statherins, PRPs, and other salivary proteins that at first sight appear more resistant to proteolysis have been identified in WS.[22] Prior to the release of glandular salivary proteins into the oral cavity, some proteolysis of primary gene products already occurs within the gland as part of the posttranslational proteolytic process. Typical examples of proteins generated through such proteolytic posttranslational modifications include histatin 5, which is formed by a chymotryptic-like cleavage event after Tyr 24 in histatin 3,[23,24] and PRP-3 and PRP-4, which are formed from PRP-1 and PRP-2, respectively, through tryptic-like cleavage after the Arg 106 residue.[25] Proteolysis occurring in the gland appears to be under biological control, as evidenced by the fact that the extent and mode of glandular protein processing is specific, virtually flow-rate independent, and shows surprisingly small differences among healthy subjects.[26] Once glandular secretions are released into the oral cavity, their proteins are exposed to a host of additional enzymes that are derived from bacteria, epithelial cells, and other host cells.[27–29] The proteolytic activity of WS is orders of

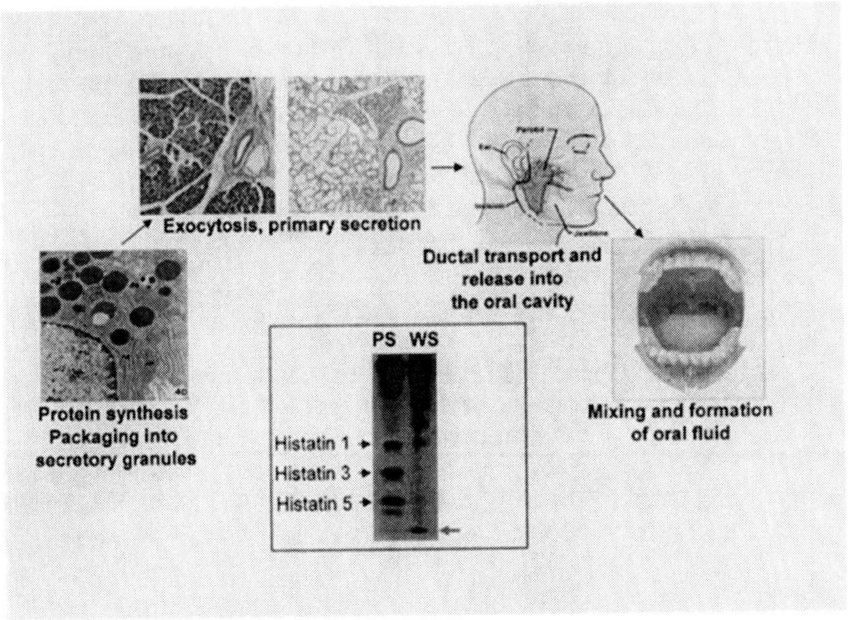

FIGURE 1. Origin and fate of glandular salivary proteins in the oral cavity. After protein synthesis, exocytosis, ductal transport and release of proteins into the oral cavity, mixing with a variety of WS-associated enzymes, occurs. The *inset* shows the differential concentrations of histatins 1, 3, and 5 in PS and WS in a cationic gel. The *arrow* points toward a variety of smaller histatin fragments that are present mainly in WS and migrate with the buffer front.

magnitude higher than that of glandular secretions.[18,28] Multiple small degradation fragments originating from histatins, statherin, and PRPs (acidic and basic) can be generated upon exposure of pure proteins to WS, and protease inhibitors prevent the degradation of these proteins.[10,17] Thus proteolysis in WS seems to be in large part responsible for the significantly lower levels of some of the salivary proteins in their intact form in WS.

IMPACT OF WS PROTEOLYSIS ON FUNCTION

Extensive proteolysis of salivary proteins in the oral cavity is likely to affect the functional activities associated with these proteins. As such, it can be hypothesized that the mode and extent of degradation of proteins with host-protective functions may be predictive of or related to a particular oral disease state. To date, the effect of WS-associated proteolysis on the functional activity of otherwise well-characterized salivary proteins, such as PRPs has hardly been addressed. In structure–function analyses of salivary proteins it has been shown that in most cases not the entire protein is needed for activity.[1] For example, the bioactive domain of acidic PRP resides in the

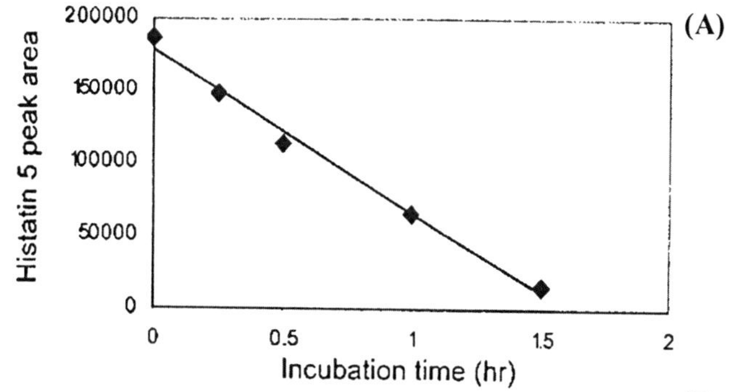

(B)

Histatin 5: DSHAKRHHGYKRKFHEKHHSHRGY

FIGURE 2. Histatin 5 degradation in WS supernatant. (**A**) Histatin 5 (200 μg) was added to 1 mL of WS supernatant (WSS) and incubated at 37°C. The degradation of histatin 5 was monitored by reversed-phase high-performance liquid chromatography (HPLC) by determining the peak area of intact histatin 5 after various incubation times. Values were corrected for the small histatin 5 peak that was naturally present in WSS (<5 μg). (**B**) Primary cleavage sites in histatin 5 determined from the amino acid sequences of identified N- and C-terminal fragments.[33]

N-terminal 30 residues,[30] and for histatins in the middle region comprising 12–14 amino acid residues.[31,32] The identification of bioactive domains in PRPs, histatins, and statherin has provided information on the structure–function relationship of these proteins. However, the question relevant to oral health is whether these *in vitro* designed and assayed fragments are actually being generated and functional in WS. In view of the dynamic nature of the degradation process, we recently initiated studies to monitor the proteolysis of purified histatin 5 added to diluted WS by characterizing the degradation products by mass spectrometry, and monitoring the biological activity of the histatin 5 degradation mixture over time. Histatin 5 is degraded by soluble WS enzymes at an average rate of approximately 100–125 μg/h at saturating substrate concentrations (FIG. 2A), partly explaining the low levels of this protein in WS. We identified at least 19 degradation fragments, the majority of which were formed through single cleavage events.[33] These preferential cleavage sites in histatin 5 are indicated with arrows in FIGURE 2B and appear to be concentrated in the middle region of the molecule. The degradation of histatin 5 has no immediate effect on its functional activity, which is consistent with the reported *in vitro* bioactivity of smaller histatin fragments.[31,32] In general, the magnitude to which WS proteolysis affects glandular salivary protein activity is dependent on the extent of protein degradation and on the functional activity of the early degradation fragments.

WS PROTEOLYSIS AND DIAGNOSTICS

It can be speculated that the mode and extent of salivary proteolysis has the potential to reflect a particular disease state. The rate of proteolysis in dental plaque is indicative of the periodontal disease status.[29] This is believed to be related to proteolytic enzymes associated with bacteria that are etiologically linked to periodontal disease. In our studies with histatin 5 we observed a high consistency in the early proteolytic degradation profiles generated in WS collected from different, orally healthy, subjects.[33] Such reproducibility in profiles suggests the involvement of one or a limited set of enzymes with similar specificities in the degradation process and points to an as yet unrecognized level of conservation. The consistency in the protein degradation pattern indicates that WS proteolytic characteristics are a more common feature of oral fluids from subjects with similar oral health status than has hitherto been believed. Other studies showed that whole and glandular saliva from caries-free subjects and those who had a high caries incidence differed not only in their protein composition,[34] but also in their ability to serve as substrates to support the growth of *S. mutans*.[27] These results indicated that the utilization and degradation of specific salivary proteins by certain microorganisms may be associated with or predictive of a particular oral disease. The commonality of these proteolytic WS characteristics among subjects with similar disease phenotypes and differences between clinical phenotypes clearly points toward the potential of exploiting this feature for diagnostic purposes.

CONCLUDING REMARKS

WS proteolysis is an ongoing process that has major impact on glandular salivary protein structure and function. It should be recognized, however, that WS proteolysis is counterbalanced by the continuous secretion of *de novo* synthesized proteins into the oral cavity. Host protection by WS is rooted in physiological as well as biochemical characteristics unique to the oral cavity. The physiological parameters relate to the fact that the oral cavity is an open-ended system where there is a constant influx and efflux of secretions leading at least to a transient supply of newly synthesized glandular salivary protein. The biochemical parameters relate to the fact that the turnover rate of these proteins into peptides is fast. We propose a concept where oral fluid provides an environment with unique dynamics that continuously modify and alter glandular salivary proteins. It is likely that there is a relative steady state between proteolytic breakdown of salivary protein and the salivary flow-driven replenishment with intact glandular proteins. Knowledge into the factors that influence this equilibrium and insights into the functional aspects of glandular protein processing are of key importance in understanding the protective role of

WS in oral disease. The current inroads that have been made with enzymes and mass spectrometric characterization of WS have already laid the groundwork to explore oral diagnostics based on entirely novel salivary parameters.

ACKNOWLEDGMENTS

The author is grateful for the contributions to this project of Xiuli Sun, Walter Siqueira, Melanie Campese, Weimin Zhang, Erdjan Salih, and Frank Oppenheim. Support from the NIH, NIDCR (Grants DE05672, DE07652, and DE14950) is gratefully acknowledged.

REFERENCES

1. LAMKIN, M.S. & F.G. OPPENHEIM. 1993. Structural features of salivary function. Crit. Rev. Oral Biol. **4:** 251–259.
2. SCHENKELS, L.C. *et al.* 1995. Biochemical composition of human saliva in relation to other mucosal fluids. Crit. Rev. Oral Biol. Med. **6:** 161–175.
3. YAO, Y. *et al.* 2003. Identification of protein components in human acquired enamel pellicle and whole saliva using novel proteomics approaches. J. Biol. Chem. **278:** 5300–5308.
4. WILMARTH, P.A. *et al.* 2004. Two-dimensional liquid chromatography study of the human whole saliva proteome. J. Proteome Res. **3:** 1017–1023.
5. HU, S. *et al.* 2005. Large-scale identification of proteins in human salivary proteome by liquid chromatography(mass spectrometry and two-dimensional gel electrophoresis-mass spectrometry. Proteomics **5:** 1714–1728.
6. VITORINO, R. *et al.* 2004. Identification of human whole saliva protein components using proteomics. Proteomics **4:** 1109–1115.
7. HARDT, M. *et al.* 2005. Toward defining the human parotid gland salivary proteome and peptidome: identification and characterization using 2D SDS-PAGE, ultrafiltration, HPLC, and mass spectrometry. Biochemistry **44:** 2885–2899.
8. OPPENHEIM, F.G. 1970. Preliminary observations on the presence and origin of serum albumin in human saliva. Helv. Odont. Acta **14:** 10–17.
9. BEELEY, J.A. *et al.* 1991. Sodium dodecyl sulphate polyacrylamide gel electrophoresis of human parotid salivary proteins. Electrophoresis **12:** 1032–1041.
10. BAUM, B.J. *et al.* 1976. Studies on histidine-rich polypeptides from human parotid saliva. Arch. Biochem. Biophys. **177:** 427–436.
11. MALAMUD, D. 2006. Salivary diagnostics: the future is now. J. Am. Dent. Assoc. **137:** 284–286.
12. WONG, D.T. 2006. Salivary diagnostics powered by nanotechnologies, proteomics and genomics. J. Am. Dent. Assoc. **137:** 313–321.
13. BAUM, B.J. *et al.* 1977. Polyacrylamide gel electrophoresis of human salivary histidine-rich-polypeptides. J. Dent. Res. **56:** 1115–1118.
14. KOUSVELARI, E.E. *et al.* 1980. Immunochemical identification and determination of proline-rich proteins in salivary secretions, enamel pellicle, and glandular tissue specimens. J. Dent. Res. **59:** 1430–1438.

15. JENSEN, J.L. *et al.* 1994. Physiological regulation of the secretion of histatins and statherins in human parotid saliva. Physiological regulation of the secretion of histatins and statherins in human parotid saliva. J. Dent. Res. **73:** 1811–1817.
16. LI, J. *et al.* 2004. Statherin is an *in vivo* pellicle constituent: identification and immuno-quantification. Arch. Oral Biol. **49:** 379–385.
17. MINAGUCHI, K. *et al.* 1988. The presence and origin of phosphopeptides in human saliva. Biochem. J. **250:** 171–177.
18. PAYNE, J.B. *et al.* 1991. Selective effects of histidine-rich polypeptides on the aggregation and viability of *Streptococcus mutans* and *Streptococcus sanguis*. Oral Microbiol. Immunol. **6:** 169–176.
19. MAKINEN, K.K. 1966. Studies on oral enzymes: I. Fractionation and characterization of aminopeptidases of human saliva. Acta Odont. Scand. **24:** 579.
20. SÖDER, P.O. 1972. Proteolytic activity in the oral cavity: proteolytic enzymes from human saliva and dental plaque material. J. Dent. Res. **51:** 389–393.
21. GERMAINE, G.R. *et al.* 1978. Whole saliva proteases: development of methods for determination of origins. Adv. Exp. Med. Biol. **107:** 849–858.
22. VITORINO, R. *et al.* 2004. Analysis of salivary peptides using HPLC-electrospray mass spectrometry. Biomed. Chomatogr. **18:** 570–575.
23. OPPENHEIM, F.G. *et al.* 1988. Histatins, a novel family of histidine-rich proteins in human parotid secretion. Isolation, characterization, primary structure, and fungistatic effects on *Candida albicans*. J. Biol. Chem. **263:** 7472–7477.
24. SABATINI, L.M. & E. A. AZEN. 1989. Histatins, a family of salivary histidine-rich proteins, are encoded by at least two loci (HIS1 and HIS2). Biochem. Biophys. Res. Commun. **160:** 495–502.
25. HAY, D.I. *et al.* 1988. The primary structures of six human salivary acidic proline-rich proteins (PRP-1, PRP-2, PRP-3, PRP-4, PIF-s and PIF-f). Biochem. J. **255:** 15–21.
26. DABBAGH, W.K. *et al.* 1994. Immunological and densitometric determination of histatin concentrations in saliva. J. Dent. Res. **73:** 150.
27. COWMAN, R.A. *et al.* 1979. Differential utilization of proteins in saliva from caries-active and caries-free subjects as growth substrates by plaque-forming streptococci. J. Dent. Res. **58:** 2019–2027.
28. WATANABE, T. *et al.* 1981. Correlation between the protease activities and the number of epithelial cells in human saliva. J. Dent. Res. **60:** 1039–1044.
29. BRETZ, W.A. & W.J. LOESCHE. 1987. Characteristics of trypsin-like activity in subgingival plaque samples. J. Dent. Res. **66:** 1668–1672.
30. AOBE, T. *et al.* 1984. Inhibition of apatite crystal growth by the amino-terminal segment of human salivary acidic proline-rich proteins. Calcif. Tissue Int. **36:** 651–658.
31. XU, L. *et al.* 1993. Salivary proteolysis of histidine-rich polypeptides and the antifungal activity of peptide degradation products. Arch. Oral Biol. **38:** 277–283.
32. RAJ, P.A. *et al.* 1990. Salivary histatin 5: dependence of sequence, chain length, and helical conformation for candidacidal activity. J. Biol. Chem. **265:** 3898–3905.
33. HELMERHORST, E.J. *et al.* 2006. Oral fluid proteolytic effects on histatin 5 structure and function. Arch. Oral Biol. **51:** 1061–1070.
34. AYAD, M. *et al.* 2000. The association of basic proline-rich peptides from human parotid gland secretions with caries experience. J. Dent. Res. **79:** 976–982.

Saliva-Based HIV Testing among Secondary School Students in Tanzania using the OraQuick® Rapid HIV1/2 Antibody Assay

CAROL HOLM-HANSEN,[a] BALTHAZAR NYOMBI,[b,c] AND MRAMBA NYINDO[d]

[a]*Norwegian Institute of Public Health, NO-0403 Oslo, Norway*

[b]*KCMC Station, Tumaini University, Moshi, Tanzania*

[c]*University of Oslo, NO-0318 Oslo, Norway*

[d]*Muslim University of Morogoro, Morogoro, Tanzania*

ABSTRACT: HIV prevalence and knowledge concerning HIV prevention among secondary school students in Tanzania was investigated. Approximately 50% of all secondary school students in Hai district and Moshi town were included in the study. Saliva samples were obtained using the OraQuick® rapid HIV-1/2 antibody assay. Forty-one (1.0%) and 211 (5.5%) students at the rural and urban schools, respectively, tested positive for HIV antibodies in saliva. HIV knowledge and beliefs varied significantly. Noninvasive saliva sample collection for HIV testing was highly acceptable. HIV infection is considerably more widespread among students attending urban rather than rural schools in the population investigated.

KEYWORDS: saliva testing; HIV; secondary school students

INTRODUCTION

The testing of oral fluids offers several advantages over blood, including ease of collection, better compliance for sample acquisition, reduced occupational risk associated with needlestick injuries and low load of infectious virus in oral fluids, and safer disposal of waste materials.[1] Noninvasive sample collection is of especial importance for ethical reasons when testing children and youth.

Address for correspondence: Carol Holm-Hansen, Norwegian Institute of Public Health, Division of Infectious Disease Control, P.O. Box 4404 Nydalen, NO-0403 Oslo, Norway. Voice: +47-22042283; fax: +47-22042301.
carol.holm-hansen@fhi.no

Ann. N.Y. Acad. Sci. 1098: 461–466 (2007).
doi: 10.1196/annals.1384.036

Few studies to date have addressed the prevalence of HIV infection among youth in developing countries. According to the National AIDS Control Programme of the Ministry of Health in Tanzania, HIV infections among secondary school students 15–19 years of age account for 3.2% of the total HIV cases reported in Tanzania.[2] While HIV testing has become widely available through voluntary counseling and testing (VCT) centers throughout the world, many people are opposed to having blood drawn for personal, cultural, and/or ethical reasons. Saliva testing for HIV may provide an attractive, cost-efficient alternative to invasive blood sample collection and testing.

HIV prevalence and knowledge concerning HIV prevention among secondary school students in Kilimanjaro region, Tanzania, was investigated in the present study. Both rural and urban schools in northern Tanzania were included in the study. The field logistics were organized by the local health and education authorities, and ethical approval was secured from the Kilimanjaro Christian Medical College Ethics Committee.

METHODS

Approximately 50% of all secondary school students in Hai district and Moshi town, Kilimanjaro region were included in the study. The headmasters, staff, and students at all participating schools were well informed prior to the study. Information concerning the aim and relevance of the study was included in the sensitization meetings. At each school the students were assembled in groups of 30–150 students, depending on the facilities available. The team again explained the purpose of the study and answered questions from the students. Written informed consent was obtained from every participant. Anonymous coded questionnaires were answered and collected prior to saliva sample collection. Anonymous coded saliva samples were obtained using the OraQuick® rapid HIV-1/2 antibody assay (OraSure Technologies, Bethlehem, PA, USA). The students received instruction concerning the correct use of OraQuick (FIG. 1) and thereafter collected their sample. The team members collected all saliva samples from the students, placed the test device into the developer solution and returned to a makeshift field laboratory, where the results were recorded.

OraQuick® is a combined sample collection device and rapid membrane enzyme immunoassay for the determination of HIV antibodies. The assay utilizes lateral flow technology and can be performed using whole blood, serum, or saliva samples. All necessary reagents are contained "within" the device. The test requires no laboratory equipment and results are provided within 20 min (FIG. 2).

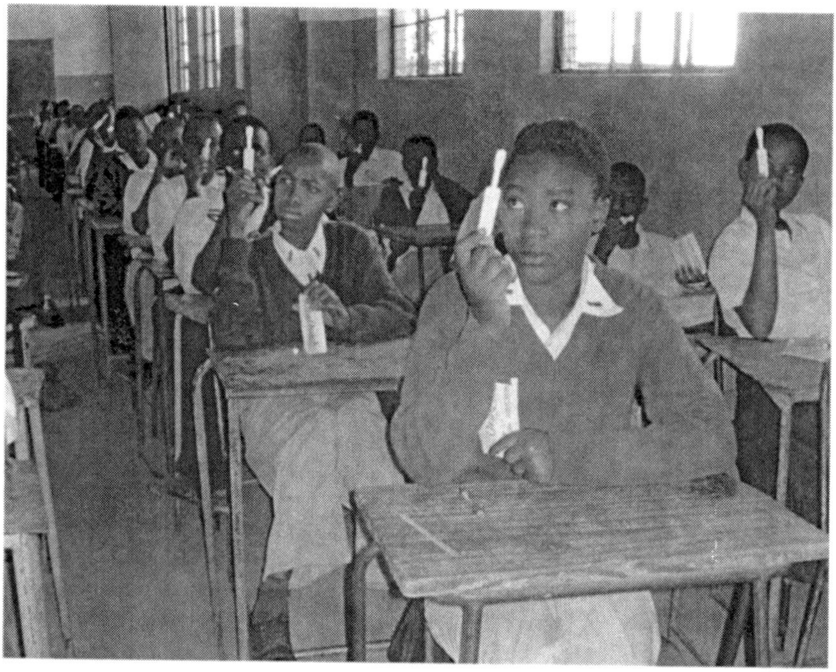

FIGURE 1. Instruction for sample collection (photo courtesy of C. Holm-Hansen).

RESULTS

A total of 3,945 and 3,825 students aged 12–24 years from 13 rural and 11 urban schools, respectively, completed questionnaires and provided anonymous saliva samples. HIV antibodies were detected in saliva samples from 41 (1.0%) and 211 (5.5%) students at the rural and urban schools, respectively. HIV knowledge and beliefs varied significantly among the students.

DISCUSSION

In the population investigated, HIV infection is considerably more widespread among students attending urban than rural schools (TABLE 1). HIV infection is most prevalent among students aged 14–18 years (TABLE 2). The prevalence of HIV is higher among female than male students of the same age group.

The difference in HIV prevalence between students attending rural and urban secondary schools may reflect different life styles. Nearly all students in Hai district attend boarding schools in this rural region of Tanzania. After classes are completed, students are expected to work in the school vegetable garden, wash clothes, and clean their dormitory rooms in addition to completing homework assignments. The investigators believe that the lifestyle

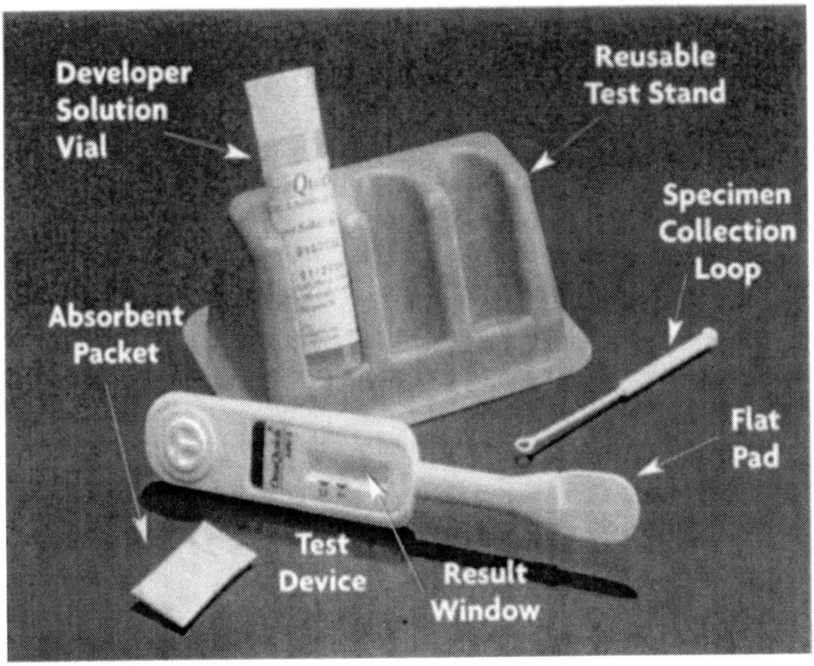

FIGURE 2. The OraQuick rapid HIV-1/2 antibody assay (photo courtesy of OraSure Technologies, Inc.).

provided by boarding schools in Hai District may in itself prevent the spread of HIV infection in that students are occupied with household tasks and lessons after school hours. In contrast, the majority of students in Moshi town are day students. After school the students frequently visit the town center.

Many students wanted to receive their HIV test results even though they knew that samples were collected anonymously. Students wishing to receive their test results were advised to seek VCT at the local health facilities. Many students and some teachers felt that the health facilities were too far away from the schools and requested that HIV testing be provided at the schools in the future. In addition, the students did not want to visit VCT centers unless saliva testing for HIV was available.

TABLE 1. HIV distribution by gender in Hai district and Moshi town secondary schools; HIV-positive test results, total number of participants, sex, and prevalence in %

Location	HIV+/total	HIV+/male	HIV+/female	HIV+/NA[a]
Hai district (rural)	41/3934[b] (1.0%)	12/1679 (0.7%)	28/2206 (1.3%)	1/49 (2.0%)
Moshi town (urban)	211/3819[b] (5.5%)	85/1808 (4.7%)	121/1908 (6.3%)	5/103 (4.8%)

[a]NA = gender not given.
[b]Invalid assays removed from totals.

TABLE 2. HIV distribution by age and gender in Hai district and Moshi town secondary schools: HIV-positive test results, total number of participants, sex, and prevalence in %

Age	Male Hai	Female Hai	NA[a] Hai	Male Moshi	Female Moshi	NA[a] Moshi
0	0/11	0/8	0/1	1/12	3/16	1/10
12	1/5	0/12	0/1	0/1	1/2	–
13	1/28 (3.6%)	0/118 (0%)	0/2	0/6 (0%)	4/45 (8.9%)	0/1
14	2/179 (1.1%)	2/363 (0.6%)	0/8	10/132 (7.6%)	21/272 (7.7%)	0/2
15	2/253 (0.8%)	8/451 (1.8%)	0/7	22/348 (6.3%)	34/413 (8.2%)	1/16
16	2/356 (0.8%)	9/469 (1.9%)	1/15	16/314 (5.1%)	24/357 (6.7%)	2/14
17	1/313 (0.3%)	4/347 (1.2%)	0/8	12/288 (4.2%)	15/266 (5.6%)	0/25
18	2/245 (0.8%)	4/232 (1.7%)	0/3	15/278 (5.4%)	16/269 (5.9%)	0/19
19	0/109 (0%)	1/95 (1.0%)	0/1	3/195 (1.5%)	3/177 (1.7%)	1/13
20	1/79	0/61	0/1	4/140	0/69	0/3
21	0/44	0/36	0/2	0/46	0/16	–
22	0/32	0/10	–	0/29	0/4	–
23	0/19	0/3	–	2/13	–	–
24+	0/6	0/1	–	0/6	0/2	–
ND	-/4	-/7	–	-/2	-/4	–
Total	12/1683 0.7%	28/2213 1.3%	1/49	85/1810 4.7%	121/1912 6.3%	5/103

[a]NA = gender not given.
ND = invalid test.

Information regarding HIV/AIDS obtained from the questionnaires was of little value. Most students claimed that they had heard of HIV/AIDS and knew how to protect themselves. However, discussions with students indicated a number of misconceptions. Many students firmly believed that condoms, tampons, and female sanitary napkins are "infected" with HIV. Frighteningly, teachers also had misconceptions concerning HIV/AIDS. At one school, teachers believed that the expiration date on condoms indicated the date at which HIV in the condoms was no longer infectious. In contrast, a number of students asked questions concerning the prevalence of different HIV subtypes and resistance to antiretroviral drugs. In general, the results of the questionnaires indicate a need for improved educational materials addressing HIV/AIDS. Information should be made available at schools and in the community.

Noninvasive saliva sample collection and testing for HIV testing using the OraQuick® rapid HIV-1/2 antibody assay was highly acceptable among the students. In accordance with recommendations from the students, saliva-based testing for HIV should be offered at schools, local health clinics, and VCT centers. The results have been reported to the local and national health authorities in Tanzania.

ACKNOWLEDGMENTS

Timothy Shuma and Christopher Temu organized the logistics in Hai district and Moshi town, respectively. Excellent technical assistance was provided by

Christopher Panda, Johana Kazoka, Inger Lise Haugen, and Hilde Bakke. The OraQuick® rapid HIV-1/2 antibody tests were kindly provided free of charge by OraSure Technologies, Inc.

REFERENCES

1. Holm-Hansen, C., N.T. Constantine & G. Haukenes. 1993. Detection of antibodies to HIV in homologous sets of plasma, urine and oral mucosal transudate samples using rapid assays in Tanzania. Clin. Diagn. Virol. **1:** 207–214.
2. NACP-MoH. National AIDS Control Programme HIV/AIDS/STD Surveillance; 2000. Report No. 15.

Lab-on-a-Chip Technologies for Oral-Based Cancer Screening and Diagnostics

Capabilities, Issues, and Prospects

MICHAEL G. MAUK,[a] BARRY L. ZIOBER,[b] ZONGYUAN CHEN,[a] JASON A. THOMPSON,[a] AND HAIM H. BAU[a]

[a]*Department of Mechanical Engineering and Applied Science, School of Engineering and Applied Science, University of Pennsylvania, Philadelphia, Pennsylvania 19104, USA*

[b]*Department of Otorhinolaryngology, Head and Neck Surgery, Hospital of the University of Pennsylvania, Philadelphia, Pennsylvania 19104, USA*

ABSTRACT: The design of a microfluidic lab-on-a-chip system for point-of-care cancer screening and diagnosis of oral squamous cell carcinoma (OSCC) is presented. The chip is based on determining a ~30-gene transcription profile in cancer cells isolated from oral fluid samples. Microfluidic cell sorting using magnetic beads functionalized with an antibody against cancer-specific cell-surface antigens (e.g., epithelial cell adhesion molecule [EpCAM]) is described. A comprehensive cancer diagnostics chip will integrate microfluidic components for cell lysis, nucleic acid extraction, and amplification and detection of a panel of mRNA isolated from a subpopulation of cancer cells contained in a clinical specimen.

KEYWORDS: cancer diagnostics; oral squamous cell carcinoma (OSCC); microfluidics; lab-on-a-chip; EpCAM

INTRODUCTION

Oral squamous cell carcinoma (OSCC), constituting 40% of all head and neck cancers, offers both a compelling opportunity and illustrative case study for the development of new point-of-care cancer diagnostics technology. Oral cancer as a target for new cancer screening modalities is a good choice because early detection methods are sorely lacking. Despite advances in diagnostics

NOTE: M.G. Mauk and B.L. Ziober contributed equally to this manuscript.

Address for correspondence: Haim H. Bau, 237 Towne Building, School of Engineering and Applied Science, Philadelphia, PA 19104-6315. Voice: 215-898-8363; fax: 215-573-6334.
bau@seas.upenn.edu

**Ann. N.Y. Acad. Sci. 1098: 467–475 (2007). © 2007 New York Academy of Sciences.
doi: 10.1196/annals.1384.025**

and therapy, the 5-year survival rate for OSCC patients remains at about 50%.[1] Early detection of OSCC could greatly reduce morbidity by fostering more timely initiations of therapy and patient monitoring, and also would help avoid inappropriately aggressive surgical treatments that result in severe disfigurement. Moreover, oral fluid samples from OSCC patients collected by noninvasive methods are found to contain precancerous (dysplastic) and cancerous cells that (1) express specific cancer markers that serve as molecular targets for sensitive and specific detection, and (2) are amenable to more elaborate analysis (typing and staging) using gene expression profiling.

There is an urgent need for new technologies to enable inexpensive, convenient, and rapid cancer screening and diagnostics.[2] Cancer tests that could be employed at the point-of-care, for example, doctors' and dentists' offices, and operated without extensive training or expertise are of special interest. Lab-on-a-chip microfluidics[3–5]—the miniaturization of fluidic networks for chemical and biochemical processing and analysis—offers a means for mass-produced, low-cost, single-use (disposable) devices for cancer screening and diagnostics, providing easily interpreted test results in a time frame of 10 to 60 min. Ideally, these lab-on-a-chip cancer diagnostics systems would use 10 to 1,000 μL of various types of clinical specimens including oral fluids, whole blood, serum, or urine, collected by minimally invasive methods, as well as samples, such as tissue biopsies, intraductal breast fluid, bronchial lavages, and lung aspirations. In general, the anticipated benefits of microfluidics for clinical diagnostics derive from the use of small sample volumes, automated operation, short processing times, and near real-time reporting of results, reduced reagent consumption, reproducibility and consistency, reduced exposure to hazardous materials and infectious agents, minimal risk of sample contamination, convenient disposal, and low cost.

In the last decade, a diverse array of microfluidic components and systems have been developed for immunoassays and include cell sorting, detection and counting, lysis, nucleic acid and protein isolation and amplification, and detection and quantification of nucleic acids and proteins.[3–5] Lab-on-a-chip devices for detection of infectious agents and toxins are becoming well established. The microfluidic technology developed for pathogen detection can be adapted and extended for the more difficult task of cancer screening and diagnostics. In the simplest approach, an automated immunoassay of a single cancer marker or a panel of cancer markers can be implemented on a credit card–sized microfluidic cassette for point-of-care cancer screening. More robust and detailed tests can be realized by quantifying a panel of 10 to 30 mRNA or proteins for determining a cancer-specific gene transcription or expression profile. To assess a gene transcription profile, microfluidics systems need to include components for cell sorting to enrich the sample in cancer cells, cell lysis, nucleic acid isolation, multiplex reverse transcriptase-polymerase chain reaction (RT-PCR) or other analogous amplification techniques, and multiplex detection and quantification of the gene transcripts. Newly developed bio-barcode assays for multiplex detection of proteins[6] and gene transcripts[7] offer an alternative approach for nonenzymatic multiplex amplification and detection of both nucleic acids and proteins.

We assess the feasibility of a lab-on-a-chip molecular diagnostics system for OSCC screening and diagnostics using oral fluid samples. Processing steps for cancer diagnosis are identified. Supporting data from benchtop studies demonstrating methods for isolating cancer cells from oral fluids, and gene expression analysis identifying an OSCC-related transcription profile, serve as the basis of the microfluidic cancer diagnostics system. We present a design for a lab-on-a-chip system that tests for a cancer-related gene transcription signature by assaying a panel of mRNA extracted from precancer or cancer cells isolated from an oral fluid sample.

FIGURE 1 depicts a flow process for a cancer diagnostics protocol whereby a cancer-specific gene expression profile is determined. The cancer diagnostic process comprises an initial step to remove lymphocytes that interfere with subsequent steps for isolating cancer cells from the sample. The sample is depleted of lymphocytes by immunoseparation using magnetic beads coated with anti-CD45 antibody, which binds to lymphocytes. Next, the cancer cells are sorted from the sample using magnetic beads coated with anti-epithelial cell adhesion molecule (EpCAM) antibody. EpCAM is a cell membrane glycoprotein that is aberrantly expressed on the surface of cancerous epithelial cells associated with OSCC. The separated cancer cells can be detected and (optionally) counted. The detection of EpCAM-expressing cells in the sample serves as the first screening test for cancer. The separated cancer cells are then subjected to a thermal and/or chemical lysis step, and the mRNA are isolated from the lysate using solid-phase extraction or by hybridization with magnetic beads coated with mRNA-specific oligonucleotides. Multiplex mRNA amplification by RT-PCR, linear amplification, or a bio-barcode technique is followed by detection of labeled cDNA, aRNA, or bio-barcodes using either fluorescence with fiber optic sensors or electrochemical sensors.

SUPPORTING STUDIES

Cancer cells are isolated from the sample by immunoseparation using paramagnetic beads coated with antibodies that bind to cell membrane proteins specific to cell types (including lymphocytes and cancer cells) making up the heterogeneous sample. The relevant findings may be summarized as follows:

1. Western blots with Ber-EP4 (monoclonal antibody to EpCAM) indicate that EpCAM is expressed only in OSCC cells and is not detectable in normal cells or fibroblasts. Our findings are consistent with a growing body of literature documenting high EpCAM expression in various cancer cells of epithelial origin.[8]

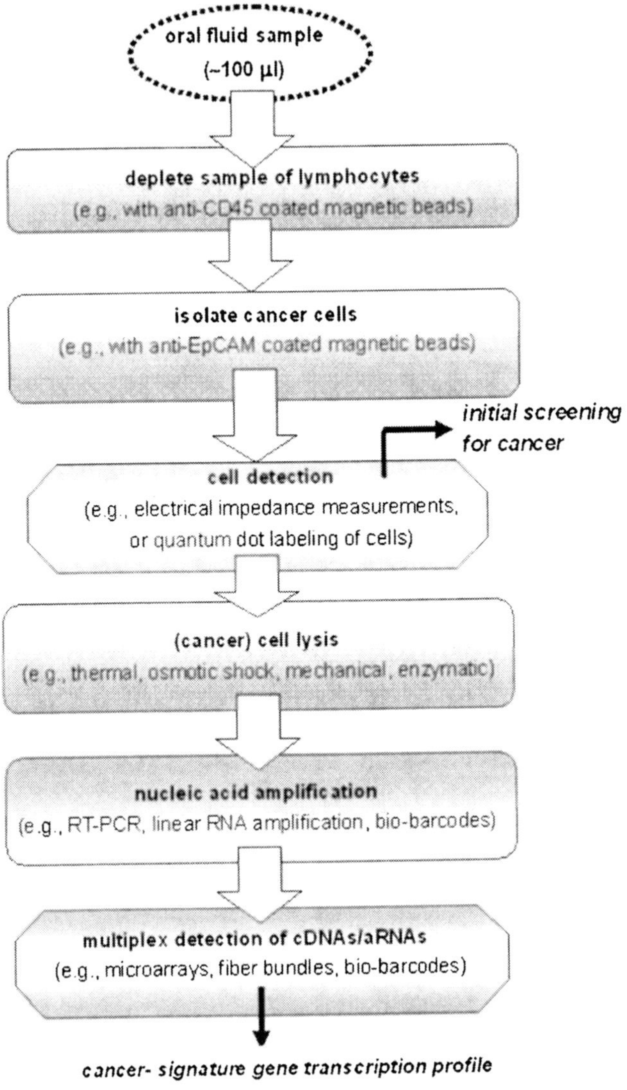

FIGURE 1. Sample processing steps for isolating cancer cells from saliva and assaying a panel of mRNAs.

2. Magnetic beads functionalized with Ber-EP4 separate cancer cells from a suspension containing known quantities of labeled cancer cells (from various cell lines) and normal cells. In a negative control, magnetic beads functionalized with anti-IgG failed to bind to either cancer or normal cells. FIGURE 2 shows a cancer cell tagged with 4.5-μm diameter magnetic beads functionalized with anti-EpCAM antibody.

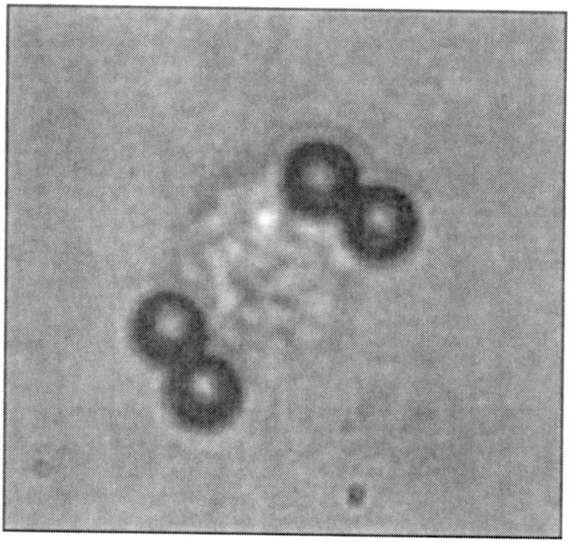

FIGURE 2. OSCC cancer cell bound with four 4.5-μm diameter superparamagnetic beads (Dynal, Invitrogen, Carlsbad, CA, USA).

3. Magnetic beads conjugated with antibodies to CD45 separated lymphocytes from a mixture containing 2×10^7 lymphocytes and various quantities of stained cancer cells. Subsequently, the cancer cells were isolated (with greater than 80% efficiency) using magnetic beads functionalized with Ber-EP4.
4. Subsequent to the removal of lymphocytes with magnetic beads functionalized with antibodies to CD45, magnetic beads functionalized with Ber-EP4 isolated more than 9,000 cells/mL from unstimulated whole saliva from T4 patients, more than 1,000 cells/mL from T1 patients, and less than 15 cells/mL from healthy patients (FIG. 3). T refers to tumor stage of the tumor node metastasis (TNM) classification system, and 1–4 denotes tumor size (1 smallest, 4 largest). The epithelial origin of the isolated cells was demonstrated by staining them with pan-cytokeratin. The isolated cells were further identified as OSCC cells by labeling with antibody to HSP-47 (clone M10.1061). HSP-47 was shown to be singularly expressed on OSCC cells.
5. HSP-47 has been identified as another protein that is uniquely expressed on OSCC cell membranes and that can be used for labeling and discriminating isolated cells.[9]
6. A 25-gene transcription signature for OSCC can classify normal and OSCC specimens. This 25-gene predictor was 96% accurate on cross-validation, averaging 87% accuracy using three independent validation tests.[10]

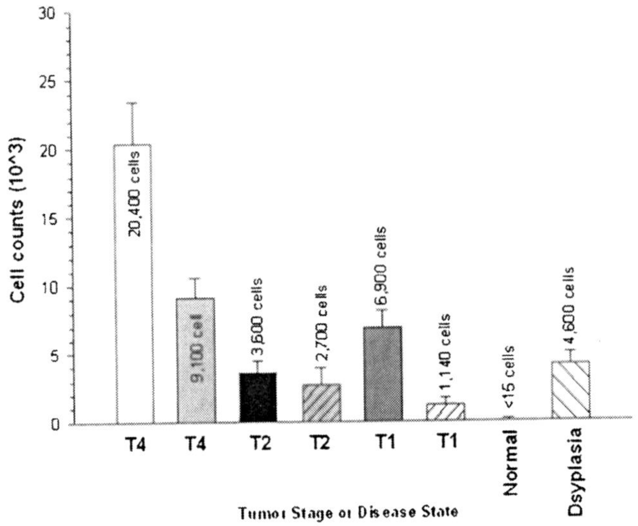

FIGURE 3. Isolation of tumor cells in OSCC patient saliva. A process using negative selection first with magnetic beads bound to the antibody CD45 (the major lymphocyte marker) followed by positive selection with magnetic beads functionalized with Ber-EP4 antibodies recovered the tumors. Total tumor cells isolated from each tumor are shown. Essentially no cells were isolated from normal patients' saliva. These numbers of cells should be sufficient for obtaining enough RNA for complete analysis.

MICROFLUIDIC IMPLEMENTATIONS AND DESIGN APPROACHES

FIGURE 4 shows a microfluidic device for continuous sorting of cells using the principle of magnetic field flow fractionation (MFFF).[11] A mixed population of cells (cancerous and noncancerous) are incubated with 4.5-μm diameter superparamagnetic beads (Dynel Cellection™ Epithelial Enrich, Invitrogen, Carlsbad, CA, USA) coated with anti-EpCAM antibody to selectively tag the target cancer cells with magnetic beads. The incubated sample is injected into a flow channel of a polycarbonate chip, and hydrodynamically focused with a surrounding sheath of flowing buffer. Both sample and buffer are propelled by programmed syringe pumps. To allow flow visualization, dyes are added to the fluids. The sample stream remained confined axially within the buffer sheath along the chamber's entire length (FIG. 4A). The sample stream is subjected to the field of an external permanent magnet positioned as indicated. FIGURE 4B is a histogram of the transverse distances (y) traveled by the bead–OSCC cell complexes, unlabeled cells, and free beads. The measurements were taken a short distance downstream of the injection point (dashed circle in FIG. 4A) with or without the external magnetic field. In the absence of a magnetic field, the cells and beads remained in the core of the flow and were kept separated from

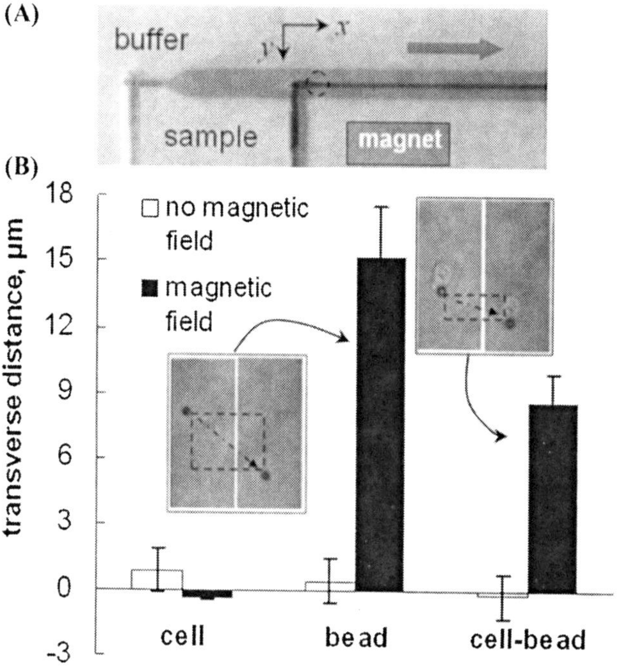

FIGURE 4. (**A**) MFFF device for separating magnetic bead-bound cells from unbound cells and unbound beads. Cancer cells (expressing the surface protein EpCAM) and normal cells (no EpCAM) are mixed with 4.5-μm diameter superparamagnetic beads (Dynal Cellection Epithelial Enrich, Invitrogen) and injected into a flow channel of a polycarbonate microfluidic chip. The sample is hydrodynamically focused with a sheath of buffer solution surrounding the sample stream. (**B**) Histogram showing deviation of flow path due to the applied magnetic field. With no applied field, the beads, unbound cells, and bead–cell complexes follow an unimpeded axial trajectory in the flow channel. Application of a magnetic field with an external permanent magnet causes the beads and bead–cell complexes to deviate from the axial flow and follow a characteristic trajectory, resulting in separation of beads, unbound cells, and bead–cell complexes.

the chamber walls by a surrounding buffer. When a magnetic field was applied, the magnetic beads and labeled cells diverted from the sample stream with the free beads moving faster and at a sharper angle with respect to the axis than the labeled cells (FIG. 4B). The unlabeled cells maintained their axial trajectory. Thus, bead–cell complexes can be separated (and collected) according to their distinct trajectories resulting from application of magnetic field.

FIGURE 5 depicts a schematic plan view for a microfluidic cassette that performs the steps outlined in FIGURE 1. The crucial function of isolating cancer cells from a heterogeneous clinical specimen, such as oral fluid by microfluidic immunoseparation with magnetic beads, appears feasible. Microfluidic components for lysis, nucleic acid isolation, multiplex RT-PCR and detection,

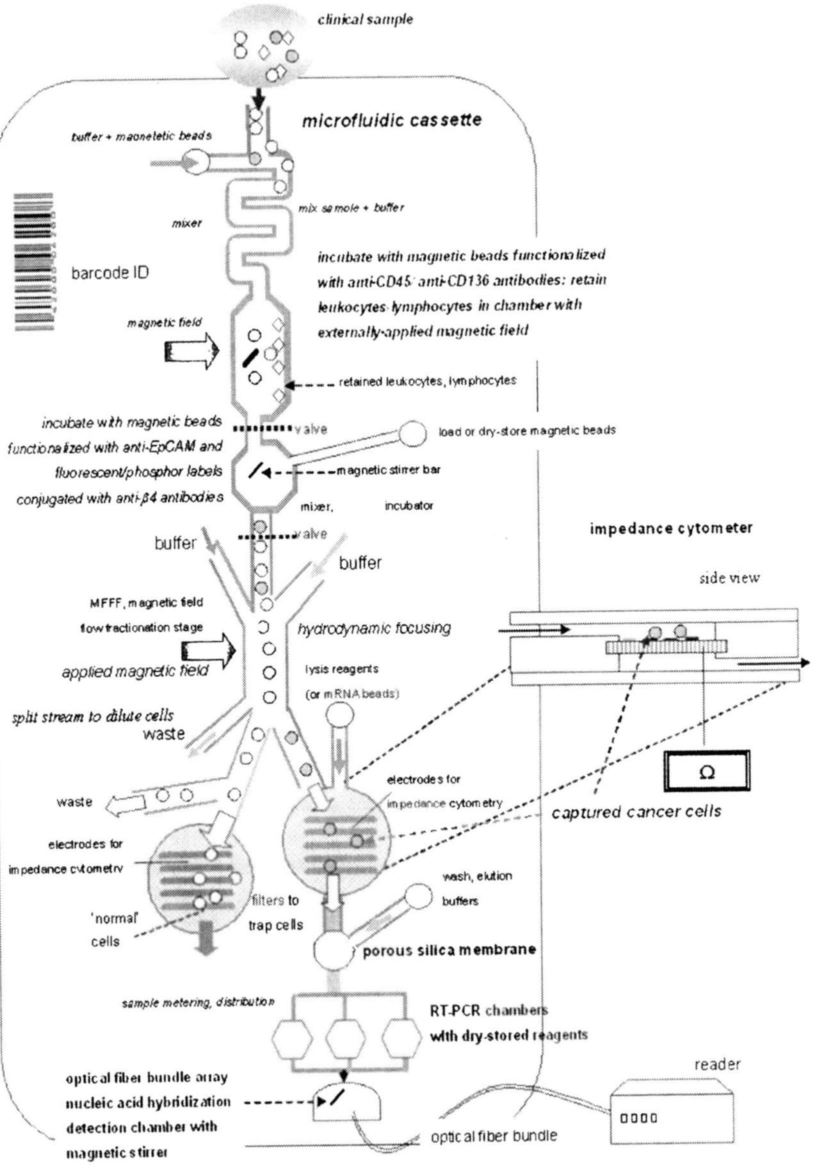

FIGURE 5. Plan view schematic of a comprehensive cancer diagnostics lab-on-a-chip integrating microfluidic components for lymphocyte depletion, cancer cell isolation and lysis, mRNA isolation, multiplex amplification, and detection of a panel of mRNA.

or analogous processes, such as bio-barcode signal amplification, have been demonstrated by numerous groups—see, for example, Ref.12. The outstanding challenge now is to seamlessly integrate the appropriate microfluidic components into a comprehensive, cost-effective lab-on-a-chip system for automated operation that provides easily interpreted, statistically significant cancer diagnostics data in a timely manner.

ACKNOWLEDGMENTS

This work was supported by a grant from the Penn Genomics Institute of the University of Pennsylvania, and the NIH Grants UO1DE0114964 and UO1DE017855.

REFERENCES

1. WEINBERG, M.A. & D.J. ESTEFAN. 2002. Assessing oral malignancies. Am. Family Phys. **65:** 1379–1384.
2. SOPER, S.A. *et al.* 2006. Point of care biosensor systems for cancer diagnostics/prognostics. Biosens. Bioelectron. **21:** 1932–1942.
3. SELVAGANAPATHY, P.R., E.T. CARLEN & C.H. MASTRANGELO. 2003. Recent progress in microfluidic devices for nucleic acid and antibody assays. Proc. IEEE **91:** 954–975.
4. VILKNER, T., D. JANASEK & A. MANZ. 2004. Micro total analysis systems. Recent developments. Anal. Chem. **76:** 3373–3386.
5. TONER, M. & D. IRIMIA. 2005. Blood-on-a-chip. Annu. Rev. Biomed. Engng. **7:** 77–103.
6. BAO, Y.P. *et al.* 2006. Detection of protein analytes via nanoparticle-based bio bar code technology. Anal. Chem. **78:** 2055–2059.
7. THAXTON, C.S., D.G. GEORGANOPOULOU & C.A. MIRKIN. 2006. Gold nanoparticle probes for the detection of nucleic acid targets. Clinica Chimica Acta **363:** 120–126.
8. WENT, P.T. *et al.* 2004. Frequent EpCAM protein expression in human carcinomas. Human Pathol. **35:** 122–128.
9. SAUK, J.J., N. NIKITAKIS & H. SIAVASH. 2005. Hsp 47, a novel collagen binding serpin chaperone, autoantigen, and therapeutic agent. Front Biosci. **10:** 107–118.
10. ZIOBER, A.F. *et al.* 2006. Identification of a gene signature for rapid screening of oral squamous cell carcinoma. Clin. Cancer Res. **20:** 5960–5971.
11. BERTHIER, J. & P. SILBERZAN. 2006. Microfluidics for Biotechnology. Artech House. Norwood, MA.
12. WEIGL, B.H. *et al.* 2006. Fully integrated multiplexed lab-on-a-card assay for enteric pathogens. Microfluidics, BioMEMS, and Medical Microsystems. Proc. SPIE **6112:** 611202.

Evaluation of UP*link*–RSV

Prototype Rapid Antigen Test for Detection of Respiratory Syncytial Virus Infection

VIJAYA K. MOKKAPATI,[a] R. SAM NIEDBALA,[b] KEITH KARDOS,[a] RONELITO J. PEREZ,[a] MING GUO,[a] HANS J. TANKE,[c] AND PAUL L. A. M. CORSTJENS[c]

[a]*OraSure Technologies Inc., Bethlehem, Pennsylvania, USA*

[b]*Department of Chemistry, Lehigh University, Bethlehem, Pennsylvania, USA*

[c]*Department of Molecular Cell Biology, Leiden University Medical Center, Leiden, the Netherlands*

ABSTRACT: A prototype rapid antigen test for the on-site detection of respiratory syncytial virus (RSV) infection was developed and evaluated. The platform uses instrumented assay analysis, eliminating potential operator bias in the interpretation of the test result that may occur with visually interpreted rapid antigen assays. The device was tested as the first point-of-care (POC) infectious disease application of novel reporter up-converting phosphor technology (UPT) using a specifically designed portable UPT reader (UP*link*™)®. Assays were performed by mixing nasopharyngeal specimen with RSV-specific UPT reporter particles and addition of the mixture to a disposable cassette containing a lateral flow (LF) strip with RSV capture antibodies. UPT reporters bound on the specific capture zone were analyzed with the UP*link* reader. Reproducibility testing of the UP*link*–RSV (UPR) test by naïve users confirmed the potential of UP*link* for POC applications where testing is not always performed by highly trained medical staff. The performance of UPR was further evaluated with clinical nasopharyngeal specimens. A prospective study at an independent test site demonstrated clinical parameters of 90% sensitivity and 98.3% specificity with an overall correlation of 96.2% as compared to viral culture with RT-PCR verification. These results are in agreement with in-house retrospective studies and results obtained with other available commercial rapid antigen assays.

KEYWORDS: up-converting phosphor; UP*link*; lateral flow; RSV

Address for correspondence: Paul L. A. M. Corstjens, Department of Molecular Cell Biology, Leiden University Medical Center, P.O. Box 9600, 2300 RC Leiden, the Netherlands. Voice: +31-71-5269209; fax: +31-71-5268270.

Corstjens@LUMC.NL

Ann. N.Y. Acad. Sci. 1098: 476–485 (2007).
doi: 10.1196/annals.1384.021

INTRODUCTION

Respiratory syncytial virus (RSV) is an enveloped, RNA virus of the *Paramyxoviridae* family. It is a frequent cause of serious lower respiratory tract disease in young children.[1] RSV bronchiolitis is a severe illness caused by RSV; if not treated it can become life-threatening in infants (as compared to adults) with small peripheral airways. In addition, RSV is a major cause of severe pneumonia in young children. RSV outbreaks occur yearly, usually during winter months, and it is recommended that health care providers consider RSV as the cause of acute respiratory symptoms. RSV is contagious and easy to transfer through casual contact, making hospital staff frequent vectors for viral transmission. Thus rapid diagnosis of patients requiring hospital admission is important not only to guide in therapeutic decisions, but also to prevent nosocomial RSV transmission and thus eventually reduce medical costs.[2,3]

Cell culture is still considered the gold standard for detection of RSV. However, for rapid diagnosis, cultivation of the virus is impractical because of the relatively long incubation needed. In addition, diagnosis of acute infection may be difficult on account of the insensitivity of viral culture.[4] In this respect, reverse transcriptase-polymerase chain reaction (RT-PCR) is a better alternative.[5] The various RT-PCR methods available for clinical laboratories can provide results within a few hours.[6] However, despite the much shorter turnaround time as compared to that of viral culture, RT-PCR has its limitations for point-of-care (POC) testing. In general, specimens requiring RT-PCR analysis are collected and batch-processed or are sent to commercial laboratories. Consequently, RT-PCR as well as viral culture is merely used to confirm the results from rapid antigenz tests.

Currently, there are a few FDA-cleared rapid RSV antigen assays available in the United States. These immunoassay-based POC devices are standardized assays convenient for analysis of single samples in the POC environment. However, the current rapid assays require visual interpretation of the test result, whereas assays that minimize operator interpretation are preferred because they reduce the chance of human error. In this article we describe a prototype UP*link* assay for the detection of RSV- in nasopharyngeal wash specimens (UP*link*–RSV, UPR). The UP*link* system comprises a portable reader with a built-in IR (infrared) laser and immunoassay devices (disposable cassettes containing test-specific LF strips). For UPR, the LF strips are RSV-specific. The reader scans LF strips in the immunoassay devices and displays the results such that subjective interpretation of the test by the operator is eliminated. The basis of the UP*link* platform is the up-converting phosphor technology (UPT), an ultrasensitive reporter particle technology.[7] These reporters, upon excitation by IR light, up-convert the energy (two-photon up-conversion) to give a visible 550-nm green emission. Since no biological specimen in nature up-converts low-energy IR light, UPT applications are unaffected by specimen background and display excellent signal-to-noise ratios.[8]

The objective of this study was to evaluate the performance of UPR (UP*link*–RSV) compared to viral culture. Samples that generated discrepant results were also evaluated by RT-PCR. The RT-PCR results were considered decisive as this method is accepted as more sensitive than viral culture.[9] For comparison, the same samples were also analyzed with Directigen RSV (DIR; Becton Dickinson Franklin Lakes, NJ, USA), another rapid antigen detection assay.

MATERIALS AND METHODS

*UP*link *Platform–Detection of UPT Reporters*

The existing UP*link* analyzer (OraSure Technologies, Bethlehem, PA, USA) was originally developed for on-site testing of drugs of abuse in oral samples. It uses a low-power (1.0-watt) IR laser that interrogates UPT-deposited lateral flow (LF) strips. The LF strip is an integrated part of the disposable plastic cassette, a self-contained immunoassay device that is inserted into the UP*link* analyzer (FIG. 1A). The analyzer is a general UPT reader with integrated software that can be modified to suit various test applications. It detects localized deposition of the UPT reporters by reading the visual light emitted upon IR excitation; the emission signal is measured and computed as relative fluorescence units (RFU). The UP*link* analyzer is provided with a bar code scanner that identifies the unique assay associated with each cassette.

*UP*link*–RSV (UPR) Assay Development*

LF strips for UPR were stripped with a test line composed of RSV monoclonal antibodies against the F, G, and N antigens (FIG. 1B). A flow-control line containing a noncontagious RSV isolate (FRhKy, Viral Antigens Inc., Memphis, TN, USA) was stripped downstream from the test line to verify migration of the UPT conjugate through the LF strip. The strips were assembled as described earlier.[10] A set of anti-F and anti-N antibodies different from the antibodies stripped on the test line were conjugated with UPT reporter particles. These reporter antibodies allow effective immuno-sandwich formation with the antibodies on the test line through RSV antigen. Conjugations were performed according to standard procedures as described earlier.[11] Freeze-dried pellets of UPT conjugates, so-called lyospheres (Biolyph LLC, Hopkins, MN, USA) were produced and stored in vacuum-sealed glass vials (FIG. 1C).

UPR Assay Procedure

The UPR test was performed by mixing assay buffer and specimen with the lyophilized reporter (lyosphere) in a glass vial followed by addition of

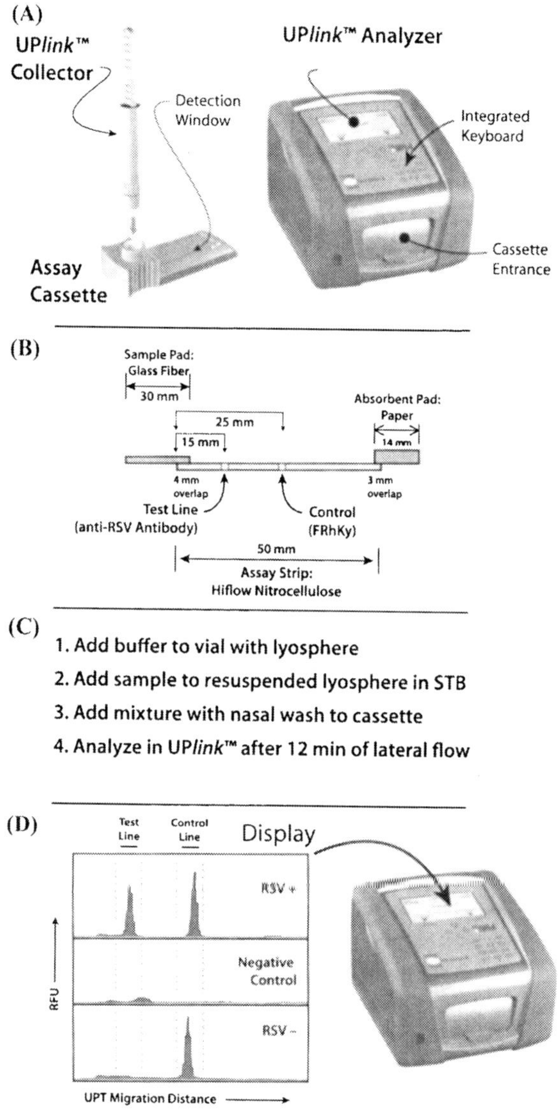

FIGURE 1. Description of the UP*link* platform and the UP*link*–RSV (UPR) assay. (**A**) The UP*link* analyzer (UPT reader containing analysis software) with matching assay cassette. (**B**) The UP*link* assay cassette with RSV-specific LF strip. Components of the UP*link*–RSV (UPR) LF strip, from left to right: a sample pad for application of a mixture of nasal wash and $UPT^{\alpha RSV}$ reporters; a nitrocellulose assay pad containing a test line with αRSV antibodies and a flow-control line with RSV antigens; an absorbent pad. (**C**) The UPR test procedure. (**D**) Emission profile (scan result) of three different UPR assays: positive, invalid, and negative. Scan profiles show signals in RFU (*y* axis). The data from the actual scan profiles are saved in memory; the displays respectively show RSV+, invalid control, and RSV–.

the mixture to the test cassette (FIG. 1C). The cassette is inserted into the portable UP*link* analyzer to scan and interpret the test result. The overall time of the assay is 15 min. In detail, the UPR test procedure consists of four steps: (1) addition of 100-μL sample treatment buffer (STB at pH 8.2 containing Tris-HCl, NaCl, EDTA, specific detergents, and ProClin 950) to a vial containing one UPT-lyosphere; (2) addition of 200-μL sample (nasopharyngeal wash) to the above mixture; (3) transfer of the entire mixture to the sample reservoir of the UPR cassette to initiate lateral LF and allow the signal to develop for 12 min; and (4) scanning of the cassette using the UP*link* reader. All specimens generating signals above the predetermined cut-off value were considered RSV-positive. The cut-off threshold value was determined on the basis of 86 frozen specimens, which were RSV-negative by viral culture. The resulting cut-off threshold value of 9,801 RFU was calculated as the average test peak area value plus twice the standard deviation. Tests were determined invalid if there was no signal generated at the flow-control line (FIG. 1D). Data were stored in the UP*link* reader and can be transferred to a personal computer. The reader display shows the result as RSV+, RSV–, or invalid control.

Blinded UPR Reproducibility Study with Inexperienced Operators

To study the accuracy and the reproducibility of the UPR assay, four specimens spiked with different levels of RSV (a mixture of ATCC VR-1302 and VR-1400) were tested by three naïve operators from different sites over three consecutive days. For every operator, three sets (one set for each day of testing) of specimens were prepared. Each set was composed of: (1) M4 transport medium (Remel. Inc., Lenexa, KS, USA) without RSV; (2) M4 spiked with RSV at 2.8×10^2 $TCID_{50}$/mL; (3) M4 spiked with RSV at 5.6×10^2 $TCID_{50}$/mL; and, (4) M4 spiked with RSV at 1.1×10^3 $TCID_{50}$/mL. These samples were designated as negative, low positive, medium positive, and high positive, respectively. Specimens in the first set (test day 1) were labeled 1–4, specimens in the second set (test day 2) were labeled 5–8, and the specimens in the third set (test day 3) were labeled 9–12. The order of the negative, low-, medium-, and high-positive sample was randomized for the 3 days of testing. All tests were performed in replicates of five.

Evaluation of UPR versus DIR at an Independent Clinical Test Site (Prospective Study)

Sample Collection

A series of 78 samples was collected and tested at the Providence Health System (PHS) in Portland, Oregon. Samples were collected and immediately analyzed over a period of 3 months (January through March). Phosphate-buffered saline nasopharyngeal wash specimens were collected from patients suspected of having RSV infection following the standard collection method as described in the package insert of BD Directigen RSV. Samples were transported on ice to the laboratory for testing. Only specimens that were refrigerated at 2–8°C for <12 h were evaluated. A portion of each specimen was aliquoted immediately after collection and transferred into M4 media for viral culture testing. Hence, samples were never frozen prior to rapid antigen testing and initiation of viral culture. Unused portions of all specimens were labeled and frozen at −20°C in case a nucleic acid verification test was required.

Laboratory Methods

Laboratory methods included antigen testing with UP*link*–RSV (UPR, OraSure Technologies), Directigen-RSV (DIR, Becton Dickinson), and viral culture with DFA detection. Viral culture was performed according to standard protocols established at the test site and antigen assays were performed according to the protocol provided by the manufacturers. Interpretation of the DIR test results was carried out according to the manufacturer's guidelines. In this method, specimens were reported positive, negative, or indeterminate (invalid control) by visual reading of the test result. For UPR, UP*link*-displayed test results (RSV+, RSV–, or invalid control) were used.

RT-PCR Verification

RT-PCR was performed on frozen specimens. All samples requiring verification were transported on dry ice to special facilities at the Children's Hospital Medical Center (CHMCC) in Cincinnati. A standard RT-PCR detection protocol suited for the detection of RSV A and B strains was used to analyze the specimens.

Statistical Analysis

UPR and DIR test results were analyzed independently. Clinical samples were considered true positives or true negatives in all cases where antigen assay results were in agreement with viral culture. When assay results and viral culture were discrepant, samples were further tested by RT-PCR. The RT-PCR results were regarded as decisive.

RESULTS

The automated analysis of the UPR assay eliminates subjective interpretation of the test results, since the software integrated in the UP*link* reader determines whether the test result is positive, negative, or invalid. Reproducibility of UPR was evaluated by repetitive testing of blinded specimens

TABLE 1. UPR reproducibility study (blinded)

Operator	RSV level	Day 1	Day 2	Day 3	Total	Agreement
1	Negative	5/5	5/5	5/5	60/60	100%
	Low pos.	5/5	5/5	5/5		
	Med. pos.	5/5	5/5	5/5		
	High pos.	5/5	5/5	5/5		
2	Negative	5/5	3/5	5/5	54/60	90%
	Low pos.	5/5	4/5	3/5		
	Med. pos.	4/5	5/5	5/5		
	High pos.	5/5	5/5	5/5		
3	Negative	5/5	5/5	4/5	57/60	95%
	Low pos.	5/5	5/5	3/5		
	Med. pos.	5/5	5/5	5/5		
	High pos.	5/5	5/5	5/5		
Agreement						95%

with variable amounts of spiked RSV (four concentration levels). Experiments performed by three independent naïve operators indicated an overall concordance of 95% (TABLE 1). The operators performed 60 tests each over a period of three consecutive days (20 tests a day), with a concordance of 100%, 90%, and 95% for operators 1, 2, and 3, respectively. The day-to-day concordance was 98.3%, 95%, and 91.7% for the first, second, and third day, respectively. Each of the four RSV concentration levels was tested 45 times by the three operators together with a concordance of 93.3%, 88.9%, 97.8%, and 100% for the negative, low-positive, medium-positive, and high-positive samples, respectively.

For an independent clinical evaluation of UPR (prospective study), a total of 78 fresh samples was gathered and tested at the PHS in Portland, Oregon. All samples were collected from patients suspected of having RSV infection; 56% of the samples were obtained from infants or children below 2 years of age. After collection, the specimens were tested within 12 h (specimens were not frozen but kept refrigerated until testing) using UPR, DIR, and viral culture.

The results from UPR and DIR tests were independently compared against the viral culture test. The comparison indicated 9 discrepant results; 4 discrepancies for UPR versus culture and 5 discrepancies for DIR versus culture (TABLE 2). These 9 discrepancies relate back to 6 viral cultures: 3 culture results discrepant with both rapid assays, 2 culture results discrepant with DIR only, and 1 culture result discrepant with UPR only. The six cultures with discrepant rapid assay results were subjected to RT-PCR verification. Among these samples only one specimen showed disagreement between culture and RT-PCR results: a negative culture result tested positive upon RT-PCR verification. After resolving the status of the discrepant samples, the clinical parameters for both UPR and DIR analysis were calculated (TABLE 2).

TABLE 2. Prospective study: performance of UPR and DIR

UPR	Pos.	Neg.	DIR	Pos.	Neg.
Culture					
Pos.	17	2	Pos.	15	1
Neg.	2	57	Neg.	4	58
Culture/RT-PCR					
Pos.	18	1	Pos.	16	0
Neg.	2	57	Neg.	4	58

Clinical parameters (after RT-PCR verification)	UPR	DIR
Sensitivity	90.0%	80.0%
Specificity	98.3%	100%
Positive Predictive value	94.7%	100%
Negative predictive value	96.6%	93.5%
Correlation	96.2%	94.8%

DISCUSSION

UPR is the first UPT/UP*link* application. The LF-based assay screens for active RSV infections using nasopharyngeal specimens and is capable of detecting viable as well as nonviable RSV fragments from A and B type RSV strains. Sensitive and chemically inert 400-nm phosphorescent UPT particles are applied as reporter in UPR. The reporter is readily detected with UP*link*, a fully automated inexpensive portable reader developed especially for on-site and POC testing. This automation entails that UPR assay results are presented as "RSV+" "RSV–," or "invalid control." It omits potentially operator-biased assay interpretation that can occur with DIR and other commonly used rapid antigen tests that rely on visual interpretation of the assay.

Reproducibility testing of the UPR assay and its developed standard test protocol was performed by three naïve operators and resulted in an overall concordance of 95%. This demonstrated that the whole procedure worked well enough to allow utilization by inexperienced users. The lowest level of agreement (88.9%) was obtained when testing a low-positive sample. This could be the result of small procedural alterations or mistakes, which will have the relatively largest effect when testing low-positive samples as their signals are closest to the predetermined threshold. Further acquaintance and experience with the whole procedure (including transport, storage, and the actual UPR assay) can be expected to have a positive impact on concordance. Testing of the high-positive sample resulted in 100% concordance.

An evaluation with 78 fresh clinical samples (56% from infants below the age of 6 months) was performed at an independent test site. RT-PCR-resolved analysis demonstrated clinical parameters of 90%, 98.3%, 94.7%, 96.6%, and

96.2% for UPR against 80%, 100%, 100%, 93.5%, and 94.8% for DIR, regarding sensitivity, specificity, positive predictive value, negative predictive value, and overall correlation, respectively. In fact, the performance of both assays with this set of fresh samples was almost similar. Out of 78 clinical samples tested, DIR scored 4 false negatives, whereas for UPR the score included 2 false negatives and 1 false positive.

CONCLUSION

On the basis of data presented in this article, we conclude that the UPR test as described here has considerable potential in the rapid diagnosis of RSV bronchiolitis and RSV-caused pneumonia. A large clinical study is required to determine whether the overall performance of UPR is better than other rapid RSV antigen-based immunoassays. However, significant added value of UPR is found in the automated analysis of the test (provided by the UP*link* platform) eliminating any operator-biased interpretation of the result. The UPR test protocol is simple and robust, allowing application in POC settings by less-trained naïve personnel. Moreover, future versions of the test platform will provide fully integrated devices that only require operator interaction for addition of the clinical specimen to the disposable test device.

ACKNOWLEDGMENT

We thank Mike Lamos (OraSure) for modifying the UP*link* reader software, Corinne Shucavage (OraSure) for technical assistance, and Shang Li (OraSure) for synthesizing the UPT reporter particles. Part of this work was supported by NIH Grant U01-DE-014964.

REFERENCES

1. HALL, C.B. 1999. Respiratory syncytial virus: a continuing culprit and conundrum. J. Pediatr. **135:** 2–7.
2. MACARTNEY, K.K. *et al.* 2000. Nosocomial respiratory syncytial virus infections: the cost-effectiveness and cost-benefit of infection control. Pediatrics **106:** 520–526.
3. MADGE, P. *et al.* 1992. Prospective controlled study of four infection-control procedures to prevent nosocomial infection with respiratory syncytial virus. Lancet **340:** 1079–1083.
4. AHLUWALIA, G. *et al.* 1987. Comparison of nasopharyngeal aspirate and nasopharyngeal swab specimens for respiratory syncytial virus diagnosis by cell culture, indirect immunofluorescence assay, and enzyme-linked immunosorbent assay. J. Clin. Microbiol. **25:** 763–767.

5. FALSEY, A.R. *et al.* 2003. Comparison of quantitative reverse transcription-PCR to viral culture for assessment of respiratory syncytial virus shedding. J. Clin. Microbiol. **41:** 4160–4165.
6. ABELS, S. *et al.* 2001. Reliable detection of respiratory syncytial virus infection in children for adequate hospital infection control management. J. Clin. Microbiol. **39:** 3135–3139.
7. NIEDBALA, R.S. *et al.* 2001. Detection of analytes by immunoassay using up-converting phosphor technology. Anal. Biochem. **293:** 22–30.
8. CORSTJENS, P.L.A.M. 2003. Lateral-flow and up-converting phosphor reporters to detect single-stranded nucleic acids in a sandwich-hybridization assay. Anal. Biochem. **312:** 191–200.
9. FALSEY, A.R. *et al.* 2002. Diagnosis of respiratory syncytial virus infection: comparison of reverse transcription-PCR to viral culture and serology in adults with respiratory illness. J. Clin. Microbiol. **40:** 817–820.
10. CORSTJENS, P.L.A.M. *et al.* 2001. Use of up-converting phosphor reporters in lateral-flow assays to detect specific nucleic acid sequences: a rapid, sensitive DNA test to identify human papillomavirus type 16 infection. Clin. Chem. **47:** 1885–1893.
11. NIEDBALA, R.S. *et al.* 2001. Detection of analytes by immunoassay using up-converting phosphor technology. Anal. Biochem. **293:** 22–30.

Immunoassay-Based Diagnostic Point-of-Care Technology for Oral Specimen

SARVAN MUNJAL,[a] PETER MIETHE,[b] LUTZ NETUSCHIL,[a] FRIEDHELM STRUCK,[a] KURT MAIER,[a] AND CLAUS BAUERMEISTER[c]

[a]*Dentognostics GmbH, Jena 07743, Germany*

[b]*FZMB, Bad Langensalza, Germany*

[c]*Laboratories for Oral Microbiology, Moers, Germany*

ABSTRACT: We have outlined our progress in developing a novel point-of-care platform to quantify micro-organisms causing dental infections and/or inflammatory markers reflecting an oral disease status. This system is based on a sandwich immunoassay technology known as ABICAP (Antibody Immuno Column for Analytical Processes) using poly-horseradish peroxidase conjugates. This assay enabled us to quantify 500 colony-forming units of *Streptococcus sobrinus* per milliliter of saliva. The platform allows rapid and convenient performance chairside of such tests by a dentist or dental hygienist within 20 minutes at the dental office.

KEYWORDS: caries risk test; chairside; immunoassay; oral specimen; point-of-care; *Streptococcus sobrinus*

As a novel point-of-care (PoC) platform, a Dento*Analyzer* (FIG. 1), was developed that renders possible the quantitative detection of various microorganisms as well as inflammatory markers (proteins) in order to monitor an oral disease status. Considering the importance of noninvasive sampling and PoC platform for dental offices,[1,2] the Dento*Analyzer* was designed as a portable user-friendly bench-top instrument. It automatically conducts the whole assay process, that is, steps like liquid handling as well as readout based on a software program and a robust algorithm.

The key component is a cartridge (FIG. 2) consisting of a liquid-handling module containing all the relevant reagents for immunological reactions like clinical sample, conjugate, wash buffers, and substrate (precipitating TMB); a reaction chamber containing six filters (including positive and negative controls), where the immunological reactions take place; and a waste container.

Address for correspondence: Dr. Lutz Netuschil, Dentognostics GmbH, Winzerlaer Str. 2, D-07743, Jena, Germany. Voice: +49-(0)3641-508 503; fax: +49-(0)3641-508 509.
netuschil@dentognostics.de

Ann. N.Y. Acad. Sci. 1098: 486–489 (2007).
doi: 10.1196/annals.1384.017

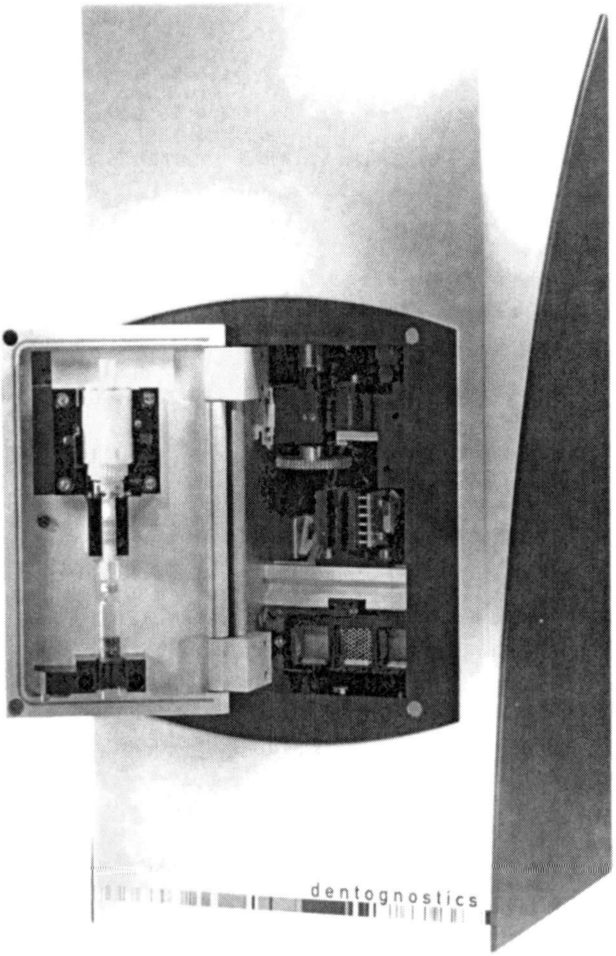

FIGURE 1. Dento *Analyzer* containing a cartridge at the inner side of its door (see FIGURE 2).

Two antibodies directed against specific epitopes of the antigens are used in a sandwich-based immunoassay technology known as antibody immuno column for analytical processes (ABICAP), which is based on an immunoaffinity filter design using flow-through solid-phase filters with extremely high binding potential.[3–6]

Oral samples including saliva and sulcus fluid have been tested for determination of intact bacteria (e.g., *S. mutans*, *S. sobrinus*) and marker enzymes (e.g., matrix metalloproteinase-8).[7] The samples or elution products were diluted in an appropriate dilution buffer, transferred into the Dento*Test* disposable and placed in the Dento*Analyzer*. The immuno-captured analyte is detected by the

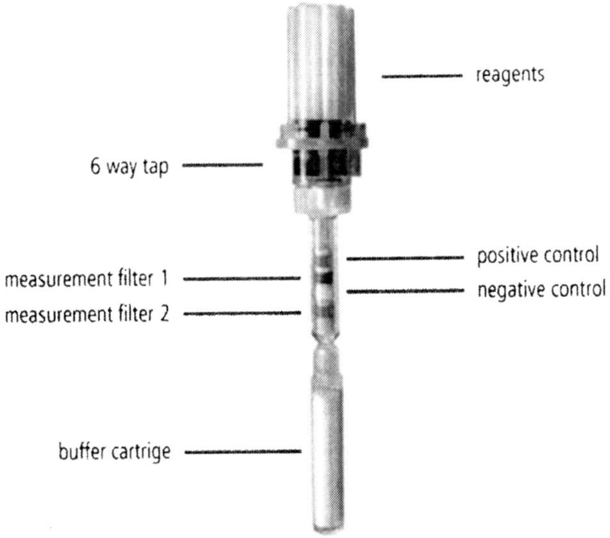

FIGURE 2. Cartridge for automated immuno-analysis.

addition of the detection antibody conjugated to poly-horseradish peroxidase (Poly-HRP, SDT GmbH, Baesweiler, Germany). The enzymatic reaction yields a blue precipitate on the filter surface (FIG. 2) with an optical density proportional to the analyte concentration. The quantitative measurement is based on kinetic analysis (FIG. 3) of TMB dye formation.[8]

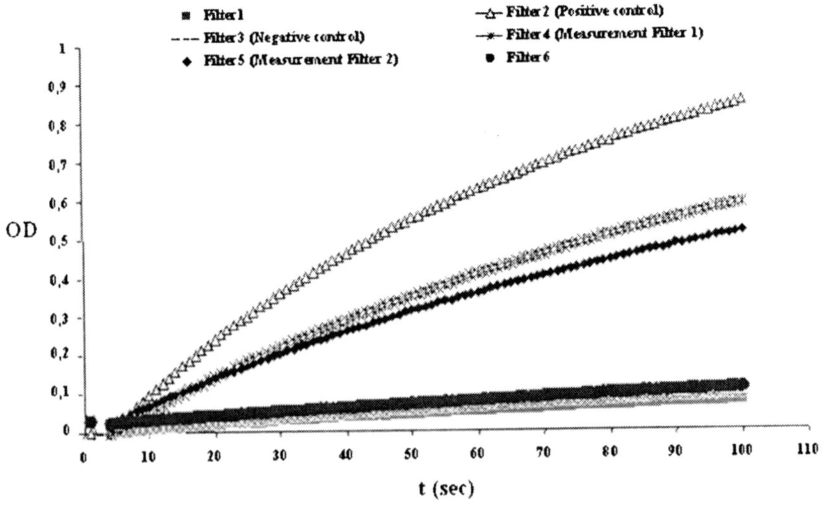

FIGURE 3. Signal course of an assay over time.

The poly-HRP-based conjugates display a much higher enzymatic activity in comparison to conventional conjugates, which makes the test ultrasensitive. For example, use of ABICAP columns and poly-HRP technology enabled us to quantify 500 colony-forming units of *Streptococus sobrinus* per milliliter of saliva (unpublished data).

This platform may become a unique chairside test analyzer for dentists, allowing quantitative detection of various dental-related bacteria and inflammatory markers. The device allows rapid and convenient performance of such tests by a dentist or dental hygienist within 20 min chairside at the dental office.

REFERENCES

1. HOLM-HANSEN, C., G. TONG, C. DAVIS, *et al.* 2006. Comparison of oral fluid collectors for use in a rapid point-of-care diagnostic device. Clin. Diagn. Lab. Immunol. **11:** 909–912.
2. MALAMUD, D., W.R. ABRAMS, H. BAU, *et al.* 2006. Oral-based techniques for the diagnosis of infectious diseases. J. Calif. Dent. Assoc. **34:** 297–301.
3. HARTMANN, H., B. LUBBERS, M. CASARETTO, *et al.* 1993. Rapid quantification of C3a and C5a using a combination of chromatographic and immunoassay procedures. J. Immunol. Methods **166:** 35–44.
4. STOVE, S., A. KLOS, W. BAUTSCH & J. KOHL.1995. Re-evaluation of the storage conditions for blood samples which are used for determination of complement activation. J. Immunol. Methods **182:** 1–5.
5. CAVUSLU, S., O. ONCUL, H. ALTUNAY, *et al.* 2003. Seroprevalence of tetanus antibody in Turkish population and effectiveness of single-dose tetanus toxoid. Eur. J. Microbiol. Infect. Dis. **22:** 431–433.
6. MEYER, M.H., M. STEHR, S. BHUJU, *et al.* 2006. Magnetic biosensor for the detection of *Yersinia pestis*. J. Microbiol. Methods: In press[Epub ahead of print] PMD: 17011649.
7. PRESCHER, N., K. MAIER, S.K. MUNJAL, *et al.* 2007. Rapid quantitative chairside test for active MMP-8 in gingival crevicular fluid – first clinical data. Ann. N. Y. Acad. Sci. This volume.
8. ZHAO, S. & R.D. FERNALD. 2005. Comprehensive algorithm for quantitative real-time polymerase chain reaction. J. Comput. Biol. **12**: 1047–1064.

Evaluation of Immunoassay-Based MMP-8 Detection in Gingival Crevicular Fluid on a Point-of-Care Platform

S. K. MUNJAL,[a] N. PRESCHER,[b] F. STRUCK,[a] T. SORSA,[c] K. MAIER,[a] AND L. NETUSCHIL[a]

[a]*Dentognostics GmbH, D-07743, Jena, Germany*

[b]*Dental Clinic, Friedrich Schiller University, Jena, Germany*

[c]*Department of Oral and Maxillofacial Diseases, Institute of Dentistry, University of Helsinki, Helsinki, Finland*

ABSTRACT: A novel immunology-based point-of-care test has been designed to assess the activated form of matrix metalloproteinase-8 (aMMP-8) for diagnosis and monitoring of periodontal diseases. The test has been automated using an analyzer, which quantitatively measures aMMP-8 in 18 min in gingival crevicular fluid (GCF). Fluid samples were collected from healthy, gingivitis-, and periodontitis-affected teeth. The test results from the analyzer were compared with quantitative aMMP-8 immunofluorometric assay (IFMA) and in-house enzyme-linked immunosorbent assay (ELISA) as well as with the periodontal state. Preliminary results of analyzer measurements of these 34 clinical samples showed a good agreement with the results from IFMA and in-house ELISA and with the clinical picture.

KEYWORDS: chairside; Dento*Analyzer*; immunoassay; MMP8; periodontitis; point-of-care

Polymorphonuclear leukocyte-derived matrix metalloproteinase-8 (MMP-8) is predominantly present in periodontitis- and peri-implantitis-affected gingival crevicular fluid (GCF).[1–4] Analysis of GCF could provide a useful noninvasive means to assess and monitor the pathophysiological status of the periodontium and dental peri-implant tissues in a site-specific manner.[4,5] A novel immunological point-of-care test was designed to detect the activated form of MMP-8 in GCF for diagnosis and monitoring of periodontal diseases.[6] Two monoclonal antibodies[3,7] were used in a sandwich-based immunoassay system using antibody immuno column for analytical processes (ABICAP) filters and poly-horseradish peroxidase (HRP) labeled conjugate.[6] The corresponding test

Address for correspondence: Dr. Sarvan Kumar Munjal, Dentognostics GmbH, Winzerlaer Str. 2, D-07743 Jena, Germany. Voice: +49-0-3641-508-525; fax: +49-0-3641-508-509.
s.munjal@dentognostics.de

Ann. N.Y. Acad. Sci. 1098: 490–492 (2007).
doi: 10.1196/annals.1384.018

was automatically carried out using an analyzer, which quantitatively measures aMMP-8 in 18–20 min in GCF.

The objective of the study was to evaluate the Dento*Analyzer* test format by comparing its results with two standard methods: in-house enzyme-linked immunosorbent assay (ELISA) and immunofluorometric assay (IFMA). GCF was collected with standard procedures (collection strips[5,8]) from healthy, gingivitis-, and periodontitis-affected teeth. The clinical periodontal status of the teeth (sites) was recorded. After testing an aliquot with the Dento*Analyzer* in the dental office, the rest of the GCF samples were divided, frozen, and transported on ice to the laboratory. One part was stored at −20°C and later tested with in-house ELISA and a second time with the analyzer. The other part was lyophilized and air-mailed to the University of Helsinki, Finland, for testing with IFMA using the same antibodies.[5,7,8] When results from all three tests were available, comparative analysis of the data was performed on samples using linear correlation analysis.

Thus, a preliminary correlation study could be performed with 15 healthy, 7 gingivitis, and 12 periodontitis samples ($n = 34$). Although the concentrations detected by IFMA were slightly higher in comparison to the Dento*Analyzer*, the test values were in good agreement with each other ($r = 0, 94$) (FIG. 1). As well, the in-house ELISA data correlated with the Dento*Analyzer* values ($r = 0, 92$) (FIG. 2).

In summary, the Dento*Analyzer* was proven as a sound platform for detection of aMMP-8 in GCF for diagnosis and monitoring of periodontal diseases. The point-of-care Dento*Analyzer* is a reliable alternative for time-consuming and labor-oriented methods like ELISA or IFMA.

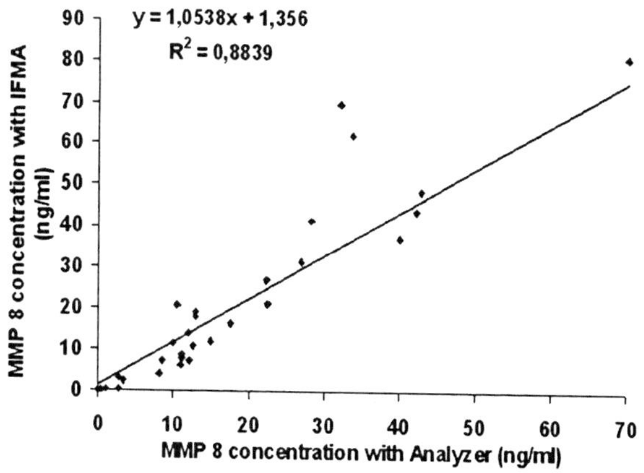

FIGURE 1. Correlation between analyzer and IFMA data.

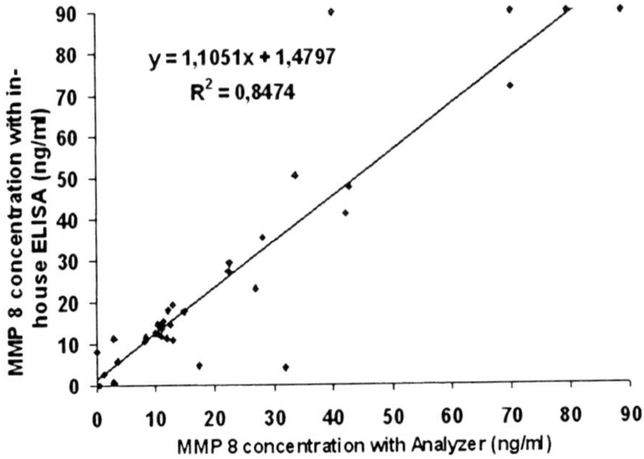

FIGURE 2. Correlation between analyzer and in-house ELISA data.

REFERENCES

1. REYNOLDS, J.J., R.M. HEMBRY & M.C. MEIKLE. 1994. Connective tissue degradation in health and periodontal disease and the roles of matrix metalloproteinases and their natural inhibitors. Adv. Dent. Res. **8:** 312–319.
2. KINANE, D.F. 2000. Regulators of tissue destruction and homeostasis as diagnostic aids in periodontology. Periodontology **24:** 215–225.
3. MA, J., U. KITTI, O. TERONEN, *et al.* 2000. Collagenases in different categories of peri-implant vertical bone loss. J. Dent. Res. **79:** 1870–1873.
4. SORSA, T., L. TJADERHANE, T. SALO. 2004. Matrix metalloproteinases (MMPs) in oral diseases. Oral Dis. **10:** 311–318.
5. SORSA, T., P. MÄNTYLÄ, H. RONKA, *et al.* 1999. Scientific basis of a matrix metalloproteinase-8 specific chair-side test for monitoring periodontal and peri-implant health and disease. Ann. N. Y. Acad. Sci. **878:** 130–140.
6. MUNJAL, S.K., P. MIETHE, L. NETUSCHIL, *et al.* 2007. Immunoassay-based diagnostic point-of-care technology for oral specimen. Ann. N. Y. Acad. Sci. This volume.
7. HANEMAAIJER, R., T. SORSA, Y.T. KONTTINEN, *et al.* 1997. Matrix metalloproteinase-8 is expressed in rheumatoid synovial fibroblasts and endothelial cells. Regulation by tumor necrosis factor-alpha and doxycycline. J. Biol. Chem. **272:** 31504–31509.
8. MÄNTYLÄ, P., M. STENMAN, D.F. KINANE, *et al.* 2003. Gingival crevicular fluid collagenase-2 (MMP-8) test stick for chair-side monitoring of periodontitis. J. Periodontal Res. **38:** 436–439.

Rapid Quantitative Chairside Test for Active MMP-8 in Gingival Crevicular Fluid

First Clinical Data

N. PRESCHER,[a] K. MAIER,[b] S. K. MUNJAL,[b] T. SORSA,[c]
C.-D. BAUERMEISTER,[d] F. STRUCK,[b] AND L. NETUSCHIL[b]

[a]*Dental Clinic, Friedrich Schiller University, Jena, Germany*

[b]*Dentognostics GmbH, Jena, Germany*

[c]*Department of Oral and Maxillofacial Diseases, Institute of Dentistry, University of Helsinki, Helsinki, Finland*

[d]*Laboratories for Oral Microbiology, Düsseldorf, Germany*

ABSTRACT: In a first pilot field study 64 gingival crevicular fluid (GCF) samples were collected from patients of dental practitioners. The dentists (one orthodontist one periodontist, and one general practitioner) were asked to monitor the respective clinical status of the sites of sampling and to collect, if possible, sulcus fluid samples from healthy as well as affected sites from the same patient. The concentration of activated matrix metalloproteinase-8 (aMMP-8) in the GCF was recorded using a set of monoclonal antibodies and a novel Dento*Analyzer*. From all three dental offices the distribution of the aMMP-8 values in GCF showed a congruent pattern, where healthy and periodontitis-affected inflamed sites were clearly disparate.

KEYWORDS: chairside; Dento*Analyzer*; immunoassay; MMP8; periodontitis; point-of-care

In the last decade it became a textbook knowledge that matrix metalloproteinases (MMPs) are responsible for tissue degradation and bone resorption.[1,2] Of these, MMP-8 in GCF[3] is the most prominent collagenase (collagenase 2) associated with periodontitis.[4–6]

It was the aim of this pilot study (1) to collect GCF samples in different dental offices, from healthy, doubtful (e.g., gingivitis), and inflamed (e.g., periodontitis-affected) sites; (2) to assess the concentration of activated

Address for correspondence: Dr. Lutz Netuschil, Dentognostics GmbH, Winzerlaer Str. 2, D-07743 Jena, Germany. Voice: +49-3641-508 503; fax: +49-3641-508 509.
l.netuschil@dentognostics.de

Ann. N.Y. Acad. Sci. 1098: 493–495 (2007).
doi: 10.1196/annals.1384.019

TABLE 1. Values (ng) of activated matrix metalloproteinase-8 (aMMP-8) per milliliter eluate of GCF

Status	Healthy	Doubtful	Periodontitis
n	21	18	25
Median	1.0	6.3	14.3
25% 75% quartile	0.2–2.9	3.5–14.2	10.9–33.7
10–90%	0.0–6.7	2.2–18.2	8.4–51.7
Range	0.0–7.4	0.0–27.1	5.7–64.6

matrix metalloproteinase-8 (aMMP-8) in the GCF with the aid of monoclonal antibodies and a novel Dento*Analyzer*[7,8]; (3) to compare the concentrations of aMMP-8 with the clinical status; and (4) to compare our data with reports in the literature.

GCF was collected with a standard technique using MMP-8 collection strips.[9,10] Concomitantly, the pocket depth was assessed and clinical status was recorded by the attending dentist. The aMMP-8 was immediately eluted from the strips and, after appropriate dilution, quantitatively assessed with the Dento*Analyzer*.[7,8] By this assay the concentration of aMMP-8 was calculated as nanogram of aMMP-8 per milliliter of eluate.

A total of 64 samples were collected, 23 by orthodontists, 18 by periodontists, and 23 in the practice of a general dentist. According to the dentists' judgment, 21 sites were estimated as healthy, 18 as doubtful (e.g., gingivitis; adjacent to brackets, etc.), and 25 as periodontically affected. In all three offices a congruent pattern of aMMP-8 values was recorded: Low values were concerned with healthy periodontium, high values reflected periodontitis-affected sites, and the doubtful cases lay in between. The summarized outcome is presented in TABLE 1.

Healthy sites (median 1.0, range 0.0–7.4 ng aMMP-8/mL eluate) were clearly different from inflammatory sites (median 14.3, range 5.7–64.6). Low values of aMMP-8 in periodontitis cases represented in *all* cases treated, nonactive disease. Whenever healthy and periodontitis sites could be compared in the same patient ($n = 30$, i.e., 15 pairs), without exception the affected sites showed higher aMMP-8 values than the healthy control sites. The aMMP-8 values assessed with the Dento*Analyzer* were found to be similar to results described previously.[9,11,12]

CONCLUSIONS

We conclude that the assessment of aMMP-8 in GCF is a noninvasive method to assess and monitor the pathophysiological status of the periodontium. The determination of aMMP-8 with the Dento*Analyzer* enables the dentist to distinguish between the healthy and the periodontitis-affected sites within 20 min.

REFERENCES

1. REYNOLDS, J.J., R.M. HEMBRY & M.C. MEIKLE. 1994. Connective tissue degradation in health and periodontal disease and the roles of matrix metalloproteinases and their natural inhibitors. Adv. Dent. Res. **8:** 312–319.
2. KINANE, D.F. 2000. Regulators of tissue destruction and homeostasis as diagnostic aids in periodontology. Periodontology **24:** 215–225.
3. SUOMALAINEN, K., T. SORSA, L. SAXEN, *et al.* 1991. Collagenase activity in gingival crevicular fluid of patients with juvenile periodontitis. Oral Microbiol. Immunol. **6:** 24–29.
4. BIRKEDAL-HANSEN, H. 1993. Role of matrix metalloproteinases in human periodontal diseases. J. Periodontol. **64:** 474–484.
5. MCCULLOCH C.A. 1994. Host enzymes in gingival crevicular fluid as diagnostic indicators of periodontitis. J. Clin. Periodontol. **21:** 497–506.
6. SORSA, T., L. TJÄDERHANE & T. SALO. 2004. Matrix metalloproteinases (MMPs) in oral diseases. Oral Dis. **10:** 311–318.
7. MUNJAL, S.K., P. MIETHE, L. NETUSCHIL, *et al.* 2007. Immunoassay-based diagnostic point-of-care technology for oral specimen. Ann. N. Y. Acad. Sci. This volume.
8. MUNJAL, S.K., N. PRESCHER, F. STRUCK, *et al.* 2007. Evaluation of immunoassay-based MMP-8 assessment in gingival crevicular fluid on a point-of-care platform. Ann. N. Y. Acad. Sci. This volume.
9. MÄNTYLÄ, P., M. STENMAN, D.F. KINANE, *et al.* 2003. Gingival crevicular fluid collagenase-8 (MMP-8) test stick for chair-side monitoring of periodontitis. J. Periodontol. Res. **38:** 436–439.
10. MÄNTYLÄ, P., M. STENMAN, D. KINANE, *et al.* 2006. Monitoring periodontal disease status in smokers and non-smokers using a gingival crevicular fluid matrix metalloproteinase-8 (MMP-8) specific chair-side test. J. Periodontal Res. **41:** 503–512.
11. CHEN, H.Y., S.W. COX, B.M. ELEY, *et al.* 2000. Matrix metalloproteinase-8 levels and elastase activities in gingival crevicular fluid from chronic adult periodontitis patients. J. Clin. Periodontol. **27:** 366–369.
12. KINANE, D.F., I.B. DARBY, S. SAID, *et al.* 2003. Changes in gingival crevicular fluid matrix metalloproteinase-8 levels during periodontal treatment and maintenance. J. Periodont. Res. **38:** 400–404.

Salivary Biomarkers Associated with Alveolar Bone Loss

F.A. SCANNAPIECO,[a] PBY NG,[a] K. HOVEY,[b] E. HAUSMANN,[a] A. HUTSON,[c] AND J. WACTAWSKI-WENDE[b]

[a]*Departments of Oral Biology, School of Dental Medicine, and Departments of* [b]*Social and Preventive Medicine and* [c]*Biostatistics, School of Public Health and Health Professions, State University at Buffalo, Buffalo, New York*

ABSTRACT: A longitudinal case–control study was performed to measure the association of salivary biomarkers with alveolar bone loss from a sub-sample of 1,256 post-menopausal women enrolled in the Buffalo Women's Health Initiative. From this cohort, 40 subjects with significant alveolar bone loss over a 5-year period were compared to 40 age-matched control subjects having no alveolar bone loss. Several biomarkers were quantitated in saliva collected at baseline by immunoassay. A positive association was noted between alveolar bone loss and salivary concentrations of hepatocyte growth factor, and interleukin-1 beta, while a negative association was noted for alveolar bone loss and salivary osteonectin. This study provides preliminary evidence that several salivary biomarkers measured at baseline may serve to predict future alveolar bone loss.

KEYWORDS: saliva; cytokines; periodontitis

OBJECTIVES

The goal of this longitudinal, case–control study was to evaluate the association, if any, between alveolar bone loss and the presence of host-derived bone resorptive factors and/or markers of bone turnover in saliva collected at baseline from a cohort of women followed for 5 years.

METHODS

Subjects were selected from 1,256 postmenopausal women who enrolled in the Buffalo Women's Health Initiative "Risk Factors for Osteoporosis and

Address for correspondence: F.A. Scannapieco, Departments of Oral Biology, School of Dental Medicine, State University at Buffalo, Buffalo, NY 14214. Voice: 716-829-3373; fax: 716-829-3942. fas1@buffalo.edu

Ann. N.Y. Acad. Sci. 1098: 496–497 (2007). © 2007 New York Academy of Sciences.
doi: 10.1196/annals.1384.034

Oral Bone Loss in Postmenopausal Women" study. From these, 40 subjects with the most significant alveolar bone loss over a 5-year follow-up period, as judged from intra-oral X-rays, and a control group of 40 age-matched subjects having no alveolar bone loss, were studied. Whole saliva, collected from each subject at baseline, was assessed by immunobioassay for IFN-γ, IL-1-β, TNF-α, hepatocyte growth factor, IL-6, IL-4, IL-8, osteonectin, and ICTP.

RESULTS

Conditional logistic regression analysis compared dichotomous saliva biomarker data cut at the median to alveolar bone loss in matched cases and controls. A positive association was noted between alveolar bone loss and salivary concentrations of hepatocyte growth factor [OR = 7.27, 95% CI 2.09–25.26; $P = 0.0018$] and IL-1β 4.06 [OR = 4.06, 95% CI 1.19–13.86, $P = 0.0254$], after adjustment for age and baseline number of teeth. A negative association was noted for alveolar bone loss and salivary concentration of osteonectin [OR = 0.30 95% CI 0.11–0.83, $P = 0.0198$], adjusted for age and baseline mean alveolar crestal height.

CONCLUSIONS

This study provides preliminary evidence that the salivary biomarkers HGF, IL-1β, and osteonectin measured at baseline may serve to predict future alveolar bone loss.

ACKNOWLEDGEMENTS

This work was supported by USPHS-NIH Grants R21 DE15854, N01 WH32122, and R01 DE13505.

Salivary Protein/Peptide Profiling with SELDI-TOF-MS

RAYMOND SCHIPPER,[a,b] ARNOUD LOOF,[d] JOLAN DE GROOT,[a,b]
LUCIEN HARTHOORN,[a,c] WAANDER VAN HEERDE,[d]
AND ERIC DRANSFIELD[a,c]

[a] *Wageningen Center for Food Sciences, Wageningen, the Netherlands*

[b] *Wageningen University, Food Chemistry Group, Wageningen, the Netherlands*

[c] *Department of Consumer and Market Insight, Agrotechnology and Food Innovations, Wageningen, the Netherlands*

[d] *University Medical Centre St. Radboud, Nijmegen, the Netherlands*

ABSTRACT: **In this study, large-scale profiling of salivary proteins and peptides ranging from 2 to 100 kDa was demonstrated using surface-enhanced laser desorption/ionization–time of flight–mass spectrometry (SELDI-TOF-MS). Results show that chip surface type and sample type critically affect the amount and composition of detected salivary proteins. Delayed processing time resulted in both increase and decrease of peak numbers consistent with proteolysis. SELDI-TOF-MS profiles also changed, depending on storage temperature, although sample processing by centrifugation and numbers of freeze–thaw cycles had a minimal impact. In conclusion, SELDI-TOF-MS offers a simple, rapid, high-throughput technique for profiling low-mass (<10 kDa) saliva proteins/peptides. We wish to use this technique to gain insight into the human saliva proteome composition and its changes over time in response to food consumption.**

KEYWORDS: **saliva; proteomics; SELDI-TOF-MS; preanalytical procedures**

INTRODUCTION

State-of-the art proteomic methods are currently being applied to the analysis of salivary peptides/proteins[1] as part of an increasing interest in exploring saliva as a diagnostic fluid.[2] Proteomics has relied heavily on two-dimensional gel electrophoresis for the separation and visualization of proteins, but this approach is insensitive to small molecular weight proteins, hydrophobic proteins,

Address for correspondence: Raymond Schipper, Ph.D., Wageningen Centre for Food Sciences/TI Food and Nutrition, Laboratory of Food Chemistry, Wageningen University, P.O. Box 8129, 6700 EV, the Netherlands. Voice: +31-317-482101; fax: +31-317-484893.
raymond.schipper@wur.nl

Ann. N.Y. Acad. Sci. 1098: 498–503 (2007).
doi: 10.1196/annals.1384.010

and low-abundance proteins. Other reports have shown that these molecules, for example, histatins, cystatins, and defensins, have important antimicrobial functions within the oral cavity.[7,8] We have explored surface-enhanced laser desorption/ionization–time of flight–mass spectrometry (SELDI-TOF-MS) as a complementary, rapid, and high-throughput proteomic approach to profile salivary proteins and peptides, and in addition we have analyzed the effect of preanalytical variables, including oral fluid sample type, processing centrifugation speed, time required for processing, storage temperature, and number of freeze–thaw cycles.

EXPERIMENTAL PROCEDURES

Whole saliva (oral fluid) was obtained from healthy nonsmoking subjects in the morning at least 2 h after eating and after rinsing the mouth with water. For nonstimulated samples, a cotton wool roll (Salivette®, Sarstedt B.V., Etten-Leur, the Netherlands) was placed in the mouth and, when saturated, the roll was transferred to a Salivette tube. For a stimulated sample, the subject chewed for 2 min on a piece of parafilm before drooling into a sterile tube. Both unstimulated and stimulated specimens were centrifuged for 5 min at 1000 × *g* or at 10,000 × *g* at 4°C. Samples were divided into 250-μL aliquots and stored either on ice at −20°C or at −80°C. Aliquots were taken after 0 or 3 h from the samples stored on ice and processed for SELDI-TOF-MS analysis. In order to study the effect of freezing and thawing cycles, frozen aliquots were thawed either (*a*) immediately, (*b*) after 1 month, or (*c*) after 6 months, and then analyzed by SELDI-TOF-MS.

Four chip surface chemistries were examined: strong anion exchanger (Q10), weak cation exchanger (CM10), reversed phase (H4), and immobilized metal (copper) affinity capture array (IMAC-Cu). Chips were processed according to the manufacturer's instructions (Ciphergen Biosystems, Fremont, CA, USA).

RESULTS AND DISCUSSION

All four surface chemistries effectively bound salivary proteins and peptides (TABLE 1) with different mass selectivity. Chips with anionic (CM10)-treated

TABLE 1. Comparison of the different chip surface types with respect to the number of peaks

Chip surface type	2–10 kDa	10–20 kDa	20–100 kDa
CM10	39	17	14
Q10	80	21	7
H4	81	15	3
IMAC-Cu	78	19	19

surfaces were effective in binding salivary proteins in the range from 6,000 to 16,000 *m/z*, while the cationic (Q10) and hydrophobic (H4) chips characterized saliva peptides and proteins in the lowest molecular (from 1,000 to 6,000 *m/z*) range. The metal affinity-binding (IMAC-Cu) chip produced the largest number of resolved peaks in a wide molecular range, with the majority of masses in the weight range between 1 and 10 kDa. These results are consistent with other reports that show whole saliva is rich in peptides in the range of 1 to 6 kDa.[3–6] In this study, masses with prominent peak intensities were found between 1,000 and 10,000 *m/z*, which are consistent with approximately identical masses of salivary peptides known as cystatins, histatins, defensins, and calgranulins. Of particular interest is calgranulin A, which responds to sensory stimulation by specific taste stimuli.[9] Other protein candidates for the detected peaks are lysozyme, fatty acid-binding protein, histatins, kallikrein, lysozyme, proloine-rich proteins, superoxide dismutase, and thymosin β-4, which also have been identified by other proteomic techniques.[10–18]

A screening advantage of SELDI-TOF-MS over MALDI-TOF-MS is its ability to perform a miniaturized on-chip prefractionation of saliva samples. Optimization of buffer systems can potentially increase resolution and the results help in the design of purification schemes to isolate specific salivary proteins with classical chromatographic methods.

Considerable overlap was observed for the resolved peaks (*m/z*) between the different array surfaces. The different surface chemistries provided complementary spectra, with considerable overlap between the observed peaks, but each surface contributed unique peaks. The IMAC-Cu chip was deemed the optimal surface since for our purposes it produced the largest number of peaks over a wide molecular range. However, to reveal specific salivary biomarkers, a combination of chip array surfaces would provide the most comprehensive mass spectra.

The usefulness of saliva analysis by SELDI-TOF-MS depends on a compatible preanalytical protocol for saliva. Once the saliva sample is collected, it is important that handling procedures do not affect the composition of the sample. The effects of the different pretreatments on saliva protein/peptide profiling are summarized in TABLE 2. Experiments examining the effect of delayed processing time indicate that new peptide fragments appear from 1 up to 3 h post-saliva donation, and quantitative analyses indicate relative intensity of other proteins and peptides changing with time. The addition of protease inhibitors did not completely block the proteolysis of some proteins. We observed that storage of saliva at −80°C maintained the integrity of the sample for SELDI-TOF-MS testing.

The goal for this study was to develop a protocol to allow analysis of a large number of samples necessary to reveal unique biomarkers. Examples of the use of SELDI-TOF-MS for detecting salivary biomarkers have recently been published.[19,20] Ryu and co-workers compared parotid salivas of subjects suffering

TABLE 2. Effect of pretreatments on SELDI-TOF-MS profiles of saliva samples

	Centrifugation		Storage conditions			
			+4°C		Frozen	
Sample type	1,000 × *g*	10,000 × *g*	0 h	3 h	–20°C	–80°C
Unstimulated						
High molecular weight	+	+	+	–/+	–/+	+
Low molecular weight	–/+	+	+	–/+	–	+
			+			
Stimulated						+
High molecular weight	+	+	+	–/+	+	+
Low molecular weight	+	+	+	+	–/+	+

+, no effect on profile; –/+, increase and/or decrease of certain protein/peptide mass spectra; –, destabilization of mass spectra.

from Sjögren's syndrome (SS) with non-SS subjects and found 10 biomarkers, 3 of which have previously not been associated with SS. An explorative study comparing protein profiles in salivary samples from a group of breast cancer patients and known healthy controls is described by Streckfus *et al.*[20] These authors identified five mass peaks that were increased more than twofold in the cancer patients.

Our research group is focused on gaining insight into the human saliva proteome composition and its changes over time in response to food consumption (Harthoorn *et al.*, submitted for publication). A lunch-induced hunger-satiety shift trial of 4.75 hours in 18 subjects (9 men and 9 women) collected whole saliva every 15 min. Saliva was analyzed for peptide and protein masses by SELDI-TOF-MS, and relative mass intensities of these spectra were related to subjectively rated scores for satiety across time and body mass index (BMI). Results show that 29 peaks showed a strong correlation to subjective satiety ratings, and 7 peaks showed significant association with body mass. Such noninvasive profiling allows the identification of novel peptide biomarkers for perceived satiety.

CONCLUSION

The goal is to characterize salivary composition for use as a diagnostic tool to monitor the physiological, health, and/or disease status of individuals. In particular we wish to associate specific salivary biomarkers with satiety. Protein microarrays, such as SELDI-TOF-MS spectra, allow a significant number of samples to be processed and compared in a relatively short time with very little sample preparation or sophisticated preanalysis chromatography. We anticipate that this proteomic technique will be a valuable tool in projects ranging from identification of diagnostic biomarkers for a variety of diseases or description of whole-body physiological changes as well as to the study of modification and biomolecular interactions of salivary peptides and proteins.

REFERENCES

1. AMADO, F.M.L. *et al.* 2005. Analysis of the human saliva proteome. Exp. Rev. Proteom. **2:** 521–539.
2. STRECKFUS, C.F. & L.R. BIGLER. 2002. Saliva as a diagnostic fluid. Oral Dis. **8:** 69–76.
3. CASTAGNOLA, M. *et al.* 2004. A cascade of 24 histatins (histatin 3 fragments) in human saliva: suggestions for a pre-secretory sequential cleavage pathway. J. Biol. Chem. **279:** 41436–41443.
4. DISTLER, W. & A. KRONCKE. 1987. Biologically-active, low-molecular-weight peptides in human saliva. J. Dent. Res. **66:** 603–604.
5. LUPI, A. *et al.* 2003. Identification of the human salivary cystatin complex by the coupling of high-performance liquid chromatography and ion-trap mass spectrometry. Proteomics **3:** 461–467.
6. PERINPANAYAGAM, H.E. *et al.* 1995. Characterization of low-molecular-weight peptides in human parotid saliva. J. Dent. Res. **74:** 345–350.
7. VAN NIEUW AMERONGEN, A., J.G.M. BOLSCHER & E.C.I. VEERMAN. 2004. Salivary proteins: protective and diagnostic value in cariology? Caries Res. **38:** 247–253.
8. TROXLER, R.F. *et al.* 1990. Structural relationship between human salivary histatins. J. Dent. Res. **69:** 2–6.
9. NEYRAUD, E. *et al.* 2006. Proteomic analysis of human whole and parotid salivas following stimulation by different tastes. J. Proteome Res. **5:** 2474–2480.
10. GUO, T. *et al.* 2006. Characterization of the human salivary proteome by capillary isoelectric focusing/nanoreversed-phase liquid chromatography coupled with ESI-tandem MS. J. Proteome Res. **5:** 1469–1478.
11. HARDT, M. *et al.* 2005. Toward defining the human parotid gland salivary proteome and peptidome: identification and characterization using 2D SDS-PAGE, ultrafiltration, HPLC, and mass spectrometry. Biochemistry **44:** 2885–2899.
12. HIRTZ, C. *et al.* 2005. Complexity of the human whole saliva proteome. J. Physiol. Biochem. **61:** 469–480.
13. HU, S. *et al.* 2005. Large-scale identification of proteins in human salivary proteome by liquid chromatography/mass spectrometry and two-dimensional gel electrophoresis-mass spectrometry. Proteomics **5:** 1714–1728.
14. HUANG, C.-M. 2004. Comparative proteomic analysis of human whole saliva. Arch. Oral Biol. **49:** 951–962.
15. VITORINO, R. *et al.* 2004. Identification of human whole saliva protein components using proteomics. Proteomics **4:** 1109–1115.
16. WALZ, A. *et al.* 2006. Proteome analysis of glandular parotid and submandibular-sublingual saliva in comparison to whole human saliva by two-dimensional gel electrophoresis. Proteomics **6:** 1631–1639.
17. WILMARTH, P.A. *et al.* 2004. Two-dimensional liquid chromatography study of the human whole saliva proteome. J. Proteome Res. **3:** 1017–1023.
18. YAO, Y. *et al.* 2003. Identification of protein components in human acquired enamel pellicle and whole saliva using novel proteomics approaches. J. Biol. Chem. **278:** 5300–5308.

19. Ryu, O.H. *et al.* 2006. Identification of parotid salivary biomarkers in Sjögren's syndrome by surface-enhanced laser desorption/ionization time-of-flight mass spectrometry and two-dimensional difference gel electrophoresis. Rheumatology **45:** 1077–1086.
20. Streckfus, C.F., L.R. Bigler & M. Zwick. 2006. The use of surface-enhanced laser desorption/ionization time-of-flight mass spectrometry to detect putative breast cancer markers in saliva: a feasibility study. J. Oral Pathol. Med. **35:** 292–300.

Acquired Enamel Pellicle and Its Potential Role in Oral Diagnostics

W. L. SIQUEIRA, E. J. HELMERHORST, W. ZHANG, E. SALIH, AND F. G. OPPENHEIM

Department of Periodontology and Oral Biology, Goldman School of Dental Medicine, Boston University, Boston, Massachusetts, USA

ABSTRACT: The acquired enamel pellicle (AEP) is a protein film with unique composition and properties, which is formed by the selective adsorption of a variety of oral fluid–derived proteins onto tooth enamel surfaces. Since events leading to caries and periodontal disease occur in close proximity to the tooth surface, pellicle constituents are likely to contain biomarkers valuable for diagnostic applications. Despite the importance of this oral structure, progress in understanding its formation and composition has been slow because of difficulties in efficient pellicle collection methods and limitations of biochemical techniques for the characterization of microgram amounts of proteins/peptides. Recent developments in both pellicle collection methods and nanoscale sensing technologies have brought the exploitation of pellicle analysis into the realm of point-of-care oral diagnostics.

KEYWORDS: acquired enamel pellicle; saliva; protein; proteomics; mass spectrometry

The acquired enamel pellicle (AEP) is an acellular biofilm formed by proteins, carbohydrates, and lipids that adsorb onto the enamel surface.[1] Nasmyth was first to describe a tooth integument and postulated that it was of embryologic origin.[2] In 1926, Chase[3] also described tooth integuments that were present not only on enamel surfaces, but also on surfaces of amalgam restorations and dentures. This finding indicated that the structures he described could not be of embryologic origin. Further explorations of these integuments by Frank in 1949 led to the distinction between two types of enamel membranes.[4] The first membrane structure is formed prior to tooth eruption and remains intact only transiently after eruption in the form of ameloblastic cellular remnants. This primary enamel pellicle is quickly succeeded by a secondary pellicle, which consists mainly of proteins that adsorb to enamel after exposure to the oral

Address for correspondence: Frank G. Oppenheim, D.M.D., Ph.D., Department of Periodontology and Oral Biology, Goldman School of Dental Medicine, Boston University, 700 Albany Street, Suite W-201, Boston, MA 02118. Voice: 617-638-4756; fax: 617-638-4924.
fropp@bu.edu

Ann. N.Y. Acad. Sci. 1098: 504–509 (2007). © 2007 New York Academy of Sciences.
doi: 10.1196/annals.1384.023

environment. The term "acquired enamel pellicle" was first proposed for this latter structure in a review of nomenclature of enamel surface integuments by Dawes and co-workers in 1963.[5] This *in vivo*–formed AEP has been the focal point of many investigations as its potential importance in maintaining the integrity of tooth surfaces had been suspected. It has become widely accepted that AEP is defined as the protein layer that is formed on enamel surfaces of teeth exposed to the oral environment. A full understanding of the formation and function of *in vivo* AEP comprised approaches to gain insights into protein adsorption, mineralization/demineralization, and bacterial adherence as well as antimicrobial activity. Despite these efforts, serious limitations have hampered progress in this area of oral biology research, limitations that are predominantly related to the small amounts of material (μg levels) that can be obtained from the enamel surfaces. Recent developments of sensitive proteomics methodologies have opened new avenues for the characterization of very small amounts of organic material, including proteins and peptides. Furthermore, this development pointed to the necessity of adequate, consistent, and reproducible AEP harvesting techniques. In view of the value of AEP analysis as a potentially important adjunct in salivary diagnostics, it is crucial that the method used to collect pellicle provide an optimal yield and ideally remove all organic material present on the tooth surface.

An efficient collection method is also vital for obtaining AEP material without contamination by any other elements present in the oral cavity. Several different collection methods of the AEP have been described over the past 40 years, each with its advantages and disadvantages. It is therefore not surprising that the heterogeneity of collection procedures yielded data reflecting qualitative and quantitative differences in AEP compositions. Almost all preferred collection methods of AEP material use the buccal/labial tooth surface for sampling on account of easy access and low risk of saliva contamination during the harvesting procedure. Originally, the *in vivo* AEP was obtained from extracted teeth, placing them in 5% HCl for 5 min before scaling them off with a small glass knife.[6] The AEP obtained by this method displayed an amino acid composition with high levels of glutamic acid and alanine, negligible amounts of methionine and cysteine, no hydroxyproline or hydroxylysine, and significant levels of hexosamines. A significant contribution to this field of research was made by Mayhall,[7] who used extracted crowns of teeth, and remounted them in a palatal appliance which was worn by subjects for 1 h. The AEP formed on these specimens was removed by immersion in 2% HCl. The composition of these pellicles showed appreciable amounts of glutamic acid, serine, and glycine, accounting for almost one-half of the total amino acid pool, but proline was significantly lower than what was found in salivary secretions. Subsequently, Sonju and Rolla[8] introduced new procedures allowing the collection of truly *in vivo*-developed AEP. These investigators performed first a thorough pumicing of natural tooth surfaces of volunteers to remove pellicle. Then the subjects were asked to refrain from any food consumption for 2 h.

After this 2-h period their teeth were isolated with cotton rolls, rinsed with water, and air-dried, and AEP was harvested. The latter procedure consisted of tooth scaling with a Pasteur pipette tip attached to a suction device. The pipette contained a glass wool plug to retain pellicle material loosened by the scaling action. Data with this *in vivo* pellicle collection showed that the amount of AEP increased over the first 1.5 h and was adequate in amount for compositional analysis. In 1982, Eggen and Rolla[9] modified their mechanical technique by adding a chemical solubilization step using 0.2 M EDTA. Employing this method, the authors discovered that the main AEP components are anionic in nature and eluted between albumin and lysozyme on a sepharose gel filtration column. Their compositional results also showed abundant amounts of serine, glycine, and glutamic acid. The fact that the amino acid compositions reported by various laboratories are very similar indicates that AEP is a universal biological entity of the oral cavity. Also noteworthy is the consistently low level of proline found in AEP preparations when compared to those of saliva, providing strong evidence that AEP formation is a specific and nonrandom process of protein adsorption.[10,11] Embery *et al.*[12] applied a two-stage solubilization step using first 2 M $CaCl_2$ and then 5% EDTA, each followed by scaling. The first chemical extraction yielded mostly high-molecular-weight glycoproteins while the second extraction contained mostly small-molecular-weight components. Electrophoretic analysis of the sample revealed the presence of phosphoproteins of low molecular weight that seemed to disappear as the pellicle was allowed to mature. In 1990 Rykke *et al.*[13] further modified their mechanical/chemical AEP collection method using a different trapping technique. Instead of glass wool, polyethylene tubes filled with distilled water were used to trap pellicle material. The amino acid profiles obtained from the analyses of the saliva samples were different from the pellicle profiles, illustrating the selective nature of the AEP formation. More recently, the use of scalers to remove pellicle was replaced by the application of a variety of membranes. The first membrane type used consisted of polyvinylidene difluoride (PVDF).[14] This study focused on the composition of *in vivo* AEP from permanent and deciduous teeth. Interestingly, the results showed distinct differences in amino acid composition, rate of formation, and ultrastructural appearance between AEP obtained from primary and permanent teeth. Further refinements of this method have been used in which teeth were scrubbed with a dental mini-sponge applicator soaked in 2% sodium dodecyl sulfate to solubilize and collect AEP.[15] In contrast to most studies, these investigations showed both inter- and intraindividual variations and very small differences between saliva and pellicle compositions.

Our laboratory has developed a mechanicochemical harvesting technique using a distinctly hydrophilic PVDF membrane soaked in 0.5 M sodium bicarbonate.[10] Pellicle was collected by rubbing the membrane against the coronal portion of the tooth surface, avoiding contact with the gingival margin. This technique in conjunction with proteomic analyses has expanded our knowledge regarding AEP composition.[10,16] To streamline and simplify the AEP recovery we recently showed that the buffer-soaked PVDF membrane can be replaced by dry electrode wick papers, which eliminates long dialysis procedures to remove the buffer salts. This method allows the direct transfer of proteins from the wick paper in an electric field into a polyacrylamide stacking gel followed by electrophoretic separation.[17] The advantage of the method using dry wick papers for AEP collection is that it reduces sample manipulation and eliminates the need for centrifugation, vortexing, and dialysis steps between AEP collection and analysis. This markedly reduces loss of protein, improves reproducibility, and even provides adequate amounts of material to determine the AEP proteome of a single subject. So far, this collection approach has led to the identification of as many as 130 different proteins/peptides associated with the AEP.[17] The AEP sampling technique has been simplified to the extent

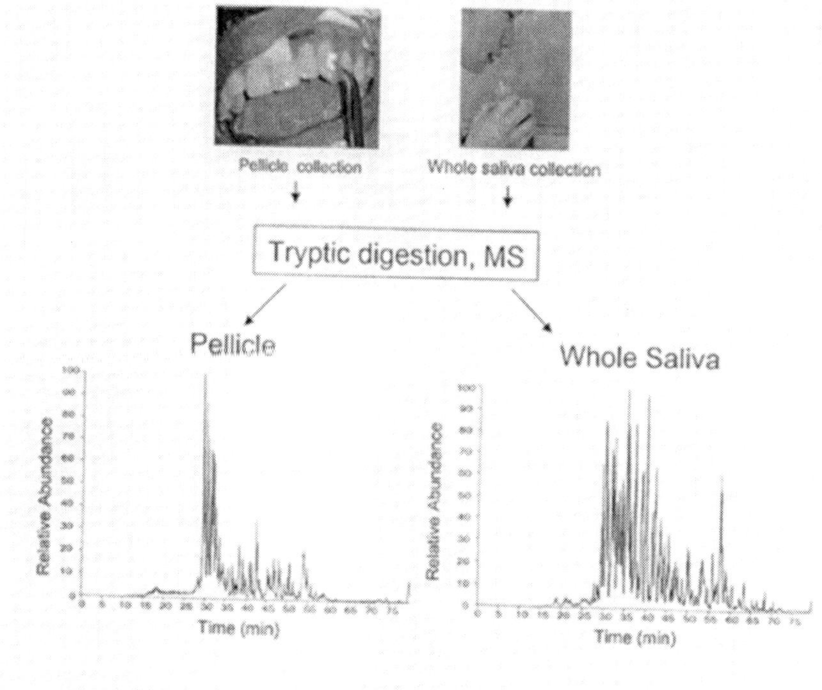

FIGURE 1. Mass spectrometric analysis of AEP and whole saliva collected from the same subjects. Note simplified collection of clinical specimens suitable for point-of-care sampling (*top*). The total ion current profiles obtained with a ProtoemEx LTQ mass spectrometer (Finnigan, San Jose, CA, USA) show distinct differences between the AEP and whole saliva samples (*bottom*).

that it could easily be used for point-of-care diagnostics, provided miniaturized multi-analyte sensing techniques can be developed. Such sensing techniques could produce "fingerprints" of AEP samples differing between health and disease states. The principle of such a fingerprinting approach is shown in FIGURE 1, which demonstrates significant differences between two types of oral specimens.

Progress made recently with proteomics allowing protein/peptide characterizations down to femtomole levels in combination with novel AEP collection approaches has generated new interest in exploring this important area of oral biology research. It is abundantly clear that there is a close relationship between saliva and AEP compositions. This relationship is important to investigate in order to understand functional parameters of oral health and disease. Both saliva as well as pellicle could reflect changes in their constituents provoked by different oral conditions, such as caries or periodontal disease. The current evidence supports the concept that with efficient AEP collection methods, and taking advantage of novel, sensitive, and sophisticated analytical tools such as mass spectrometry, a comprehensive understanding of *in vivo* AEP composition can be achieved. These timely developments are promising in complementing salivary diagnostic avenues.

ACKNOWLEDGMENTS

This study was supported by NIH/NIDCR Grants DE05672, DE07652, and DE14950.

REFERENCES

1. HANNIG, M. & A. JOINER. 2006. The structure, function and properties of the acquired pellicle. Monogr. Oral Sci. **19:** 29–64.
2. NASMYTH, A. 1839. On the structure, physiology, and pathology of the persistent capsular investments and pulp of the tooth. Trans. R. Med. Chir. Soc. Lond. **22:** 310–328.
3. CHASE, W.B. 1926. The origin, structure, and duration of Nasmyth's membrane. Anat. Rec. **33:** 357–376.
4. FRANK, R. 1949. Researchers sur la membrane de Nasmyth. Compt. Rend. Seances Soci. Biol. **143:** 1243–1245.
5. DAWES, C., G.N. JENKINS & C.H. TONGE. 1963. The nomenclature of the integuments of the enamel surface of the teeth. Br. Dent. J. **115:** 65–68.
6. ARMSTRONG, W.G. 1966. Amino-acid composition of the acquired pellicle of human tooth enamel. Nature **210:** 197–198.
7. MAYHALL, C.W. 1970. Concerning the composition and source of the acquired enamel pellicle of human teeth. Arch. Oral Biol. **15:** 1327–1341.
8. SONJU, T. & G. ROLLA. 1973. Chemical analysis of the acquired pellicle formed in two hours on cleaned human teeth *in vivo*. Rate of formation and amino acid analysis. Caries Res. **7:** 30–38.

9. EGGEN, K.H. & G. ROLLA. 1982. Gel filtration, ion exchange chromatography and chemical analysis of macromolecules present in acquired enamel pellicle (2-hour-pellicle). Scand. J. Dent. Res. **90:** 182–188.
10. YAO, Y., J. GROGAN, M. ZEHNDER, *et al.* 2001. Compositional analysis of human acquired enamel pellicle by mass spectrometry. Arch. Oral Biol. **46:** 293–303.
11. AL-HASHIMI, I. & M.J. LEVINE. 1989. Characterization of *in vivo* salivary-derived enamel pellicle. Arch. Oral Biol. **34:** 289–295.
12. EMBERY, G., T.G. HEANEY & J.B. STANBURY. 1986. Studies on the organic polyanionic constituents of human acquired dental pellicle. Arch. Oral Biol. **31:** 623–625.
13. RYKKE, M., T. SONJU & G. ROLLA. 1990. Interindividual and longitudinal studies of amino acid composition of pellicle collected *in vivo*. Scand. J. Dent. Res. **98:** 129–134.
14. SONJU CLASEN, A.B., M. HANNIG, K. SKJORLAND, *et al.* 1997. Analytical and ultrastructural studies of pellicle on primary teeth. Acta Odontol. Scand. **55:** 339–343.
15. CARLEN, A., A.C. BORJESSON, K. NIKDEL, *et al.* 1998. Composition of pellicles formed *in vivo* on tooth surfaces in different parts of the dentition, and *in vitro* on hydroxyapatite. Caries Res. **32:** 447–455.
16. YAO, Y., E.A. BERG, C.E. COSTELLO, *et al.* 2003. Identification of protein components in human acquired enamel pellicle and whole saliva using novel proteomics approaches. J. Biol. Chem. **278:** 5300–5308.
17. SIQUEIRA, W.L., W. ZHANG, E.J. HELMERHORST, *et al.* Identification of protein components in *in vivo* human acquired enamel pellicle using LC-MS/MS. Submitted for publication.

Gender-Specific Differences in Salivary Biomarker Responses to Acute Psychological Stress

NORIYASU TAKAI,[a] MASAKI YAMAGUCHI,[b] TOSHIAKI ARAGAKI,[a] KENJI ETO,[a] KENJI UCHIHASHI,[a] AND YASUO NISHIKAWA[a]

[a]*Department of Physiology, Osaka Dental University, Osaka, Japan*

[b]*Faculty of Engineering, University of Toyama, Toyama, Japan*

ABSTRACT: The stress response is regulated by two primary neuroendocrine systems, the hypothalamus-pituitary-adrenocortical (HPA) and sympathetic adrenomedullary (SAM) systems. This study investigated gender differences in the activities of these two systems in response to acute psychological stress. Subjects were categorized according to their score in Spielberger's Trait Anxiety Inventory (STAI), which assesses the predisposition to personal anxiety. High (STAI score $\geq$55)- and low (STAI score $\leq$45)-anxiety groups were selected. A video of corneal surgery was served as the stressor for 15 min. Salivary cortisol and amylase levels were used as indices of the HPA and SAM activities, respectively. β-endorphin was also assayed as a possible index of HPA activity. There were no differences in the resting salivary parameters among the groups. As expected, cortisol and amylase levels were significantly increased in all groups after the stressful video viewing. There were no gender differences in amylase levels in either the high- or low-anxiety groups. However, cortisol levels in highly anxious females were significantly lower than those in highly anxious males. Our findings show that highly anxious females exhibited lower cortisol release than highly anxious males, suggesting that high trait anxiety in females may be associated with an inability to respond with sufficient activation of HPA under acute psychological stress.

KEYWORDS: gender difference; psychological stress; saliva

INTRODUCTION

Two primary neuroendocrine systems have been of specific interest in the study of human stress; the hypothalamus-pituitary-adrenocortical (HPA)

Address for correspondence: Noriyasu Takai, Department of Physiology, Osaka Dental University, 8-1, Kuzuha-hanazono-cho, Hirakata, Osaka 573-1121, Japan. Voice: +81-72-864-3054; fax: +81-72-864-3154.
takai@cc.osaka-dent.ac.jp

Ann. N.Y. Acad. Sci. 1098: 510–515 (2007).
doi: 10.1196/annals.1384.014

system, with the secretion of cortisol, and the sympathetic adrenomedullary (SAM) system, with the secretion of catecholamine. In the HPA system, cortisol secretion is regulated by the adrenocorticotropic hormone (ACTH) from the pituitary gland. Salivary cortisol levels are closely correlated to blood cortisol levels and therefore reliably reflect HPA activity.[1] Many reports have shown that various kinds of psychological stress activate the HPA system and consequently induce significant increases in salivary cortisol levels.[1] In the SAM system, direct measurements of salivary catecholamine do not reflect SAM activity.[2] Alpha-amylase is one of the major salivary enzymes in humans, and is secreted from the salivary glands in response to sympathetic stimuli.[3] Currently, it is considered that measurement of this salivary enzyme is a useful tool for evaluating activation of the SAM system.[4,5]

Previous studies suggest that gender influences the hormonal responses to stress via the HPA axis. In animal studies, ACTH and corticosterone levels in response to stress have been shown to be consistently greater in females compared with males.[6,7] However, in human studies, no such clear-cut gender differences have been established.

Our study was designed to determine whether there are gender differences in the activation patterns of HPA and SAM in response to acute psychological stress. Saliva sampling has the advantage that it is noninvasive, making multiple sampling easy and stress-free. To avoid the confounding effects of additional stress induced by blood or urine collection, salivary cortisol and amylase were assayed as indices of the HPA and SAM system, respectively.

METHODS

A total of 83 healthy volunteers, ranging in age from 20 to 27 years (mean age: 23.3 [males], and 23.8 years [females]), participated in the study. Questionnaires administered prior to the experiment indicated that no volunteers had a physical or mental illness, were pregnant, or were taking oral contraceptives. The aims of the study and the procedures involved were explained to the subjects and written consent was obtained. The ethical research committee of Osaka Dental University approved the study protocol.

Spielberger's State-Trait Anxiety Inventory (STAI) was used to assess personal anxiety. High (STAI score $\geq$ 55)- and low (STAI score $\leq$ 45)-anxiety subjects were selected. The high-anxiety group consisted of 18 males and 14 females, and the low-anxiety group consisted of 8 males and 4 females.

After thoroughly rinsing their mouths, subjects sat unrestrained in a comfortable chair with lumbar support opposite a 19-inch TV monitor placed 100 cm away at eye level. A video recording of corneal transplant surgery, which included scenes of injection into the eyeball and incision of the cornea with scissors, served as the psychological stressor for 15 min.

subjects were instructed to tilt their heads slightly forward without taking their eyes off the TV monitor, and to accumulate saliva in the floor of the mouth before spitting into a plastic vial. A prestress saliva sample was collected 5 min before the video viewing, and the stress saliva sample was collected just after the video viewing. Saliva samples were centrifuged and the supernatant was stored at −20°C until analyzed.

Salivary cortisol and β-endorphin were assayed using ELISA kits (cortisol: Correlate-EIA kit, Assay Designs Inc. USA; β-endorphin: S-1134 kit, Peninsula Lab. Inc. USA). Salivary α-amylase activity was assayed using a kit (Neo-Amylase test, Daiichi Pure Chemicals, Japan) that used blue starch as the substrate. Human pancreatic α-amylase (Humylase Control, Pharmacia and Upjohn, Sweden) was used as the standard, and the enzyme activity was expressed as international units per milliliter of saliva (U/mL).

RESULTS

There were very wide ranges in resting basal levels of cortisol, amylase, and β-endorphin in the saliva between different individuals. No significant difference between males and females was observed in any of the pre-stress salivary parameters for both high- and low-anxiety groups (TABLE 1).

Following the video viewing, an increase in salivary amylase and cortisol levels were observed in all subjects in both groups. However, an increase in β-endorphin was observed in only 22 of 44 (50.0%) subjects.

FIGURE 1 shows the normalized values of the percentage of prestress levels for amylase, cortisol, and β-endorphin after the stressful video viewing. There were no differences in amylase levels between males and females in either the high- or low-anxiety groups. However, cortisol levels in the high-anxiety females were significantly lower than in the high-anxiety males. The β-endorphin level tended to be higher in anxious subjects, but there were no significant gender differences.

TABLE 1. Pre-stress basal levels of amylase, cortisol, and β-endorphin

	Low-anxiety group		High-anxiety group	
	Male	Female	Male	Female
Sample size	18	14	8	4
α-amylase (U/mL)	121.5 ± 57.9	132.1 ± 51.0	141.1 ± 36.7	119.1 ± 48.2
Cortisol (pg/mL)	996.6 ± 236.5	1036.4.6 ± 213.9	1094.2 ± 118.4	989.6 ± 236.5
β-endorphin (pg/mL)	2230.5 ± 624.4	2013.6 ± 713.9	1998.6 ± 489.7	2156.4 ± 848.3

Mean ± SD.

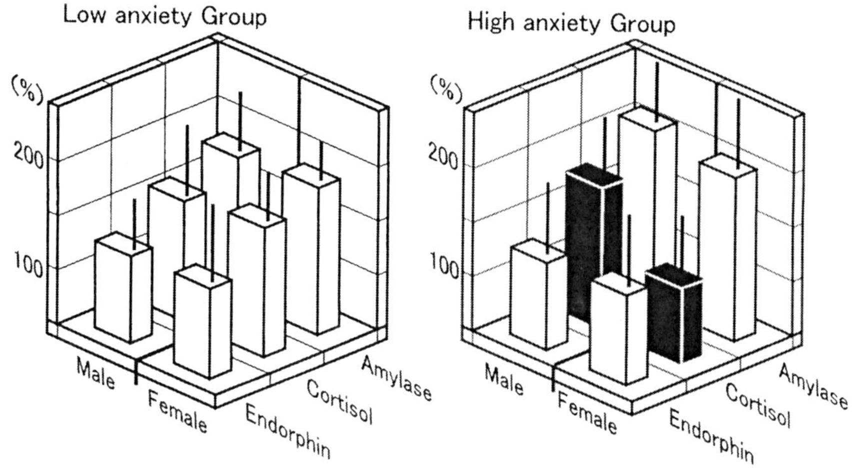

FIGURE 1. Salivary parameters after stress. Normalized values are plotted as a percentage of the resting pre-stress levels. Cortisol levels of highly anxious females were significantly lower than those of highly anxious males (*black bars*; $P < 0.01$, unpaired t-test). The data are presented as mean + SD.

DISCUSSION

In this study, we observed that there were no significant gender differences in basal levels of salivary indices for HPA (cortisol) or SAM (amylase), among the high- and low-anxiety groups. Additionally, no gender differences were seen in the levels of β-endorphin, a possible marker of the HPA system. A clear-cut gender difference in HPA activity has been demonstrated in rats. Corticosterone is functionally equivalent to cortisol in humans, and female rats show higher baseline corticosterone levels and enhanced stress responses.[8] However, in human studies, no such clear-cut gender differences have been established. While some studies have reported higher basal cortisol levels in men,[9,10] no differences were found in other studies.[11,12] Similarly, in response to psychological stressors in young subjects, certain studies have shown higher cortisol responses in male subjects compared to females,[13] whereas other studies have not revealed any gender differences.[14]

In our study, in contrast to basal HPA and SAM activities, exposure to psychological stress resulted in some sex-specific endocrine-response profiles. In the high-anxiety group, the cortisol response to the stressor was significantly lower in females compared to males, while in the low-anxiety group, there was no gender difference in the cortisol response. Gender differences in amylase responses were not observed in either the high- or low-anxiety groups. To date, the attempts to correlate personality traits have focused mainly on cortisol release. Kirschbaum *et al.*[15] divided their subjects into low and high cortisol responders, who showed different response kinetics to repeated psychological stresses. The psychological profile of high cortisol responders included the occurrence of depressive mood and low self-esteem. More recently, Schommer *et al.*[16] investigated a large sample of subjects and found no evidence of a close relationship between personality traits and circadian cortisol rhythm, or a single cortisol stress response. However, gender differences were not addressed in these studies.

In conclusion, the major finding of our study was that females with high trait anxiety exhibit lower HPA system activation with psychosocial stress than males with high trait anxiety. This suggests that high trait anxiety in females may be associated with an inability to respond to acute stress with adequate HPA hormone release.

REFERENCES

1. KIRSCHBAUM, C. & D.H. HELLHAMMER. 1994. Salivary cortisol in psychoneuroendocrine research: recent developments and applications. Psychoneuroendocrinology **19:** 313–333.
2. SCHWAB, K.O. *et al.* 1992. Free epinephrine, norepinephrine and dopamine in saliva and plasma of healthy adults. Eur. J. Clin. Chem. Clin. Biochem. **30:** 541–544.
3. GALLACHER, D.V. & O.H. PETERSEN. 1983. Stimulus-secretion coupling in mammalian salivary glands. Int. Rev. Physiol. **28:** 1–52.
4. TAKAI, N. *et al.* 2004. Effect of psychological stress on the salivary cortisol and amylase levels in healthy young adults. Arch. Oral Biol. **49:** 963–968.
5. VAN STEGERENA, A. *et al.* 2006. Salivary alpha amylase as marker for adrenergic activity during stress: effect of betablockade. Psychoneuroendocrinology **31:** 137–141.
6. HANDA, R.J. *et al.* 1994. Gonadal steroid hormone receptors and sex differences in the hypothalamo–pituitary–adrenal axis. Horm. Behav. **28:** 464–476.
7. ARMARIO, A. *et al.* 1995. Comparison of the behavioural and endocrine response to forced swimming stress in five inbred strains of rats. Psychoneuroendocrinology **20:** 879–890.
8. HEINSBROEK, R.P. *et al.* 1991. Sex- and time-dependent changes in neurochemical and hormonal variables induced by predictable and unpredictable footshock. Physiol. Behav. **49:** 1251–1256.
9. OLSSON, T. *et al.* 1989. Hormones in 'young' and 'old' elderly: pituitary-thyroid and pituitary-adrenal axes. Gerontology **35:** 144–152.
10. TERSMAN, Z. *et al.* 1991. Cardiovascular responses to psychological and physiological stressors during the menstrual cycle. Psychosom. Med. **53:** 185–197.
11. HORROCKS, P.M. *et al.* 1990. Patterns of ACTH and cortisol pulsatility over twenty-four hours in normal males and females. Clin. Endocrinol. **32:** 127–134.
12. BRANDSTADTER, J. *et al.* 1991. Developmental and personality correlates of adrenocortical activity as indexed by salivary cortisol: observations in the age range of 35 to 65 years. J. Psychosom. Res. **35:** 173–185.
13. KIRSCHBAUM, C. *et al.* 1995. Sex-specific effects of social support on cortisol and subjective responses to acute psychological stress. Psychosom. Med. **57:** 23–31.

14. COLLINS, A. & M. FRANKENHAEUSER. 1978. Stress responses in male and female engineering students. J. Human Stress **4:** 43–48.
15. KIRSCHBAUM, C. *et al.* 1995. Persistent high cortisol responses to repeated psychological stress in a subpopulation of healthy men. Psychosom. Med. **57:** 468–474.
16. SCHOMMER, N.C. *et al.* 1999. No evidence for a close relationship between personality traits and circadian cortisol rhythm or a single cortisol stress response. Psychol. Rep. **84:** 840–842.

Index of Contributors

Printed in the United States
119108LV00006B/1-24/P

9 781573 316613